7701
GAL
D0301606

$C(a)$	$\{g \in G \mid ga = ag\}$, the centralizer of a in G, 66
$C(H)$	$\{x \in G \mid xh = hx$ for all $h \in H\}$, the centralizer of H, 68
$\phi(n)$	Euler phi function of n, 80
$N(H)$	$\{x \in G \mid xHx^{-1} = H\} = \{x \in G \mid Hx = xH\}$, the normalizer of H in G, 90
$\text{cl}(a)$	conjugacy class of a, 90
G^n	$\{g^n \mid g \in G\}$, 91
S_n	group of one-to-one functions from $\{1, 2, \ldots, n\}$ to itself, 96
A_n	alternating group of degree n, 104
$G \approx \overline{G}$	G and $\overline{G}$ are isomorphic, 121
ϕ_a	mapping given by $\phi_a(x) = axa^{-1}$ for all x, 129
$\text{Aut}(G)$	group of automorphisms of the group G, 129
$\text{Inn}(G)$	group of inner automorphisms of G, 129
aH	$\{ah \mid h \in H\}$, 137
aHa^{-1}	$\{aha^{-1} \mid h \in H\}$, 137
$\|G:H\|$	the index of H in G, 141
$\text{stab}_G(i)$	$\{\phi \in G \mid \phi(i) = i\}$, the stabilizer of i under the permutation group G, 144
$\text{orb}_G(i)$	$\{\phi(i) \mid \phi \in G\}$, the orbit of i under the permutation group G, 144
$G_1 \oplus G_2 \oplus \cdots \oplus G_n$	external direct product of groups $G_1, G_2, \ldots, G_n$, 153
$U_k(n)$	$\{x \in U(n) \mid x \bmod k = 1\}$, 157
G'	commutator subgroup, 174
$H \lhd G$	H is a normal subgroup of G, 177
G/H	factor group, 179
HK	$\{hk \mid h \in H, k \in K\}$, 187
$H \times K$	internal direct product of H and K, 187
$H_1 \times H_2 \times \cdots \times H_n$	internal direct product of $H_1, \ldots, H_n$, 189
$\text{Ker}\,\phi$	kernel of the homomorphism ϕ, 199
$\phi^{-1}(g')$	inverse image of g' under ϕ, 201
$\phi^{-1}(\overline{K})$	inverse image of $\overline{K}$ under ϕ, 202
$Z[x]$	ring of polynomials with integer coefficients, 236
$M_2(Z)$	ring of all 2×2 matrices with integer entries, 236
$R_1 \oplus R_2 \oplus \cdots \oplus R_n$	direct sum of rings, 237
nZ	ring of multiples of n, 239
$Z[i]$	ring of Gaussian integers, 239
$U(R)$	group of units of the ring R, 242
$Z_n[i]$	ring of Gaussian integers modulo n, 244
$\text{char}\,R$	characteristic of R, 252
R/A	factor ring, 262
$\langle a \rangle$	principal ideal generated by a, 262
$\langle a_1, a_2, \ldots, a_n \rangle$	ideal generated by $a_1, a_2, \ldots, a_n$, 262
$A + B$	sum of ideals A and B, 269
AB	product of ideals A and B, 269
$\text{Ann}(A)$	annihilator of A, 270
$N(A)$	nil radical of A, 270
$F(x)$	field of quotients of $F[x]$, 285
$R[x]$	ring of polynomials over R, 291

barcode next page →

SIXTH EDITION

Contemporary Abstract Algebra

Joseph A. Gallian

University of Minnesota Duluth

Houghton Mifflin Company

Boston   New York

Dedicated to the memory of my mother and father

Publisher: *Jack Shira*
Sponsoring Editor: *Lauren Schultz*
Associate Editor: *Lisa Collette*
Editorial Associate: *Kasey McGarrigle*
Project Editor: *Marilyn Rothenberger*
Art and Design Manager: *Gary Crespo*
Senior Photo Editor: *Jennifer Meyer Dare*
Senior Composition Buyer: *Sarah Ambrose*
Senior Marketing Manager: *Ben Rivera*
Marketing Assistant: *Lisa Lawler*

Cover image: © *Michael Field*

Printed in the U.S.A.

Library of Congress Control Number: 2004114157

ISBN: 0-618-51471-6

56789-EB-08 07 06

Contents

iii

**PART 5
Special Topics 399**

Preface

Dear Sir or Madam, will you read my book, it took me years to write, will you take a look?

JOHN LENNON AND PAUL McCARTNEY,
Paperback Writer, single

Although I wrote the first edition of this book twenty years ago, my goals for it remain the same. I want students to receive a solid introduction to the traditional topics. I want readers to come away with the view that abstract algebra is a contemporary subject—that its concepts and methodologies are being used by working computer scientists, physicists, and chemists. I want students to enjoy reading the book. To this end, I have included lines from popular songs, poems, quotations, biographies, historical notes, dozens of photographs, hundreds of figures, numerous tables and charts, and reproductions of stamps and currency that honor mathematicians. I want students to be able to do computations and to write proofs. Accordingly, I have included an abundance of exercises to develop both skills.

Changes for the sixth edition include freshening of the exercises, quotations, biographies, references, and the suggested readings. The major additions for the sixth edition are at the website: www.d.umn.edu/~jgallian. Many new computer exercises have been included and many others have been expanded and improved. Some of the exercises illustrate the concepts from the chapter where they appear, some anticipate concepts that are in later chapters, and others ask the students to make conjectures based on data generated by the software. Because so much of abstract algebra is based on concrete examples, these exercises help students to see underlying structures. Also new to the website is a lengthy essay that provides advice to students on how to learn the subject and how to do proofs. The website also includes true/false questions that test students' understanding of the concepts.

The Houghton Mifflin companion website at www.college.hmco.com provides additional teaching resources that are password protected. These resources include an Instructor's Resource Manual that contains solutions to all even-numbered exercises and a bank of sample tests with solutions, an Instructor's Answer Key for the computer exercises that ask users to make conjectures, and a laboratory manual, written by Julianne Rainbolt and the author, for instructors who want to incorporate a computer algebra system into their courses. The lab manual contains exercises that are designed to be completed with the free software GAP. Also available at www.d.umn.edu/~jgallian is a link to Alec Mihailovs's website that has a Maple manual keyed to this book.

For the first time, there is a Student Solutions Manual for the odd-numbered exercises available from the publisher.

I wish to express my appreciation to Lisa Collette, Lauren Schultz, and Kasey McGarrigle from Houghton Mifflin as well as to the publisher's production staff.

Petr Kovar assisted me with word processing, Craig Kirkpatrick was the copy editor, Paul Lorczak was the accuracy reviewer, and Marilyn Rothenberger served as freelance project editor. I am grateful to each of them.

I greatly valued the careful reading of the manuscript and thoughtful input of the following people, who kindly served as reviewers for the sixth edition: Jerry R. Beehler, Tri-State University; Clifford Bergman, Iowa State University; Cherlyn Converse, California State University Fullerton; David J. DeVries, Georgia College & State University; Jason Douma, University of Sioux Falls; Ken Levasseur, University of Massachusetts Lowell; Benny Lo, DeVry University; Alec Mihailovs, Shepherd College; Paul C. Pasles, Villanova University; David Shannon, Transylvania University; Christopher Terry, Augusta State University; Jianqiang Zhao, Eckerd College.

Many users of the text kindly sent comments and suggestions to me about the sixth edition. I welcome them for the seventh edition. My address is jgallian@d.umn.edu.

Joseph A. Gallian

PART 1

Integers and Equivalence Relations

 For online student resources,
visit this textbook's website at
math.college.hmco.com/students

0 Preliminaries

The whole of science is nothing more than a refinement of everyday thinking.

ALBERT EINSTEIN, *Physics and Reality*

PROPERTIES OF INTEGERS

Much of abstract algebra involves properties of integers and sets. In this chapter we collect the properties we need for future reference.

An important property of the integers, which we will often use, is the so-called Well Ordering Principle. Since this property cannot be proved from the usual properties of arithmetic, we will take it as an axiom.

Well Ordering Principle

Every nonempty set of positive integers contains a smallest member.

The concept of divisibility plays a fundamental role in the theory of numbers. We say a nonzero integer t is a *divisor* of an integer s if there is an integer u such that $s = tu$. In this case, we write $t \mid s$ (read "t divides s"). When t is not a divisor of s, we write $t \nmid s$. A *prime* is a positive integer greater than 1 whose only positive divisors are 1 and itself. We say an integer s is a *multiple* of an integer t if there is an integer u such that $s = tu$.

As our first application of the Well Ordering Principle, we establish a fundamental property of integers that we will use often.

Theorem 0.1   **Division Algorithm**

> Let a and b be integers with $b > 0$. Then there exist unique
> integers q and r with the property that $a = bq + r$, where
> $0 \leq r < b$.

PROOF. We begin with the existence portion of the theorem. Consider the set $S = \{a - bk \mid k \text{ is an integer and } a - bk \geq 0\}$. If $0 \in S$, then b divides a and we may obtain the desired result with $q = a/b$ and $r = 0$. Now assume $0 \notin S$. Since S is nonempty [if $a > 0$, $a - b \cdot 0 \in S$; if $a < 0$, $a - b(2a) = a(1 - 2b) \in S$; $a \neq 0$ since $0 \notin S$], we may apply the Well Ordering Principle to conclude that S has a smallest member, say $r = a - bq$. Then $a = bq + r$ and $r \geq 0$, so all that remains to be proved is that $r < b$.

If $r \geq b$, then $a - b(q + 1) = a - bq - b = r - b \geq 0$, so that $a - b(q + 1) \in S$. But $a - b(q + 1) < a - bq$, and $a - bq$ is the *smallest* member of S. So, $r < b$.

To establish the uniqueness of q and r, let us suppose that there are integers q, q', r, and r' such that

$$a = bq + r, \quad 0 \leq r < b \quad \text{and} \quad a = bq' + r', \quad 0 \leq r' < b.$$

For convenience, we may also suppose that $r' \geq r$. Then $bq + r = bq' + r'$ and $b(q - q') = r' - r$. So, b divides $r' - r$ and $0 \leq r' - r \leq r' < b$. It follows that $r' - r = 0$, and therefore $r' = r$ and $q = q'$. ∎

The integer q in the division algorithm is called the *quotient* upon dividing a by b; the integer r is called the *remainder* upon dividing a by b.

EXAMPLE 1   For $a = 17$ and $b = 5$, the division algorithm gives $17 = 5 \cdot 3 + 2$; for $a = -23$ and $b = 6$, the division algorithm gives $-23 = 6(-4) + 1$. ◆

Several states use linear functions to encode the month and date of birth into a three-digit number that is incorporated into driver's license numbers. If the encoding function is known, the division algorithm can be used to recapture the month and date of birth from the three-digit number. For instance, the last three digits of a Florida male driver's license number are those given by the formula $40(m - 1) + b$, where m is the number of the month of birth and b is the day of birth. Thus, since $177 = 40 \cdot 4 + 17$, a person with these last three digits was born on May 17. For New York licenses issued prior to September of 1992, the last two digits indicate the year of birth, and the three preceding digits code the month and date of birth. For a male driver, these three digits are $63m + 2b$, where m denotes the number of the month of birth and b is the date of birth. So, since $701 = 63 \cdot 11 + 2 \cdot 4$, a license that ends with 70174 indicates that the holder is a male born on November 4, 1974. (In cases where the formula for the driver's license number yields the same result for two or more people, a "tie-breaking" digit is inserted before the two digits for the year of birth.) Incidentally, Wisconsin uses the same method as Florida to encode birth information, but the numbers immediately precede the last pair of digits.

DEFINITIONS *Greatest Common Divisor, Relatively Prime Integers*

> The *greatest common divisor* of two nonzero integers a and b is the largest of all common divisors of a and b. We denote this integer by gcd(a, b). When gcd$(a, b) = 1$, we say a and b are *relatively prime.*

The following property of the greatest common divisor of two integers plays a critical role in abstract algebra. The proof provides an application of the division algorithm and our second application of the Well Ordering Principle.

Theorem 0.2 **GCD Is a Linear Combination**

> *For any nonzero integers a and b, there exist integers s and t such that gcd$(a, b) = as + bt$. Moreover, gcd(a, b) is the smallest positive integer of the form $as + bt$.*

PROOF. Consider the set $S = \{am + bn \mid m, n$ are integers and $am + bn > 0\}$. Since S is obviously nonempty (if some choice of m and n makes $am + bn < 0$, then replace m and n by $-m$ and $-n$), the Well Ordering Principle asserts that S has a smallest member, say, $d = as + bt$. We claim that $d = $ gcd(a, b). To verify this claim, use the division algorithm to write $a = dq + r$, where $0 \le r < d$. If $r > 0$, then $r = a - dq = a - (as + bt)q = a - asq - btq = a(1 - sq) + b(-tq) \in S$, contradicting the fact that d is the smallest member of S. So, $r = 0$ and d divides a. Analogously (or, better yet, by symmetry), d divides b as well. This proves that d is a common divisor of a and b. Now suppose d' is another common divisor of a and b and write $a = d'h$ and $b = d'k$. Then $d = as + bt = (d'h)s + (d'k)t = d'(hs + kt)$, so that d' is a divisor of d. Thus, among all common divisors of a and b, d is the greatest. ■■

EXAMPLE 2 $\gcd(4, 15) = 1$; $\gcd(4, 10) = 2$; $\gcd(2^2 \cdot 3^2 \cdot 5, 2 \cdot 3^3 \cdot 7^2) = 2 \cdot 3^2$. Note that 4 and 15 are relatively prime, whereas 4 and 10 are not. Also, $4 \cdot 4 + 15(-1) = 1$ and $4(-2) + 10 \cdot 1 = 2$. ◆

Although the result above is a powerful theoretical tool, it does not provide a practical method for calculating $\gcd(a, b)$ or specific values for s and t. The next example shows how this can be done.

EXAMPLE 3   Euclidean Algorithm (Proposition 2 of Book Seven of the *Elements*)
For any pair of positive integers a and b, we may find $\gcd(a, b)$ by repeated use of division to produce a decreasing sequence of integers $r_1 > r_2 > \cdots$ as follows.

$$
\begin{aligned}
a &= bq_1 + r_1 & 0 &< r_1 < b, \\
b &= r_1q_2 + r_2 & 0 &< r_2 < r_1, \\
r_1 &= r_2q_3 + r_3 & 0 &< r_3 < r_2, \\
&\vdots \\
r_{k-3} &= r_{k-2}q_{k-1} + r_{k-1} & 0 &< r_{k-1} < r_{k-2}, \\
r_{k-2} &= r_{k-1}q_k + r_k & 0 &< r_k < r_{k-1}, \\
r_{k-1} &= r_kq_{k+1} + 0.
\end{aligned}
$$

(It is a consequence of the Well Ordering Principle that this process must eventually result in a remainder of 0.) Then r_k, the last nonzero remainder, is $\gcd(a, b)$.

To see that $r_k = \gcd(a, b)$, observe that since $r_k \mid r_{k-1}$, the next-to-last equation implies that $r_k \mid r_{k-2}$. This, in turn, implies $r_k \mid r_{k-3}$. Continuing to work backwards in this fashion, we see that $r_k \mid b$ and $r_k \mid a$. This proves that r_k is a common divisor of a and b. Now if r is another common divisor of a and b, the first equation above shows that $r \mid r_1$. The second equation then shows that $r \mid r_2$. Continuing, we see that $r \mid r_k$ and indeed r_k is the greatest of all common divisors of a and b.

Let's apply this process to compute gcd(2520, 154).

$$2520 = 154 \cdot 16 + 56,$$
$$154 = 56 \cdot 2 + 42,$$
$$56 = 42 \cdot 1 + 14,$$
$$42 = 14 \cdot 3.$$

Since 14 is the last nonzero remainder, gcd(2520, 154) = 14.

The preceding equations also provide a method of expressing gcd(a, b) as a linear combination of a and b. For instance, we may use the equations to write the remainders in terms of 2520 and 154. Namely,

$$56 = 2520 + 154(-16),$$
$$42 = 154 + 56(-2),$$
$$14 = 56 + 42(-1)$$
$$= 56 + [154 + 56(-2)](-1)$$
$$= 56 \cdot 3 + 154(-1)$$
$$= [2520 + 154(-16)]3 + 154(-1)$$
$$= 2520 \cdot 3 + 154(-49). \qquad \blacklozenge$$

The next lemma is frequently used. It appeared in Euclid's *Elements*.

Euclid's Lemma $p \mid ab$ implies $p \mid a$ or $p \mid b$

> *If p is a prime that divides ab, then p divides a or p divides b.*

PROOF. Suppose p is a prime that divides ab but does not divide a. We must show that p divides b. Since p does not divide a, there are integers s and t such that $1 = as + pt$. Then $b = abs + ptb$, and since p divides the right-hand side of this equation, p also divides b. ∎

Note that Euclid's Lemma may fail when p is not a prime, since $6 \mid 4 \cdot 3$ but $6 \nmid 4$ and $6 \nmid 3$.

Our next property shows that the primes are the building blocks for all integers. We will often use this property without explicitly saying so.

***Theorem 0.3*   Fundamental Theorem of Arithmetic**

> *Every integer greater than 1 is a prime or a product of primes. This product is unique, except for the order in which the factors appear. Thus, if $n = p_1 p_2 \cdots p_r$ and $n = q_1 q_2 \cdots q_s$, where the p's and q's are primes, then $r = s$ and, after renumbering the q's, we have $p_i = q_i$ for all i.*

We will prove the existence portion of Theorem 0.3 later in this chapter. The uniqueness portion is a consequence of Euclid's Lemma (Exercise 25).

Another concept that frequently arises is that of the least common multiple of two integers.

DEFINITION   *Least Common Multiple*

The least common multiple of two nonzero integers a and b is the smallest positive integer that is a multiple of both a and b. We will denote this integer by lcm(a, b).

We leave it as an exercise (Exercise 12) to prove that every common multiple of a and b is a multiple of lcm(a, b).

EXAMPLE 4   lcm(4, 6) $= 12$; lcm(4, 8) $= 8$; lcm(10, 12) $= 60$; lcm(6, 5) $= 30$; lcm($2^2 \cdot 3^2 \cdot 5$, $2 \cdot 3^3 \cdot 7^2$) $= 2^2 \cdot 3^3 \cdot 5 \cdot 7^2$.   ◆

MODULAR ARITHMETIC

Another application of the division algorithm that will be important to us is modular arithmetic. Modular arithmetic is an abstraction of a method of counting that you often use. For example, if it is now September, what month will it be 25 months from now? Of course, the answer is October, but the interesting fact is that you didn't arrive at the answer by starting with September and counting off 25 months. Instead, without even thinking about it, you simply observed that $25 = 2 \cdot 12 + 1$, and you added 1 month to September. Similarly, if it is now Wednesday, you know that in 23 days it will be Friday. This time, you arrived at your answer by noting that $23 = 7 \cdot 3 + 2$, so you added 2 days to Wednesday instead of counting off 23 days. If your electricity is off for 26 hours, you must advance your clock 2 hours, since $26 = 2 \cdot 12 + 2$. Surprisingly, this simple idea has numerous important applications in mathematics and computer science. You will see a few of them in this section. The following notation is convenient.

When $a = qn + r$, where q is the quotient and r is the remainder upon dividing a by n, we write $a \bmod n = r$ or $a = r \bmod n$. Thus,

$$3 \bmod 2 = 1 \text{ since } 3 = 1 \cdot 2 + 1,$$
$$6 \bmod 2 = 0 \text{ since } 6 = 3 \cdot 2 + 0,$$
$$11 \bmod 3 = 2 \text{ since } 11 = 3 \cdot 3 + 2,$$
$$62 \bmod 85 = 62 \text{ since } 62 = 0 \cdot 85 + 62,$$
$$-2 \bmod 15 = 13 \text{ since } -2 = (-1)15 + 13.$$

In general, if a and b are integers and n is a positive integer, then $a \bmod n = b \bmod n$ if and only if n divides $a - b$ (Exercise 9).

In our applications, we will use addition and multiplication mod n. When you wish to compute $ab \bmod n$ or $(a + b) \bmod n$ and a or b is greater than n, it is easier to "mod first." For example, to compute $(27 \cdot 36) \bmod 11$, we note that $27 \bmod 11 = 5$ and $36 \bmod 11 = 3$, so $(27 \cdot 36) \bmod 11 = (5 \cdot 3) \bmod 11 = 4$. (See Exercise 11.)

Modular arithmetic is often used in assigning an extra digit to identification numbers for the purpose of detecting forgery or errors. We present two such applications.

EXAMPLE 5   The United States Postal Service money order shown in Figure 0.1 has an identification number consisting of 10 digits together with an extra digit called a *check*. The check digit is the 10-digit number modulo 9. Thus, the number 3953988164 has the check digit 2, since $3953988164 \bmod 9 = 2$.[†] If the number 39539881642 were incorrectly entered into a computer (programmed to calculate the check digit) as, say, 39559881642 (an error in the fourth position), the machine would calculate the check digit as 4, whereas the entered check digit would be 2. Thus the error would be detected.   ◆

EXAMPLE 6   Airline companies, United Parcel Service, and the rental car companies Avis and National use the modulo 7 values of

---

[†]The value of $N \bmod 9$ is easy to compute with a calculator. If $N = 9q + r$, where r is the remainder upon dividing N by 9, then on a calculator screen $N \div 9$ appears as $q.rrrr\ldots$, so the first decimal digit is the check digit. For example, $3953988164 \div 9 = 439332018.222$, so 2 is the check digit. If N has too many digits for your calculator, replace N by the sum of its digits and divide that number by 9. Thus, $3953988164 \bmod 9 = 56 \bmod 9 = 2$.

Figure 0.1

identification numbers to assign check digits. Thus, the identification number 00121373147367 (see Figure 0.2) has the check digit 3 appended to it because 121373147367 mod 7 = 3. Similarly, the UPS pickup record number 768113999, shown in Figure 0.3, has the check digit 2 appended to it. ◆

The methods used by the Postal Service and the airline companies do not detect all single-digit errors (see Exercises 33 and 37). However, detection of all single-digit errors, as well as nearly all errors involving the transposition of two adjacent digits, is easily achieved. One method that does this is the one used to assign the so-called Universal Product Code (UPC) to most retail items (see Figure 0.4). A UPC identification number has 12 digits. The first six digits identify the manufacturer, the next five identify the product, and the last is a check. (For many items, the 12th digit is

Figure 0.2

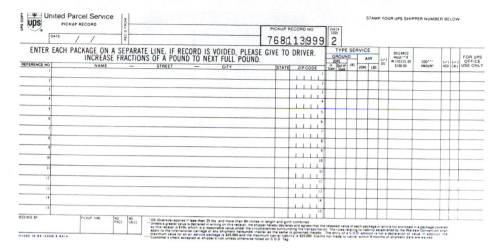

Figure 0.3

not printed, but it is always bar-coded.) In Figure 0.4, the check digit is 8.

To explain how the check digit is calculated, it is convenient to introduce the dot product notation for two k-tuples:

$$(a_1, a_2, \ldots, a_k) \cdot (w_1, w_2, \ldots, w_k) = a_1 w_1 + a_2 w_2 + \cdots + a_k w_k.$$

An item with the UPC identification number $a_1 a_2 \cdots a_{12}$ satisfies the condition

$$(a_1, a_2, \ldots, a_{12}) \cdot (3, 1, 3, 1, \ldots, 3, 1) \bmod 10 = 0.$$

To verify that the number in Figure 0.4 satisfies the condition above, we calculate

$$(0 \cdot 3 + 2 \cdot 1 + 1 \cdot 3 + 0 \cdot 1 + 0 \cdot 3 + 0 \cdot 1 + 6 \cdot 3 + 5 \cdot 1$$
$$+ 8 \cdot 3 + 9 \cdot 1 + 7 \cdot 3 + 8 \cdot 1) \bmod 10 = 90 \bmod 10 = 0.$$

The fixed k-tuple used in the calculation of check digits is called the *weighting vector.*

Now suppose a single error is made in entering the number in Figure 0.4 into a computer. Say, for instance, that 021000958978 is entered (notice that the seventh digit is incorrect). Then the computer calculates

$$0 \cdot 3 + 2 \cdot 1 + 1 \cdot 3 + 0 \cdot 1 + 0 \cdot 3 + 0 \cdot 1 + 9 \cdot 3$$
$$+ 5 \cdot 1 + 8 \cdot 3 + 9 \cdot 1 + 7 \cdot 3 + 8 \cdot 1 = 99.$$

Since 99 mod 10 $\neq$ 0, the entered number cannot be correct.

In general, any single error will result in a sum that is not 0 modulo 10.

Figure 0.4

The advantage of the UPC scheme is that it will detect nearly all errors involving the transposition of two adjacent digits as well as all errors involving one digit. For doubters, let us say that the identification number given in Figure 0.4 is entered as 021000658798. Notice that the last two digits preceding the check digit have been transposed. But by calculating the dot product, we obtain 94 mod 10 $\neq$ 0, so we have detected an error. In fact, the only undetected transposition errors of adjacent digits a and b are those where $|a - b| = 5$. To verify this, we observe that a transposition error of the form

$$a_1 a_2 \cdots a_i a_{i+1} \cdots a_{12} \rightarrow a_1 a_2 \cdots a_{i+1} a_i \cdots a_{12}$$

is undetected if and only if

$$(a_1, a_2, \ldots, a_{i+1}, a_i, \ldots, a_{12}) \cdot (3, 1, 3, 1, \ldots, 3, 1) \bmod 10 = 0.$$

That is, the error is undetected if and only if

$$(a_1, a_2, \ldots, a_{i+1}, a_i, \ldots, a_{12}) \cdot (3, 1, 3, 1, \ldots, 3, 1) \bmod 10$$

$$= (a_1, a_2, \ldots, a_i, a_{i+1}, \ldots, a_{12}) \cdot (3, 1, 3, 1, \ldots, 3, 1) \bmod 10.$$

This equality simplifies to either

$$(3a_{i+1} + a_i) \bmod 10 = (3a_i + a_{i+1}) \bmod 10$$

or

$$(a_{i+1} + 3a_i) \bmod 10 = (a_i + 3a_{i+1}) \bmod 10$$

depending on whether i is even or odd. Both cases reduce to $2(a_{i+1} - a_i) \bmod 10 = 0$. It follows that $|a_{i+1} - a_i| = 5$, if $a_{i+1} \neq a_i$.

In 2005 United States companies began to phase in the use of a 13th digit to be in conformance with the 13-digit product indentification numbers used in Europe. The weighing vector for 13 digit numbers is $(1, 3, 1, 3, \ldots, 3, 1)$.

Identification numbers printed on bank checks (on the bottom left between the two colons) consist of an eight-digit number $a_1 a_2 \cdots a_8$ and a check digit a_9, so that

$$(a_1, a_2, \ldots, a_9) \cdot (7, 3, 9, 7, 3, 9, 7, 3, 9) \bmod 10 = 0.$$

As is the case for the UPC scheme, this method detects all single-digit errors and all errors involving the transposition of adjacent digits a and b except when $|a - b| = 5$. But it also detects most errors of the form $\cdots abc \cdots \rightarrow \cdots cba \cdots$, whereas the UPC method detects no errors of this form.

In Chapter 5, we will examine more sophisticated means of assigning check digits to numbers.

What about error correction? Suppose you have a number such as 73245018 and you would like to be sure that even if a single mistake were made in entering this number into a computer, the computer would nevertheless be able to determine the correct number. (Think of it. You could make a mistake in dialing a telephone number but still get the correct phone to ring!) This is possible using two check digits. One of the check digits determines the magnitude of any single-digit error, while the other check digit locates the position of the error. With these two pieces of information, you can fix the error. To illustrate the idea, let us say that we have the eight-digit identification number $a_1 a_2 \cdots a_8$. We assign two check digits a_9 and a_{10} so that

$$(a_1 + a_2 + \cdots + a_9 + a_{10}) \bmod 11 = 0$$

and

$$(a_1, a_2, \ldots, a_9, a_{10}) \cdot (1, 2, 3, \ldots, 10) \bmod 11 = 0$$

are satisfied.

Let's do an example. Say our number before appending the two check digits is 73245018. Then a_9 and a_{10} are chosen to satisfy

$$(7 + 3 + 2 + 4 + 5 + 0 + 1 + 8 + a_9 + a_{10}) \bmod 11 = 0 \qquad (1)$$

and

$$(7 \cdot 1 + 3 \cdot 2 + 2 \cdot 3 + 4 \cdot 4 + 5 \cdot 5 + 0 \cdot 6 + \qquad (2)$$
$$1 \cdot 7 + 8 \cdot 8 + a_9 \cdot 9 + a_{10} \cdot 10) \bmod 11 = 0.$$

Since $7 + 3 + 2 + 4 + 5 + 0 + 1 + 8 = 30$ and $30 \bmod 11 = 8$, Equation (1) reduces to

$$(8 + a_9 + a_{10}) \bmod 11 = 0. \qquad (1')$$

Likewise, since $(7 \cdot 1 + 3 \cdot 2 + 2 \cdot 3 + 4 \cdot 4 + 5 \cdot 5 + 0 \cdot 6 + 1 \cdot 7 + 8 \cdot 8) \bmod 11 = 10$, Equation (2) reduces to

$$(10 + 9a_9 + 10a_{10}) \bmod 11 = 0. \qquad (2')$$

Since we are using mod 11, we may rewrite Equation (2′) as

$$(-1 - 2a_9 - a_{10}) \bmod 11 = 0$$

and add this to Equation (1′) to obtain $7 - a_9 = 0$. Thus $a_9 = 7$. Now substituting $a_9 = 7$ into Equation (1′) or Equation (2′), we obtain $a_{10} = 7$ as well. So, the number is encoded as 7324501877.

Now let us suppose that this number is erroneously entered into a computer programmed with our encoding scheme as 7824501877 (an error in position 2). Since the sum of the digits of the received number mod 11 is 5, we know that some digit is 5 too large (assuming only one error has been made). But which one? Say the error is in position i. Then the second dot product has the form $a_1 \cdot 1 + a_2 \cdot 2 + \cdots + (a_i + 5)i + a_{i+1} \cdot (i + 1) + \cdots + a_{10} \cdot 10 = (a_1, a_2, \cdots, a_{10}) \cdot (1, 2, \cdots, 10) + 5i$. So, $(7, 8, 2, 4, 5, 0, 1, 8, 7, 7) \cdot (1, 2, 3, 4, 5, 6, 7, 8, 9, 10) \bmod 11 = 5i \bmod 11$. Since the left-hand side mod 11 is 10, we see that $i = 2$. Our conclusion: The digit in position 2 is 5 too large. We have successfully corrected the error.

MATHEMATICAL INDUCTION

There are two forms of proof by mathematical induction that we will use. Both are equivalent to the Well Ordering Principle. The explicit formulation of the method of mathematical induction came in the 16th century. Francisco Maurolycus (1494–1575), a teacher of Galileo, used it in 1575 to prove that $1 + 3 + 5 + \cdots + (2n - 1) = n^2$, and Blaise Pascal (1623–1662) used it when he presented what we now call Pascal's triangle for the coefficients of the binomial expansion. The term *mathematical induction* was coined by Augustus De Morgan.

Theorem 0.4 **First Principle of Mathematical Induction**

> *Let S be a set of integers containing a. Suppose S has the property that whenever some integer $n \geq a$ belongs to S, then the integer $n + 1$ also belongs to S. Then, S contains every integer greater than or equal to a.*

PROOF.   The proof is left as an exercise (Exercise 27).   ■ ■

So, to use induction to prove that a statement involving positive integers is true for every positive integer, we must first verify that the statement is true for the integer 1. We then *assume* the statement is true for the integer n and use this assumption to prove that the statement is true for the integer $n + 1$.

Our next example uses some facts about plane geometry. Recall that given a straightedge and compass we can construct a right angle.

EXAMPLE 7   We use induction to prove that given a straightedge, a compass, and a unit length, we can construct a line segment of length $\sqrt{n}$ for every positive integer n. The case when $n = 1$ is given. Now we assume that we can construct a line segment of length $\sqrt{n}$. Then use the straightedge and compass to construct a right triangle with height 1 and base $\sqrt{n}$. The hypotenuse of the triangle has length $\sqrt{n+1}$. So, by induction, we can construct a line segment of length $\sqrt{n}$ for every positive integer n.   ◆

EXAMPLE 8   DeMoivre's Theorem
We use induction to prove that for every positive integer n and every real number θ, $(\cos\theta + i\sin\theta)^n = \cos n\theta + i\sin n\theta$, where i is the complex number $\sqrt{-1}$. Obviously, the statement is true for $n = 1$. Now assume it is true for n. We must prove that $(\cos\theta + i\sin\theta)^{n+1} = \cos(n + 1)\theta + i\sin(n + 1)\theta$. Observe that

$$(\cos\theta + i\sin\theta)^{n+1} = (\cos\theta + i\sin\theta)^n(\cos\theta + i\sin\theta)$$
$$= (\cos n\theta + i\sin n\theta)(\cos\theta + i\sin\theta)$$
$$= \cos n\theta\cos\theta + i(\sin n\theta\cos\theta +$$
$$\sin\theta\cos n\theta) - \sin n\theta\sin\theta.$$

Now, using trigonometric identities for $\cos(\alpha + \beta)$ and $\sin(\alpha + \beta)$, we see that this last term is $\cos(n + 1)\theta + i\sin(n + 1)\theta$. So, by induction, the statement is true for all positive integers.   ◆

In many instances, the assumption that a statement is true for an integer n does not readily lend itself to a proof that the statement is true for the integer $n + 1$. In such cases, the following equivalent form of induction may be more convenient. Some authors call this formulation the *strong form* of induction.

***Theorem 0.5*   Second Principle of Mathematical Induction**

Let S be a set of integers containing a. Suppose S has the property that n belongs to S whenever every integer less than n and greater than or equal to a belongs to S. Then, S contains every integer greater than or equal to a.

PROOF.   The proof is left to the reader.   ■ ■

To use this form of induction, we first show that the statement is true for the integer a. We then *assume* that the statement is true for *all* integers that are greater than or equal to a and less than n, and use this assumption to prove that the statement is true for n.

EXAMPLE 9   We will use the Second Principle of Mathematical Induction with $a = 2$ to prove the existence portion of the Fundamental Theorem of Arithmetic. Let S be the set of integers greater than 1 that are primes or products of primes. Clearly, $2 \in S$. Now we assume that for some integer n, S contains all integers k with $2 \leq k < n$. We must show that $n \in S$. If n is a prime, then $n \in S$ by definition. If n is not a prime, then n can be written in the form ab, where $1 < a < n$ and $1 < b < n$. Since we are assuming that both a and b belong to S, we know that each of them is a prime or a product of primes. Thus, n is also a product of primes. This completes the proof.   ◆

Notice that it is more natural to prove the Fundamental Theorem of Arithmetic with the Second Principle of Mathematical Induction than with the First Principle. Knowing that a particular integer factors as a product of primes does not tell you anything about factoring the next larger integer. (Does knowing that 5280 is a product of primes help you to factor 5281 as a product of primes?)

The following problem appeared in the "Brain Boggler" section of the January 1988 issue of the science magazine *Discover.*

EXAMPLE 10   The Quakertown Poker Club plays with blue chips worth $5.00 and red chips worth $8.00. What is the largest bet that cannot be made?

To gain insight into this problem, we try various combinations of blue and red chips and obtain 5, 8, 10, 13, 15, 16, 18, 20, 21, 23, 24, 26, 28, 29, 30, 31, 32, 33, 34, 35, 36, 37, 38, 39, 40. It appears that the answer is 27. But how can we be sure? Well, we need only prove that every integer greater than 27 can be written in the form $a \cdot 5 + b \cdot 8$, where a and b are nonnegative integers. This will solve the problem, since a represents the number of blue chips and b the number of red chips needed to make a bet of $a \cdot 5 + b \cdot 8$. For the purpose of contrast, we will give two proofs—one using the First Principle of Mathematical Induction and one using the Second Principle.

Let S be the set of all integers of the form $a \cdot 5 + b \cdot 8$, where a and b are nonnegative. Obviously, $28 \in S$. Now assume that some integer $n \in S$, say, $n = a \cdot 5 + b \cdot 8$. We must show that $n + 1 \in S$. First, note that since $n \geq 28$, we cannot have both a and b less than 3. If $a \geq 3$, then

$$n + 1 = (a \cdot 5 + b \cdot 8) + (-3 \cdot 5 + 2 \cdot 8)$$
$$= (a - 3) \cdot 5 + (b + 2) \cdot 8.$$

(Regarding chips, this last equation says that we may increase a bet from n to $n + 1$ by removing three blue chips from the pot and adding two red chips.) If $b \geq 3$, then

$$n + 1 = (a \cdot 5 + b \cdot 8) + (5 \cdot 5 - 3 \cdot 8)$$
$$= (a + 5) \cdot 5 + (b - 3) \cdot 8.$$

(The bet can be increased by 1 by removing three red chips and adding five blue chips.) This completes the proof.

To prove the same statement by the Second Principle, we note that each of the integers 28, 29, 30, 31, and 32 is in S. Now assume that for some integer $n > 32$, S contains all integers k with $28 \leq k < n$. We must show that $n \in S$. Since $n - 5 \in S$, there are nonnegative integers a and b such that $n - 5 = a \cdot 5 + b \cdot 8$. But then $n = (a + 1) \cdot 5 + b \cdot 8$. Thus n is in S. ◆

EQUIVALENCE RELATIONS

In mathematics, things that are considered different in one context may be viewed as equivalent in another context. We have already seen one such example. Indeed, the sums $2 + 1$ and $4 + 4$ are certainly different in ordinary arithmetic, but are the same under modulo 5 arithmetic. Congruent triangles that are situated differently in the plane are not the same, but they are often considered to be the same in plane geometry. In physics, vectors of the same magnitude and direction can produce different effects—a 10-pound weight placed 2 feet from a fulcrum produces a different effect than a 10-pound weight placed 1 foot from a fulcrum. But in linear algebra, vectors of the same magnitude and direction are considered to be the same. What is needed to make these distinctions precise is an appropriate generalization of the notion of equality; that is, we need a formal mechanism for specifying whether or not two quantities are the same in a given setting. This mechanism is an equivalence relation.

DEFINITION   *Equivalence Relation*

An *equivalence relation* on a set S is a set R of ordered pairs of elements of S such that

1. $(a, a) \in R$ for all $a \in S$ (reflexive property).
2. $(a, b) \in R$ implies $(b, a) \in R$ (symmetric property).
3. $(a, b) \in R$ and $(b, c) \in R$ imply $(a, c) \in R$ (transitive property).

When R is an equivalence relation on a set S, it is customary to write aRb instead of $(a, b) \in R$. Also, since an equivalence relation is just a generalization of equality, a suggestive symbol such as $\approx$, $\equiv$, or $\sim$ is usually used to denote the relation. Using this notation, the three conditions for an equivalence relation become $a \sim a$; $a \sim b$ implies $b \sim a$; and $a \sim b$ and $b \sim c$ imply $a \sim c$. If $\sim$ is an equivalence relation on a set S and $a \in S$, then the set $[a] = \{x \in S \mid x \sim a\}$ is called the *equivalence class of S containing a.*

EXAMPLE 11   Let S be the set of all triangles in a plane. If $a, b \in S$, define $a \sim b$ if a and b are similar—that is, if a and b have corresponding angles that are the same. Then, $\sim$ is an equivalence relation on S.   ◆

EXAMPLE 12   Let S be the set of all polynomials with real coefficients. If $f, g \in S$, define $f \sim g$ if $f' = g'$, where f' is the derivative of f. Then, $\sim$ is an equivalence relation on S. Since two functions with equal derivatives differ by a constant, we see that for any f in S, $[f] = \{f + c \mid c$ is real$\}$.   ◆

EXAMPLE 13   Let S be the set of integers and let n be a positive integer. If $a, b \in S$, define $a \equiv b$ if a mod $n = b$ mod n (that is, if $a - b$ is divisible by n). Then, $\equiv$ is an equivalence relation on S and $[a] = \{a + kn \mid k \in S\}$. Since this particular relation is important in abstract algebra, we will take the trouble to verify that it is indeed an equivalence relation. Certainly, $a - a$ is divisible by n, so that $a \equiv a$ for all a in S. Next, assume that $a \equiv b$, say, $a - b = rn$. Then, $b - a = (-r)n$, and therefore $b \equiv a$. Finally, assume that $a \equiv b$ and $b \equiv c$, say, $a - b = rn$ and $b - c = sn$. Then, we have $a - c = (a - b) + (b - c) = rn + sn = (r + s)n$, so that $a \equiv c$.   ◆

EXAMPLE 14   Let $\equiv$ be as in Example 13 and let $n = 7$. Then we have $16 \equiv 2$; $9 \equiv -5$; and $24 \equiv 3$. Also, $[1] = \{\ldots, -20, -13, -6, 1, 8, 15, \ldots\}$ and $[4] = \{\ldots, -17, -10, -3, 4, 11, 18, \ldots\}$.   ◆

EXAMPLE 15   Let $S = \{(a, b) \mid a, b$ are integers, $b \neq 0\}$. If $(a, b), (c, d) \in S$, define $(a, b) \approx (c, d)$ if $ad = bc$. Then $\approx$ is an equivalence relation on S. [The motivation for this example comes from fractions. In fact, the pairs (a, b) and (c, d) are equivalent if the fractions a/b and c/d are equal.]

To verify that $\approx$ is an equivalence relation on S, note that $(a, b) \approx (a, b)$ requires that $ab = ba$, which is true. Next, we assume that $(a, b) \approx (c, d)$, so that $ad = bc$. We have $(c, d) \approx (a, b)$ provided that $cb = da$, which is true from commutativity of multiplication. Finally, we assume that $(a, b) \approx (c, d)$ and $(c, d) \approx (e, f)$ and prove that $(a, b) \approx (e, f)$. This amounts to using $ad = bc$ and $cf = de$ to show that $af = be$. Multiplying both sides of $ad = bc$ by f and replacing cf by de, we obtain $adf = bcf = bde$. Since $d \neq 0$, we can cancel d from the first and last terms.   ◆

DEFINITION   *Partition*

A *partition* of a set S is a collection of nonempty disjoint subsets of S whose union is S. Figure 0.5 illustrates a partition of a set into four subsets.

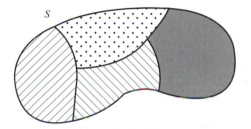

Figure 0.5   Partition of S into four subsets.

EXAMPLE 16   The sets $\{0\}$, $\{1, 2, 3, \ldots\}$, and $\{\ldots, -3, -2, -1\}$ constitute a partition of the set of integers.   ◆

EXAMPLE 17   The set of nonnegative integers and the set of nonpositive integers do not partition the integers, since both contain 0.   ◆

The next theorem reveals that equivalence relations and partitions are intimately intertwined.

Theorem 0.6   Equivalence Classes Partition

The equivalence classes of an equivalence relation on a set S constitute a partition of S. Conversely, for any partition P of S, there is an equivalence relation on S whose equivalence classes are the elements of P.

PROOF.   Let $\sim$ be an equivalence relation on a set S. For any $a \in S$, the reflexive property shows that $a \in [a]$. So, $[a]$ is nonempty and the union of all equivalence classes is S. Now, suppose that $[a]$ and $[b]$ are distinct equivalence classes. We must show that $[a] \cap [b] = \emptyset$. On the contrary, assume $c \in [a] \cap [b]$. We will show that $[a] \subseteq [b]$. To this end, let $x \in [a]$. We then have $c \sim a$, $c \sim b$, and $x \sim a$. By the symmetric property, we also have $a \sim c$. Thus, by transitivity, $x \sim c$, and by transitivity again, $x \sim b$. This proves $[a] \subseteq [b]$. Analogously, $[b] \subseteq [a]$. Thus, $[a] = [b]$, in contradiction to our assumption that $[a]$ and $[b]$ are distinct equivalence classes.

To prove the converse, let P be a collection of nonempty disjoint subsets of S whose union is S. Define $a \sim b$ if a and b belong to the same subset in the collection. We leave it to the reader to show that $\sim$ is an equivalence relation on S (Exercise 51).   ■ ■

FUNCTIONS (MAPPINGS)

Although the concept of a function plays a central role in nearly every branch of mathematics, the terminology and notation associated with functions vary quite a bit. In this section, we establish ours.

DEFINITION *Function (Mapping)*

> A *function* (or *mapping*) ϕ *from a set A to a set B* is a rule that assigns to each element a of A exactly one element b of B. The set A is called the *domain of* ϕ, and B is called the *range of* ϕ. If ϕ assigns b to a, then b is called the *image of a under* ϕ. The subset of B comprising all the images of elements of A is called the *image of A under* ϕ.

We use the shorthand $\phi: A \rightarrow B$ to mean that ϕ is a mapping from A to B. We will write $\phi(a) = b$ or $\phi: a \rightarrow b$ to indicate that ϕ carries a to b.

DEFINITION *Composition of Functions*

> Let $\phi: A \rightarrow B$ and $\psi: B \rightarrow C$. The *composition* $\psi\phi$ is the mapping from A to C defined by $(\psi\phi)(a) = \psi(\phi(a))$ for all a in A. The composition function $\psi\phi$ can be visualized as in Figure 0.6.

In calculus courses, the composition of f with g is written $(f \circ g)(x)$ and is defined as $f(g(x))$. When we compose functions, we omit the "circle."

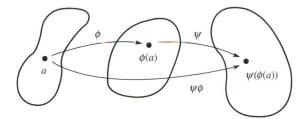

Figure 0.6 Composition of functions ϕ and ψ.

There are several kinds of functions that occur often enough to be given names.

DEFINITION *One-to-One Function*

A function ϕ from a set A is called *one-to-one* if for every a_1, a_2 $\in A$, $\phi(a_1) = \phi(a_2)$ implies $a_1 = a_2$.

The term *one-to-one* is suggestive, since the definition ensures that one element of B can be the image of only one element of A. Alternatively, ϕ is one-to-one if $a_1 \neq a_2$ implies $\phi(a_1) \neq \phi(a_2)$. That is, different elements of A map to different elements of B. See Figure 0.7.

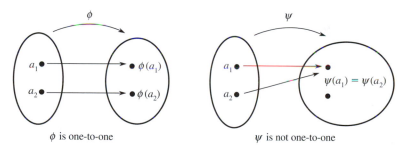

Figure 0.7

DEFINITION *Function from A onto B*

A function ϕ from a set A to a set B is said to be *onto B* if each element of B is the image of at least one element of A. In symbols, $\phi: A \rightarrow B$ is onto if for each b in B there is at least one a in A such that $\phi(a) = b$. See Figure 0.8.

The next theorem summarizes the facts about functions we will need.

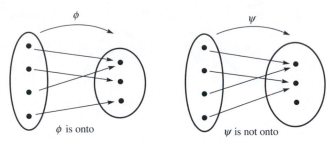

ϕ ψ

ϕ is onto ψ is not onto

Figure 0.8

Theorem 0.7 **Properties of Functions**

> *Given functions $\alpha: A \rightarrow B$, $\beta: B \rightarrow C$, and $\gamma: C \rightarrow D$, then*
> **1.** $\gamma(\beta\alpha) = (\gamma\beta)\alpha$ *(associativity).*
> **2.** *If α and β are one-to-one, then $\beta\alpha$ is one-to-one.*
> **3.** *If α and β are onto, then $\beta\alpha$ is onto.*
> **4.** *If α is one-to-one and onto, then there is a function α^{-1}*
> *from B onto A such that $(\alpha^{-1}\alpha)(a) = a$ for all a in A and*
> *$(\alpha\alpha^{-1})(b) = b$ for all b in B.*

PROOF.   We prove only part 1. The remaining parts are left as exercises (Exercise 53). Let $a \in A$. Then $(\gamma(\beta\alpha))(a) = \gamma((\beta\alpha)(a)) = \gamma(\beta(\alpha(a)))$. On the other hand, $((\gamma\beta)\alpha)(a) = (\gamma\beta)(\alpha(a)) = \gamma(\beta(\alpha(a)))$. So, $\gamma(\beta\alpha) = (\gamma\beta)\alpha$.   ∎

EXAMPLE 18   Let **Z** denote the set of integers, **R** the set of real numbers, and **N** the set of nonnegative integers. The following table illustrates the properties of one-to-one and onto.

Domain	Range	Rule	One-to-one	Onto
Z	**Z**	$x \rightarrow x^3$	Yes	No
R	**R**	$x \rightarrow x^3$	Yes	Yes
Z	**N**	$x \rightarrow |x|$	No	Yes
Z	**Z**	$x \rightarrow x^2$	No	No

To verify that $x \rightarrow x^3$ is one-to-one in the first two cases, notice that if $x^3 = y^3$, we may take the cube roots of both sides of the equation to obtain $x = y$. Clearly, the mapping from **Z** to **Z** given by $x \rightarrow x^3$ is not onto, since 2 is the cube of no integer. However, $x \rightarrow x^3$ defines an onto function from **R** to **R**, since every real number is the cube of its cube root (that is, $\sqrt[3]{b} \rightarrow b$). The remaining verifications are left to the reader.   ◆

EXERCISES

If you really want something in this life, you have to work for it—Now quiet, they're about to announce the lottery numbers!

HOMER SIMPSON

1. For $n = 5, 8, 12, 20$, and 25, find all positive integers less than n and relatively prime to n.

2. Determine $\gcd(2^4 \cdot 3^2 \cdot 5 \cdot 7^2, 2 \cdot 3^3 \cdot 7 \cdot 11)$ and $\text{lcm}(2^3 \cdot 3^2 \cdot 5, 2 \cdot 3^3 \cdot 7 \cdot 11)$.

3. Determine 51 mod 13, 342 mod 85, 62 mod 15, 10 mod 15, $(82 \cdot 73)$ mod 7, $(51 + 68)$ mod 7, $(35 \cdot 24)$ mod 11, and $(47 + 68)$ mod 11.

4. Find integers s and t such that $1 = 7 \cdot s + 11 \cdot t$. Show that s and t are not unique.

5. In Florida, the fourth and fifth digits from the end of a driver's license number give the year of birth. The last three digits for a male with birth month m and birth date b are represented by $40(m - 1) + b$. For females the digits are $40(m - 1) + b + 500$. Determine the dates of birth of people who have last five digits 42218 and 53953.

6. For driver's license numbers issued in New York prior to September of 1992, the three digits preceding the last two of the number of a male with birth month m and birth date b are represented by $63m + 2b$. For females the digits are $63m + 2b + 1$. Determine the dates of birth and sex(es) corresponding to the numbers 248 and 601.

7. Show that if a and b are positive integers, then $ab = \text{lcm}(a, b) \cdot \gcd(a, b)$.

8. Suppose a and b are integers that divide the integer c. If a and b are relatively prime, show that ab divides c. Show, by example, that if a and b are not relatively prime, then ab need not divide c.

9. If a and b are integers and n is a positive integer, prove that a mod $n = b$ mod n if and only if n divides $a - b$.

10. Let $d = \gcd(a, b)$. If $a = da'$ and $b = db'$, show that $\gcd(a', b') = 1$.

11. Let n be a fixed positive integer greater than 1. If a mod $n = a'$ and b mod $n = b'$, prove that $(a + b)$ mod $n = (a' + b')$ mod n and (ab) mod $n = (a'b')$ mod n. (This exercise is referred to in Chapters 6, 8, and 15.)

12. Let a and b be positive integers and let $d = \gcd(a, b)$ and $m = \text{lcm}(a, b)$. If t divides both a and b, prove that t divides d. If s is a multiple of both a and b, prove that s is a multiple of m.

13. Let n and a be positive integers and let $d = \gcd(a, n)$. Show that the equation ax mod $n = 1$ has a solution if and only if $d = 1$. (This exercise is referred to in Chapter 2.)

14. Show that $5n + 3$ and $7n + 4$ are relatively prime for all n.

15. Let a, b, s, and t be integers. If $a \bmod st = b \bmod st$, show that $a \bmod s = b \bmod s$ and $a \bmod t = b \bmod t$. What condition on s and t is needed to make the converse true? (This exercise is referred to in Chapter 8.)

16. Use the Euclidean algorithm to find gcd(34, 126) and write it as a linear combination of 34 and 126.

17. Show that $\gcd(a, bc) = 1$ if and only if $\gcd(a, b) = 1$ and $\gcd(a, c) = 1$. (This exercise is referred to in Chapter 8.)

18. Let $p_1, p_2, \ldots, p_n$ be primes. Show that $p_1 p_2 \cdots p_n + 1$ is divisible by none of these primes.

19. Prove that there are infinitely many primes. (*Hint:* Use Exercise 18.)

20. For every positive integer n, prove that $1 + 2 + \cdots + n = n(n + 1)/2$.

21. For every positive integer n, prove that a set with exactly n elements has exactly 2^n subsets (counting the empty set and the entire set).

22. Prove that $2^n 3^{2n} - 1$ is always divisible by 17.

23. Prove that there is some positive integer n such that $n, n + 1, n + 2, \ldots, n + 200$ are all composite.

24. (Generalized Euclid's Lemma) If p is a prime and p divides $a_1 a_2 \cdots a_n$, prove that p divides a_i for some i.

25. Use the Generalized Euclid's Lemma (see Exercise 24) to establish the uniqueness portion of the Fundamental Theorem of Arithmetic.

26. What is the largest bet that cannot be made with chips worth $7.00 and $9.00? Verify that your answer is correct with both forms of induction.

27. Prove that the First Principle of Mathematical Induction is a consequence of the Well Ordering Principle.

28. The Fibonacci numbers are: 1, 1, 2, 3, 5, 8, 13, 21, 34, ... In general, the Fibonacci numbers are defined by $f_1 = 1$, $f_2 = 1$, and for $n \geq 3$, $f_n = f_{n-1} + f_{n-2}$. Prove that the nth Fibonacci number f_n satisfies $f_n < 2^n$.

29. In the cut "As" from *Songs in the Key of Life*, Stevie Wonder mentions the equation $8 \times 8 \times 8 \times 8 = 4$. Find all integers n for which this statement is true, modulo n.

30. Prove that for every integer n, $n^3 \bmod 6 = n \bmod 6$.

31. If it were 2:00 A.M. now, what time would it be 3736 hours from now?

32. Determine the check digit for a money order with identification number 7234541780.

33. Suppose that in one of the noncheck positions of a money order number, the digit 0 is substituted for the digit 9 or vice versa. Prove that this error will not be detected by the check digit. Prove that all other errors involving a single position are detected.

34. Suppose that a money order identification number and check digit of 21720421168 is erroneously copied as 27750421168. Will the check digit detect the error?

35. A transposition error involving distinct adjacent digits is one of the form $\ldots ab \ldots \rightarrow \ldots ba \ldots$ with $a \neq b$. Prove that the money order check digit scheme will not detect such errors unless the check digit itself is transposed.

36. Determine the check digit for the Avis rental car with identification number 540047. (See Example 6.)

37. Show that a substitution of a digit a_i' for the digit a_i ($a_i' \neq a_i$) in a noncheck position of a UPS number is detected if and only if $|a_i - a_i'| \neq 7$.

38. Determine which transposition errors involving adjacent digits are detected by the UPS check digit.

39. Use the UPC scheme to determine the check digit for the number 07312400508.

40. Explain why the check digit for a money order for the number N is the repeated decimal digit in the real number $N \div 9$.

41. The International Standard Book Number (ISBN) $a_1a_2a_3a_4a_5a_6a_7a_8a_9a_{10}$ has the property $(a_1, a_2, \ldots, a_{10}) \cdot (10, 9, 8, 7, 6, 5, 4, 3, 2, 1) \bmod 11 = 0$. The digit a_{10} is the check digit. When a_{10} is required to be 10 to make the dot product 0, the character X is used as the check digit. Verify the check digit for the ISBN assigned to this book.

42. Suppose that an ISBN has a smudged entry where the question mark appears in the number 0-716?-2841-9. Determine the missing digit.

43. Suppose three consecutive digits abc of an ISBN are scrambled as bca. Which such errors will go undetected?

44. The ISBN 0-669-03925-4 is the result of a transposition of two adjacent digits not involving the first or last digit. Determine the correct ISBN.

45. Suppose the weighting vector for ISBNs was changed to (1, 2, 3, 4, 5, 6, 7, 8, 9, 10). Explain how this would affect the check digit.

46. Use the two-check-digit error-correction method described in this chapter to append two check digits to the number 73445860.

47. Suppose that an eight-digit number has two check digits appended using the error-correction method described in this chapter and it is incorrectly transcribed as 4302511568. If exactly one digit is incorrect, determine the correct number.

48. Let S be the set of real numbers. If $a, b \in S$, define $a \sim b$ if $a - b$ is an integer. Show that $\sim$ is an equivalence relation on S. Describe the equivalence classes of S.

49. Let S be the set of integers. If $a, b \in S$, define aRb if $ab \geq 0$. Is R an equivalence relation on S?

50. Let S be the set of integers. If $a, b \in S$, define aRb if $a + b$ is even. Prove that R is an equivalence relation and determine the equivalence classes of S.

51. Complete the proof of Theorem 0.6 by showing that $\sim$ is an equivalence relation on S.

52. Prove that none of the integers 11, 111, 1111, 11111, . . . is a square of an integer.

53. Complete the proof of Theorem 0.7.

COMPUTER EXERCISES ▬▬▬▬▬▬▬▬▬▬▬▬▬▬▬▬▬▬▬▬ ▪▪

There is nothing more practical than a good theory.

LEONID BREZHNEV

Software for the computer exercises in this chapter is available at the website:

http://www.d.umn.edu/~jgallian

1. This software checks the validity of a Postal Service money order number. Use it to verify that 39539881642 is valid. Now enter the same number with one digit incorrect. Was the error detected? Enter the number with the 9 in position 2 replaced with a 0. Was the error detected? Explain why or why not. Enter the number with two digits transposed. Was the error detected? Explain why or why not.

2. This software checks the validity of a UPC number. Use it to verify that 090146003386 is valid. Now enter the same number with one digit incorrect. Was the error detected? Enter the number with two consecutive digits transposed. Was the error detected? Enter the number with the second 3 and the 8 transposed. Was the error detected? Explain why or why not. Enter the number with the 9 and the 1 transposed. Was the error detected? Explain why or why not.

3. This software checks the validity of a UPS number. Use it to verify that 8733456723 is valid. Now enter the same number with one digit incorrect. Was the error detected? Enter the number with two consecutive digits transposed. Was the error detected? Enter the number with the 8 replaced by 1. Was the error detected? Explain why or why not.

4. This software checks the validity of an identification number on a bank check. Use it to verify that 091902049 is valid. Now enter the same number with one digit incorrect. Was the error detected? Enter the number with two consecutive digits transposed. Was the error detected? Enter the number with the 2 and the 4 transposed. Was the error detected? Explain why or why not.

5. This software checks the validity of an ISBN. Use it to verify that 0395872456 is valid. Now enter the same number with one digit incorrect. Was the error detected? Enter the number with two digits transposed (they need not be consecutive). Was the error detected?

6. This software determines the two check digits for the mod 11 decimal error-correction scheme discussed in this chapter. Run the program with the input 21355432, 20965744, 10033456. Then enter these numbers with the two check digits appended with one digit incorrect. Was the error corrected?

Suggested Readings

Linda Deneen, "Secret Encryption with Public Keys," *The UMAP Journal* 8 (1987): 9–29.

> This well-written article describes several ways in which modular arithmetic can be used to code secret messages. They range from a simple scheme used by Julius Caesar to a highly sophisticated scheme invented in 1978 and based on modular n arithmetic, where n has more than 200 digits.

J. A. Gallian, "Assigning Driver's License Numbers," *Mathematics Magazine* 64 (1991): 13–22.

> This article describes various methods used by the states to assign driver's license numbers. Several include check digits for error detection.

J. A. Gallian, "The Mathematics of Identification Numbers," *The College Mathematics Journal* 22 (1991): 194–202.

> This article is a comprehensive survey of check digit schemes that are associated with identification numbers.

J. A. Gallian and S. Winters, "Modular Arithmetic in the Marketplace," *The American Mathematical Monthly* 95 (1988): 548–551.

> This article provides a more detailed analysis of the check digit schemes presented in this chapter. In particular, the error detection rates for the various schemes are given.

PART 2

Groups

For online student resources,
visit this textbook's website at
math.college.hmco.com/students

1 | Introduction to Groups

SYMMETRIES OF A SQUARE

Suppose we remove a square region from a plane, move it in some way, then put the square back into the space it originally occupied. Our goal in this chapter is to describe in some reasonable fashion all possible ways in which this can be done. More specifically, we want to describe the possible relationships between the starting position of the square and its final position in terms of motions. However, we are interested in the net effect of a motion, rather than the motion itself. Thus, for example, we consider a 90° rotation and a 450° rotation as equal, since they have the same effect on every point. With this simplifying convention, it is an easy matter to achieve our goal.

To begin, we can think of the square region as being transparent (glass, say), with the corners marked on one side with the colors blue, white, pink, and green. This makes it easy to distinguish between motions that have different effects. With this marking scheme, we are now in a position to describe, in simple fashion, all possible ways in which a square object can be repositioned. See Figure 1.1. We now claim that any motion—no matter how complicated—is equivalent to one of these eight. To verify this claim, observe that the final position of the square is completely determined by the location and orientation (that is, face up or face down) of any particular corner. But, clearly, there are only four locations and two orientations for a given corner, so there are exactly eight distinct final positions for the corner.

R_0 = Rotation of 0° (no change in position)

R_{90} = Rotation of 90° (counterclockwise)

R_{180} = Rotation of 180°

R_{270} = Rotation of 270°

H = Flip about a horizontal axis

V = Flip about a vertical axis

D = Flip about the main diagonal

D' = Flip about the other diagonal

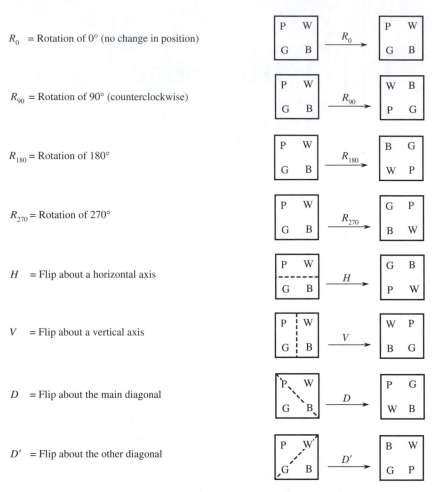

Figure 1.1

Let's investigate some consequences of the fact that every motion is equal to one of the eight listed in Figure 1.1. Suppose a square is repositioned by a rotation of 90° followed by a flip about the horizontal axis of symmetry. In pictures,

Thus, we see that this pair of motions—taken together—is equal to the single motion D. This observation suggests that we can compose two motions to obtain a single motion. And indeed we can, since the eight motions may be viewed as functions from the square region to itself, and as such we can combine them using function composition.

With this in mind, we may now write $HR_{90} = D$. The eight motions R_0, R_{90}, R_{180}, R_{270}, H, V, D, and D', together with the operation composition, form a mathematical system called the *dihedral group of order 8* (the order of a group is the number of elements it contains). It is denoted by D_4. Rather than introduce the formal definition of a group here, let's look at some properties of groups by way of the example D_4.

To facilitate future computations, we construct an *operation table* or *Cayley table* (so named in honor of the prolific English mathematician Arthur Cayley, who first introduced them in 1854) for D_4 below. The circled entry represents the fact that $D = HR_{90}$. (In general, ab denotes the entry at the intersection of the row with a at the left and the column with b at the top.)

	R_0	R_{90}	R_{180}	R_{270}	H	V	D	D'
R_0	R_0	R_{90}	R_{180}	R_{270}	H	V	D	D'
R_{90}	R_{90}	R_{180}	R_{270}	R_0	D'	D	H	V
R_{180}	R_{180}	R_{270}	R_0	R_{90}	V	H	D'	D
R_{270}	R_{270}	R_0	R_{90}	R_{180}	D	D'	V	H
H	H	$\textcircled{D}$	V	D'	R_0	R_{180}	R_{90}	R_{270}
V	V	D'	H	D	R_{180}	R_0	R_{270}	R_{90}
D	D	V	D'	H	R_{270}	R_{90}	R_0	R_{180}
D'	D'	H	D	V	R_{90}	R_{270}	R_{180}	R_0

Notice how orderly this table looks! This is no accident. Perhaps the most important feature of this table is that it has been completely filled in without introducing any new motions. Of course, this is because, as we have already pointed out, any sequence of motions turns out to be the same as one of these eight. Algebraically, this says that if A and B are in D_4, then so is AB. This property is called *closure*, and it is one of the requirements for a mathematical system to be a group. Next, notice that if A is any element of D_4, then $AR_0 = R_0A = A$. Thus, combining any element A on either side with R_0 yields A back again. An element R_0 with this property is called an *identity*, and every group must have one. Moreover, we see that for each element A in D_4, there is exactly one element B in D_4 such that $AB = BA = R_0$. In this case, B is said to be the *inverse* of A and vice versa. For example, R_{90} and R_{270} are inverses of each other, and H is its own inverse. The term *inverse* is a descriptive one, for if A and B are inverses of each other, then B "undoes" whatever A "does," in the sense that A and B

taken together in either order produce R_0, representing no change. Another striking feature of the table is that every element of D_4 appears exactly once in each row and column. This feature is something that all groups must have, and, indeed, it is quite useful to keep this fact in mind when constructing the table in the first place.

Another property of D_4 deserves special comment. Observe that $HD \neq DH$ but $R_{90}R_{180} = R_{180}R_{90}$. Thus, in a group, AB may or may not be the same as BA. If it happens that $AB = BA$ for *all* choices of group elements A and B, we say the group is *commutative* or—better yet—*Abelian* (in honor of the great Norwegian mathematician Niels Abel). Otherwise, we say the group is *non-Abelian*.

Thus far, we have illustrated, by way of D_4, three of the four conditions that define a group—namely, closure, existence of an identity, and existence of inverses. The remaining condition required for a group is *associativity*; that is, $(ab)c = a(bc)$ for all a, b, c in the set. To be sure that D_4 is indeed a group, we should check this equation for each of the $8^3 = 512$ possible choices of a, b, and c in D_4. In practice, however, this is rarely done! Here, for example, we simply observe that the eight motions are functions and the operation is function composition. Then, since function composition is associative, we do not have to check the equations.

THE DIHEDRAL GROUPS

The analysis carried out above for a square can similarly be done for an equilateral triangle or regular pentagon or, indeed, any regular n-gon ($n \geq 3$). The corresponding group is denoted by D_n and is called the *dihedral group of order 2n*.

The dihedral groups arise frequently in art and nature. Many of the decorative designs used on floor coverings, pottery, and buildings have one of the dihedral groups as a group of symmetry. Corporation logos are rich sources of dihedral symmetry [1]. Chrysler's logo has D_5 as a symmetry group, and that of Mercedes-Benz has D_3. The ubiquitous five-pointed star has symmetry group D_5. The phylum Echinodermata contains many sea animals such as starfish, sea cucumbers, feather stars, and sand dollars that exhibit patterns with D_5 symmetry.

Chemists classify molecules according to their symmetry. Moreover, symmetry considerations are applied in orbital calculations, in determining energy levels of atoms and molecules, and in the study of molecular vibrations. The symmetry group of a pyramidal molecule such as ammonia (NH_3), depicted in Figure 1.2, has symmetry group D_3.

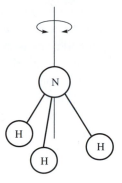

Figure 1.2   A pyramidal molecule with symmetry group D_3.

Mineralogists determine the internal structures of crystals (that is, rigid bodies in which the particles are arranged in three-dimensional repeating patterns—table salt and table sugar are two examples) by studying two-dimensional x-ray projections of the atomic makeup of the crystals. The symmetry present in the projections reveals the internal symmetry of the crystals themselves. Commonly occurring symmetry patterns are D_4 and D_6 (see Figure 1.3). Interestingly, it is mathematically impossible for a crystal to possess a D_n symmetry pattern with $n = 5$ or $n > 6$.

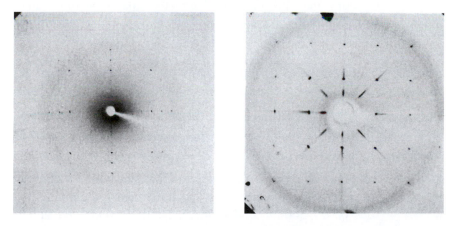

Figure 1.3   X-ray diffraction photos revealing D_4 symmetry patterns in crystals.

The dihedral group of order $2n$ is often called the *group of symmetries of a regular n-gon*. A *plane symmetry* of a figure F in a plane is a function from the plane to itself that carries F onto F and preserves distances; that is, for any points p and q in the plane, the distance from the image of p to the image of q is the same as the distance from p to q. (The term *symmetry* is from the Greek word *symmetros*, meaning "of like measure.") The *symmetry group* of a plane figure is the set of all symmetries of the figure. Symmetries in three dimensions are defined analogously. Obviously, a rotation of a plane about a point in the plane is a symmetry of the plane and a rotation about a line in three dimensions is a symmetry in three-dimensional space. Similarly, any translation of a plane or of three-dimensional space is a symmetry. A *reflection across a line L* is that function that leaves every point of L fixed and takes any point q, not on L, to the point q' so that L is the perpendicular bisector of the line segment joining q and q' (see Figure 1.4). A reflection across a plane in three dimensions is defined analogously. Notice that the restriction of a 180° rotation about a line L in three dimensions to a plane containing L is a reflection across L in the plane. Thus, in the dihedral groups, the motions that we described as flips about axes of symmetry in three dimensions (for example, H, V, D, D') are reflections across lines in two dimensions. Just as a reflection across a line is a plane symmetry that cannot be achieved by a physical motion of the plane in two dimensions, a reflection across a plane is a three-dimensional symmetry that cannot be achieved by a physical motion of three-dimensional space. A cup, for instance, has reflective symmetry across the plane bisecting the cup, but this symmetry cannot be duplicated with a physical motion in three dimensions.

Figure 1.4

Many objects and figures have rotational symmetry but not reflective symmetry. A symmetry group consisting of the rotational symmetries of 0°, 360°/n, 2(360°)/n, . . . , $(n - 1)360°/n$, and no other symmetries is called a *cyclic rotation group of order n* and is denoted by $\langle R_{360/n} \rangle$. Cyclic rotation groups, along with dihedral groups, are favorites of artists, designers, and nature. Figure 1.5 illustrates with corporate logos the cyclic rotation groups of orders 2, 3, 4, 5, 6, 8, 16, and 20.

Further examples of the occurrence of dihedral groups and cyclic groups in art and nature can be found in the references. A study of symmetry in greater depth is given in Chapters 27 and 28.

Figure 1.5 Logos with cyclic rotation symmetry groups.

EXERCISES

The only way to learn mathematics is to do mathematics.

PAUL HALMOS,
Hilbert Space Problem Book

1. With pictures and words, describe each symmetry in D_3 (the set of symmetries of an equilateral triangle).
2. Write out a complete Cayley table for D_3.
3. Is D_3 Abelian?
4. Describe in pictures or words the elements of D_5 (symmetries of a regular pentagon).
5. For $n \geq 3$, describe the elements of D_n. (*Hint:* You will need to consider two cases—n even and n odd.) How many elements does D_n have?
6. In D_n, explain geometrically why a reflection followed by a reflection must be a rotation.
7. In D_n, explain geometrically why a rotation followed by a rotation must be a rotation.
8. In D_n, explain geometrically why a rotation and a reflection taken together in either order must be a reflection.
9. Associate the number $+1$ with a rotation and the number -1 with a reflection. Describe an analogy between multiplying these two numbers and multiplying elements of D_n.

10. If r_1, r_2, and r_3 represent rotations from D_n and f_1, f_2, and f_3 represent reflections from D_n, determine whether $r_1r_2 f_1r_3 f_2 f_3r_3$ is a rotation or a reflection.

11. Find elements A, B, and C in D_4 such that $AB = BC$ but $A \neq C$. (Thus, "cross cancellation" is not valid.)

12. Explain what the following diagram proves about the group D_n.

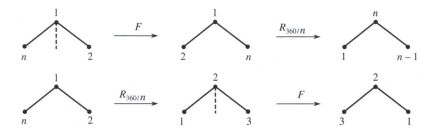

13. Describe the symmetries of a nonsquare rectangle. Construct the corresponding Cayley table.

14. Describe the symmetries of a parallelogram that is neither a rectangle nor a rhombus. Describe the symmetries of a rhombus that is not a rectangle.

15. Describe the symmetries of a noncircular ellipse. Do the same for a hyperbola.

16. Consider an infinitely long strip of equally spaced H's:

$$\cdots H H H H \cdots$$

Describe the symmetries of this strip. Is the group of symmetries of the strip Abelian?

17. For each of the snowflakes in the figure, find the symmetry group and locate the axes of reflective symmetry (disregard imperfections).

Photographs of snowflakes from the Bentley and Humphrey atlas.

18. Does the agitator of a washing machine have a cyclic symmetry group or a dihedral symmetry group?

19. Does an airplane propeller have a cyclic symmetry group or a dihedral symmetry group?

20. Bottle caps that are pried off typically have 22 ridges around the rim. Find the symmetry group of such a cap.

21. Determine the symmetry group of the outer shell of the cross section of the human immunodeficiency virus (HIV) shown below.

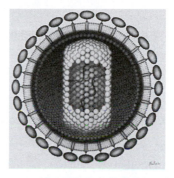

22. For each design below, determine the symmetry group (ignore imperfections).

23. What would the effect be if a six-bladed ceiling fan were designed so that the centerlines of two of the blades were at a 70° angle while all the other blades were set at a 58° angle?

Reference

1. B. B. Capitman, *American Trademark Designs*, New York: Dover, 1976.

Suggested Reading

Michael Field and Martin Golubitsky, *Symmetry in Chaos*, Oxford University Press, 1992.

This book has many beautiful symmetric designs that arise in chaotic dynamic systems.

Suggested Website

http://ccins.camosun.bc.ca/~jbritton/jbsymteslk.htm

This spectacular website on symmetry and tessellations has numerous activities and links to many other sites on related topics. It is a wonderful website for K–12 teachers and students.

Niels Abel

He [Abel] has left mathematicians something to keep them busy for five hundred years.

CHARLES HERMITE

This stamp was issued in 1929 to commemorate the 100th anniversary of Abel's death.

A 500-kroner bank note first issued by Norway in 1948.

NIELS HENRIK ABEL, one of the foremost mathematicians of the 19th century, was born in Norway on August 5, 1802. At the age of 16, he began reading the classic mathematical works of Newton, Euler, Lagrange, and Gauss. When Abel was 18 years old, his father died, and the burden of supporting the family fell upon him. He took in private pupils and did odd jobs, while continuing to do mathematical research. At the age of 19, Abel solved a problem that had vexed leading mathematicians for hundreds of years. He proved that, unlike the situation for equations of degree 4 or less, there is no finite (closed) formula for the solution of the general fifth-degree equation.

Although Abel died long before the advent of the subjects that now make up abstract algebra, his solution to the quintic problem laid the groundwork for many of these subjects. In addition to his work in the theory of equations, Abel made outstanding contributions to the theory of elliptic functions, elliptic integrals, Abelian integrals, and infinite series. Just when his work was beginning to receive the attention it deserved, Abel contracted tuberculosis. He died on April 6, 1829, at the age of 26.

To find more information about Abel, visit:

http://www-groups.dcs.st-and.ac.uk/~history/

2

Groups

A good stock of examples, as large as possible, is indispensable for a thorough understanding of any concept, and when I want to learn something new, I make it my first job to build one.

PAUL R. HALMOS

DEFINITION AND EXAMPLES OF GROUPS

The term *group* was used by Galois around 1830 to describe sets of one-to-one functions on finite sets that could be grouped together to form a closed set. As is the case with most fundamental concepts in mathematics, the modern definition of a group that follows is the result of a long evolutionary process. Although this definition was given by both Heinrich Weber and Walter von Dyck in 1882, it did not gain universal acceptance until the twentieth century.

DEFINITION *Binary Operation*

Let G be a set. A *binary operation* on G is a function that assigns each ordered pair of elements of G an element of G.

A binary operation on a set G, then, is simply a method (or formula) by which the members of an ordered pair from G combine to yield a new member of G. This condition is called *closure*. The most familiar binary operations are ordinary addition, subtraction, and multiplication of integers. Division of integers is not a binary operation on the integers because an integer divided by an integer need not be an integer.

The binary operations addition modulo n and multiplication modulo n on the set $\{0, 1, 2, \ldots, n - 1\}$, which we denote by Z_n, play an extremely important role in abstract algebra. In certain situations we will want to combine the elements of Z_n by addition modulo n only; in other situations we will want to use both addition modulo n and multiplication modulo n to combine the elements. It will be clear from the context whether we are using addition only or addition and multiplication. For example, when multiplying matrices with entries from Z_n, we will need both addition modulo n and multiplication modulo n.

DEFINITION *Group*

Let G be a nonempty set together with a binary operation (usually called multiplication) that assigns to each ordered pair (a, b) of elements of G an element in G denoted by ab. We say G is a *group* under this operation if the following three properties are satisfied.

1. *Associativity.* The operation is associative; that is, $(ab)c = a(bc)$ for all a, b, c in G.
2. *Identity.* There is an element e (called the *identity*) in G such that $ae = ea = a$ for all a in G.
3. *Inverses.* For each element a in G, there is an element b in G (called an *inverse* of a) such that $ab = ba = e$.

In words, then, a group is a set together with an associative operation such that there is an identity, every element has an inverse, and any pair of elements can be combined without going outside the set. Be sure to verify closure when testing for a group (see Example 5). Notice that if a is the inverse of b, then b is the inverse of a.

If a group has the property that $ab = ba$ for every pair of elements a and b, we say the group is *Abelian*. A group is *non-Abelian* if there is some pair of elements a and b for which $ab \neq ba$. When encountering a particular group for the first time, one should determine whether or not it is Abelian.

Now that we have the formal definition of a group, our first job is to build a good stock of examples. These examples will be used throughout the text to illustrate the theorems. (The best way to grasp the meat of a theorem is to see what it says in specific cases.) As we progress, the reader is bound to have hunches and conjectures that can be tested against the stock of examples. To develop a better understanding of the following examples, the reader should supply the missing details.

EXAMPLE 1   The set of integers Z (so denoted because the German word for numbers is *Zahlen*), the set of rational numbers Q (for quotient), and the set of real numbers **R** are all groups under ordinary addition. In each case, the identity is 0 and the inverse of a is $-a$.   ◆

EXAMPLE 2   The set of integers under ordinary multiplication is not a group. Since the number 1 is the identity, property 3 fails. For example, there is no *integer b* such that $5b = 1$.   ◆

EXAMPLE 3   The subset $\{1, -1, i, -i\}$ of the complex numbers is a group under complex multiplication. Note that -1 is its own inverse, whereas the inverse of i is $-i$, and vice versa.   ◆

EXAMPLE 4   The set Q^+ of positive rationals is a group under ordinary multiplication. The inverse of any a is $1/a = a^{-1}$.   ◆

EXAMPLE 5   The set S of positive irrational numbers together with 1 under multiplication satisfies the three properties given in the definition of a group but is not a group. Indeed, $\sqrt{2} \cdot \sqrt{2} = 2$, so S is not closed under multiplication.   ◆

EXAMPLE 6   A rectangular array of the form $\begin{bmatrix} a & b \\ c & d \end{bmatrix}$ is called a 2×2 *matrix*. The set of all 2×2 matrices with real entries is a group under componentwise addition. That is,

$$\begin{bmatrix} a_1 & b_1 \\ c_1 & d_1 \end{bmatrix} + \begin{bmatrix} a_2 & b_2 \\ c_2 & d_2 \end{bmatrix} = \begin{bmatrix} a_1 + a_2 & b_1 + b_2 \\ c_1 + c_2 & d_1 + d_2 \end{bmatrix}$$

The identity is $\begin{bmatrix} 0 & 0 \\ 0 & 0 \end{bmatrix}$, and the inverse of $\begin{bmatrix} a & b \\ c & d \end{bmatrix}$ is $\begin{bmatrix} -a & -b \\ -c & -d \end{bmatrix}$.   ◆

EXAMPLE 7   The set $Z_n = \{0, 1, \ldots, n - 1\}$ for $n \geq 1$ is a group under addition modulo n. For any $j > 0$ in Z_n, the inverse of j is $n - j$. This group is usually referred to as the *group of integers modulo n.*   ◆

As we have seen, the real numbers, the 2×2 matrices with real entries, and the integers modulo n are all groups under the appropriate addition. But what about multiplication? In each case, the existence of some elements that do not have inverses prevents the set from being a group under the usual multiplication. However, we can form a group in each case by simply throwing out the rascals. Examples 8, 9, and 11 illustrate this.

EXAMPLE 8   The set **R*** of nonzero real numbers is a group under ordinary multiplication. The identity is 1. The inverse of a is $1/a$.   ◆

EXAMPLE 9[†]   The *determinant* of the 2 × 2 matrix $\begin{bmatrix} a & b \\ c & d \end{bmatrix}$ is the number $ad - bc$. If A is a 2 × 2 matrix, det A denotes the determinant of A. The set

$$GL(2, \mathbf{R}) = \left\{ \begin{bmatrix} a & b \\ c & d \end{bmatrix} \middle| a, b, c, d \in \mathbf{R}, ad - bc \neq 0 \right\}$$

of 2 × 2 matrices with real entries and nonzero determinant is a non-Abelian group under the operation

$$\begin{bmatrix} a_1 & b_1 \\ c_1 & d_1 \end{bmatrix} \begin{bmatrix} a_2 & b_2 \\ c_2 & d_2 \end{bmatrix} = \begin{bmatrix} a_1a_2 + b_1c_2 & a_1b_2 + b_1d_2 \\ c_1a_2 + d_1c_2 & c_1b_2 + d_1d_2 \end{bmatrix}.$$

The first step in verifying that this set is a group is to show that the product of two matrices with nonzero determinant also has nonzero determinant. This follows from the fact that for any pair of 2 × 2 matrices A and B, det (AB) = (det A)(det B).

Associativity can be verified by direct (but cumbersome) calculations. The identity is $\begin{bmatrix} 1 & 0 \\ 0 & 1 \end{bmatrix}$; the inverse of $\begin{bmatrix} a & b \\ c & d \end{bmatrix}$ is

$$\begin{bmatrix} \dfrac{d}{ad - bc} & \dfrac{-b}{ad - bc} \\ \dfrac{-c}{ad - bc} & \dfrac{a}{ad - bc} \end{bmatrix}$$

(explaining the requirement that $ad - bc \neq 0$). This very important non-Abelian group is called the *general linear group* of 2 × 2 matrices over **R**.   ◆

EXAMPLE 10   The set of all 2 × 2 matrices with real number entries is not a group under the operation defined in Example 9. Inverses do not exist when the determinant is 0.   ◆

[†]For simplicity, we have restricted our matrix examples to the 2 × 2 case. However, readers who have had linear algebra can readily generalize to $n \times n$ matrices.

Now that we have shown how to make subsets of the real numbers and subsets of the set of 2 × 2 matrices into multiplicative groups, we next consider the integers under multiplication modulo n.

EXAMPLE 11   (L. Euler, 1761)
By Exercise 13 in Chapter 0, an integer a has a multiplicative inverse modulo n if and only if a and n are relatively prime. So, for each $n > 1$, we define $U(n)$ to be the set of all positive integers less than n and relatively prime to n. Then $U(n)$ is a group under multiplication modulo n. (We leave it to the reader to check that this set is closed under this operation.)

For $n = 10$, we have $U(10) = \{1, 3, 7, 9\}$. The Cayley table for $U(10)$ is

mod 10	1	3	7	9
1	1	3	7	9
3	3	9	1	7
7	7	1	9	3
9	9	7	3	1

(Recall that $ab \bmod n$ is the unique integer r with the property $a \cdot b = nq + r$, where $0 \le r < n$ and $a \cdot b$ is ordinary multiplication.) In the case that n is a prime, $U(n) = \{1, 2, \ldots, n - 1\}$.   ◆

In his classic book *Lehrbuch der Algebra*, published in 1899, Heinrich Weber gave an extensive treatment of the groups $U(n)$ and described them as the most important examples of finite Abelian groups.

EXAMPLE 12   The set $\{0, 1, 2, 3\}$ is not a group under multiplication modulo 4. Although 1 and 3 have inverses, the elements 0 and 2 do not.   ◆

EXAMPLE 13   The set of integers under subtraction is not a group, since the operation is not associative.   ◆

With the examples given thus far as a guide, it is wise for the reader to pause here and think of his or her own examples. Study actively! Don't just read along and be spoon-fed by the book.

EXAMPLE 14   For all integers $n \ge 1$, the set of complex nth roots of unity

$$\left\{ \cos \frac{k \cdot 360°}{n} + i \sin \frac{k \cdot 360°}{n} \,\middle|\, k = 0, 1, 2, \ldots, n - 1 \right\}$$

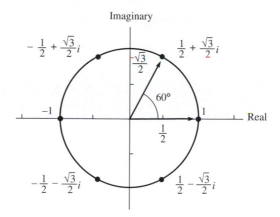

Figure 2.1

(i.e., complex zeros of $x^n - 1$) is a group under multiplication. (See DeMoivre's Theorem—Example 8 in Chapter 0.) Compare this group with the one in Example 3. ◆

The complex number $a + bi$ can be represented geometrically as the point (a, b) in a plane coordinatized by a horizontal real axis and a vertical i or imaginary axis. The distance from the point $a + bi$ to the origin is $\sqrt{a^2 + b^2}$ and is often denoted by $|a + bi|$. For any angle θ, the line segment joining the complex number $\cos \theta + i \sin \theta$ and the origin forms an angle of θ with the positive real axis. Thus, the six complex zeros of $x^6 = 1$ are located at points around the circle of radius 1, 60° apart, as shown in Figure 2.1.

EXAMPLE 15   The set $\mathbf{R}^n = \{(a_1, a_2, \ldots, a_n) \mid a_1, a_2, \ldots, a_n \in \mathbf{R}\}$ is a group under componentwise addition [i.e., $(a_1, a_2, \ldots, a_n) + (b_1, b_2, \ldots, b_n) = (a_1 + b_1, a_2 + b_2, \ldots, a_n + b_n)$]. ◆

EXAMPLE 16   For a fixed point (a, b) in $\mathbf{R}^2$, define $T_{a,b}: \mathbf{R}^2 \rightarrow \mathbf{R}^2$ by $(x, y) \rightarrow (x + a, y + b)$. Then $G = \{T_{a,b} \mid a, b \in \mathbf{R}\}$ is a group under function composition. Straightforward calculations show that $T_{a,b}T_{c,d} = T_{a+c,b+d}$. From this formula we may observe that G is closed, $T_{0,0}$ is the identity, the inverse of $T_{a,b}$ is $T_{-a,-b}$, and G is Abelian. Function composition is always associative. The elements of G are called *translations*. ◆

EXAMPLE 17   The set of all 2×2 matrices with determinant 1 with entries from Q (rationals), $\mathbf{R}$ (reals), $\mathbf{C}$ (complex numbers), or Z_p (p a prime) is a non-Abelian group under matrix multiplication. This group is called the *special linear group* of 2×2 matrices over Q, $\mathbf{R}$, $\mathbf{C}$, or Z_p, respectively. If the entries are from F, where F is any of the above, we denote this group by $SL(2, F)$. For the group $SL(2, F)$, the formula given in Example 9 for the inverse of $\begin{bmatrix} a & b \\ c & d \end{bmatrix}$ simplifies to $\begin{bmatrix} d & -b \\ -c & a \end{bmatrix}$. When the matrix entries are from Z_p, we use modulo p arithmetic to compute determinants, matrix products, and inverses. To illustrate the case $SL(2, Z_5)$, consider the element $A = \begin{bmatrix} 3 & 4 \\ 4 & 4 \end{bmatrix}$. Then $\det A = 3 \cdot 4 - 4 \cdot 4 = -4 \bmod 5 = 1$ and the inverse of A is $\begin{bmatrix} 4 & -4 \\ -4 & 3 \end{bmatrix} = \begin{bmatrix} 4 & 1 \\ 1 & 3 \end{bmatrix}$. Note that $\begin{bmatrix} 3 & 4 \\ 4 & 4 \end{bmatrix} \begin{bmatrix} 4 & 1 \\ 1 & 3 \end{bmatrix} = \begin{bmatrix} 1 & 0 \\ 0 & 1 \end{bmatrix}$ when the arithmetic is done modulo 5. ◆

Example 9 is a special case of the following general construction.

EXAMPLE 18   Let F be any of Q, R, C, or Z_p (p a prime). The set $GL(2, F)$ of all 2×2 matrices with nonzero determinants and entries from F is a non-Abelian group under matrix multiplication. As in Example 17, when F is Z_p, modulo p arithmetic is used to calculate determinants, the matrix products, and inverses. The formula given in Example 9 for the inverse of $\begin{bmatrix} a & b \\ c & d \end{bmatrix}$ remains valid for elements from $GL(2, Z_p)$ provided we interpret division by $ad - bc$ as multiplication by the inverse of $ad - bc$ modulo p. For example, in $GL(2, Z_7)$, consider $\begin{bmatrix} 4 & 5 \\ 6 & 3 \end{bmatrix}$. Then the determinant $(ad - bc) \bmod 7$ is $(12 - 30) \bmod 7 = -18 \bmod 7 = 3 \bmod 7$ and the inverse of 3 is 5 [since $(3 \cdot 5) \bmod 7 = 1$]. So, the inverse of $\begin{bmatrix} 4 & 5 \\ 6 & 3 \end{bmatrix}$ is $\begin{bmatrix} 3 \cdot 5 & 2 \cdot 5 \\ 1 \cdot 5 & 4 \cdot 5 \end{bmatrix} = \begin{bmatrix} 1 & 3 \\ 5 & 6 \end{bmatrix}$. [The reader should check that $\begin{bmatrix} 4 & 5 \\ 6 & 3 \end{bmatrix} \begin{bmatrix} 1 & 3 \\ 5 & 6 \end{bmatrix} = \begin{bmatrix} 1 & 0 \\ 0 & 1 \end{bmatrix}$ in $GL(2, Z_7)$]. ◆

EXAMPLE 19   The set $\{1, 2, \ldots, n - 1\}$ is a group under multiplication modulo n if and only if n is prime. ◆

EXAMPLE 20   The set of all symmetries of the infinite ornamental pattern in which arrowheads are spaced uniformly a unit apart along a line is an Abelian group under composition. Let T denote a translation to the right by one unit, T^{-1} a translation to the left by one unit, and H a reflection across the horizontal line of the figure. Then, every member of the group is of the form $x_1 x_2 \cdots x_n$, where each $x_i \in \{T, T^{-1}, H\}$. In this case, we say that T, T^{-1}, and H *generate* the group. ◆

Table 2.1 summarizes many of the specific groups that we have presented thus far.

As the examples above demonstrate, the notion of a group is a very broad one indeed. The goal of the axiomatic approach is to find properties general enough to permit many diverse examples having these properties and specific enough to allow one to deduce many interesting consequences.

The goal of abstract algebra is to discover truths about algebraic systems (that is, sets with one or more binary operations) that are independent of the specific nature of the operations. All one knows or needs to know is that these operations, whatever they may be,

TABLE 2.1   Summary of Group Examples (F can be any of Q, $\mathbf{R}$, $\mathbf{C}$, or Z_p; L is a reflection)

Group	Operation	Identity	Form of Element	Inverse	Abelian
Z	Addition	0	k	$-k$	Yes
Q^+	Multiplication	1	m/n, $m, n > 0$	n/m	Yes
Z_n	Addition mod n	0	k	$n - k$	Yes
$\mathbf{R}^*$	Multiplication	1	x	$1/x$	Yes
$GL(2, F)$	Matrix multiplication	$\begin{bmatrix} 1 & 0 \\ 0 & 1 \end{bmatrix}$	$\begin{bmatrix} a & b \\ c & d \end{bmatrix}$, $ad - bc \neq 0$	$\begin{bmatrix} \dfrac{d}{ad - bc} & \dfrac{-b}{ad - bc} \\ \dfrac{-c}{ad - bc} & \dfrac{a}{ad - bc} \end{bmatrix}$	No
$U(n)$	Multiplication mod n	1	k, $\gcd(k, n) = 1$	Solution to $kx \bmod n = 1$	Yes
$\mathbf{R}^n$	Componentwise addition	$(0, 0, \ldots, 0)$	$(a_1, a_2, \ldots, a_n)$	$(-a_1, -a_2, \ldots, -a_n)$	Yes
$SL(2, F)$	Matrix multiplication	$\begin{bmatrix} 1 & 0 \\ 0 & 1 \end{bmatrix}$	$\begin{bmatrix} a & b \\ c & d \end{bmatrix}$, $ad - bc = 1$	$\begin{bmatrix} d & -b \\ -c & a \end{bmatrix}$	No
D_n	Composition	R_0	R_α, L	$R_{360 - \alpha}$, L	No

have certain properties. We then seek to deduce consequences of these properties. This is why this branch of mathematics is called *abstract* algebra. It must be remembered, however, that when a specific group is being discussed, a specific operation must be given (at least implicitly).

ELEMENTARY PROPERTIES OF GROUPS

Now that we have seen many diverse examples of groups, we wish to deduce some properties that they share. The definition itself raises some fundamental questions. Every group has *an* identity. Could a group have more than one? Every group element has *an* inverse. Could an element have more than one? The examples suggest not. But examples can only suggest. One cannot prove that every group has a unique identity by looking at examples, because each example inherently has properties that may not be shared by all groups. We are forced to restrict ourselves to the properties that all groups have; that is, we must view groups as abstract entities rather than argue by example. The next three theorems illustrate the abstract approach.

Theorem 2.1   **Uniqueness of the Identity**

> *In a group G, there is only one identity element.*

PROOF.   Suppose both e and e' are identities of G. Then,

1. $ae = a$ for all a in G, and
2. $e'a = a$ for all a in G.

The choices of $a = e'$ in (1) and $a = e$ in (2) yield $e'e = e'$ and $e'e = e$. Thus, e and e' are both equal to $e'e$ and so are equal to each other.   ■ ■

Because of this theorem, we may unambiguously speak of "the identity" of a group and denote it by "e" (because the German word for identity is *Einheit*).

Theorem 2.2   **Cancellation**

> *In a group G, the right and left cancellation laws hold; that is, ba = ca implies b = c, and ab = ac implies b = c.*

PROOF.   Suppose $ba = ca$. Let a' be an inverse of a. Then, multiplying on the right by a' yields $(ba)a' = (ca)a'$. Associativity yields $b(aa') = c(aa')$. Then, $be = ce$ and, therefore, $b = c$ as desired. Similarly, one can prove that $ab = ac$ implies $b = c$ by multiplying by a' on the left.   ■ ■

A consequence of the cancellation property is the fact that in a Cayley table for a group, each group element occurs exactly once in each row and column (see Exercise 23). Another consequence of the cancellation property is the uniqueness of inverses.

Theorem 2.3   **Uniqueness of Inverses**

> For each element a in a group G, there is a unique element b in G such that $ab = ba = e$.

PROOF.   Suppose b and c are both inverses of a. Then $ab = e$ and $ac = e$, so that $ab = ac$. Now cancel a.   ■ ■

As was the case with the identity element, it is reasonable, in view of Theorem 2.3, to speak of "the inverse" of an element g of a group; in fact, we may unambiguously denote it by g^{-1}. This notation is suggested by that used for ordinary real numbers under multiplication. Similarly, when n is a positive integer, the associative law allows us to use g^n to denote the unambiguous product

$$\underbrace{gg \cdots g}_{n \text{ factors}}.$$

We define $g^0 = e$. When n is negative, we define $g^n = (g^{-1})^{|n|}$ [for example, $g^{-3} = (g^{-1})^3$]. Unlike for real numbers, in an abstract group we do not permit noninteger exponents such as $g^{1/2}$. With this notation, the familiar laws of exponents hold for groups; that is, for all integers m and n and any group element g, we have $g^m g^n = g^{m+n}$ and $(g^m)^n = g^{mn}$. Although the way one manipulates the group expressions $g^m g^n$ and $(g^m)^n$ coincides with the laws of exponents for real numbers, the laws of exponents fail to hold for expressions involving two group elements. Thus, for groups in general, $(ab)^n \neq a^n b^n$ (see Exercise 15). Also, one must be careful with this notation when dealing with a specific group whose binary operation is addition and is denoted by "+." In this case, the definitions and group properties expressed in multiplicative notation must be translated to additive notation. For example, the inverse of g is written as $-g$. Likewise, for example, g^3

TABLE 2.2

	Multiplicative Group		Additive Group
$a \cdot b$ or ab	Multiplication	$a + b$	Addition
e or 1	Identity or one	0	Zero
a^{-1}	Multiplicative inverse of a	$-a$	Additive inverse of a
a^n	Power of a	na	Multiple of a
ab^{-1}	Quotient	$a - b$	Difference

means $g + g + g$ and is usually written as $3g$, whereas g^{-3} means $(-g) + (-g) + (-g)$ and is written as $-3g$. When additive notation is used, do not interpret "ng" as combining n and g under the group operation; n may not even be an element of the group! Table 2.2 shows the common notation and corresponding terminology for groups under multiplication and groups under addition. As is the case for real numbers, we use $a - b$ as an abbreviation for $a + (-b)$.

Because of the associative property, we may unambiguously write the expression abc, for this can be reasonably interpreted as only $(ab)c$ or $a(bc)$, which are equal. In fact, by using induction and repeated application of the associative property, one can prove a general associative property that essentially means that parentheses can be inserted or deleted at will without affecting the value of a product involving any number of group elements. Thus,

$$a^2(bcdb^2) = a^2b(cd)b^2 = (a^2b)(cd)b^2 = a(abcdb)b,$$

and so on.

HISTORICAL NOTE

We conclude this chapter with a bit of history concerning the non-commutativity of matrix multiplication. In 1925, quantum theory was replete with annoying and puzzling ambiguities. It was Werner Heisenberg who recognized the cause. He observed that the product of the quantum-theoretical analogs of the classical Fourier series did not necessarily commute. For all his boldness, this shook Heisenberg. As he later recalled [2, p. 94]:

> In my paper the fact that XY was not equal to YX was very disagreeable to me. I felt this was the only point of difficulty in the whole scheme, otherwise I would be perfectly happy. But this difficulty had worried me and I was not able to solve it.

Heisenberg asked his teacher, Max Born, if his ideas were worth publishing. Born was fascinated and deeply impressed by Heisenberg's new approach. Born wrote [1, p. 217]:

> After having sent off Heisenberg's paper to the *Zeitschrift für Physik* for publication, I began to ponder over his symbolic multiplication, and was soon so involved in it that I thought about it for the whole day and could hardly sleep at night. For I felt there was something fundamental behind it, the consummation of our endeavors of many years. And one morning, about the 10 July 1925, I suddenly saw light: Heisenberg's symbolic multiplication was nothing but the matrix calculus, well-known to me since my student days from Rosanes' lectures in Breslau.

Born and his student, Pascual Jordan, reformulated Heisenberg's ideas in terms of matrices, but it was Heisenberg who was credited with the formulation. In his autobiography, Born laments [1, p. 219]:

> Nowadays the textbooks speak without exception of Heisenberg's matrices, Heisenberg's commutation law, and Dirac's field quantization.
>
> In fact, Heisenberg knew at that time very little of matrices and had to study them.

Upon learning in 1933 that he was to receive the Nobel Prize with Dirac and Schrödinger for this work, Heisenberg wrote to Born [1, p. 220]:

> If I have not written to you for such a long time, and have not thanked you for your congratulations, it was partly because of my rather bad conscience with respect to you. The fact that I am to receive the Nobel Prize alone, for work done in Göttingen in collaboration—you, Jordan, and I—this fact depresses me and I hardly know what to write to you. I am, of course, glad that our common efforts are now appreciated, and I enjoy the recollection of the beautiful time of collaboration. I also believe that all good physicists know how great was your and Jordan's contribution to the structure of quantum mechanics—and this remains unchanged by a wrong decision from outside. Yet I myself can do nothing but thank you again for all the fine collaboration, and feel a little ashamed.

The story has a happy ending, however, because Born received the Nobel Prize in 1954 for his fundamental work in quantum mechanics.

EXERCISES

"For example," is not proof.

JEWISH PROVERB

1. Give two reasons why the set of odd integers under addition is not a group.

2. Referring to Example 13, verify the assertion that subtraction is not associative.

3. Show that $\{1, 2, 3\}$ under multiplication modulo 4 is not a group but that $\{1, 2, 3, 4\}$ under multiplication modulo 5 is a group.

4. Show that the group $GL(2, \mathbf{R})$ of Example 9 is non-Abelian by exhibiting a pair of matrices A and B in $GL(2, \mathbf{R})$ such that $AB \neq BA$.

5. Find the inverse of the element $\begin{bmatrix} 2 & 6 \\ 3 & 5 \end{bmatrix}$ in $GL(2, Z_{11})$.

6. Give an example of group elements a and b with the property that $a^{-1}ba \neq b$.

7. Translate each of the following multiplicative expressions into its additive counterpart. Assume that addition is commutative.
   a. a^2b^3
   b. $a^{-2}(b^{-1}c)^2$
   c. $(ab^2)^{-3}c^2 = e$

8. Show that the set $\{5, 15, 25, 35\}$ is a group under multiplication modulo 40. What is the identity element of this group? Can you see any relationship between this group and $U(8)$?

9. (From the GRE Practice Exam) Let p and q be distinct primes. Suppose that H is a proper subset of the integers and H is a group under addition that contains exactly three elements of the set $\{p, p + q, pq, p^q, q^p\}$. Determine which of the following are the three elements in H.
   a. pq, p^q, q^p
   b. $p + q, pq, p^q$
   c. $p, p + q, pq$
   d. p, p^q, q^p
   e. p, pq, p^q

10. In the notation of Example 16, verify that $T_{a,b}T_{c,d} = T_{a+c,b+d}$.

11. Prove that the set of all 2×2 matrices with entries from $\mathbf{R}$ and determinant $+1$ is a group under matrix multiplication.

12. For any integer $n > 2$, show that there are at least two elements in $U(n)$ that satisfy $x^2 = 1$.

13. An abstract algebra teacher intended to give a typist a list of nine integers that form a group under multiplication modulo 91. Instead, one of the nine integers was inadvertently left out, so that the list appeared as 1, 9, 16, 22, 53, 74, 79, 81. Which integer was left out? (This really happened!)

14. Let G be a group with the following property: Whenever a, b, and c belong to G and $ab = ca$, then $b = c$. Prove that G is Abelian. ("Cross" cancellation implies commutativity.)

15. (Law of Exponents for Abelian Groups) Let a and b be elements of an Abelian group and let n be any integer. Show that $(ab)^n = a^nb^n$. Is this also true for non-Abelian groups?

16. (Socks-Shoes Property) In a group, prove that $(ab)^{-1} = b^{-1}a^{-1}$. Find an example that shows that it is possible to have $(ab)^{-2} \neq b^{-2}a^{-2}$. Find distinct nonidentity elements a and b from a non-Abelian group with the property that $(ab)^{-1} = a^{-1}b^{-1}$. Draw an analogy between the statement $(ab)^{-1} = b^{-1}a^{-1}$ and the act of putting on and taking off your socks and shoes.

17. Prove that a group G is Abelian if and only if $(ab)^{-1} = a^{-1}b^{-1}$ for all a and b in G.

18. In a group, prove that $(a^{-1})^{-1} = a$ for all a.

19. For any elements a and b from a group and any integer n, prove that $(a^{-1}ba)^n = a^{-1}b^na$.

20. If $a_1, a_2, \ldots, a_n$ belong to a group, what is the inverse of $a_1a_2 \cdots a_n$?

21. The integers 5 and 15 are among a collection of 12 integers that form a group under multiplication modulo 56. List all 12.

22. Give an example of a group with 105 elements. Give two examples of groups with 44 elements.

23. Prove that every group table is a *Latin square*[†]; that is, each element of the group appears exactly once in each row and each column. (This exercise is referred to in this chapter.)

24. Construct a Cayley table for $U(12)$.

25. Suppose the table below is a group table. Fill in the blank entries.

	e	a	b	c	d
e	e	—	—	—	—
a	—	b	—	—	e
b	—	c	d	e	—
c	—	d	—	a	b
d	—	—	—	—	—

26. Prove that if $(ab)^2 = a^2b^2$ in a group G, then $ab = ba$.

27. Let a, b, and c be elements of a group. Solve the equation $axb = c$ for x. Solve $a^{-1}xa = c$ for x.

28. Prove that the set of all rational numbers of the form 3^m6^n, where m and n are integers, is a group under multiplication.

29. Let G be a finite group. Show that the number of elements x of G such that $x^3 = e$ is odd. Show that the number of elements x of G such that $x^2 \neq e$ is even.

[†]Latin squares are useful in designing statistical experiments. There is also a close connection between Latin squares and finite geometries.

30. Give an example of a group with elements a, b, c, d, and x such that $axb = cxd$ but $ab \neq cd$. (Hence "middle cancellation" is not valid in groups.)

31. Suppose that G is a group with the property that for every choice of elements in G, $axb = cxd$ implies $ab = cd$. Prove that G is Abelian. ("Middle cancellation" implies commutativity.)

32. In D_n, let $r = R_{360/n}$ and let f be any reflection. Use a diagram to verify that $frf = r^{-1}$. Use this relation to write the following elements in the form r^i or $r^i f$, where $0 \leq i < n$.

    a. In D_4, $fr^{-2}fr^5$
    b. In D_5, $r^{-3}fr^4fr^{-2}$
    c. In D_6, $fr^5fr^{-2}f$

33. Prove that if G is a group with the property that the square of every element is the identity, then G is Abelian. (This exercise is referred to in Chapter 26.)

34. Prove that the set of all 3×3 matrices with real entries of the form

$$\begin{bmatrix} 1 & a & b \\ 0 & 1 & c \\ 0 & 0 & 1 \end{bmatrix}$$

    is a group. (Multiplication is defined by

$$\begin{bmatrix} 1 & a & b \\ 0 & 1 & c \\ 0 & 0 & 1 \end{bmatrix} \begin{bmatrix} 1 & a' & b' \\ 0 & 1 & c' \\ 0 & 0 & 1 \end{bmatrix} = \begin{bmatrix} 1 & a+a' & b'+ac'+b \\ 0 & 1 & c'+c \\ 0 & 0 & 1 \end{bmatrix}.$$

    This group, sometimes called the *Heisenberg group* after the Nobel Prize–winning physicist Werner Heisenberg, is intimately related to the Heisenberg Uncertainty Principle of quantum physics.)

35. Prove the assertion made in Example 19 that the set $\{1, 2, \ldots, n - 1\}$ is a group under multiplication modulo n if and only if n is prime.

36. In a finite group, show that the number of nonidentity elements that satisfy the equation $x^5 = e$ is a multiple of 4.

37. Let $G = \left\{ \begin{bmatrix} a & a \\ a & a \end{bmatrix} \,\middle|\, a \in \mathbf{R}, a \neq 0 \right\}$. Show that G is a group under matrix multiplication. Explain why each element of G has an inverse even though the matrices have 0 determinant. (Compare with Example 10.)

COMPUTER EXERCISES ▬▬▬▬▬▬▬▬▬▬▬▬▬ ▪▪

Almost immediately after the war, Johnny [Von Neumann] and I also began to discuss the possibilities of using computers heuristically to try to obtain insights into questions of pure mathematics. By producing examples and by observing the properties of special mathematical objects, one could hope to obtain clues as to the behavior of general statements which have been tested on examples.

s. m. ULAM, Adventures of a Mathematician

Software for the computer exercises in this chapter is available at the website:

http://www.d.umn.edu/~jgallian

1. This software prints the elements of $U(n)$ and the inverse of each element.

2. This software determines the size of $U(k)$. Run the program for $k = 9$, 27, 81, 243, 25, 125, 49, 121. On the basis of this output, try to guess a formula for the size of $U(p^n)$ as a function of the prime p and the integer n. Run the program for $k = 18, 54, 162, 486, 50, 250, 98, 242$. Make a conjecture about the relationship between the size of $U(2p^n)$ and the size of $U(p^n)$, where p is a prime greater than 2.

3. This software computes the inverse of any element in $GL(2, Z_p)$, where p is a prime.

4. This software determines the number of elements in $GL(2, Z_p)$ and $SL(2, Z_p)$. (The technical term for the number of elements in a group is the *order* of the group.) Run the program for $p = 3, 5, 7$, and 11. Do you see a relationship between the orders of $GL(2, Z_p)$ and $SL(2, Z_p)$ and $p - 1$? Does this relationship hold for $p = 2$? Based on these examples, does it appear that p always divides the order of $SL(2, Z_p)$? What about $p - 1$? What about $p + 1$? Guess a formula for the order of $SL(2, Z_p)$. Guess a formula for the order of $GL(2, Z_p)$.

References

1. Max Born, *My Life: Recollections of a Nobel Laureate*, New York: Charles Scribner's Sons, 1978.
2. J. Mehra and H. Rechenberg, *The Historical Development of Quantum Theory*, Vol. 3, New York: Springer-Verlag, 1982.

Suggested Readings

Marcia Ascher, *Ethnomathematics*, Pacific Grove, CA: Brooks/Cole, 1991.
    Chapter 3 of this book describes how the dihedral group of order 8 can be used to encode the social structure of the kin system of family relationships among a tribe of native people of Australia.

Arie Bialostocki, "An Application of Elementary Group Theory to Central Solitaire," *The College Mathematics Journal*, May 1998: 208–212.

> The author uses properties of groups to analyze the peg board game central solitaire (which also goes by the name peg solitaire).

J. E. White, "Introduction to Group Theory for Chemists," *Journal of Chemical Education* 44 (1967): 128–135.

> Students interested in the physical sciences may find this article worthwhile. It begins with easy examples of groups and builds up to applications of group theory concepts and terminology to chemistry.

3 Finite Groups; Subgroups

In our own time, in the period 1960–1980, we have seen particle physics emerge as the playground of group theory.

FREEMAN DYSON

TERMINOLOGY AND NOTATION

As we will soon discover, finite groups—that is, groups with finitely many elements—have interesting arithmetic properties. To facilitate the study of finite groups, it is convenient to introduce some terminology and notation.

DEFINITION   *Order of a Group*

The number of elements of a group (finite or infinite) is called its *order*. We will use $|G|$ to denote the order of G.

Thus, the group Z of integers under addition has infinite order, whereas the group $U(10) = \{1, 3, 7, 9\}$ under multiplication modulo 10 has order 4.

DEFINITION   *Order of an Element*

The *order* of an element g in a group G is the smallest positive integer n such that $g^n = e$. (In additive notation, this would be $ng = 0$.) If no such integer exists, we say that g has *infinite order*. The order of an element g is denoted by $|g|$.

So, to find the order of a group element g, you need only compute the sequence of products $g, g^2, g^3, \ldots$, until you reach the identity for the first time. The exponent of this product (or coefficient if the operation is addition) is the order of g. If the identity never appears in the sequence, then g has infinite order.

EXAMPLE 1   Consider $U(15) = \{1, 2, 4, 7, 8, 11, 13, 14\}$ under multiplication modulo 15. This group has order 8. To find the order of the element 7, say, we compute the sequence $7^1 = 7$, $7^2 = 4$, $7^3 = 13$, $7^4 = 1$, so $|7| = 4$. To find the order of 11, we compute $11^1 = 11$, $11^2 = 1$, so $|11| = 2$. Similar computations show that $|1| = 1$, $|2| = 4$, $|4| = 2$, $|8| = 4$, $|13| = 4$, $|14| = 2$. [Here is a trick that makes these calculations easier. Rather than compute the sequence 13^1, 13^2, 13^3, 13^4, we may observe that $13 = -2 \bmod 15$, so that $13^2 = (-2)^2 = 4$, $13^3 = -2 \cdot 4 = -8$, $13^4 = (-2)(-8) = 1$.]   ◆

EXAMPLE 2   Consider Z_{10} under addition modulo 10. Since $1 \cdot 2 = 2$, $2 \cdot 2 = 4$, $3 \cdot 2 = 6$, $4 \cdot 2 = 8$, $5 \cdot 2 = 0$, we know that $|2| = 5$. Similar computations show that $|0| = 1$, $|7| = 10$, $|5| = 2$, $|6| = 5$.   ◆

EXAMPLE 3   Consider Z under ordinary addition. Here every nonzero element has infinite order, since the sequence a, $2a$, $3a$, . . . never includes 0 when $a \neq 0$.   ◆

The perceptive reader may have noticed among our examples of groups in Chapter 2 that some are subsets of others with the same binary operation. The group $SL(2, \mathbf{R})$ in Example 17, for instance, is a subset of the group $GL(2, \mathbf{R})$ in Example 9. Similarly, the group of complex numbers $\{1, -1, i, -i\}$ under multiplication is a subset of the group described in Example 14 for n equal to any multiple of 4. This situation arises so often that we introduce a special term to describe it.

DEFINITION   *Subgroup*

> If a subset H of a group G is itself a group under the operation of G, we say that H is a *subgroup* of G.

We use the notation $H \leq G$ to mean that H is a subgroup of G. If we want to indicate that H is a subgroup of G, but not equal to G itself, we write $H < G$. Such a subgroup is called a *proper subgroup*. The subgroup $\{e\}$ is called the *trivial subgroup* of G; a subgroup that is not $\{e\}$ is called a *nontrivial subgroup* of G.

Notice that Z_n under addition modulo n is *not* a subgroup of Z under addition, since addition modulo n is not the operation of Z.

SUBGROUP TESTS

When determining whether or not a subset H of a group G is a subgroup of G, one need not directly verify the group axioms. The next three results provide simple tests that suffice to show that a subset of a group is a subgroup.

Theorem 3.1 **One-Step Subgroup Test**

> *Let G be a group and H a nonempty subset of G. If ab^{-1} is in H whenever a and b are in H, then H is a subgroup of G. (In additive notation, if a − b is in H whenever a and b are in H, then H is a subgroup of G.)*

PROOF. Since the operation of H is the same as that of G, it is clear that this operation is associative. Next, we show that e is in H. Since H is nonempty, we may pick some x in H. Then, letting $a = x$ and $b = x$ in the hypothesis, we have $e = xx^{-1} = ab^{-1}$ is in H. To verify that x^{-1} is in H whenever x is in H, all we need to do is to choose $a = e$ and $b = x$ in the statement of the theorem. Finally, the proof will be complete when we show that H is closed; that is, if x, y belong to H, we must show that xy is in H also. Well, we have already shown that y^{-1} is in H whenever y is; so letting $a = x$ and $b = y^{-1}$, we have $xy = x(y^{-1})^{-1} = ab^{-1}$ is in H. ∎

Although we have dubbed Theorem 3.1 the "One-Step Subgroup Test," there are actually four steps involved in applying the theorem. (After you gain some experience, the first three steps will be routine.) Notice the similarity between the last three steps listed below and the three steps involved in the Principle of Mathematical Induction.

1. Identify the property P that distinguishes the elements of H; that is, identify a defining condition.
2. Prove that the identity has property P. (This verifies that H is nonempty.)
3. *Assume* that two elements a and b have property P.
4. Use the assumption about a and b to show that ab^{-1} has property P.

The procedure is illustrated in Examples 4 and 5.

EXAMPLE 4   Let G be an Abelian group with identity e. Then $H = \{x \in G \mid x^2 = e\}$ is a subgroup of G. Here, the defining property of H is the condition $x^2 = e$. So, we first note that $e^2 = e$, so that H is nonempty. Now we assume that a and b belong to H. This means that $a^2 = e$ and $b^2 = e$. Finally, we must show that $(ab^{-1})^2 = e$. Since G is Abelian, $(ab^{-1})^2 = ab^{-1}ab^{-1} = a^2(b^{-1})^2 = a^2(b^2)^{-1} = ee^{-1} = e$. Therefore, ab^{-1} belongs to H and, by the One-Step Subgroup Test, H is a subgroup of G.   ◆

In many instances, a subgroup will consist of all elements that have a particular form. Then the property P is that the elements have that particular form. This is illustrated in the following example.

EXAMPLE 5   Let G be an Abelian group under multiplication with identity e. Then $H = \{x^2 \mid x \in G\}$ is a subgroup of G. (In words, H is the set of all "squares.") Since $e^2 = e$, the identity has the correct form. Next, we write two elements of H in the correct form, say, a^2 and b^2. We must show that $a^2(b^2)^{-1}$ also has the correct form; that is, $a^2(b^2)^{-1}$ is the square of some element. Since G is Abelian, we may write $a^2(b^2)^{-1}$ as $(ab^{-1})^2$, which is the correct form. Thus, H is a subgroup of G.   ◆

Beginning students often prefer to use the next theorem instead of Theorem 3.1.

Theorem 3.2   **Two-Step Subgroup Test**

> *Let G be a group and let H be a nonempty subset of G. If ab is in H whenever a and b are in H (H is closed under the operation), and a^{-1} is in H whenever a is in H (H is closed under taking inverses), then H is a subgroup of G.*

PROOF.   By Theorem 3.1, it suffices to show that $a, b \in H$ implies $ab^{-1} \in H$. So, we suppose that $a, b \in H$. Since H is closed under taking inverses, we also have $b^{-1} \in H$. Thus, $ab^{-1} \in H$ by closure under multiplication.   ∎

How do you prove that a subset of a group is *not* a subgroup? Here are three possible ways, any one of which guarantees that the subset is not a subgroup:

1. Show that the identity is not in the set.
2. Exhibit an element of the set whose inverse is not in the set.
3. Exhibit two elements of the set whose product is not in the set.

EXAMPLE 6 Let G be the group of nonzero real numbers under multiplication, $H = \{x \in G \mid x = 1$ or x is irrational$\}$ and $K = \{x \in G \mid x \geq 1\}$. Then H is not a subgroup of G, since $\sqrt{2} \in H$ but $\sqrt{2} \cdot \sqrt{2} = 2 \notin H$. Also, K is not a subgroup, since $2 \in K$ but $2^{-1} \notin K$. ◆

When dealing with finite groups, it is easier to use the following subgroup test.

Theorem 3.3 Finite Subgroup Test

> *Let H be a nonempty finite subset of a group G. If H is closed under the operation of G, then H is a subgroup of G.*

PROOF. In view of Theorem 3.2, we need only prove that $a^{-1} \in H$ whenever $a \in H$. If $a = e$, then $a^{-1} = a$ and we are done. If $a \neq e$, consider the sequence $a, a^2, \ldots$. By closure, all of these elements belong to H. Since H is finite, not all of these elements are distinct. Say $a^i = a^j$ and $i > j$. Then, $a^{i-j} = e$; and since $a \neq e$, $i - j > 1$. Thus, $aa^{i-j-1} = a^{i-j} = e$ and, therefore, $a^{i-j-1} = a^{-1}$. But, $i - j - 1 \geq 1$ implies $a^{i-j-1} \in H$ and we are done. ■ ■

EXAMPLES OF SUBGROUPS

The proofs of the next few theorems show how our subgroup tests work. We first introduce an important notation. For any element a from a group, we let $\langle a \rangle$ denote the set $\{a^n \mid n \in Z\}$. In particular, observe that the exponents of a include all negative integers as well as 0 and the positive integers (a^0 is defined to be the identity).

Theorem 3.4 $\langle a \rangle$ Is a Subgroup

> *Let G be a group, and let a be any element of G. Then, $\langle a \rangle$ is a subgroup of G.*

PROOF. Since $a \in \langle a \rangle$, $\langle a \rangle$ is not empty. Let a^n, $a^m \in \langle a \rangle$. Then, $a^n(a^m)^{-1} = a^{n-m} \in \langle a \rangle$; so, by Theorem 3.1, $\langle a \rangle$ is a subgroup of G. ■ ■

The subgroup $\langle a \rangle$ is called the *cyclic subgroup of G generated by a*. In the case that $G = \langle a \rangle$, we say that G is *cyclic* and a is a *generator of G*. (A cyclic group may have many generators.) Notice that although the list $\ldots, a^{-2}, a^{-1}, a^0, a^1, a^2, \ldots$ has infinitely many entries, the set $\{a^n \mid n \in Z\}$ might have only finitely many elements. Also note that, since $a^i a^j = a^{i+j} = a^{j+i} = a^j a^i$, every cyclic group is Abelian.

EXAMPLE 7   In $U(10)$, $\langle 3 \rangle = \{3, 9, 7, 1\} = U(10)$, for $3^1 = 3$, $3^2 = 9$, $3^3 = 7$, $3^4 = 1$, $3^5 = 3^4 \cdot 3 = 1 \cdot 3$, $3^6 = 3^4 \cdot 3^2 = 9, \ldots; 3^{-1} = 7$ (since $3 \cdot 7 = 1$), $3^{-2} = 9$, $3^{-3} = 3$, $3^{-4} = 1$, $3^{-5} = 3^{-4} \cdot 3^{-1} = 1 \cdot 7$, $3^{-6} = 3^{-4} \cdot 3^{-2} = 1 \cdot 9 = 9, \ldots$. ◆

EXAMPLE 8   In Z_{10}, $\langle 2 \rangle = \{2, 4, 6, 8, 0\}$. Remember, a^n means na when the operation is addition. ◆

EXAMPLE 9   In Z, $\langle -1 \rangle = Z$. Here each entry in the list $\ldots$, $-2(-1), -1(-1), 0(-1), 1(-1), 2(-1), \ldots$ represents a distinct group element. ◆

EXAMPLE 10   In D_n, the dihedral group of order $2n$, let R denote a rotation of $360/n$ degrees. Then,

$$R^n = R_{360°} = e, \qquad R^{n+1} = R, \qquad R^{n+2} = R^2, \ldots.$$

Similarly, $R^{-1} = R^{n-1}, R^{-2} = R^{n-2}, \ldots$, so that $\langle R \rangle = \{e, R, \ldots, R^{n-1}\}$. We see, then, that the powers of R "cycle back" periodically with period n. Visually, raising R to successive positive powers is the same as moving counterclockwise around the following circle one node at a time, whereas raising R to successive negative powers is the same as moving around the circle clockwise one node at a time.

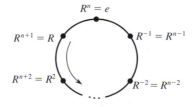

In Chapter 4 we will show that $|\langle a \rangle| = |a|$; that is, the order of the subgroup generated by a is the order of a itself. (Actually, the definition of $|a|$ was chosen to ensure the validity of this equation.)

We next consider one of the most important subgroups.

DEFINITION   *Center of a Group*

The *center*, $Z(G)$, of a group G is the subset of elements in G that commute with every element of G. In symbols,

$$Z(G) = \{a \in G \mid ax = xa \text{ for all } x \text{ in } G\}.$$

[The notation $Z(G)$ comes from the fact that the German word for center is *Zentrum*. The term was coined by J. A. de Seguier in 1904.]

Theorem 3.5   Center Is a Subgroup

> *The center of a group G is a subgroup of G.*

PROOF.   For variety, we shall use Theorem 3.2 to prove this result. Clearly, $e \in Z(G)$, so $Z(G)$ is nonempty. Now, suppose $a, b \in Z(G)$. Then, $(ab)x = a(bx) = a(xb) = (ax)b = (xa)b = x(ab)$ for all x in G; and, therefore, $ab \in Z(G)$.

Next, assume that $a \in Z(G)$. Then, we have $ax = xa$ for all x in G. What we want is $a^{-1}x = xa^{-1}$ for all x in G. Informally, all we need do to obtain the second equation from the first one is simultaneously to bring the a's across the equal sign:

$$ax = xa$$

becomes $xa^{-1} = a^{-1}x$. (Be careful here; groups need not be commutative. The a on the left comes across as a^{-1} on the left, and the a on the right comes across as a^{-1} on the right.) Formally, the desired equation can be obtained from the original one by multiplying it on the left and right by a^{-1}, like so:

$$a^{-1}(ax)a^{-1} = a^{-1}(xa)a^{-1},$$

$$(a^{-1}a)xa^{-1} = a^{-1}x(aa^{-1}),$$

$$exa^{-1} = a^{-1}xe,$$

$$xa^{-1} = a^{-1}x.$$

This shows that $a^{-1} \in Z(G)$ whenever a is.   ■ ■

For practice, let's determine the centers of the dihedral groups.

EXAMPLE 11   For $n \geq 3$,

$$Z(D_n) = \begin{cases} \{R_0, R_{180}\} & \text{when } n \text{ is even,} \\ \{R_0\} & \text{when } n \text{ is odd.} \end{cases}$$

To verify this, first observe that since every rotation in D_n is a power of $\langle R_{360/n} \rangle$, rotations commute with rotations. We now investigate when a rotation commutes with a reflection. Let R be any rotation in D_n and let F be any reflection in D_n. Observe that since RF is a reflection we have $RF = (RF)^{-1} = F^{-1}R^{-1} = FR^{-1}$. Thus it follows that R and F commute if and only if $FR = RF = FR^{-1}$. By cancellation, this holds if and only if $R = R^{-1}$. But $R = R^{-1}$ only when $R = R_0$ or $R = R_{180}$, and R_{180} is in D_n only when n is even. So, we have proved that $Z(D_n) = \{R_0\}$ when n is odd and $Z(D_n) = \{R_0, R_{180}\}$ when n is even.   ◆

Although an element from a non-Abelian group does not necessarily commute with every element of the group, there are always some elements with which it will commute. For example, every element a commutes with all powers of a. This observation prompts the next definition and theorem.

DEFINITION   *Centralizer of a in G*

Let a be a fixed element of a group G. The *centralizer of a in G*, $C(a)$, is the set of all elements in G that commute with a. In symbols, $C(a) = \{g \in G \mid ga = ag\}$.

EXAMPLE 12   In D_4, we have the following centralizers:

$$C(R_0) = D_4 = C(R_{180}),$$

$$C(R_{90}) = \{R_0, R_{90}, R_{180}, R_{270}\} = C(R_{270}),$$

$$C(H) = \{R_0, H, R_{180}, V\} = C(V),$$

$$C(D) = \{R_0, D, R_{180}, D'\} = C(D').$$ ◆

Notice that each of the centralizers in Example 12 is actually a subgroup of D_4. The next theorem shows that this was not a coincidence.

Theorem 3.6 $C(a)$ Is a Subgroup

For each a in a group G, the centralizer of a is a subgroup of G.

PROOF.   A proof similar to that of Theorem 3.5 is left to the reader to supply (Exercise 19).   ■ ■

Notice that for every element a of a group G, $Z(G) \subseteq C(a)$. Also, observe that G is Abelian if and only if $C(a) = G$ for all a in G.

EXERCISES

The purpose of proof is to understand, not to verify.
ARNOLD ROSS

1. For each group in the following list, find the order of the group and the order of each element in the group. What relation do you see between the orders of the elements of a group and the order of the group?

$$Z_{12}, \qquad U(10), \qquad U(12), \qquad U(20), \qquad D_4$$

2. Let Q be the group of rational numbers under addition and let Q^* be the group of nonzero rational numbers under multiplication. In Q, list the elements in $\langle \frac{1}{2} \rangle$. In Q^*, list the elements in $\langle \frac{1}{2} \rangle$.

3. Let Q and Q^* be as in Exercise 2. Find the order of each element in Q and in Q^*.

4. Prove that in any group, an element and its inverse have the same order.

5. Without actually computing the orders, explain why the two elements in each of the following pairs of elements from Z_{30} must have the same order: $\{2, 28\}, \{8, 22\}$. Do the same for the following pairs of elements from $U(15)$: $\{2, 8\}, \{7, 13\}$.

6. Let x belong to a group. If $x^2 \neq e$ and $x^6 = e$, prove that $x^4 \neq e$ and $x^5 \neq e$. What can we say about the order of x?

7. Show that if a is an element of a group G, then $|a| \leq |G|$.

8. Show that $U(14) = \langle 3 \rangle = \langle 5 \rangle$. [Hence, $U(14)$ is cyclic.] Is $U(14) = \langle 11 \rangle$?

9. Show that $U(20) \neq \langle k \rangle$ for any k in $U(20)$. [Hence, $U(20)$ is not cyclic.]

10. Prove that an Abelian group with two elements of order 2 must have a subgroup of order 4.

11. Find a group that contains elements a and b such that $|a| = |b| = 2$ and
    a. $|ab| = 3$,   **b.** $|ab| = 4$,   **c.** $|ab| = 5$.
    Can you see any relationship among $|a|$, $|b|$, and $|ab|$?

12. Suppose that H is a proper subgroup of Z under addition and H contains 18, 30, and 40. Determine H.

13. For each divisor $k > 1$ of n, let $U_k(n) = \{x \in U(n) \mid x \bmod k = 1\}$. [For example, $U_3(21) = \{1, 4, 10, 13, 16, 19\}$ and $U_7(21) = \{1, 8\}$.] List the

elements of $U_4(20)$, $U_5(20)$, $U_5(30)$, and $U_{10}(30)$. Prove that $U_k(n)$ is a subgroup of $U(n)$. Let $H = \{x \in U(10) \mid x \bmod 3 = 1\}$. Is H a subgroup of $U(10)$? (This exercise is referred to in Chapter 8.)

14. If H and K are subgroups of G, show that $H \cap K$ is a subgroup of G. (Can you see that the same proof shows that the intersection of any number of subgroups of G, finite or infinite, is again a subgroup of G?)

15. Let G be a group. Show that $Z(G) = \cap_{a \in G} C(a)$. [This means the intersection of *all* subgroups of the form $C(a)$.]

16. Let G be a group, and let $a \in G$. Prove that $C(a) = C(a^{-1})$.

17. Suppose G is the group defined by the following Cayley table.

	1	2	3	4	5	6	7	8
1	1	2	3	4	5	6	7	8
2	2	1	8	7	6	5	4	3
3	3	4	5	6	7	8	1	2
4	4	3	2	1	8	7	6	5
5	5	6	7	8	1	2	3	4
6	6	5	4	3	2	1	8	7
7	7	8	1	2	3	4	5	6
8	8	7	6	5	4	3	2	1

a. Find the centralizer of each member of G.
b. Find $Z(G)$.
c. Find the order of each element of G. How are these orders arithmetically related to the order of the group?

18. If a and b are distinct group elements, prove that either $a^2 \neq b^2$ or $a^3 \neq b^3$.

19. Prove Theorem 3.6.

20. If H is a subgroup of G, then by the *centralizer* $C(H)$ of H we mean the set $\{x \in G \mid xh = hx \text{ for all } h \in H\}$. Prove that $C(H)$ is a subgroup of G.

21. Must the centralizer of an element of a group be Abelian?

22. Must the center of a group be Abelian?

23. Let G be an Abelian group with identity e and let n be some fixed integer. Prove that the set of all elements of G that satisfy the equation $x^n = e$ is a subgroup of G. Give an example of a group G in which the set of all elements of G that satisfy the equation $x^2 = e$ does not form a subgroup of G. (This exercise is referred to in Chapter 11.)

24. Suppose a belongs to a group and $|a| = 5$. Prove that $C(a) = C(a^3)$. Find an element a from some group such that $|a| = 6$ and $C(a) \neq C(a^3)$.

25. Determine all finite subgroups of $\mathbf{R}^*$, the group of nonzero real numbers under multiplication.

26. Suppose n is an even positive integer and H is a subgroup of Z_n. Prove that either every member of H is even or exactly half of the members of H are even.

27. Suppose a group contains elements a and b such that $|a| = 4$, $|b| = 2$, and $a^3b = ba$. Find $|ab|$.

28. Consider the elements $A = \begin{bmatrix} 0 & -1 \\ 1 & 0 \end{bmatrix}$ and $B = \begin{bmatrix} 0 & 1 \\ -1 & -1 \end{bmatrix}$ from $SL(2, \mathbf{R})$.

Find $|A|$, $|B|$, and $|AB|$. Does your answer surprise you?

29. Consider the element $A = \begin{bmatrix} 1 & 1 \\ 0 & 1 \end{bmatrix}$ in $SL(2, \mathbf{R})$. What is the order of A? If

we view $A = \begin{bmatrix} 1 & 1 \\ 0 & 1 \end{bmatrix}$ as a member of $SL(2, Z_p)$ (p is a prime), what is

the order of A?

30. For any positive integer n and any angle θ, show that in the group $SL(2, \mathbf{R})$,

$$\begin{bmatrix} \cos \theta & -\sin \theta \\ \sin \theta & \cos \theta \end{bmatrix}^n = \begin{bmatrix} \cos n\theta & -\sin n\theta \\ \sin n\theta & \cos n\theta \end{bmatrix}.$$

Use this formula to find the order of

$$\begin{bmatrix} \cos 60° & -\sin 60° \\ \sin 60° & \cos 60° \end{bmatrix} \text{ and } \begin{bmatrix} \cos \sqrt{2}° & -\sin \sqrt{2}° \\ \sin \sqrt{2}° & \cos \sqrt{2}° \end{bmatrix}.$$

(Geometrically, $\begin{bmatrix} \cos \theta & -\sin \theta \\ \sin \theta & \cos \theta \end{bmatrix}$ represents a rotation of the plane θ

degrees.)

31. Let G be the symmetry group of a circle. Show that G has elements of every finite order as well as elements of infinite order.

32. Let x belong to a group and $|x| = 6$. Find $|x^2|$, $|x^3|$, $|x^4|$, and $|x^5|$. Let y belong to a group and $|y| = 9$. Find $|y^i|$ for $i = 2, 3, \ldots, 8$. Do these examples suggest any relationship between the order of the power of an element and the order of the element?

33. D_4 has seven cyclic subgroups. List them. Find a subgroup of D_4 of order 4 that is not cyclic.

34. $U(15)$ has six cyclic subgroups. List them.

35. Prove that a group of even order must have an element of order 2.

36. Suppose G is a group that has exactly eight elements of order 3. How many subgroups of order 3 does G have?

37. Let H be a subgroup of a finite group G. Suppose that g belongs to G and n is the smallest positive integer such that $g^n \in H$. Prove that n divides $|g|$.

38. Compute the orders of the following groups.
    a. $U(3)$, $U(4)$, $U(12)$
    b. $U(5)$, $U(7)$, $U(35)$
    c. $U(4)$, $U(5)$, $U(20)$
    d. $U(3)$, $U(5)$, $U(15)$
    On the basis of your answers, make a conjecture about the relationship among $|U(r)|$, $|U(s)|$, and $|U(rs)|$.

39. Let $\mathbf{R}^*$ be the group of nonzero real numbers under multiplication and let $H = \{x \in \mathbf{R}^* \mid x^2 \text{ is rational}\}$. Prove that H is a subgroup of $\mathbf{R}^*$. Can the exponent 2 be replaced by any positive integer and still have H be a subgroup?

40. Compute $|U(4)|$, $|U(10)|$, and $|U(40)|$. Do these groups provide a counterexample to your answer to Exercise 38? If so, revise your conjecture.

41. Find a cyclic subgroup of order 4 in $U(40)$.

42. Find a noncyclic subgroup of order 4 in $U(40)$.

43. Let $G = \left\{ \begin{bmatrix} a & b \\ c & d \end{bmatrix} \middle| a, b, c, d \in Z \right\}$ under addition. Let $H = \left\{ \begin{bmatrix} a & b \\ c & d \end{bmatrix} \in G \middle| a + b + c + d = 0 \right\}$. Prove that H is a subgroup of G. What if 0 is replaced by 1?

44. Let $H = \{A \in GL(2, \mathbf{R}) \mid \det A \text{ is a power of } 2\}$. Show that H is a subgroup of $GL(2, \mathbf{R})$.

45. Let H be a subgroup of $\mathbf{R}$ under addition. Let $K = \{2^a \mid a \in H\}$. Prove that K is a subgroup of $\mathbf{R}^*$ under multiplication.

46. Let G be a group of functions from $\mathbf{R}$ to $\mathbf{R}^*$, where the operation of G is multiplication of functions. Let $H = \{f \in G \mid f(2) = 1\}$. Prove that H is a subgroup of G. Can 2 be replaced by any real number?

47. Let $G = GL(2, \mathbf{R})$ and let $H = \left\{ \begin{bmatrix} a & 0 \\ 0 & b \end{bmatrix} \middle| a \text{ and } b \text{ are nonzero integers} \right\}$ under the operation of matrix multiplication. Prove or disprove that H is a subgroup of $GL(2, \mathbf{R})$.

48. Let $H = \{a + bi \mid a, b \in \mathbf{R}, ab \geq 0\}$. Prove or disprove that H is a subgroup of $\mathbf{C}$ under addition.

49. Let $H = \{a + bi \mid a, b \in \mathbf{R}, a^2 + b^2 = 1\}$. Prove or disprove that H is a subgroup of $\mathbf{C}^*$ under multiplication. Describe the elements of H geometrically.

50. The smallest subgroup containing a collection of elements S is the subgroup H with the property that if K is any subgroup containing S then K

also contains H. (So, the smallest subgroup containing S is contained in every subgroup that contains S.) The notation for this subgroup is $\langle S \rangle$. In the group Z, find

a. $\langle 8, 14 \rangle$
b. $\langle 8, 13 \rangle$
c. $\langle 6, 15 \rangle$
d. $\langle m, n \rangle$
e. $\langle 12, 18, 45 \rangle$.

In each part, find an integer k such that the subgroup is $\langle k \rangle$.

51. Let $G = GL(2, \mathbf{R})$.

a. Find $C\left(\begin{bmatrix} 1 & 1 \\ 1 & 0 \end{bmatrix}\right)$.

b. Find $C\left(\begin{bmatrix} 0 & 1 \\ 1 & 0 \end{bmatrix}\right)$.

c. Find $Z(G)$.

52. Let G be a finite group with more than one element. Show that G has an element of prime order.

53. Let a belong to a group and $|a| = m$. If n is relatively prime to m, show that a can be written as the nth power of some element in the group.

54. Let G be a finite Abelian group and let a and b belong to G. Prove that the set $\langle a, b \rangle = \{a^i b^j \mid i, j \in Z\}$ is a subgroup of G. What can you say about $|\langle a, b \rangle|$ in terms of $|a|$ and $|b|$?

COMPUTER EXERCISES ■■

A Programmer's Lament

I really hate this damned machine;
I wish that they would sell it
It never does quite what I want
but only what I tell it.

DENNIE L. VAN TASSEL,
The Compleat Computer

Software for the computer exercises in this chapter is available at the website:

http://www.d.umn.edu/~jgallian

1. This software determines the cyclic subgroups of $U(n)$ generated by each k in $U(n)$ ($n < 100$). Run the program for $n = 12, 15,$ and 30. Compare the order of the subgroups with the order of the group itself. What arithmetic relationship do these integers have?

2. This exercise repeats Exercise 1 for the group Z_n. The program lists the elements of Z_n that generate all of Z_n—that is, those elements k, $0 \leq k$

$\leq n - 1$, for which $Z_n = \langle k \rangle$. How does this set compare with $U(n)$? Make a conjecture.

3. This software does the following: For each pair of elements a and b from $U(n)$ ($n < 100$), it prints $|a|$, $|b|$, and $|ab|$ on the same line. Run the program for several values of n. Is there an arithmetic relationship between $|ab|$ and $|a|$ and $|b|$?

4. This exercise repeats Exercise 3 for Z_n using $a + b$ in place of ab.

5. This software computes the order of elements in $GL(2, Z_p)$. Enter several choices for matrices A and B. The software returns $|A|$, $|B|$, $|AB|$, $|BA|$, $|A^{-1}BA|$, and $|B^{-1}AB|$. Do you see any relationship between $|A|$, $|B|$ and $|AB|$? Do you see any relationship between $|AB|$ and $|BA|$? Make a conjecture about this relationship. Test your conjecture for several other choices for A and B. Do you see any relationship between $|B|$ and $|A^{-1}BA|$? Do you see any relationship between $|A|$ and $|B^{-1}AB|$? Make a conjecture about this relationship. Test your conjecture for several other choices for A and B.

Suggested Readings

J. Gallian and M. Reid, "Abelian Forcing Sets," *American Mathematical Monthly* 100 (1993): 580–582.

> A set S is called *Abelian forcing* if the only groups that satisfy $(ab)^n = a^n b^n$ for all a and b in the group and all n in S are the Abelian ones. This paper characterizes the Abelian forcing sets.

Gina Kolata, "Perfect Shuffles and Their Relation to Math," *Science* 216 (1982): 505–506.

> This is a delightful nontechnical article that discusses how group theory and computers were used to solve a difficult problem about shuffling a deck of cards. Serious work on the problem was begun by an undergraduate student as part of a programming course.

Suggested Software

Ellen Maycock Parker, *Laboratory Experiences in Group Theory*, Washington, DC: Mathematical Association of America, 1996.

> This workbook consists of 15 laboratories designed to be used with the software *Exploring Small Groups* as a supplement to a course in abstract algebra. The book includes the *Exploring Small Groups* software on a 3.5″ HD PC-compatible disk that can be run on Windows. To order, call 1-800-331-1622.

4 Cyclic Groups

The notion of a "group," viewed only 30 years ago
as the epitome of sophistication, is today one of the
mathematical concepts most widely used in physics,
chemistry, biochemistry, and mathematics itself.

ALEXEY SOSINSKY, 1991

PROPERTIES OF CYCLIC GROUPS

Recall from Chapter 3 that a group G is called *cyclic* if there is an element a in G such that $G = \{a^n \mid n \in Z\}$. Such an element a is called a *generator* of G. In view of the notation introduced in the preceding chapter, we may indicate that G is a cyclic group generated by a by writing $G = \langle a \rangle$.

In this chapter, we examine cyclic groups in detail and determine their important characteristics. We begin with a few examples.

EXAMPLE 1 The set of integers Z under ordinary addition is cyclic. Both 1 and -1 are generators. (Recall that, when the operation is addition, 1^n is interpreted as

$$\underbrace{1 + 1 + \cdots + 1}_{n \text{ terms}}$$

when n is positive and as

$$\underbrace{(-1) + (-1) + \cdots + (-1)}_{|n| \text{ terms}}$$

when n is negative.) ◆

EXAMPLE 2   The set $Z_n = \{0, 1, \ldots, n - 1\}$ for $n \geq 1$ is a cyclic group under addition modulo n. Again, 1 and $-1 = n - 1$ are generators.   ◆

Unlike Z, which has only two generators, Z_n may have many generators (depending on which n we are given).

EXAMPLE 3 $Z_8 = \langle 1 \rangle = \langle 3 \rangle = \langle 5 \rangle = \langle 7 \rangle$. To verify, for instance, that $Z_8 = \langle 3 \rangle$, we note that $\langle 3 \rangle = \{3, 3 + 3, 3 + 3 + 3, \ldots\}$ is the set $\{3, 6, 1, 4, 7, 2, 5, 0\} = Z_8$. Thus, 3 is a generator of Z_8. On the other hand, 2 is not a generator, since $\langle 2 \rangle = \{0, 2, 4, 6\} \neq Z_8$.   ◆

EXAMPLE 4   (See Example 11 in Chapter 2.)
$U(10) = \{1, 3, 7, 9\} = \{3^0, 3^1, 3^3, 3^2\} = \langle 3 \rangle$. Also, $\{1, 3, 7, 9\} = \{7^0, 7^3, 7^1, 7^2\} = \langle 7 \rangle$. So both 3 and 7 are generators for $U(10)$.   ◆

Quite often in mathematics, a "nonexample" is as helpful in understanding a concept as an example. With regard to cyclic groups, $U(8)$ serves this purpose; that is, $U(8)$ is not a cyclic group. How can we verify this? Well, note that $U(8) = \{1, 3, 5, 7\}$. But,

$$\langle 1 \rangle = \{1\}$$
$$\langle 3 \rangle = \{3, 1\}$$
$$\langle 5 \rangle = \{5, 1\}$$
$$\langle 7 \rangle = \{7, 1\}$$

so that $U(8) \neq \langle a \rangle$ for any a in $U(8)$.

With these examples under our belts, we are now ready to tackle cyclic groups in an abstract way and state their key properties.

Theorem 4.1   **Criterion for $a^i = a^j$**

> *Let G be a group, and let a belong to G. If a has infinite order, then all distinct powers of a are distinct group elements. If a has finite order, say, n, then $\langle a \rangle = \{e, a, a^2, \ldots, a^{n-1}\}$ and $a^i = a^j$ if and only if n divides $i - j$.*

PROOF.   If a has infinite order, there is no nonzero n such that a^n is the identity. Since $a^i = a^j$ implies $a^{i-j} = e$, we must have $i - j = 0$, and the first statement of the theorem is proved.

Now assume that $|a| = n$. We will prove that $\langle a \rangle = \{e, a, \ldots, a^{n-1}\}$. Certainly, the elements $e, a, \ldots, a^{n-1}$ are distinct. For if $a^i = a^j$ with $0 \leq j < i \leq n - 1$, then $a^{i-j} = e$. But this contradicts the fact that n is the least positive integer such that a^n is the identity.

Now, suppose that a^k is an arbitrary member of $\langle a \rangle$. By the division algorithm, there exist integers q and r such that

$$k = qn + r \quad \text{with} \quad 0 \leq r < n.$$

Then $a^k = a^{qn+r} = a^{qn}a^r = (a^n)^q a^r = ea^r = a^r$, so that $a^k \in \{e, a, a^2, \ldots, a^{n-1}\}$. This proves that $\langle a \rangle = \{e, a, a^2, \ldots, a^{n-1}\}$.

Next, we assume that $a^i = a^j$ and prove that n divides $i - j$. We begin by observing that $a^i = a^j$ implies $a^{i-j} = e$. Again, by the division algorithm, there are integers q and r such that

$$i - j = qn + r \quad \text{with} \quad 0 \leq r < n.$$

Then, $a^{i-j} = a^{qn+r}$, and therefore $e = a^{i-j} = a^{qn+r} = (a^n)^q a^r = e^q a^r = ea^r = a^r$. Since n is the least positive integer such that a^n is the identity, we must have $r = 0$, so that n divides $i - j$.

Conversely, if $i - j = nq$, then $a^{i-j} = a^{nq} = e^q = e$, so that $a^i = a^j$. ∎

Theorem 4.1 reveals the reason for the dual use of the notation and terminology for the order of an element and the order of a group.

Corollary 1 $|a| = |\langle a \rangle|$

> *For any group element a, $|a| = |\langle a \rangle|$.*

One special case of Theorem 4.1 occurs so often that it deserves singling out.

Corollary 2 $a^k = e$ **Implies that** $|a|$ **Divides** k

> *Let G be a group and let a be an element of order n in G. If $a^k = e$, then n divides k.*

PROOF. Since $a^k = e = a^0$, we know by Theorem 4.1 that n divides $k - 0$. ∎

Theorem 4.1 and its corollaries for the case $|a| = 6$ are illustrated in Figure 4.1.

What is important about Theorem 4.1 in the finite case is that it says that multiplication in $\langle a \rangle$ is essentially done by *addition* modulo n. That is, if $(i + j) \bmod n = k$, then $a^i a^j = a^k$. Thus, no matter what group G is, or how the element a is chosen, multiplication in $\langle a \rangle$ works the same as addition in Z_n whenever $|a| = n$. Similarly, if a has infinite order, then multiplication in $\langle a \rangle$ works the same as addition in Z, since $a^i a^j = a^{i+j}$ and no modular arithmetic is done.

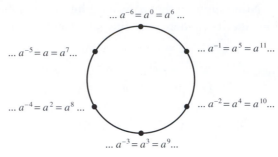

... $a^{-6} = a^0 = a^6$ ...

... $a^{-5} = a = a^7$...

... $a^{-1} = a^5 = a^{11}$...

... $a^{-4} = a^2 = a^8$ ...

... $a^{-2} = a^4 = a^{10}$...

... $a^{-3} = a^3 = a^9$...

Figure 4.1

For these reasons, the cyclic groups Z_n and Z serve as prototypes for all cyclic groups, and algebraists say that there is essentially only one cyclic group of each order. What is meant by this is that, although there may be many different sets of the form $\{a^n \mid n \in Z\}$, there is essentially only one way to operate on these sets. Algebraists do not really care what the elements of a set are; they care only about the algebraic properties of the set—that is, the ways in which the elements of a set can be combined. We will return to this theme in the chapter on isomorphisms (Chapter 6).

The next theorem provides a simple method for computing $|a^k|$ knowing only $|a|$, and its first corollary provides a simple way to tell when $\langle a^i \rangle = \langle a^j \rangle$.

Theorem 4.2 $\langle a^k \rangle = \langle a^{\gcd(n,k)} \rangle$

> *Let a be an element of order n in a group and let k be a positive integer. Then $\langle a^k \rangle = \langle a^{\gcd(n,k)} \rangle$ and $|a^k| = n/\gcd(n,k)$.*

PROOF.   To simplify the notation, let $d = \gcd(n,k)$ and let $k = dr$. Since $a^k = (a^d)^r$, we have by closure that $\langle a^k \rangle \subseteq \langle a^d \rangle$. By Theorem 0.2 (the gcd theorem), there are integers s and t such that $d = ns + kt$. So, $a^d = a^{ns + kt} = a^{ns}a^{kt} = (a^n)^s(a^k)^t = e(a^k)^t = (a^k)^t \in \langle a^k \rangle$. This proves $\langle a^d \rangle \subseteq \langle a^k \rangle$. So, we have verified that $\langle a^k \rangle = \langle a^{\gcd(n,k)} \rangle$.

We prove the second part of the theorem by showing first that $|a^d| = n/d$ for any divisor d of n. Clearly, $(a^d)^{n/d} = a^n = e$, so that $|a^d| \leq n/d$. On the other hand, if i is a positive integer less than n/d, then $(a^d)^i \neq e$ by definition of $|a|$. We now apply this fact with $d = \gcd(n,k)$ to obtain $|a^k| = |\langle a^k \rangle| = |\langle a^{\gcd(n,k)} \rangle| = |a^{\gcd(n,k)}| = n/\gcd(n,k)$. ■ ■

The advantage of Theorem 4.2 is that it allows us to replace one generator of a cyclic subgroup with a more convenient one. For example, if $|a| = 30$, we have $\langle a^{26} \rangle = \langle a^2 \rangle$, $\langle a^{23} \rangle = \langle a \rangle$, $\langle a^{22} \rangle = \langle a^2 \rangle$, $\langle a^{21} \rangle = \langle a^3 \rangle$. From this we can easily see that $|a^{23}| = 30$ and $|a^{22}| = 15$. Moreover, if one wants to list the elements of, say, $\langle a^{21} \rangle$, it is easier to list the elements of $\langle a^3 \rangle$ instead. (Try it doing it both ways!).

Corollary 1 **Criterion for $\langle a^i \rangle = \langle a^j \rangle$**

> *Let $|a| = n$. Then $\langle a^i \rangle = \langle a^j \rangle$ if and only if $gcd(n,i) = gcd(n, j)$.*

PROOF. Theorem 4.2 shows that $\langle a^i \rangle = \langle a^{gcd(n,i)} \rangle$ and $\langle a^j \rangle = \langle a^{gcd(n,j)} \rangle$, so that the proof reduces to proving that $\langle a^{gcd(n,i)} \rangle = \langle a^{gcd(n,j)} \rangle$ if and only if $gcd(n,i) = gcd(n,j)$. Certainly, $gcd(n,i) = gcd(n,j)$ implies that $\langle a^{gcd(n,i)} \rangle = \langle a^{gcd(n,j)} \rangle$. On the other hand, $\langle a^{gcd(n,i)} \rangle = \langle a^{gcd(n,j)} \rangle$ implies that $|a^{gcd(n,i)}| = |a^{gcd(n,j)}|$, so that by the second conclusion of Theorem 4.2, we have $n/gcd(n,i) = n/gcd(n,j)$, and therefore $gcd(n,i) = gcd(n,j)$. ∎

Two special cases of Corollary 1 arise so frequently that they deserve to be singled out.

Corollary 2 **Generators of Cyclic Groups**

> *Let $G = \langle a \rangle$ be a cyclic group of order n. Then $G = \langle a^k \rangle$ if and only if $gcd(n,k) = 1$.*

Corollary 3 **Generators of Z_n**

> *An integer k in Z_n is a generator of Z_n if and only if $gcd(n,k) = 1$.*

The value of Corollary 2 is that once one generator of a cyclic group has been found, all generators of the cyclic group can easily be determined. For example, consider the subgroup of all rotations in D_6. Clearly, one generator is R_{60}. And, since $|R_{60}| = 6$, we see by Corollary 2 that the only other generator is $(R_{60})^5 = R_{300}$. Of course, we could have readily deduced this information without the aid of Corollary 2 by direct calculations. So, to illustrate the real power of Corollary 2, let us use it to find all generators of the cyclic group $U(50)$. First, note that direct computations show that $|U(50)| = 20$ and that 3 is one of its generators. Thus, in view of Corollary 2, the complete list of generators for $U(50)$ is

$$3 \bmod 50 = 3, \qquad 3^{11} \bmod 50 = 47,$$
$$3^3 \bmod 50 = 27, \qquad 3^{13} \bmod 50 = 23,$$
$$3^7 \bmod 50 = 37, \qquad 3^{17} \bmod 50 = 13,$$
$$3^9 \bmod 50 = 33, \qquad 3^{19} \bmod 50 = 17.$$

Admittedly, we had to do some arithmetic here, but it certainly entailed much less work than finding all the generators by simply determining the order of each element of $U(50)$ one by one.

CLASSIFICATION OF SUBGROUPS OF CYCLIC GROUPS

The next theorem tells us how many subgroups a finite cyclic group has and how to find them.

Theorem 4.3   **Fundamental Theorem of Cyclic Groups**

> *Every subgroup of a cyclic group is cyclic. Moreover, if*
> $|\langle a \rangle| = n$, *then the order of any subgroup of $\langle a \rangle$ is a divisor*
> *of n; and, for each positive divisor k of n, the group $\langle a \rangle$ has*
> *exactly one subgroup of order k—namely, $\langle a^{n/k} \rangle$.*

Before we prove this theorem, let's see what it means. Understanding what a theorem means is a prerequisite to understanding its proof. Suppose $G = \langle a \rangle$ and G has order 30. The first and second parts of the theorem say that if H is any subgroup of G, then H has the form $\langle a^{30/k} \rangle$ for some k that is a divisor of 30. The third part of the theorem says that G has one subgroup of each of the orders 1, 2, 3, 5, 6, 10, 15, and 30—and no others. The proof will also show how to find these subgroups.

PROOF.   Let $G = \langle a \rangle$ and suppose that H is a subgroup of G. We must show that H is cyclic. If it consists of the identity alone, then clearly H is cyclic. So we may assume that $H \neq \{e\}$. We now claim that H contains an element of the form a^t, where t is positive. Since $G = \langle a \rangle$, every element of H has the form a^t; and when a^t belongs to H with $t < 0$, then a^{-t} belongs to H also and $-t$ is positive. Thus, our claim is verified. Now let m be the least positive integer such that $a^m \in H$. By closure, $\langle a^m \rangle \subseteq H$. We next claim that $H = \langle a^m \rangle$. To prove this claim, it suffices to let b be an arbitrary member of H and show that b is in $\langle a^m \rangle$.

Since $b \in G = \langle a \rangle$, we have $b = a^k$ for some k. Now, apply the division algorithm to k and m to obtain integers q and r such that $k = mq + r$ where $0 \leq r < m$. Then $a^k = a^{mq+r} = a^{mq}a^r$, so that $a^r = a^{-mq}a^k$. Since $a^k = b \in H$ and $a^{-mq} = (a^m)^{-q}$ is in H also, $a^r \in H$. But, m is the *least* positive integer such that $a^m \in H$, and $0 \leq r < m$, so r must be 0. Therefore, $b = a^k = a^{mq} = (a^m)^q \in \langle a^m \rangle$. This proves the assertion of the theorem that every subgroup of a cyclic group is cyclic.

To prove the next portion of the theorem, suppose that $|\langle a \rangle| = n$ and H is any subgroup of $\langle a \rangle$. We have already shown that $H = \langle a^m \rangle$, where m is the least positive integer such that $a^m \in H$. Using $e = b = a^n$ as in the preceding paragraph, we have $n = mq$.

Finally, let k be any positive divisor of n. We will show that $\langle a^{n/k} \rangle$ is the one and only subgroup of $\langle a \rangle$ of order k. From Theorem 4.2, we see that $\langle a^{n/k} \rangle$ has order $n/\gcd(n,n/k) = n/(n/k) = k$. Now let H be any subgroup of $\langle a \rangle$ of order k. We have already shown above that $H = \langle a^m \rangle$, where m is a divisor of n. Then $m = \gcd(n,m)$ and $k = |a^m| = |a^{\gcd(n,m)}| = n/\gcd(n,m) = n/m$. Thus, $m = n/k$ and $H = \langle a^{n/k} \rangle$. ∎

Returning for a moment to our discussion of the cyclic group $\langle a \rangle$, where a has order 30, we may conclude from Theorem 4.3 that the subgroups of $\langle a \rangle$ are precisely those of the form $\langle a^m \rangle$, where m is a divisor of 30. Moreover, if k is a divisor of 30, the subgroup of order k is $\langle a^{30/k} \rangle$. So the list of subgroups of $\langle a \rangle$ is:

$$
\begin{aligned}
\langle a \rangle &= \{e, a, a^2, \ldots, a^{29}\} && \text{order } 30, \\
\langle a^2 \rangle &= \{e, a^2, a^4, \ldots, a^{28}\} && \text{order } 15, \\
\langle a^3 \rangle &= \{e, a^3, a^6, \ldots, a^{27}\} && \text{order } 10, \\
\langle a^5 \rangle &= \{e, a^5, a^{10}, a^{15}, a^{20}, a^{25}\} && \text{order } 6, \\
\langle a^6 \rangle &= \{e, a^6, a^{12}, a^{18}, a^{24}\} && \text{order } 5, \\
\langle a^{10} \rangle &= \{e, a^{10}, a^{20}\} && \text{order } 3, \\
\langle a^{15} \rangle &= \{e, a^{15}\} && \text{order } 2, \\
\langle a^{30} \rangle &= \{e\} && \text{order } 1.
\end{aligned}
$$

In general, if $\langle a \rangle$ has order n and k divides n, then $\langle a^{n/k} \rangle$ is the unique subgroup of order k.

Taking the group in Theorem 4.3 to be Z_n and a to be 1, we obtain the following important special case.

Corollary   Subgroups of Z_n

> *For each positive divisor k of n, the set $\langle n/k \rangle$ is the unique subgroup of Z_n of order k; moreover, these are the only subgroups of Z_n.*

EXAMPLE 5   The list of subgroups of Z_{30} is

$$\langle 1 \rangle = \{0, 1, 2, \dots, 29\} \qquad \text{order } 30,$$
$$\langle 2 \rangle = \{0, 2, 4, \dots, 28\} \qquad \text{order } 15,$$
$$\langle 3 \rangle = \{0, 3, 6, \dots, 27\} \qquad \text{order } 10,$$
$$\langle 5 \rangle = \{0, 5, 10, 15, 20, 25\} \qquad \text{order } 6,$$
$$\langle 6 \rangle = \{0, 6, 12, 18, 24\} \qquad \text{order } 5,$$
$$\langle 10 \rangle = \{0, 10, 20\} \qquad \text{order } 3,$$
$$\langle 15 \rangle = \{0, 15\} \qquad \text{order } 2,$$
$$\langle 30 \rangle = \{0\} \qquad \text{order } 1. \qquad \blacklozenge$$

By combining Theorems 4.2 and 4.3, we can easily count the number of elements of each order in a finite cyclic group. For convenience, we introduce an important number-theoretic function called the *Euler phi function*. Let $\phi(1) = 1$, and for any integer $n > 1$, let $\phi(n)$ denote the number of positive integers less than n and relatively prime to n. Notice that by definition of the group $U(n)$, $|U(n)| = \phi(n)$. The first 12 values of $\phi(n)$ are given in Table 4.1.

TABLE 4.1   Values of $\phi(n)$

n	1	2	3	4	5	6	7	8	9	10	11	12
$\phi(n)$	1	1	2	2	4	2	6	4	6	4	10	4

Theorem 4.4   **Number of Elements of Each Order in a Cyclic Group**

> *If d is a positive divisor of n, the number of elements of order d in a cyclic group of order n is $\phi(d)$.*

PROOF.   By Theorem 4.3, the group has exactly one subgroup of order d—call it $\langle a \rangle$. Then every element of order d also generates the subgroup $\langle a \rangle$ and, by Corollary 2 of Theorem 4.2, an element a^k generates $\langle a \rangle$ if and only if $\gcd(k, d) = 1$. The number of such elements is precisely $\phi(d)$.   ∎

Notice that for a finite cyclic group of order n, the number of elements of order d for any divisor d of n depends only on d. Thus, Z_8, Z_{640}, and Z_{80000} each have $\phi(8) = 4$ elements of order 8.

Although there is no formula for the number of elements of each order for arbitrary finite groups, we still can say something important in this regard.

Corollary   **Number of Elements of Order d in a Finite Group**

> *In a finite group, the number of elements of order d is divisible by $\phi(d)$.*

PROOF. If a finite group has no elements of order d, the statement is true, since $\phi(d)$ divides 0. Now suppose that $a \in G$ and $|a| = d$. By Theorem 4.4, we know that $\langle a \rangle$ has $\phi(d)$ elements of order d. If all elements of order d in G are in $\langle a \rangle$, we are done. So, suppose that there is an element b in G of order d that is not in $\langle a \rangle$. Then, $\langle b \rangle$ also has $\phi(d)$ elements of order d. This means that we have found $2\phi(d)$ elements of order d in G provided that $\langle a \rangle$ and $\langle b \rangle$ have no elements of order d in common. If there is an element c of order d that belongs to both $\langle a \rangle$ and $\langle b \rangle$, then we have $\langle a \rangle = \langle c \rangle = \langle b \rangle$, so that $b \in \langle a \rangle$, which is a contradiction. Continuing in this fashion, we see that the number of elements of order d in a finite group is a multiple of $\phi(d)$. ∎

The relationships among the various subgroups of a group can be illustrated with a *subgroup lattice* of the group. This is a diagram that includes all the subgroups of the group and connects a subgroup H at one level to a subgroup K at a higher level with a sequence of line segments if and only if H is a proper subgroup of K. Although there are many ways to draw such a diagram, the connections between the subgroups must be the same. Typically one attempts to present the diagram in an eye-pleasing fashion. The lattice diagram for Z_{30} is shown in Figure 4.2. Notice that $\langle 10 \rangle$ is a subgroup of both $\langle 2 \rangle$ and $\langle 5 \rangle$, but $\langle 6 \rangle$ is not a subgroup of $\langle 10 \rangle$.

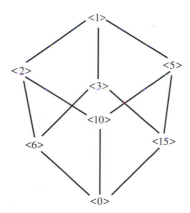

Figure 4.2   Subgroup lattice of Z_{30}.

The precision of Theorem 4.3 can be appreciated by comparing the ease with which we are able to identify the subgroups of Z_{30} with that of doing the same for, say, $U(30)$ or D_{30}. And these groups have relatively simple structures among noncyclic groups.

We will prove in Chapter 7 that a certain portion of Theorem 4.3 extends to arbitrary finite groups; namely, the order of a subgroup divides the order of the group itself. We will also see, however, that a finite group need not have exactly one subgroup corresponding to each divisor of the order of the group. For some divisors, there may be none at all, whereas for other divisors, there may be many.

   One final remark about the importance of cyclic groups is appropriate. Although cyclic groups constitute a very narrow class of finite groups, we will see in Chapter 11 that they play the role of building blocks for all finite Abelian groups in much the same way that primes are the building blocks for the integers and that chemical elements are the building blocks for the chemical compounds.

EXERCISES ━━━━━━━━━━━━━━━━━━━━━━━━━━━━━━━━━━━━━━━ ■■

It is not unreasonable to use the hypothesis.

ARNOLD ROSS

1. Find all generators of Z_6, Z_8, and Z_{20}.
2. Suppose that $\langle a \rangle$, $\langle b \rangle$, and $\langle c \rangle$ are cyclic groups of orders 6, 8, and 20, respectively. Find all generators of $\langle a \rangle$, $\langle b \rangle$, and $\langle c \rangle$.
3. List the elements of the subgroups $\langle 20 \rangle$ and $\langle 10 \rangle$ in Z_{30}.
4. List the elements of the subgroups $\langle 3 \rangle$ and $\langle 15 \rangle$ in Z_{18}.
5. List the elements of the subgroups $\langle 3 \rangle$ and $\langle 7 \rangle$ in $U(20)$.
6. What do Exercises 3, 4, and 5 have in common? Try to make a generalization that includes these three cases.
7. Find an example of a noncyclic group, all of whose proper subgroups are cyclic.
8. Let a be an element of a group and let $|a| = 15$. Compute the orders of the following elements of G.
   a. a^3, a^6, a^9, a^{12}
   b. a^5, a^{10}
   c. a^2, a^4, a^8, a^{14}
9. How many subgroups does Z_{20} have? List a generator for each of these subgroups. Suppose that $G = \langle a \rangle$ and $|a| = 20$. How many subgroups does G have? List a generator for each of these subgroups.
10. Let $G = \langle a \rangle$ and let $|a| = 24$. List all generators for the subgroup of order 8.
11. Let G be a group and let $a \in G$. Prove that $\langle a^{-1} \rangle = \langle a \rangle$.
12. Suppose that a has infinite order. Find all generators of the subgroup $\langle a^3 \rangle$.

13. Suppose that $|a| = 24$. Find a generator for $\langle a^{21} \rangle \cap \langle a^{10} \rangle$. In general, what is a generator for the subgroup $\langle a^m \rangle \cap \langle a^n \rangle$?

14. Suppose that a cyclic group G has exactly three subgroups: G itself, $\{e\}$, and a subgroup of order 7. What is $|G|$? What can you say if 7 is replaced with p where p is a prime?

15. Let G be an Abelian group and let $H = \{g \in G|\ |g|$ divides $12\}$. Prove that H is a subgroup of G. Is there anything special about 12 here? Would your proof be valid if 12 were replaced by some other positive integer? State the general result.

16. Find a collection of distinct subgroups $\langle a_1 \rangle, \langle a_2 \rangle, \ldots, \langle a_n \rangle$ of Z_{240} with the property that $\langle a_1 \rangle \subset \langle a_2 \rangle \subset \cdots \subset \langle a_n \rangle$ with n as large as possible.

17. Complete the following statement: $|a| = |a^2|$ if and only if $|a| \ldots$.

18. If a cyclic group has an element of infinite order, how many elements of finite order does it have?

19. List the cyclic subgroups of $U(30)$.

20. Suppose that G is an Abelian group of order 35 and every element of G satisfies the equation $x^{35} = e$. Prove that G is cyclic. Does your argument work if 35 is replaced with 33?

21. Let G be a group and let a be an element of G.
   a. If $a^{12} = e$, what can we say about the order of a?
   b. If $a^m = e$, what can we say about the order of a?
   c. Suppose that $|G| = 24$ and that G is cyclic. If $a^8 \neq e$ and $a^{12} \neq e$, show that $\langle a \rangle = G$.

22. Prove that a group of order 3 must be cyclic.

23. Let Z denote the group of integers under addition. Is every subgroup of Z cyclic? Why? Describe all the subgroups of Z.

24. For any element a in any group G, prove that $\langle a \rangle$ is a subgroup of $C(a)$ (the centralizer of a).

25. If d is a positive integer, $d \neq 2$, and d divides n, show that the number of elements of order d in D_n is $\phi(d)$. How many elements of order 2 does D_n have?

26. Find all generators of Z.

27. Let a be an element of a group and suppose that a has infinite order. How many generators does $\langle a \rangle$ have?

28. Let a be a group element that has infinite order. Prove that $\langle a^i \rangle = \langle a^j \rangle$ if and only if $i = \pm j$.

29. List all the elements of order 8 in $Z_{8000000}$. How do you know your list is complete?

30. Suppose a and b belong to a group, a has odd order, and $aba^{-1} = b^{-1}$. Show that $b^2 = e$.

31. Let G be a finite group. Show that there exists a fixed positive integer n such that $a^n = e$ for all a in G. (Note that n is independent of a.)

32. Determine the subgroup lattice for Z_{12}.

33. Determine the subgroup lattice for Z_{p^2q}, where p and q are distinct primes.

34. Determine the subgroup lattice for Z_8.

35. Determine the subgroup lattice for Z_{p^n}, where p is a prime and n is some positive integer.

36. Prove that a finite group is the union of proper subgroups if and only if the group is not cyclic.

37. Show that the group of positive rational numbers under multiplication is not cyclic.

38. Consider the set $\{4, 8, 12, 16\}$. Show that this set is a group under multiplication modulo 20 by constructing its Cayley table. What is the identity element? Is the group cyclic? If so, find all of its generators.

39. Give an example of a group that has exactly 6 subgroups (including the trivial subgroup and the group itself). Generalize to exactly n subgroups for any positive integer n.

40. Let m and n be elements of the group Z. Find a generator for the group $\langle m \rangle \cap \langle n \rangle$.

41. Suppose that a and b are group elements that commute and have orders m and n. If $\langle a \rangle \cap \langle b \rangle = \{e\}$, prove that the group contains an element whose order is the least common multiple of m and n. Show that this need not be true if a and b do not commute.

42. Prove that an infinite group must have an infinite number of subgroups.

43. Let p be a prime. If a group has more than $p - 1$ elements of order p, why can't the group be cyclic?

44. Suppose that G is a cyclic group and that 6 divides $|G|$. How many elements of order 6 does G have? If 8 divides $|G|$, how many elements of order 8 does G have? If a is one element of order 8, list the other elements of order 8.

45. List all the elements of Z_{40} that have order 10.

46. Let $|x| = 40$. List all the elements of $\langle x \rangle$ that have order 10.

47. Determine the orders of the elements of D_{33} and how many there are of each.

48. If G is a cyclic group and 15 divides the order of G, determine the number of solutions in G of the equation $x^{15} = e$. If 20 divides the order of G, determine the number of solutions of $x^{20} = e$. Generalize.

49. If G is an Abelian group and contains cyclic subgroups of orders 4 and 5, what other sizes of cyclic subgroups must G contain?

50. If G is an Abelian group and contains cyclic subgroups of orders 4 and 6, what other sizes of cyclic subgroups must G contain?

51. Prove that no group can have exactly two elements of order 2.

52. Given the fact that $U(49)$ is cyclic and has 42 elements, deduce the number of generators that $U(49)$ has without actually finding any of the generators.

53. Let a and b be elements of a group. If $|a| = 10$ and $|b| = 21$, show that $\langle a \rangle \cap \langle b \rangle = \{e\}$.

54. Let a and b belong to a group. If $|a|$ and $|b|$ are relatively prime, show that $\langle a \rangle \cap \langle b \rangle = \{e\}$.

55. Let a and b belong to a group. If $|a| = 24$ and $|b| = 10$, what are the possibilities for $|\langle a \rangle \cap \langle b \rangle|$?

56. Prove that $U(2^n)$ $(n \geq 3)$ is not cyclic.

57. Suppose that G is a group of order 16 and that, by direct computation, you know that G has at least nine elements x such that $x^8 = e$. Can you conclude that G is not cyclic? What if G has at least five elements x such that $x^4 = e$? Generalize.

58. Prove that Z_n has an even number of generators if $n > 2$. What does this tell you about $\phi(n)$?

59. If $|a^5| = 12$, what are the possibilities for $|a|$? If $|a^4| = 12$, what are the possibilities for $|a|$?

60. Suppose that $|x| = n$. Find a necessary and sufficient condition on r and s such that $\langle x^r \rangle \subseteq \langle x^s \rangle$.

61. Let p be a prime. Show that in a cyclic group of order $p^n - 1$, every element is a pth power (that is, every element can be written in the form a^p for some a).

62. Prove that $H = \left\{ \begin{bmatrix} 1 & n \\ 0 & 1 \end{bmatrix} \middle| n \in Z \right\}$ is a cyclic subgroup of $GL(2, \mathbf{R})$.

63. Let a and b belong to a group. If $|a| = 12$, $|b| = 22$, and $\langle a \rangle \cap \langle b \rangle \neq \{e\}$, prove that $a^6 = b^{11}$.

64. Suppose that G is a finite group with the property that every nonidentity element has prime order (for example, D_3 and D_5). If $Z(G)$ is not trivial, prove that every nonidentity element of G has the same order.

65. Let G be the set of all polynomials of the form $ax^2 + bx + c$ with coefficients from the set $\{0, 1, 2\}$. We can make G a group under addition by adding the polynomials in the usual way, except that we use modulo 3 to combine the coefficients. With this operation, prove that G is a group of order 27 that is not cyclic.

COMPUTER EXERCISES ▬▬▬▬▬▬▬▬▬▬▬▬▬▬▬▬▬▬ ■■

The nerds are running the world now.

JOE PISCOPO

Software for the computer exercises in this chapter is available at the website:

http://www.d.umn.edu/~jgallian

1. This software determines if $U(n)$ is cyclic. Run the program for $n = 8$, 32, 64, and 128. Make a conjecture. Run the program for $n = 3, 9, 27$, 81, 243, 5, 25, 125, 7, 49, 11, and 121. Make a conjecture. Run the program for $n = 12, 20, 28, 44, 52, 15, 21, 33, 39, 51, 57, 69, 35, 55, 65$, and 85. Make a conjecture.

2. For any pair of positive integers m and n, let $Z_m \oplus Z_n = \{(a, b) \mid a \in Z_m$, $b \in Z_n\}$. For any pair of elements (a, b) and (c, d) in $Z_m \oplus Z_n$, define $(a, b) + (c, d) = ((a + c) \bmod m, (b + d) \bmod n)$. [For example, in $Z_3 \oplus Z_4$, we have $(1, 2) + (2, 3) = (0, 1)$.] This software checks whether or not $Z_m \oplus Z_n$ is cyclic. Run the program for the following choices of m and n: (2, 2), (2, 3), (2, 4), (2, 5), (3, 4), (3, 5), (3, 6), (3, 7), (3, 8), (3, 9), and (4, 6). On the basis of this output, guess how m and n must be related for $Z_m \oplus Z_n$ to be cyclic.

3. In this exercise, $a, b \in U(n)$. Define $\langle a, b \rangle = \{a^i b^j \mid 0 \leq i < |a|, 0 \leq j < |b|\}$. This software computes the orders of $\langle a, b \rangle$, $\langle a \rangle$, $\langle b \rangle$, and $\langle a \rangle \cap \langle b \rangle$. Run the program for the following choices of a, b, and n: (21, 101, 550), (21, 49, 550), (7, 11, 100), (21, 31, 100), and (63, 77, 100). On the basis of your output, make a conjecture about arithmetic relationships among $|\langle a, b \rangle|$, $|\langle a \rangle|$, $|\langle b \rangle|$, and $|\langle a \rangle \cap \langle b \rangle|$.

4. For each positive integer n, this software gives the order of $U(n)$ and the order of each element in $U(n)$. Do you see any relationship between the order of $U(n)$ and the order of its elements? Run the program for $n = 8$, 16, 32, 64, and 128. Make a conjecture about the number of elements of order 2 in $U(2^k)$ when k is at least 3. Make a conjecture about the number of elements of order 4 in $U(2^k)$ when k is at least 4. Make a conjecture about the number of elements of order 8 in $U(2^k)$ when k is at least 5. Make a conjecture about the maximum order of any element in $U(2^k)$ when k is at least 3. Try to find a formula for an element of order 4 in $U(2^k)$ when k is at least 4.

5. For each positive integer n, this software lists the number of elements of $U(n)$ of each order. For each order d of some element of $U(n)$, this software lists $\phi(d)$ and the number of elements of order d. (Recall that $\phi(d)$ is the number of positive integers less than or equal to d and relatively prime to d). Do you see any relationship between the number of elements of order d and $\phi(d)$? Run the program for $n = 3, 9, 27, 81, 5, 25, 125, 7, 49$, and 343. Make a conjecture about the number of elements of order d and $\phi(d)$ when n is a power of an odd prime. Run the program for $n = 6, 18, 54, 162, 10, 50, 250, 14, 98$, and 686. Make a conjecture about the number of elements of order d and $\phi(d)$ when n is twice a power of an odd prime. Make a conjecture about the number of elements of various orders in $U(p^k)$ and $U(2p^k)$ where p is an odd prime.

6. For each positive integer n, this software gives the order of $U(n)$. Run the program for $n = 9, 27, 81$, and 243. Try to guess a formula for the order of $U(3^k)$ when k is at least 2. Run the program for $n = 18, 54, 162$, and 486. How does the order of $U(2 \cdot 3^k)$ appear to be related to the order of $U(3^k)$? Run the program for $n = 25, 125$, and 625. Try to guess a formula for the order of $U(5^k)$ when k is at least 2. Run the program for $n = 50, 250$, and 1250. How does the order of $U(2 \cdot 5^k)$ appear to be related to the order of $U(5^k)$? Run the program for $n = 49$ and 343. Try to guess a formula for the order of $U(7^k)$ when k is at least 2. Run the program for $n = 98$ and 686. How does the order of $U(2 \cdot 7^k)$ appear to be related to the order of $U(7^k)$? Based on your guesses for $U(3^k)$, $U(5^k)$, and $U(7^k)$, guess a formula for the order of $U(p^k)$ when p is an odd prime and k is at least 2. What about the order of $U(2p^k)$ when p is an odd prime and k is at least 2. Does your formula also work when k is 1?

Suggested Reading

Deborah L. Massari, "The Probability of Regenerating a Cyclic Group," *Pi Mu Epsilon Journal* 7 (1979): 3–6.

In this easy-to-read paper, it is shown that the probability of a randomly chosen element from a cyclic group being a generator of the group depends only on the set of prime divisors of the order of the group, and not on the order itself. This article, written by an undergraduate student, received first prize in a Pi Mu Epsilon Paper Contest.

J. J. Sylvester

I really love my subject.

J. J. SYLVESTER

JAMES JOSEPH SYLVESTER was the most influential mathematician in America in the 19th century. Sylvester was born on September 3, 1814, in London and showed his mathematical genius early. At the age of 14, he studied under De Morgan and won several prizes for his mathematics, and at the unusually young age of 25, he was elected a Fellow of the Royal Society.

After receiving B.A. and M.A. degrees from Trinity College in Dublin in 1841, Sylvester began a professional life that was to include academics, law, and actuarial careers. In 1876, at the age of 62, he was appointed to a prestigious position at the newly founded Johns Hopkins University. During his seven years at Johns Hopkins, Sylvester pursued research in pure mathematics, the first ever done in America, with tremendous vigor and enthusiasm. He also founded the *American Journal of Mathematics,* the first journal in America devoted to mathematical research. Sylvester returned to England in 1884 to a professorship at Oxford, a position he held until his death on March 15, 1897.

Sylvester's major contributions to mathematics were in the theory of equations, matrix theory, determinant theory, and invariant theory (which he founded with Cayley). His writings and lectures—flowery and eloquent, pervaded with poetic flights, emotional expressions, bizarre utterances, and paradoxes—reflected the personality of this sensitive, excitable, and enthusiastic man. We quote three of his students.[†] E. W. Davis commented on Sylvester's teaching methods.

> Sylvester's methods! He had none. "Three lectures will be delivered on a New Universal Algebra," he would say; then, "The course must be extended to twelve." It did last all the rest of that year. The following year the course was to be *Substitutions-Theorie,* by Netto. We all got the text. He lectured about three times, following the text closely and stopping sharp at the end of the hour. Then he began to think about matrices again. "I must give one lecture a week on those," he

---

[†]F. Cajori, *Teaching and History of Mathematics in the U.S.,* Washington, 1890, 265–266.

said. He could not confine himself to the hour, nor to the one lecture a week. Two weeks were passed, and Netto was forgotten entirely and never mentioned again. Statements like the following were not infrequent in his lectures: "I haven't proved this, but I am as sure as I can be of anything that it must be so. From this it will follow, etc." At the next lecture it turned out that what he was so sure of was false. Never mind, he kept on forever guessing and trying, and presently a wonderful discovery followed, then another and another. Afterward he would go back and work it all over again, and surprise us with all sorts of side lights. He then made another leap in the dark, more treasures were discovered, and so on forever.

Sylvester's enthusiasm for teaching and his influence on his students are captured in the following passage written by Sylvester's first student at Johns Hopkins, G. B. Halsted.

A short, broad man of tremendous vitality, . . . Sylvester's capacious head was ever lost in the highest cloud-lands of pure mathematics. Often in the dead of night he would get his favorite pupil, that he might communicate the very last product of his creative thought. Everything he saw suggested to him something new in the higher algebra. This transmutation of everything into new mathematics was a revelation to those who knew him intimately. They began to do it themselves.

Another characteristic of Sylvester, which is very unusual among mathematicians, was his apparent inability to remember mathematics! W. P. Durfee had the following to say

Sylvester had one remarkable peculiarity. He seldom remembered theorems, propositions, etc., but had always to deduce them when he wished to use them. In this he was the very antithesis of Cayley, who was thoroughly conversant with everything that had been done in every branch of mathematics.

I remember once submitting to Sylvester some investigations that I had been engaged on, and he immediately denied my first statement, saying that such a proposition had never been heard of, let alone proved. To his astonishment, I showed him a paper of his own in which he had proved the proposition; in fact, I believe the object of his paper had been the very proof which was so strange to him.

For more information about Sylvester, visit:

http://www-groups.dcs.st-and.ac.uk/~history/

SUPPLEMENTARY EXERCISES FOR CHAPTERS 1–4 ━━━ ■■

> *It is better to wear out than to rust out.*
>
> BISHOP RICHARD CUMBERLAND

True/False questions for Chapters 1–4 are available on the web at:

http://www.d.umn.edu/~jgallian/TF

1. Let G be a group and let H be a subgroup of G. For any fixed x in G, define $xHx^{-1} = \{xhx^{-1} \mid h \in H\}$. Prove the following.
   a. xHx^{-1} is a subgroup of G.
   b. If H is cyclic, then xHx^{-1} is cyclic.
   c. If H is Abelian, then xHx^{-1} is Abelian.
   The group xHx^{-1} is called a *conjugate* of H. (Note that conjugation preserves structure.)

2. Let G be a group and let H be a subgroup of G. Define $N(H) = \{x \in G \mid xHx^{-1} = H\}$. Prove that $N(H)$ (called the *normalizer* of H) is a subgroup of G.†

3. Let G be a group. For each $a \in G$, define cl$(a) = \{xax^{-1} \mid x \in G\}$. Prove that these subsets of G partition G. [cl(a) is called the *conjugacy class* of a.]

4. The group defined by the following table is called the *group of quaternions*. Use the table to determine each of the following:
   a. The center
   b. cl(a)
   c. cl(b)
   d. All cyclic subgroups

	e	a	a^2	a^3	b	ba	ba^2	ba^3
e	e	a	a^2	a^3	b	ba	ba^2	ba^3
a	a	a^2	a^3	e	ba^3	b	ba	ba^2
a^2	a^2	a^3	e	a	ba^2	ba^3	b	ba
a^3	a^3	e	a	a^2	ba	ba^2	ba^3	b
b	b	ba	ba^2	ba^3	a^2	a^3	e	a
ba	ba	ba^2	ba^3	b	a	a^2	a^3	e
ba^2	ba^2	ba^3	b	ba	e	a	a^2	a^3
ba^3	ba^3	b	ba	ba^2	a^3	e	a	a^2

---

†This very important subgroup was first used by L. Sylow in 1872 to prove the existence of certain kinds of subgroups in a group. His work is discussed in Chapter 24.

5. (Conjugation preserves order.) Prove that, in any group, $|xax^{-1}| = |a|$. (This exercise is referred to in Chapter 24.)

6. Prove that, in any group, $|ab| = |ba|$.

7. If a, b, and c are elements of a group, give an example to show that it need not be the case that $|abc| = |cba|$.

8. Let a and b belong to a group G. Prove that there is an element x in G such that $xax = b$ if and only if $ab = c^2$ for some element c in G.

9. Prove that if a is the only element of order 2 in a group, then a lies in the center of the group.

10. Let G be the plane symmetry group of the infinite strip of equally spaced H's shown below.

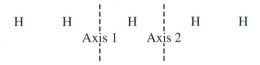

Let x be the reflection about Axis 1 and let y be the reflection about Axis 2. Calculate $|x|$, $|y|$, and $|xy|$. Must the product of elements of finite order have finite order?

11. What are the orders of the elements of D_{15}? How many elements have each of these orders?

12. Prove that a group of order 4 is Abelian.

13. Prove that a group of order 5 must be cyclic.

14. Prove that an Abelian group of order 6 must be cyclic.

15. Let G be an Abelian group and let n be a fixed positive integer. Let $G^n = \{g^n \mid g \in G\}$. Prove that G^n is a subgroup of G. Give an example showing that G^n need not be a subgroup of G when G is non-Abelian. (This exercise is referred to in Chapter 11.)

16. Let $G = \{a + b\sqrt{2}\}$, where a and b are rational numbers not both 0. Prove that G is a group under ordinary multiplication.

17. (1969 Putnam Competition) Prove that no group is the union of two proper subgroups. Does the statement remain true if "two" is replaced by "three"?

18. Prove that the subset of elements of finite order in an Abelian group forms a subgroup. (This subgroup is called the *torsion subgroup*.) Is the same thing true for non-Abelian groups?

19. Let p be a prime and let G be an Abelian group. Show that the set of all elements whose orders are powers of p is a subgroup of G.

20. Let x and y belong to a group G. Assume that $x \neq e$, $|y| = 2$, and $yxy^{-1} = x^2$. Find $|x|$.

21. Suppose that a finite group is generated by two elements a and b (that is, every element of the group can be expressed as some product of a's and b's). Given that $a^3 = b^2 = e$ and $ba^2 = ab$, construct the Cayley table for the group. We have already seen an example of a group that satisfies these conditions. Name it.

22. Suppose that a finite group is generated by two elements a and b. Given that $a^4 = b^2 = e$ and $ba = a^3b$, construct the Cayley table for the group. Give an example of a group that satisfies these conditions.

23. Let x and y belong to a group G. If $xy \in Z(G)$, prove that $xy = yx$.

24. Suppose that H and K are nontrivial subgroups of Q under addition. Show that $H \cap K$ is a nontrivial subgroup of Q. Is this true if Q is replaced by $\mathbf{R}$?

25. Let H be a subgroup of G and let g be an element of G. Prove that $N(gHg^{-1}) = gN(H)g^{-1}$. See Exercise 2 for the notation.

26. Let H be a subgroup of a group G and let $|g| = n$. If g^m belongs to H and m and n are relatively prime, prove that g belongs to H.

27. Find a group that contains elements a and b such that $|a| = 2$, $|b| = 11$, and $|ab| = 2$.

28. Suppose that G is a group with exactly eight elements of order 10. How many cyclic subgroups of order 10 does G have?

29. (1989 Putnam Competition) Let S be a nonempty set with an associative operation that is left and right cancellative ($xy = xz$ implies $y = z$, and $yx = zx$ implies $y = z$). Assume that for every a in S the set $\{a^n \mid n = 1, 2, 3, \ldots\}$ is finite. Must S be a group?

30. Let $H_1, H_2, H_3, \ldots$ be a sequence of subgroups of a group with the property that $H_1 \subseteq H_2 \subseteq H_3 \ldots$. Prove that the union of the sequence is a subgroup.

31. Let $\mathbf{R}^*$ be the group of nonzero real numbers under multiplication and let $H = \{g \in \mathbf{R}^* \mid$ some nonzero integer power of g is a rational number$\}$. Prove that H is a subgroup of $\mathbf{R}^*$.

32. Suppose that a and b belong to a group, a and b commute, and $|a|$ and $|b|$ are relatively prime. Prove that $|ab| = |a||b|$. Give an example showing that $|ab|$ need not be $|a||b|$ when a and b commute but $|a|$ and $|b|$ are not relatively prime. (Don't use $b = a^{-1}$.)

33. Let $H = \{A \in GL(2, \mathbf{R}) \mid \det A$ is rational$\}$. Prove or disprove that H is a subgroup of $GL(2, \mathbf{R})$. What if "rational" is replaced by "an integer"?

34. Suppose that G is a group that has exactly one nontrivial proper subgroup. Prove that G is cyclic and $|G| = p^2$, where p is prime.

35. Suppose that G is a group and G has exactly two nontrivial proper subgroups. Prove that G is cyclic and $|G| = pq$, where p and q are distinct primes, or that G is cyclic and $|G| = p^3$, where p is prime.

36. If $|a^2| = |b^2|$, prove or disprove that $|a| = |b|$.

37. (1995 Putnam Competition) Let S be a set of real numbers that is closed under multiplication. Let T and U be disjoint subsets of S whose union is S. Given that the product of any three (not necessarily distinct) elements of T is in T and that the product of any three elements of U is in U, show that at least one of the two subsets T and U is closed under multiplication.

38. If p is an odd prime, prove that there is no group that has exactly p elements of order p.

39. Give an example of a group G with infinitely many distinct subgroups $H_1, H_2, H_3, \ldots$ such that $H_1 \subset H_2 \subset H_3 \ldots$.

40. Suppose a and b are group elements and $b \neq e$. If $a^{-1}ba = b^2$ and $|a| = 3$, find $|b|$. What is $|b|$, if $|a| = 5$? What can you say about $|b|$ in the case where $|a| = k$?

41. Let a and b belong to a group G. Show that there is an element g in G such that $g^{-1} abg = ba$.

42. Suppose G is a group and $x^3y^3 = y^3x^3$ for every x and y in G. Let $H = \{x \in G| \ |x| \text{ is relatively prime to } 3\}$. Prove that elements of H commute with each other and that H is a subgroup of G. Is your argument valid if 3 is replaced by an arbitrary positive integer n? Explain why or why not.

43. Let G be a finite group and let S be a subset of G that contains more than half of the elements of G. Show that every element of G can be expressed in the form s_1s_2 where s_1 and s_2 belong to S.

44. Let G be a group and let f be a function from G to some set. Show that $H = \{g \in G| f(xg) = f(x) \text{ for all } x \in G\}$ is a subgroup of G. In the case that G is the group of real numbers under addition and $f(x) = \sin x$, describe H.

45. Let G be a cyclic group of order n and let H be the subgroup of order d. Show that $H = \{x \in G| \ |x| \text{ divides } d\}$.

46. Let a be an element of maximum order from a finite Abelian group G. Prove that for any element b in G, $|b|$ divides $|a|$. Show by example that this need not be true for finite non-Abelian groups.

5 | Permutation Groups

> *Wigner's discovery about the electron permutation group was just the beginning. He and others found many similar applications and nowadays group theoretical methods—especially those involving characters and representations—pervade all branches of quantum mechanics.*
>
> GEORGE MACKEY, *Proceedings of the American Philosophical Society*

DEFINITION AND NOTATION

In this chapter, we study certain groups of functions, called permutation groups, from a set A to itself. In the early and mid-19th century, groups of permutations were the only groups investigated by mathematicians. It was not until around 1850 that the notion of an abstract group was introduced by Cayley, and it took another quarter century before the idea firmly took hold.

DEFINITIONS *Permutation of A, Permutation Group of A*

A *permutation* of a set A is a function from A to A that is both one-to-one and onto. A *permutation group* of a set A is a set of permutations of A that forms a group under function composition.

Although groups of permutations of any nonempty set A of objects exist, we will focus on the case where A is finite. Furthermore, it is customary, as well as convenient, to take A to be a set of the form $\{1, 2, 3, \ldots, n\}$ for some positive integer n. Unlike in calculus, where most functions are defined on infinite sets and are given by formulas, in algebra, permutations of finite sets are usually given by an explicit listing of each element of the domain and its corresponding functional value. For example, we define a permutation α of the set $\{1, 2, 3, 4\}$ by specifying

$$\alpha(1) = 2, \qquad \alpha(2) = 3, \qquad \alpha(3) = 1, \qquad \alpha(4) = 4.$$

A more convenient way to express this correspondence is to write α in array form as

$$\alpha = \begin{bmatrix} 1 & 2 & 3 & 4 \\ 2 & 3 & 1 & 4 \end{bmatrix}.$$

Here $\alpha(j)$ is placed directly below j for each j. Similarly, the permutation β of the set $\{1, 2, 3, 4, 5, 6\}$ given by

$$\beta(1) = 5, \quad \beta(2) = 3, \quad \beta(3) = 1, \quad \beta(4) = 6, \quad \beta(5) = 2, \quad \beta(6) = 4$$

is expressed in array form as

$$\beta = \begin{bmatrix} 1 & 2 & 3 & 4 & 5 & 6 \\ 5 & 3 & 1 & 6 & 2 & 4 \end{bmatrix}.$$

Composition of permutations expressed in array notation is carried out from right to left by going from top to bottom, then again from top to bottom. For example, let

$$\sigma = \begin{bmatrix} 1 & 2 & 3 & 4 & 5 \\ 2 & 4 & 3 & 5 & 1 \end{bmatrix}$$

and

$$\gamma = \begin{bmatrix} 1 & 2 & 3 & 4 & 5 \\ 5 & 4 & 1 & 2 & 3 \end{bmatrix};$$

then

$$\gamma\sigma = \begin{bmatrix} 1 & 2 & 3 & 4 & 5 \\ 5 & 4 & 1 & 2 & 3 \end{bmatrix}\begin{bmatrix} 1 & 2 & 3 & 4 & 5 \\ 2 & 4 & 3 & 5 & 1 \end{bmatrix} = \begin{bmatrix} 1 & 2 & 3 & 4 & 5 \\ 4 & 2 & 1 & 3 & 5 \end{bmatrix}.$$

On the right we have 4 under 1, since $(\gamma\sigma)(1) = \gamma(\sigma(1)) = \gamma(2) = 4$, so $\gamma\sigma$ sends 1 to 4. The remainder of the bottom row $\gamma\sigma$ is obtained in a similar fashion.

We are now ready to give some examples of permutation groups.

EXAMPLE 1   Symmetric Group S_3

Let S_3 denote the set of all one-to-one functions from $\{1, 2, 3\}$ to itself. Then S_3, under function composition, is a group with six elements. The six elements are

$$\varepsilon = \begin{bmatrix} 1 & 2 & 3 \\ 1 & 2 & 3 \end{bmatrix}, \quad \alpha = \begin{bmatrix} 1 & 2 & 3 \\ 2 & 3 & 1 \end{bmatrix}, \quad \alpha^2 = \begin{bmatrix} 1 & 2 & 3 \\ 3 & 1 & 2 \end{bmatrix},$$

$$\beta = \begin{bmatrix} 1 & 2 & 3 \\ 1 & 3 & 2 \end{bmatrix}, \qquad \alpha\beta = \begin{bmatrix} 1 & 2 & 3 \\ 2 & 1 & 3 \end{bmatrix}, \qquad \alpha^2\beta = \begin{bmatrix} 1 & 2 & 3 \\ 3 & 2 & 1 \end{bmatrix}.$$

Note that $\beta\alpha = \begin{bmatrix} 1 & 2 & 3 \\ 3 & 2 & 1 \end{bmatrix} = \alpha^2\beta \neq \alpha\beta$, so that S_3 is non-Abelian.   ◆

The relation $\beta\alpha = \alpha^2\beta$ can be used to compute other products in S_3 without resorting to the arrays. For example, $\beta\alpha^2 = (\beta\alpha)\alpha = (\alpha^2\beta)\alpha = \alpha^2(\beta\alpha) = \alpha^2(\alpha^2\beta) = \alpha^4\beta = \alpha\beta$.

Example 1 can be generalized as follows.

EXAMPLE 2   Symmetric Group S_n

Let $A = \{1, 2, \ldots, n\}$. The set of all permutations of A is called the *symmetric group of degree n* and is denoted by S_n. Elements of S_n have the form

$$\alpha = \begin{bmatrix} 1 & 2 & \ldots & n \\ \alpha(1) & \alpha(2) & \ldots & \alpha(n) \end{bmatrix}.$$

It is easy to compute the order of S_n. There are n choices of $\alpha(1)$. Once $\alpha(1)$ has been determined, there are $n - 1$ possibilities for $\alpha(2)$ [since α is one-to-one, we must have $\alpha(1) \neq \alpha(2)$]. After choosing $\alpha(2)$, there are exactly $n - 2$ possibilities for $\alpha(3)$. Continuing along in this fashion, we see that S_n has $n(n - 1) \cdots 3 \cdot 2 \cdot 1 = n!$ elements. We leave it to the reader to prove that S_n is non-Abelian when $n \geq 3$ (Exercise 41).   ◆

The symmetric groups are rich in subgroups. The group S_4 has 30 subgroups, and S_5 has well over 100 subgroups.

EXAMPLE 3   Symmetries of a Square

As a third example, we associate each motion in D_4 with the permutation of the locations of each of the four corners of a square. For example, if we label the four corner positions as in the figure below and keep these labels fixed for reference, we may describe a 90° counterclockwise rotation by the permutation

$$\rho = \begin{bmatrix} 1 & 2 & 3 & 4 \\ 2 & 3 & 4 & 1 \end{bmatrix},$$

whereas a reflection across a horizontal axis yields

$$\phi = \begin{bmatrix} 1 & 2 & 3 & 4 \\ 2 & 1 & 4 & 3 \end{bmatrix}.$$

These two elements generate the entire group (that is, every element is some combination of the ρ's and ϕ's).

When D_4 is represented in this way, we see that it is a subgroup of S_4. ◆

CYCLE NOTATION

There is another notation commonly used to specify permutations. It is called *cycle notation* and was first introduced by the great French mathematician Cauchy in 1815. Cycle notation has theoretical advantages in that certain important properties of the permutation can be readily determined when cycle notation is used.

As an illustration of cycle notation, let us consider the permutation

$$\alpha = \begin{bmatrix} 1 & 2 & 3 & 4 & 5 & 6 \\ 2 & 1 & 4 & 6 & 5 & 3 \end{bmatrix}.$$

This assignment of values could be presented schematically as follows:

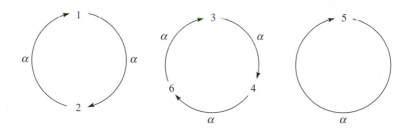

Although mathematically satisfactory, such diagrams are cumbersome. Instead, we leave out the arrows and simply write $\alpha = (1, 2)$ $(3, 4, 6)(5)$. As a second example, consider

$$\beta = \begin{bmatrix} 1 & 2 & 3 & 4 & 5 & 6 \\ 5 & 3 & 1 & 6 & 2 & 4 \end{bmatrix}.$$

In cycle notation, β can be written $(2, 3, 1, 5)(6, 4)$ or $(4, 6)(3, 1, 5, 2)$, since both of these unambiguously specify the function β. An expression of the form $(a_1, a_2, \ldots, a_m)$ is called a *cycle of length m* or an *m-cycle*.

A multiplication of cycles can be introduced by thinking of a cycle as a permutation that fixes any symbol not appearing in the cycle. Thus, the cycle (4, 6) can be thought of as representing the permutation $\begin{bmatrix} 1 & 2 & 3 & 4 & 5 & 6 \\ 1 & 2 & 3 & 6 & 5 & 4 \end{bmatrix}$. In this way, we can multiply cycles by thinking of them as permutations given in array form. Consider the following example from S_8. Let $\alpha = (13)(27)(456)(8)$ and $\beta = (1237)(648)(5)$. (When the domain consists of single-digit integers, it is common practice to omit the commas between the digits.) What is the cycle form of $\alpha\beta$? Of course, one could say that $\alpha\beta = (13)(27)(456)(8)(1237)(648)(5)$, but it is usually more desirable to express a permutation in a *disjoint* cycle form (that is, the various cycles have no number in common). Well, keeping in mind that function composition is done from right to left and that each cycle that does not contain a symbol fixes the symbol, we observe that: (5) fixes 1; (648) fixes 1; (1237) sends 1 to 2; (8) fixes 2; (456) fixes 2; (27) sends 2 to 7; and (13) fixes 7. So the net effect of $\alpha\beta$ is to send 1 to 7. Thus we begin $\alpha\beta = (17 \cdots) \cdots$. Now, repeating the entire process beginning with 7, we have, cycle by cycle, right to left, $7 \to 7 \to 7 \to 1 \to 1 \to 1 \to 1 \to 3$, so that $\alpha\beta = (173 \cdots) \cdots$. Ultimately, we have $\alpha\beta = (1732)(48)(56)$. The important thing to bear in mind when multiplying cycles is to "keep moving" from one cycle to the next from right to left. (*Warning:* Some authors compose cycles from left to right. When reading another text, be sure to determine which convention is being used.)

To be sure you understand how to switch from one notation to the other and how to multiply permutations, we will do one more example of each.

If array notations for α and β, respectively, are

$$\begin{bmatrix} 1 & 2 & 3 & 4 & 5 \\ 2 & 1 & 3 & 5 & 4 \end{bmatrix} \quad \text{and} \quad \begin{bmatrix} 1 & 2 & 3 & 4 & 5 \\ 5 & 4 & 1 & 2 & 3 \end{bmatrix}$$

then, in cycle notation, $\alpha = (12)(3)(45)$, $\beta = (153)(24)$, and $\alpha\beta = (12)(3)(45)(153)(24)$.

To put $\alpha\beta$ in disjoint cycle form, observe that (24) fixes 1; (153) sends 1 to 5; (45) sends 5 to 4; and (3) and (12) both fix 4. So, $\alpha\beta$ sends 1 to 4. Continuing in this way we obtain $\alpha\beta = (14)(253)$.

One can convert $\alpha\beta$ back to array form without converting each cycle of $\alpha\beta$ into array form by simply observing that (14) means 1 goes to 4 and 4 goes to 1; (253) means $2 \to 5, 5 \to 3, 3 \to 2$.

One final remark about cycle notation: Mathematicians prefer not to write cycles that have only one entry. In this case, it is understood that any missing element is mapped to itself. With this convention, the permutation α above can be written as (12)(45). Similarly,

$$\alpha = \begin{bmatrix} 1 & 2 & 3 & 4 & 5 \\ 3 & 2 & 4 & 1 & 5 \end{bmatrix}$$

can be written $\alpha = (134)$. Of course, the identity permutation consists only of cycles with one entry, so we cannot omit all of these! In this case, one usually writes just one cycle. For example,

$$\varepsilon = \begin{bmatrix} 1 & 2 & 3 & 4 & 5 \\ 1 & 2 & 3 & 4 & 5 \end{bmatrix}$$

could be written as $\varepsilon = (5)$ or $\varepsilon = (1)$. Just remember that missing elements are mapped to themselves.

PROPERTIES OF PERMUTATIONS

We are now ready to state several theorems about permutations and cycles. The proof of the first theorem is implicit in our discussion of writing permutations in cycle form.

Theorem 5.1   **Products of Disjoint Cycles**

> *Every permutation of a finite set can be written as a cycle or as a product of disjoint cycles.*

PROOF.   Let α be a permutation on $A = \{1, 2, \ldots, n\}$. To write α in disjoint cycle form, we start by choosing any member of A, say a_1, and let

$$a_2 = \alpha(a_1), \qquad a_3 = \alpha(\alpha(a_1)) = \alpha^2(a_1),$$

and so on, until we arrive at $a_1 = \alpha^m(a_1)$ for some m. We know that such an m exists because the sequence $a_1, \alpha(a_1), \alpha^2(a_1), \ldots$ must be finite; so there must eventually be a repetition, say $\alpha^i(a_1) = \alpha^j(a_1)$ for some i and j with $i < j$. Then $a_1 = \alpha^m(a_1)$, where $m = j - i$. We express this relationship among $a_1, a_2, \ldots, a_m$ as

$$\alpha = (a_1, a_2, \ldots, a_m) \cdots .$$

The three dots at the end indicate the possibility that we may not have exhausted the set A in this process. In such a case, we merely choose any element b_1 of A not appearing in the first cycle and proceed to create a new cycle as before. That is, we let $b_2 = \alpha(b_1)$, $b_3 = \alpha^2(b_1)$, and so on, until we reach $b_1 = \alpha^k(b_1)$ for some k. This new cycle will have no elements in common with the previously constructed cycle. For, if so, then $\alpha^i(a_1) = \alpha^j(b_1)$ for some i and j. But then $\alpha^{i-j}(a_1) = b_1$, and therefore $b_1 = a_t$ for some t. This contradicts the way b_1 was chosen. Continuing this process until we run out of elements of A, our permutation will appear as

$$\alpha = (a_1, a_2, \ldots, a_m)(b_1, b_2, \ldots, b_k) \cdots (c_1, c_2, \ldots, c_s).$$

In this way, we see that every permutation can be written as a product of disjoint cycles. ■ ■

Theorem 5.2 Disjoint Cycles Commute

> *If the pair of cycles $\alpha = (a_1, a_2, \ldots, a_m)$ and $\beta = (b_1, b_2, \ldots, b_n)$ have no entries in common, then $\alpha\beta = \beta\alpha$.*

PROOF.   For definiteness, let us say that α and β are permutations of the set

$$S = \{a_1, a_2, \ldots, a_m, b_1, b_2, \ldots, b_n, c_1, c_2, \ldots, c_k\}$$

where the c's are the members of S left fixed by both α and β. To prove that $\alpha\beta = \beta\alpha$, we must show that $(\alpha\beta)(x) = (\beta\alpha)(x)$ for all x in S. If x is one of the a elements, say a_i, then

$$(\alpha\beta)(a_i) = \alpha(\beta(a_i)) = \alpha(a_i) = a_{i+1},$$

since β fixes all a elements. (We interpret a_{i+1} as a_1 if $i = m$.) For the same reason,

$$(\beta\alpha)(a_i) = \beta(\alpha(a_i)) = \beta(a_{i+1}) = a_{i+1}.$$

Hence, the functions of $\alpha\beta$ and $\beta\alpha$ agree on the a elements. A similar argument shows that $\alpha\beta$ and $\beta\alpha$ agree on the b elements as well. Finally, suppose that x is a c element, say c_i. Then, since both α and β fix c elements, we have

$$(\alpha\beta)(c_i) = \alpha(\beta(c_i)) = \alpha(c_i) = c_i$$

and

$$(\beta\alpha)(c_i) = \beta(\alpha(c_i)) = \beta(c_i) = c_i.$$

This completes the proof.   ■ ■

In demonstrating how to multiply cycles, we showed that the product $(13)(27)(456)(8)(1237)(648)(5)$ can be written in disjoint cycle form as $(1732)(48)(56)$. Is economy in expression the only advantage to writing a permutation in disjoint cycle form? No. The next theorem shows that the disjoint cycle form has the enormous advantage of allowing us to "eyeball" the order of the permutation.

Theorem 5.3   **Order of a Permutation (Ruffini—1799)**

> *The order of a permutation of a finite set written in disjoint cycle form is the least common multiple of the lengths of the cycles.*

PROOF.   First, observe that a cycle of length n has order n. (Verify this yourself.) Next, suppose that α and β are disjoint cycles of lengths m and n, and let k be the least common multiple of m and n. It follows from Theorem 4.1 that both α^k and β^k are the identity permutation ε and, since α and β commute, $(\alpha\beta)^k = \alpha^k\beta^k$ is also the identity. Thus, we know by Corollary 2 to Theorem 4.1 ($a^k = e$ implies that $|a|$ divides k) that the order of $\alpha\beta$—let us call it t—must divide k. But then $(\alpha\beta)^t = \alpha^t\beta^t = \varepsilon$, so that $\alpha^t = \beta^{-t}$. However, it is clear that if α and β have no common symbol, the same is true for α^t and β^{-t}, since raising a cycle to a power does not introduce new symbols. But, if α^t and β^{-t} are equal and have no common symbol, they must both be the identity, because every symbol in α^t is fixed by β^{-t} and vice versa (remember that a symbol not appearing in a permutation is fixed by the permutation). It follows, then, that both m and n must divide t. This means that k, the least common multiple of m and n, divides t also. This shows that $k = t$.

Thus far, we have proved that the theorem is true in the cases where the permutation is a single cycle or a product of two disjoint cycles. The general case involving more than two cycles can be handled in an analogous way.   ■ ■

As we will soon see, a particularly important kind of permutation is a cycle of length 2—that is, a permutation of the form (ab). Many authors call these permutations *transpositions,* since the effect of (ab) is to interchange or transpose a and b.

Theorem 5.4   **Product of 2-Cycles**

> *Every permutation in S_n, $n > 1$, is a product of 2-cycles.*

PROOF.   First, note that the identity can be expressed as $(12)(12)$, and so it is a product of 2-cycles. By Theorem 5.1, we know that every permutation can be written in the form

$$(a_1 a_2 \cdots a_k)(b_1 b_2 \cdots b_t) \cdots (c_1 c_2 \cdots c_s).$$

A direct computation shows that this is the same as

$$(a_1 a_k)(a_1 a_{k-1}) \cdots (a_1 a_2)(b_1 b_t)(b_1 b_{t-1}) \cdots (b_1 b_2) \cdots$$
$$(c_1 c_s)(c_1 c_{s-1}) \cdots (c_1 c_2).$$

This completes the proof.   ■ ■

The decompositions in the following example demonstrate this technique.

EXAMPLE 4

$$(12345) = (15)(14)(13)(12)$$
$$(1632)(457) = (12)(13)(16)(47)(45)   ◆$$

The decomposition of a permutation into a product of 2-cycles given in the proof of Theorem 5.4 is not the only way a permutation can be written as a product of 2-cycles. Although the next example shows that even the *number* of 2-cycles may vary from one decomposition to another, we will prove in Theorem 5.5 (first proved by Cauchy) that there is one aspect of a decomposition that never varies.

EXAMPLE 5

$$(12345) = (54)(53)(52)(51)$$
$$(12345) = (54)(52)(21)(25)(23)(13)   ◆$$

We isolate a special case of Theorem 5.5 as a lemma.

Lemma

> If $\varepsilon = \beta_1 \beta_2 \cdots \beta_r$, where the β's are 2-cycles, then r is even.

PROOF.   Clearly, $r \neq 1$, since a 2-cycle is not the identity. If $r = 2$, we are done. So, we suppose that $r > 2$, and we proceed by induction.

Since $(ij) = (ji)$, the product $\beta_{r-1}\beta_r$ can be expressed in one of the following forms shown on the right:

$$\varepsilon = (ab)(ab)$$
$$(ab)(bc) = (ac)(ab)$$
$$(ac)(cb) = (bc)(ab)$$
$$(ab)(cd) = (cd)(ab).$$

If the first case occurs, we may delete $\beta_{r-1}\beta_r$ from the original product to obtain $\varepsilon = \beta_1\beta_2 \cdots \beta_{r-2}$, and therefore, by the Second Principle of Mathematical Induction, $r - 2$ is even. In the other three cases, we replace the form of $\beta_{r-1}\beta_r$ on the right by its counterpart on the left to obtain a new product of r 2-cycles that is still the identity, but where the rightmost occurrence of the integer a is in the second-from-the-rightmost 2-cycle of the product instead of the rightmost 2-cycle. We now repeat the procedure just described with $\beta_{r-2}\beta_{r-1}$, and, as before, we obtain a product of $(r - 2)$ 2-cycles equal to the identity or a new product of r 2-cycles, where the rightmost occurrence of a is in the third 2-cycle from the right. Continuing this process, we must obtain a product of $(r - 2)$ 2-cycles equal to the identity, because otherwise we have a product equal to the identity in which the only occurrence of the integer a is in the leftmost 2-cycle, and such a product does not fix a, whereas the identity does. Hence, by induction, $r - 2$ is even, and r is even as well. ■ ■

Theorem 5.5   Always Even or Always Odd

If a permutation α can be expressed as a product of an even (odd) number of 2-cycles, then every decomposition of α into a product of 2-cycles must have an even (odd) number of 2-cycles. In symbols, if

$$\alpha = \beta_1\beta_2 \cdots \beta_r \qquad and \qquad \alpha = \gamma_1\gamma_2 \cdots \gamma_s,$$

where the β's and the γ's are 2-cycles, then r and s are both even or both odd.

PROOF.   Observe that $\beta_1\beta_2 \cdots \beta_r = \gamma_1\gamma_2 \cdots \gamma_s$ implies

$$\varepsilon = \gamma_1\gamma_2 \cdots \gamma_s\beta_r^{-1} \cdots \beta_2^{-1}\beta_1^{-1}$$
$$= \gamma_1\gamma_2 \cdots \gamma_s\beta_r \cdots \beta_2\beta_1,$$

since a 2-cycle is its own inverse. Thus, the lemma on page 102 guarantees that $s + r$ is even. It follows that r and s are both even or both odd.   ■■

DEFINITION   *Even and Odd Permutations*

A permutation that can be expressed as a product of an even number of 2-cycles is called an *even* permutation. A permutation that can be expressed as a product of an odd number of 2-cycles is called an *odd* permutation.

Theorems 5.4 and 5.5 together show that every permutation can be unambiguously classified as either even or odd. The significance of this observation is given in Theorem 5.6.

Theorem 5.6   Even Permutations Form a Group

The set of even permutations in S_n forms a subgroup of S_n.

PROOF.   This proof is left to the reader (Exercise 13).   ■■

The subgroup of even permutations in S_n arises so often that we give it a special name and notation.

DEFINITION   *Alternating Group of Degree n*

The group of even permutations of n symbols is denoted by A_n and is called the *alternating group of degree n*.

The next result shows that exactly half of the elements of S_n ($n > 1$) are even permutations.

Theorem 5.7

For $n > 1$, A_n has order $n!/2$.

PROOF.   For each odd permutation α, the permutation $(12)\alpha$ is even and $(12)\alpha \neq (12)\beta$ when $\alpha \neq \beta$. Thus, there are at least as many even permutations as there are odd ones. On the other hand, for each even permutation α, the permutation $(12)\alpha$ is odd and $(12)\alpha \neq (12)\beta$ when $\alpha \neq \beta$. Thus, there are at least as many odd permutations as there are even ones. It follows that there are equal numbers of even and odd permutations. Since $|S_n| = n!$, we have $|A_n| = n!/2$.   ■■

The names for the symmetric group and the alternating group of degree *n* come from the study of polynomials over *n* variables. A *symmetric* polynomial in the variables $x_1, x_2, \ldots, x_n$ is one that is unchanged under any transposition of two of the variables. An *alternating* polynomial is one that changes signs under any transposition of two of the variables. For example, the polynomial $x_1 x_2 x_3$ is unchanged by any transposition of two of the three variables, whereas the polynomial $(x_1 - x_2)(x_1 - x_3)(x_2 - x_3)$ changes signs when any two of the variables are transposed. Since every member of the symmetric group is the product of transpositions, the symmetric polynomials are those that are unchanged by members of the symmetric group. Likewise, since any member of the alternating group is the product of an even number of transpositions, the alternating polynomials are those that are unchanged by members of the alternating group.

The alternating groups are among the most important examples of groups. The groups A_4 and A_5 will arise on several occasions in later chapters. In particular, A_5 has great historical significance.

A geometric interpretation of A_4 is given in Example 6, and a multiplication table for A_4 is given as Table 5.1.

EXAMPLE 6   Rotations of a Tetrahedron
The 12 rotations of a regular tetrahedron can be conveniently described with the elements of A_4. The top row of Figure 5.1 illustrates the iden-

TABLE 5.1   The Alternating Group A_4 of Even Permutations of $\{1, 2, 3, 4\}$

(In this table, the permutations of A_4 are designated as $\alpha_1, \alpha_2, \ldots, \alpha_{12}$ and an entry *k* inside the table represents α_k. For example, $\alpha_3 \alpha_8 = \alpha_6$.)

	α_1	α_2	α_3	α_4	α_5	α_6	α_7	α_8	α_9	α_{10}	α_{11}	α_{12}
$(1) = \alpha_1$	1	2	3	4	5	6	7	8	9	10	11	12
$(12)(34) = \alpha_2$	2	1	4	3	6	5	8	7	10	9	12	11
$(13)(24) = \alpha_3$	3	4	1	2	7	8	5	6	11	12	9	10
$(14)(23) = \alpha_4$	4	3	2	1	8	7	6	5	12	11	10	9
$(123) = \alpha_5$	5	8	6	7	9	12	10	11	1	4	2	3
$(243) = \alpha_6$	6	7	5	8	10	11	9	12	2	3	1	4
$(142) = \alpha_7$	7	6	8	5	11	10	12	9	3	2	4	1
$(134) = \alpha_8$	8	5	7	6	12	9	11	10	4	1	3	2
$(132) = \alpha_9$	9	11	12	10	1	3	4	2	5	7	8	6
$(143) = \alpha_{10}$	10	12	11	9	2	4	3	1	6	8	7	5
$(234) = \alpha_{11}$	11	9	10	12	3	1	2	4	7	5	6	8
$(124) = \alpha_{12}$	12	10	9	11	4	2	1	3	8	6	5	7

tity and three 180° "edge" rotations about axes joining midpoints of two edges. The second row consists of 120° "face" rotations about axes joining a vertex to the center of the opposite face. The third row consists of −120° (or 240°) "face" rotations. Notice that the four rotations in the second row can be obtained from those in the first row by left-multiplying the four in the first row by the rotation (123), whereas those in the third row can be obtained from those in the first row by left-multiplying the ones in the first row by (132). ◆

Many molecules with chemical formulas of the form AB_4, such as methane (CH_4) and carbon tetrachloride (CCl_4), have A_4 as their symmetry group. Figure 5.2 shows the form of one such molecule.

Many games and puzzles can be analyzed using permutations.

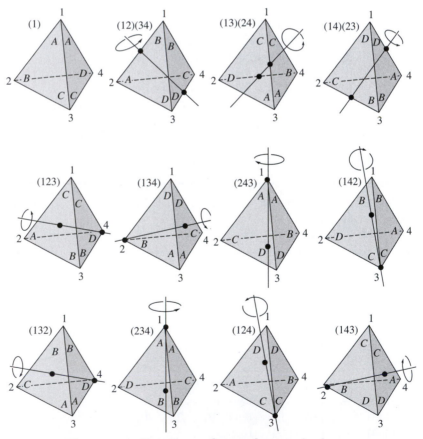

Figure 5.1   Rotations of a regular tetrahedron.

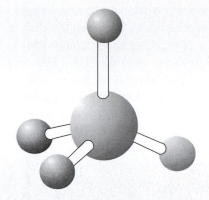

Figure 5.2   A tetrahedral AB_4 molecule.

EXAMPLE 7   (Loren Larson) A Sliding Disk Puzzle
Consider the puzzle shown below (the space in the middle is empty).

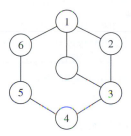

By sliding disks from one position to another along the lines indicated without lifting or jumping, can we obtain the following arrangement?

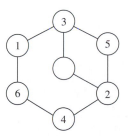

To answer this question, we view the positions as numbered in the first figure above and consider two basic operations. Let r denote the following operation: Move the disk in position 1 to the center position, then move the disk in position 6 to position 1, the disk in position 5 to position 6, the disk in position 4 to position 5, the disk in position 3 to position 4, then the disk in the middle position to position 3. Let s denote the operation: Move the disk in position 1 to the center position, then move the disk in position 2 to position 1, then move the disk in position 3 to position 2, and finally move the disk in the center to position 3. In permutation notation, we have $r = (13456)$ and $s = (132)$. The permutation for the arrangement we seek is (16523). Clearly, if we can express (16523) as a string of r's and s's, we can achieve the desired arangement. Rather than attempt to find an appropriate combination of r's and s's by hand, it is easier to employ computer software that is designed for this kind of problem. One such software program is GAP (see Suggested Software at the end of this chapter). With GAP, all we need to do is use the following commands:

gap> G := SymmetricGroup(6);
gap> r := (1,3,4,5,6); s := (1, 3, 2);
gap> K := Subgroup(G,[r,s]);
gap> Factorization(K,(1,6,5,2,3));

The first three lines inform the computer that our group is the subgroup of S_6 generated by $r = (13456)$ and $s = (132)$. The fourth line requests that (16523) be expressed in terms of r and s. The response to the command

$$\text{gap} > \text{Size } (K);$$

tells us that the order of the subgroup generated by r and s is 360. Then, observing that r and s are even permutations and that $|A_6| = 360$, we deduce that r and s can achieve any arrangement that corresponds to an even permutation.   ◆

GAP can even compute the 43,252,003,274,489,856,000 (43+ quintillion) permutations of the Rubik's Cube! Labeling the faces of the cube as shown here,

			1	2	3								
			4	top	5								
			6	7	8								
9	10	11	17	18	19	25	26	27	33	34	35
12	left	13	20	front	21	28	right	29	36	rear	37
14	15	16	22	23	24	30	31	32	38	39	40
			41	42	43								
			44	bottom	45								
			46	47	48								

the group of permutations of the cube is generated by the following rotations of the six layers:

top = (1,3,8,6)(2,5,7,4)(9,33,25,17)(10,34,26,18)(11,35,27,19)
left = (9,11,16,14)(10,13,15,12)(1,17,41,40)(4,20,44,37)(6,22,46,35)
front = (17,19,24,22)(18,21,23,20)(6,25,43,16)(7,28,42,13)(8,30,41,11)
right = (25,27,32,30)(26,29,31,28)(3,38,43,19)(5,36,45,21)(8,33,48,24)
rear = (33,35,40,38)(34,37,39,36)(3,9,46,32)(2,12,47,29)(1,14,48,27)
bottom = (41,43,48,46)(42,45,47,44)(14,22,30,38)(15,23,31,39)
(16,24,32,40)

A CHECK DIGIT SCHEME BASED ON D_5

In Chapter 0, we presented several schemes for appending a check digit to an identification number. Among these schemes, only the International Standard Book Number method was capable of detecting all single-digit errors and all transposition errors involving adjacent digits. However, recall that this success was achieved by introducing the alphabetical character X to handle the case where 10 was required to make the dot product 0 modulo 11.

In contrast, in 1969, J. Verhoeff [2] devised a method utilizing the dihedral group of order 10 that detects all single-digit errors and all transposition errors involving adjacent digits without the necessity of avoiding certain numbers or introducing a new character. To describe this method, consider the permutation $\sigma = (01589427)(36)$ and the dihedral group of order 10 as represented in Table 5.2. (Here we use 0

TABLE 5.2   Multiplication for D_5

*	0	1	2	3	4	5	6	7	8	9
0	0	1	2	3	4	5	6	7	8	9
1	1	2	3	4	0	6	7	8	9	5
2	2	3	4	0	1	7	8	9	5	6
3	3	4	0	1	2	8	9	5	6	7
4	4	0	1	2	3	9	5	6	7	8
5	5	9	8	7	6	0	4	3	2	1
6	6	5	9	8	7	1	0	4	3	2
7	7	6	5	9	8	2	1	0	4	3
8	8	7	6	5	9	3	2	1	0	4
9	9	8	7	6	5	4	3	2	1	0

through 4 for the rotations, 5 through 9 for the reflections, and $*$ for the operation of D_5.)

Verhoeff's idea is to view the digits 0 through 9 as the elements of the group D_5 and to replace ordinary addition with calculations done in D_5. In particular, to any string of digits $a_1a_2 \ldots a_{n-1}$, we append the check digit a_n so that $\sigma(a_1) * \sigma^2(a_2) * \cdots * \sigma^{n-2}(a_{n-2}) * \sigma^{n-1}(a_{n-1})$ $* \sigma^n(a_n) = 0$. [Here $\sigma^2(x) = \sigma(\sigma(x))$, $\sigma^3(x) = \sigma(\sigma^2(x))$, and so on.] Since σ has the property that $\sigma^i(a) \neq \sigma^i(b)$ if $a \neq b$, all single-digit errors are detected. Also, because

$$a * \sigma(b) \neq b * \sigma(a) \qquad \text{if } a \neq b, \tag{1}$$

as can be checked on a case-by-case basis (see Exercise 47), it follows that all transposition errors involving adjacent digits are detected [since Equation (1) implies that $\sigma^i(a) * \sigma^{i+1}(b) \neq \sigma^i(b) * \sigma^{i+1}(a)$ if $a \neq b$].

From 1990 until 2002, the German government used a minor modification of Verhoeff's check digit scheme to append a check digit to the serial numbers on German banknotes. Table 5.3 gives the values of the functions σ, σ^2, $\ldots$, σ^{10} needed for the computations. [The functional value $\sigma^i(j)$ appears in the row labeled with σ^i and the column labeled j.] Since the serial numbers on the banknotes use 10 letters of the alphabet in addition to the 10 decimal digits, it is necessary to assign numerical values to the letters to compute the check digit. This assignment is shown in Table 5.4.

To any string of digits $a_1a_2 \ldots a_{10}$ corresponding to a banknote serial number, the check digit a_{11} is chosen such that $\sigma(a_1) * \sigma^2(a_2) * \cdots$ $* \sigma^9(a_9) * \sigma^{10}(a_{10}) * a_{11} = 0$ [instead of $\sigma(a_1) * \sigma^2(a_2) * \cdots *$ $\sigma^{10}(a_{10}) * \sigma^{11}(a_{11}) = 0$ as in the Verhoeff scheme].

TABLE 5.3 Powers of σ

	0	1	2	3	4	5	6	7	8	9
σ	1	5	7	6	2	8	3	0	9	4
σ^2	5	8	0	3	7	9	6	1	4	2
σ^3	8	9	1	6	0	4	3	5	2	7
σ^4	9	4	5	3	1	2	6	8	7	0
σ^5	4	2	8	6	5	7	3	9	0	1
σ^6	2	7	9	3	8	0	6	4	1	5
σ^7	7	0	4	6	9	1	3	2	5	8
σ^8	0	1	2	3	4	5	6	7	8	9
σ^9	1	5	7	6	2	8	3	0	9	4
σ^{10}	5	8	0	3	7	9	6	1	4	2

TABLE 5.4 Letter Values

A	D	G	K	L	N	S	U	Y	Z
0	1	2	3	4	5	6	7	8	9

To trace through a specific example, consider the banknote (featuring the mathematician Gauss) shown in Figure 5.3 with the number AG8536827U7. To verify that 7 is the appropriate check digit, we observe that $\sigma(0) * \sigma^2(2) * \sigma^3(8) * \sigma^4(5) * \sigma^5(3) * \sigma^6(6) * \sigma^7(8) * \sigma^8(2) * \sigma^9(7) * \sigma^{10}(7) * 7 = 1 * 0 * 2 * 2 * 6 * 6 * 5 * 2 * 0 * 1 * 7 = 0$, as it should be. [To illustrate how to use the multiplication table for D_5, we compute $1 * 0 * 2 * 2 = (1 * 0) * 2 * 2 = 1 * 2 * 2 = (1 * 2) * 2 = 3 * 2 = 0.$]

Figure 5.3 German banknote with serial number AG8536827U and check digit 7.

One shortcoming of the German banknote scheme is that it does not distinguish between a letter and its assigned numerical value. Thus, a substitution of 7 for U (or vice versa) and the transposition of 7 and U are not detected by the check digit. Moreover, the banknote scheme does not detect all transpositions of adjacent characters involving the check digit itself. For example, the transposition of D and 8 in positions 10 and 11 is not detected. Both of these defects can be avoided by using the Verhoeff method with D_{18}, the dihedral group of order 36, to assign every letter and digit a distinct value together with an appropriate function σ (see Gallian [1]). Using this method to append a check character, all single-position errors and all transposition errors involving adjacent digits will be detected.

EXERCISES ▪▪

When you feel how depressingly
slowly you climb,
it's well to remember that
Things Take Time.

PIET HEIN, "T. T. T.,"
Grooks (1966)†

1. Find the order of each of the following permutations.
   a. (14)
   b. (147)
   c. (14762)

2. What is the order of a k-cycle $(a_1a_2 \cdots a_k)$?

3. What is the order of each of the following permutations?
   a. (124)(357)
   b. (124)(3567)
   c. (124)(35)
   d. (124)(357869)
   e. (1235)(24567)
   f. (345)(245)

4. What is the order of each of the following permutations?

   a. $\begin{bmatrix} 1 & 2 & 3 & 4 & 5 & 6 \\ 2 & 1 & 5 & 4 & 6 & 3 \end{bmatrix}$

   b. $\begin{bmatrix} 1 & 2 & 3 & 4 & 5 & 6 & 7 \\ 7 & 6 & 1 & 2 & 3 & 4 & 5 \end{bmatrix}$

5. What is the order of the product of a pair of disjoint cycles of lengths 4 and 6?

6. Show that A_8 contains an element of order 15.

7. What are the possible orders for the elements of S_6 and A_6? What about S_7 and A_7? (This exercise is referred to in Chapter 25.)

8. What is the maximum order of any element in A_{10}?

9. Determine whether the following permutations are even or odd.
   a. (135)
   b. (1356)
   c. (13567)
   d. (12)(134)(152)
   e. (1243)(3521)

---

†Hein is a Danish engineer and poet, and is the inventor of the game *Hex*.

10. Show that a function from a finite set S to itself is one-to-one if and only if it is onto. Is this true when S is infinite? (This exercise is referred to in Chapter 6.)

11. Let n be a positive integer. If n is odd, is an n-cycle an odd or an even permutation? If n is even, is an n-cycle an odd or an even permutation?

12. If α is even, prove that α^{-1} is even. If α is odd, prove that α^{-1} is odd.

13. Prove Theorem 5.6.

14. In S_n, let α be an r-cycle, β an s-cycle, and γ a t-cycle. Complete the following statements: $\alpha\beta$ is even if and only if $r + s$ is _____; $\alpha\beta\gamma$ is even if and only if $r + s + t$ is ____.

15. Let α and β belong to S_n. Prove that $\alpha\beta$ is even if and only if α and β are both even or both odd.

16. Associate an even permutation with the number $+1$ and an odd permutation with the number -1. Draw an analogy between the result of multiplying two permutations and the result of multiplying their corresponding numbers $+1$ or -1.

17. Let

$$\alpha = \begin{bmatrix} 1 & 2 & 3 & 4 & 5 & 6 \\ 2 & 1 & 3 & 5 & 4 & 6 \end{bmatrix} \quad \text{and} \quad \beta = \begin{bmatrix} 1 & 2 & 3 & 4 & 5 & 6 \\ 6 & 1 & 2 & 4 & 3 & 5 \end{bmatrix}.$$

Compute each of the following.
a. α^{-1}
b. $\beta\alpha$
c. $\alpha\beta$

18. Let

$$\alpha = \begin{bmatrix} 1 & 2 & 3 & 4 & 5 & 6 & 7 & 8 \\ 2 & 3 & 4 & 5 & 1 & 7 & 8 & 6 \end{bmatrix} \quad \text{and} \quad \beta = \begin{bmatrix} 1 & 2 & 3 & 4 & 5 & 6 & 7 & 8 \\ 1 & 3 & 8 & 7 & 6 & 5 & 2 & 4 \end{bmatrix}.$$

Write α, β, and $\alpha\beta$ as
a. products of disjoint cycles,
b. products of 2-cycles.

19. Show that if H is a subgroup of S_n, then either every member of H is an even permutation or exactly half of the members are even. (This exercise is referred to in Chapter 25.)

20. Compute the order of each member of A_4. What arithmetic relationship do these orders have with the order of A_4?

21. Do the odd permutations in S_n form a group? Why?

22. Let α and β belong to S_n. Prove that $\alpha^{-1}\beta^{-1}\alpha\beta$ is an even permutation.

23. Use Table 5.1 to compute the following.
a. The centralizer of $\alpha_3 = (13)(24)$.
b. The centralizer of $\alpha_{12} = (124)$.

24. How many elements of order 5 are in S_7?

25. How many odd permutations of order 4 does S_6 have?

26. Prove that (1234) is not the product of 3-cycles.

27. Let $\beta \in S_7$ and suppose $\beta^4 = (2143567)$. Find β.

28. Let $\beta = (123)(145)$. Write β^{99} in disjoint cycle form.

29. Find three elements σ in S_9 with the property that $\sigma^3 = (157)(283)(469)$.

30. What cycle is $(a_1 a_2 \cdots a_n)^{-1}$?

31. Let G be a group of permutations on a set X. Let $a \in X$ and define stab$(a) = \{\alpha \in G | \alpha(a) = a\}$. We call stab$(a)$ the *stabilizer of a in G* (since it consists of all members of G that leave a fixed). Prove that stab(a) is a subgroup of G. (This subgroup was introduced by Galois in 1832.) This exercise is referred to in Chapter 7.

32. Let $\beta = (1, 3, 5, 7, 9, 8, 6)(2, 4, 10)$. What is the smallest positive integer n for which $\beta^n = \beta^{-5}$?

33. Let $\alpha = (1, 3, 5, 7, 9)(2, 4, 6)(8, 10)$. If α^m is a 5-cycle, what can you say about m?

34. Let $H = \{\beta \in S_5 | \beta(1) = 1 \text{ and } \beta(3) = 3\}$. Prove that H is a subgroup of S_5. Is your argument valid when 5 is replaced by any $n \geq 3$?

35. How many elements of order 5 are there in A_6?

36. In S_4, find a cyclic subgroup of order 4 and a noncyclic subgroup of order 4.

37. Suppose that β is a 10-cycle. For which integers i between 2 and 10 is β^i also a 10-cycle?

38. In S_3, find elements α and β such that $|\alpha| = 2$, $|\beta| = 2$, and $|\alpha\beta| = 3$.

39. Find group elements α and β such that $|\alpha| = 3$, $|\beta| = 3$, and $|\alpha\beta| = 5$.

40. Represent the symmetry group of an equilateral triangle as a group of permutations of its vertices (see Example 3).

41. Prove that S_n is non-Abelian for all $n \geq 3$.

42. Let α and β belong to S_n. Prove that $\beta\alpha\beta^{-1}$ and α are both even or both odd.

43. Show that A_5 has 24 elements of order 5, 20 elements of order 3, and 15 elements of order 2. (This exercise is referred to in Chapter 25.)

44. If $\beta \in S_7$ and $|\beta^3| = 7$, prove that $|\beta| = 7$.

45. Show that every element in A_n for $n \geq 3$ can be expressed as a 3-cycle or a product of three cycles.

46. Show that for $n \geq 3$, $Z(S_n) = \{\varepsilon\}$.

47. Verify the statement made in the discussion of the Verhoeff check digit scheme based on D_5 that $a * \sigma(b) \neq b * \sigma(a)$ for distinct a and b. Use this to prove that $\sigma^i(a) * \sigma^{i+1}(b) \neq \sigma^i(b) * \sigma^{i+1}(a)$ for all i. Prove that this implies that all transposition errors involving adjacent digits are detected.

48. Use the Verhoeff check digit scheme based on D_5 to append a check digit to 45723.

49. Prove that every element of S_n $(n > 1)$ can be written as a product of elements of the form $(1k)$.

50. (Indiana College Mathematics Competition) A card-shuffling machine always rearranges cards in the same way relative to the order in which they were given to it. All of the hearts arranged in order from ace to king were put into the machine, and then the shuffled cards were put into the machine again to be shuffled. If the cards emerged in the order 10, 9, Q, 8, K, 3, 4, A, 5, J, 6, 2, 7, in what order were the cards after the first shuffle?

51. Show that a permutation with odd order must be an even permutation.

52. Let G be a group. Prove or disprove that $H = \{g^2 \mid g \in G\}$ is a subgroup of G. (Compare with Example 5 in Chapter 3.)

53. Why does the fact that the orders of the elements of A_4 are 1, 2, and 3 imply that $|Z(A_4)| = 1$?

54. Label the four locations of tires on an automobile with the labels 1, 2, 3, and 4, clockwise. Let a represent the operation of switching the tires in positions 1 and 3 and switching the tires in positions 2 and 4. Let b represent the operation of rotating the tires in positions 2, 3, and 4 clockwise and leaving the tire in position 1 as is. Let G be the group of all possible combinations of a and b. How many elements are in G?

55. Shown below are four tire rotation patterns recommended by the Dunlop Tire Company. Explain how these patterns can be represented as permutations in S_4 and find the smallest subgroup of S_4 that contains these four patterns. Is the subgroup Abelian?

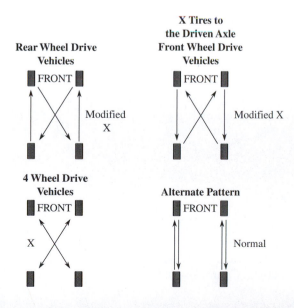

COMPUTER EXERCISES

> *Science is what we understand well enough to explain to a computer. Art is everything else we do.*
>
> DONALD KNUTH, The Art of Computer Programming, 1969

Software for Computer Exercise 1 in this chapter is available at the website:

http://www.d.umn.edu/~jgallian

1. This software determines whether the two permutations $(1x)$ and $(123 \cdots n)$ generate S_n for various choices of x and n (that is, whether every element of S_n can be expressed as some product of these permutations). For $n = 4$, run the program for $x = 2, 3,$ and 4. For $n = 5$, run the program for $x = 2, 3, 4,$ and 5. For $n = 6$, run the program for $x = 2, 3, 4, 5,$ and 6. For $n = 8$, run the program for $x = 2, 3, 4, 5, 6, 7,$ and 8. Conjecture a necessary and sufficient condition involving x and n for $(1x)$ and $(123 \cdots n)$ to generate S_n.

2. Use a software package (see Suggested Software on the following page) to express the following permutations in terms of the r and s given in Example 7. (For GAP, the prompt brk> means that the permutation entered is not in the group. In this situation, use Control-D to return to the main prompt. Be advised that GAP composes permutations from left to right as opposed to our method of right to left.)

   a. $(4,5,6)$
   b. $(2,3)$
   c. $(1,2)(3,4)$
   d. $(1,2)(3,4)(5,6)$

3. Repeat Computer Exercise 2 for the puzzle shown here.

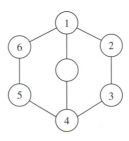

References

1. J. A. Gallian, "The Mathematics of Identification Numbers," *The College Mathematics Journal* 22 (1991): 194–202.
2. J. Verhoeff, *Error Detecting Decimal Codes,* Amsterdam: Mathematisch Centrum, 1969.

Suggested Readings

Douglas E. Ensely, "Invariants Under Actions to Amaze Your Friends," *Mathematics Magazine,* Dec. 1999: 383–387.

> This article explains some card tricks that are based on permutation groups.

Dmitry Fomin, "Getting It Together with 'Polynominoes,' " *Quantum,* Nov./Dec. 1991: 20–23.

> In this article, permutation groups are used to analyze various sorts of checkerboard tiling problems.

J. A. Gallian, "Error Detection Methods," *ACM Computing Surveys* 28 (1996): 504–517.

> This article gives a comprehensive survey of error-detection methods that use check digits.

I. N. Herstein and I. Kaplansky, *Matters Mathematical,* New York: Chelsea, 1978.

> Chapter 3 of this book discusses several interesting applications of permutations to games.

Douglas Hofstadter, "The Magic Cube's Cubies Are Twiddled by Cubists and Solved by Cubemeisters," *Scientific American* 244 (1981): 20–39.

> This article, written by a Pulitzer Prize recipient, discusses the group theory involved in the solution of the Magic (Rubik's) Cube. In particular, permutation groups, subgroups, conjugates (elements of the form xyx^{-1}), commutators (elements of the form $xyx^{-1}y^{-1}$), and the "always-even-always-odd" theorem (Theorem 5.5) are prominently mentioned. At one point, Hofstadter says, "It is this kind of marvelously concrete illustration of an abstract notion of group theory that makes the Magic Cube one of the most amazing things ever invented for teaching mathematical ideas."

John O. Kiltinen, *Oval Track & Other Permutation Puzzles & Just Enough Group Theory to Solve Them*, Mathematical Association of America, Washington, D.C., 2003.

> This book and the software that comes with it present the user with an array of computerized puzzles, plus tools to vary them in thousands of ways. The book provides the background needed to use the puzzle software to its fullest potential, and also gives the reader a gentle, not-too-technical introduction to the theory of permutation groups that is a prerequisite to a full understanding of how to solve puzzles of this type. The website http://www-instruct.nmu.edu/math_cs/kiltinen/web/mathpuzzles/ provides resources that expand upon the book. It also has news about puzzle software—modules that add functionality and fun to puzzles.

Will Oakley, "Portrait of Three Puzzle Graces," *Quantum,* Nov./Dec. 1991: 83–86.
> The author uses permutation groups to analyze solutions to the 15 puzzle, Rubik's Cube, and Rubik's Clock.

A. White and R. Wilson, "The Hunting Group," *Mathematical Gazette* 79 (1995): 5–16.
> This article explains how permutation groups are used in bell ringing.

S. Winters, "Error-Detecting Schemes Using Dihedral Groups," *UMAP Journal* 11, no. 4 (1990): 299–308.
> This article discusses error-detection schemes based on D_n for n odd. Schemes for both one and two check digits are analyzed.

Suggested Software

GAP is free for downloading. Versions are available for Unix, IBM-compatibles, and Macintosh at:

> http://www-gap.dcs.st-andrews.ac.uk/~gap/
> http://www.math.rwth-aachen.de/LDFM/GAP/
> gopher://archives.math.utk.edu:70/11/software/multi-platform/gap

GAP for DOS:

> http://archives.math.utk.edu/software/
> msdos/modern.algebra/gap/.html

Augustin Cauchy

You see that little young man? Well! He will supplant all of us in so far as we are mathematicians.

SPOKEN BY LAGRANGE
TO LAPLACE ABOUT THE
11-YEAR-OLD CAUCHY

This stamp was issued by France in Cauchy's honor.

AUGUSTIN LOUIS CAUCHY was born on August 21, 1789, in Paris. By the time he was 11, both Laplace and Lagrange had recognized Cauchy's extraordinary talent for mathematics. In school he won prizes for Greek, Latin, and the humanities. At the age of 21, he was given a commission in Napoleon's army as a civil engineer. For the next few years, Cauchy attended to his engineering duties while carrying out brilliant mathematical research on the side.

In 1815, at the age of 26, Cauchy was made Professor of Mathematics at the École Polytechnique and was recognized as the leading mathematician in France. Cauchy and his contemporary Gauss were among the last mathematicians to know the whole of mathematics as known at their time, and both made important contributions to nearly every branch, both pure and applied, as well as to physics and astronomy.

Cauchy introduced a new level of rigor into mathematical analysis. We owe our contemporary notions of limit and continuity to him. He gave the first proof of the Fundamental Theorem of Calculus. Cauchy was the founder of complex function theory and a pioneer in the theory of permutation groups and determinants. His total written output of mathematics fills 24 large volumes and is second only to that of Euler. He wrote more than 500 research papers after the age of 50. Cauchy died at the age of 67 on May 23, 1857.

For more information about Cauchy, visit:

http://www–groups.dcs.st-and.ac.uk/~history/

6 Isomorphisms

The basis for poetry and scientific discovery is the ability to comprehend the unlike in the like and the like in the unlike.

JACOB BRONOWSKI

MOTIVATION

Suppose an American and a German are asked to count a handful of objects. The American says, "One, two, three, four, five, ... ," whereas the German says "Eins, zwei, drei, vier, fünf, . . ." Are the two doing different things? No. They are both counting the objects, but they are using different terminology to do so. Similarly, when one person says: "Two plus three is five" and another says: "Zwei und drei ist fünf," the two are in agreement on the *concept* they are describing, but they are using different terminology to describe the concept. An analogous situation often occurs with groups; the same group is described with different terminology. We have seen two examples of this so far. In Chapter 1, we described the symmetries of a square in geometric terms (e.g., R_{90}), whereas in Chapter 5 we described the *same* group by way of permutations of the corners. In both cases, the underlying group was the symmetries of a square. In Chapter 4, we observed that when we have a cyclic group of order n generated by a, the operation turns out to be essentially that of addition modulo n, since $a^r a^s = a^k$, where $k = (r + s) \bmod n$. For example, each of $U(43)$ and $U(49)$ is cyclic of order 42. So, each has the form $\langle a \rangle$, where $a^r a^s = a^{(r + s) \bmod 42}$.

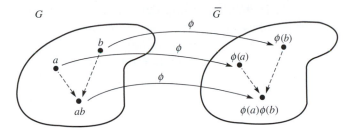

Figure 6.1

DEFINITION AND EXAMPLES

In this chapter, we give a formal method for determining whether two groups defined in different terms are really the *same*. When this is the case, we say that there is an isomorphism between the two groups. This notion was first introduced by Galois about 175 years ago. The term *isomorphism* is derived from the Greek words *isos*, meaning "same" or "equal," and *morphe*, meaning "form." R. Allenby has colorfully defined an algebraist as "a person who can't tell the difference between isomorphic systems."

DEFINITION · *Group Isomorphism*

An *isomorphism* ϕ from a group G to a group $\overline{G}$ is a one-to-one mapping (or function) from G onto $\overline{G}$ that preserves the group operation. That is,

$$\phi(ab) = \phi(a)\phi(b) \qquad \text{for all } a, b \text{ in } G.$$

If there is an isomorphism from G onto $\overline{G}$, we say that G and $\overline{G}$ are *isomorphic* and write $G \approx \overline{G}$.

This definition can be visualized as shown in Figure 6.1. The pairs of dashed arrows represent the group operations.

TABLE 6.1

G Operation	$\overline{G}$ Operation	Operation Preservation
$\cdot$	$\cdot$	$\phi(a \cdot b) = \phi(a) \cdot \phi(b)$
$\cdot$	$+$	$\phi(a \cdot b) = \phi(a) + \phi(b)$
$+$	$\cdot$	$\phi(a + b) = \phi(a) \cdot \phi(b)$
$+$	$+$	$\phi(a + b) = \phi(a) + \phi(b)$

It is implicit in the definition of isomorphism that isomorphic groups have the same order. It is also implicit in the definition of isomorphism that the operation on the left side of the equal sign is that of G, whereas the operation on the right side is that of $\overline{G}$. The four cases involving $\cdot$ and $+$ are shown in Table 6.1.

There are four separate steps involved in proving that a group G is isomorphic to a group $\overline{G}$.

STEP 1   "Mapping." Define a candidate for the isomorphism; that is, define a function ϕ from G to $\overline{G}$.

STEP 2   "1–1." Prove that ϕ is one-to-one; that is, assume that $\phi(a) = \phi(b)$ and prove that $a = b$.

STEP 3   "Onto." Prove that ϕ is onto; that is, for any element $\overline{g}$ in $\overline{G}$, find an element g in G such that $\phi(g) = \overline{g}$.

STEP 4   "O.P." Prove that ϕ is operation-preserving; that is, show that $\phi(ab) = \phi(a)\phi(b)$ for all a and b in G.

None of these steps is unfamiliar to you. The only one that may appear novel is the fourth one. It requires that one be able to obtain the same result by combining two elements and then mapping, or by mapping two elements and then combining them. Roughly speaking, this says that the two processes—operating and mapping—can be done in either order without affecting the result. This same concept arises in calculus when we say

$$\lim_{x \to a} (f(x) \cdot g(x)) = \lim_{x \to a} f(x) \lim_{x \to a} g(x)$$

or

$$\int_a^b (f + g)\, dx = \int_a^b f\, dx + \int_a^b g\, dx.$$

Before going any further, let's consider some examples.

EXAMPLE 1   Let G be the real numbers under addition and let $\overline{G}$ be the positive real numbers under multiplication. Then G and $\overline{G}$ are isomorphic under the mapping $\phi(x) = 2^x$. Certainly, ϕ is a function from G to $\overline{G}$. To prove that it is one-to-one, suppose that $2^x = 2^y$. Then $\log_2 2^x = \log_2 2^y$, and therefore $x = y$. For "onto," we must find for any positive real number y some real number x such that $\phi(x) = y$; that is, $2^x = y$. Well, solving for x gives $\log_2 y$. Finally,

$$\phi(x + y) = 2^{x+y} = 2^x \cdot 2^y = \phi(x)\phi(y)$$

for all x and y in G, so that ϕ is operation-preserving as well.   ◆

EXAMPLE 2   Any infinite cyclic group is isomorphic to Z. Indeed, if a is a generator of the cyclic group, the mapping $a^k \rightarrow k$ is an isomorphism. Any finite cyclic group $\langle a \rangle$ of order n is isomorphic to Z_n under the mapping $a^k \rightarrow k$ mod n. That these correspondences are functions and are one-to-one is the essence of Theorem 4.1. Obviously, the mappings are onto. That the mappings are operation-preserving follows from Exercise 11 in Chapter 0 in the finite case and from the definitions in the infinite case.   ◆

EXAMPLE 3   The mapping from **R** under addition to itself given by $\phi(x) = x^3$ is *not* an isomorphism. Although ϕ is one-to-one and onto, it is not operation-preserving, since it is not true that $(x + y)^3 = x^3 + y^3$ for all x and y.   ◆

EXAMPLE 4 $U(10) \approx Z_4 \approx U(5)$. To verify this, one need only observe that both $U(10)$ and $U(5)$ are cyclic of order 4. Then appeal to Example 2.   ◆

EXAMPLE 5 $U(10) \not\approx U(12)$. This is a bit trickier to prove. First, note that $x^2 = 1$ for all x in $U(12)$. Now, suppose that ϕ is an isomorphism from $U(10)$ onto $U(12)$. Then,

$$\phi(9) = \phi(3 \cdot 3) = \phi(3)\phi(3) = 1$$

and

$$\phi(1) = \phi(1 \cdot 1) = \phi(1)\phi(1) = 1.$$

Thus, $\phi(9) = \phi(1)$, but $9 \neq 1$, which is a contradiction to the supposed one-to-one character of ϕ.   ◆

EXAMPLE 6   There is no isomorphism from Q, the group of rational numbers under addition, to Q^*, the group of nonzero rational numbers under multiplication. If ϕ were such a mapping, there would be a rational number a such that $\phi(a) = -1$. But then

$$-1 = \phi(a) = \phi(\tfrac{1}{2}a + \tfrac{1}{2}a) = \phi(\tfrac{1}{2}a)\phi(\tfrac{1}{2}a) = [\phi(\tfrac{1}{2}a)]^2.$$

However, no rational number squared is -1.   ◆

EXAMPLE 7   Let $G = SL(2, \mathbf{R})$, the group of 2×2 real matrices with determinant 1. Let M be any 2×2 real matrix with determinant

1. Then we can define a mapping from G to G itself by $\phi_M(A) = MAM^{-1}$ for all A in G. To verify that ϕ_M is an isomorphism, we carry out the four steps.

STEP 1 ϕ_M is a function from G to G. Here, we must show that $\phi_M(A)$ is indeed an element of G whenever A is. This follows from properties of determinants:

$$\det(MAM^{-1}) = (\det M)(\det A)(\det M)^{-1} = 1 \cdot 1 \cdot 1^{-1} = 1.$$

Thus, MAM^{-1} is in G.

STEP 2 ϕ_M is one-to-one. Suppose that $\phi_M(A) = \phi_M(B)$. Then $MAM^{-1} = MBM^{-1}$ and, by left and right cancellation, $A = B$.

STEP 3 ϕ_M is onto. Let B belong to G. We must find a matrix A in G such that $\phi_M(A) = B$. How shall we do this? If such a matrix A is to exist, it must have the property that $MAM^{-1} = B$. But this tells us exactly what A must be! For we can solve for A to obtain $A = M^{-1}BM$.

STEP 4 ϕ_M is operation-preserving. Let A and B belong to G. Then,

$$\phi_M(AB) = M(AB)M^{-1} = MA(M^{-1}M)BM^{-1}$$
$$= (MAM^{-1})(MBM^{-1}) = \phi_M(A)\phi_M(B).$$

The mapping ϕ_M is called *conjugation* by M. ◆

CAYLEY'S THEOREM

Our first theorem is a classic result of Cayley. An important generalization of it will be given in Chapter 25.

Theorem 6.1 **Cayley's Theorem (1854)**

> *Every group is isomorphic to a group of permutations.*

PROOF. To prove this, let G be any group. We must find a group $\overline{G}$ of permutations that we believe is isomorphic to G. Since G is all we have to work with, we will have to use it to construct $\overline{G}$. For any g in G, define a function T_g from G to G by

$$T_g(x) = gx \qquad \text{for all } x \text{ in } G.$$

(In words, T_g is just multiplication by g on the left.) We leave it as an exercise (Exercise 21) to prove that T_g is a permutation on the set of elements of G. Now, let $\overline{G} = \{T_g \mid g \in G\}$. Then, $\overline{G}$ is a group under the operation of function composition. To verify this, we first observe that for any g and h in G we have $T_g T_h(x) = T_g(T_h(x)) = T_g(hx) = g(hx) = (gh)x = T_{gh}(x)$, so that $T_g T_h = T_{gh}$. From this it follows that T_e is the identity and $(T_g)^{-1} = T_{g^{-1}}$ (see Exercise 9). Since function composition is associative, we have verified all the conditions for $\overline{G}$ to be a group.

The isomorphism ϕ between G and $\overline{G}$ is now ready-made. For every g in G, define $\phi(g) = T_g$. If $T_g = T_h$, then $T_g(e) = T_h(e)$ or $ge = he$. Thus, $g = h$ and ϕ is one-to-one. By the way $\overline{G}$ was constructed, we see that ϕ is onto. The only condition that remains to be checked is that ϕ is operation-preserving. To this end, let x and y belong to G. Then

$$\phi(xy) = T_{xy} = T_x T_y = \phi(x)\phi(y). \quad \blacksquare \blacksquare$$

The group $\overline{G}$ constructed above is called the *left regular representation of G*.

EXAMPLE 8   For concreteness, let us calculate the left regular representation $\overline{U(12)}$ for $U(12) = \{1, 5, 7, 11\}$. Writing the permutations of $U(12)$ in array form, we have (remember, T_x is just multiplication by x)

$$T_1 = \begin{bmatrix} 1 & 5 & 7 & 11 \\ 1 & 5 & 7 & 11 \end{bmatrix}, \quad T_5 = \begin{bmatrix} 1 & 5 & 7 & 11 \\ 5 & 1 & 11 & 7 \end{bmatrix},$$

$$T_7 = \begin{bmatrix} 1 & 5 & 7 & 11 \\ 7 & 11 & 1 & 5 \end{bmatrix}, \quad T_{11} = \begin{bmatrix} 1 & 5 & 7 & 11 \\ 11 & 7 & 5 & 1 \end{bmatrix}.$$

It is instructive to compare the Cayley table for $U(12)$ and its left regular representation $\overline{U(12)}$.

$U(12)$	1	5	7	11
1	1	5	7	11
5	5	1	11	7
7	7	11	1	5
11	11	7	5	1

$\overline{U(12)}$	T_1	T_5	T_7	T_{11}
T_1	T_1	T_5	T_7	T_{11}
T_5	T_5	T_1	T_{11}	T_7
T_7	T_7	T_{11}	T_1	T_5
T_{11}	T_{11}	T_7	T_5	T_1

It should be abundantly clear from these tables that $U(12)$ and $\overline{U(12)}$ are only notationally different.   ◆

Cayley's Theorem is important for two contrasting reasons. One is that it allows us to represent an abstract group in a concrete way. A second is that it shows that the present-day set of axioms we have adopted for a group is the correct abstraction of its much earlier predecessor—a group of permutations. Indeed, Cayley's Theorem tells us that abstract groups are not different from permutation groups. Rather, it is the viewpoint that is different. It is this difference of viewpoint that has stimulated the tremendous progress in group theory and many other branches of mathematics in the twentieth century.

It is sometimes very difficult to prove or disprove, whichever the case may be, that two particular groups are isomorphic. For example, it requires somewhat sophisticated techniques to prove the surprising fact that the group of real numbers under addition is isomorphic to the group of complex numbers under addition. Likewise, it is not easy to prove the fact that the group of nonzero complex numbers under multiplication is isomorphic to the group of complex numbers with absolute value of 1 under multiplication. In geometric terms, this says that, as groups, the punctured plane and the unit circle are isomorphic.

PROPERTIES OF ISOMORPHISMS

Our next two theorems give a catalog of properties of isomorphisms and isomorphic groups.

Theorem 6.2 **Properties of Isomorphisms Acting on Elements**

Suppose that ϕ is an isomorphism from a group G onto a group $\overline{G}$. Then

1. *ϕ carries the identity of G to the identity of $\overline{G}$.*
2. *For every integer n and for every group element a in G, $\phi(a^n) = [\phi(a)]^n$.*
3. *For any elements a and b in G, a and b commute if and only if $\phi(a)$ and $\phi(b)$ commute.*
4. *$G = \langle a \rangle$ if and only if $\overline{G} = \langle \phi(a) \rangle$.*
5. *$|a| = |\phi(a)|$ for all a in G (isomorphisms preserve orders).*
6. *For a fixed integer k and a fixed group element b in G, the equation $x^k = b$ has the same number of solutions in G as does the equation $x^k = \phi(b)$ in $\overline{G}$.*

PROOF.   We will restrict ourselves to proving only properties 1, 2, and 4 but observe that property 5 follows from property 4. For convenience, let us denote the identity in G by e and the identity in $\overline{G}$ by $\overline{e}$. Then, since $e = ee$, we have

$$\phi(e) = \phi(ee) = \phi(e)\phi(e).$$

Also, because $\phi(e) \in \overline{G}$, we have $\phi(e) = \overline{e}\phi(e)$, as well. Thus, by cancellation, $\overline{e} = \phi(e)$. This proves property 1.

For positive integers, property 2 follows from the definition of an isomorphism and mathematical induction. If n is negative, then $-n$ is positive, and we have from property 1 and the observation about the positive integer case that $e = \phi(e) = \phi(g^n g^{-n}) = \phi(g^n)\phi(g^{-n}) = \phi(g^n)(\phi(g))^{-n}$. Thus, multiplying both sides on the right by $(\phi(g))^n$, we have $(\phi(g))^n = \phi(g^n)$.

To prove property 4, note that, by closure, $\langle\phi(a)\rangle \subseteq \overline{G}$. Because ϕ is onto, for any element b in $\overline{G}$, there is an element a^k in G such that $\phi(a^k) = b$. Thus, $b = (\phi(a))^k$ and so $b \in \langle\phi(a)\rangle$. This proves that $\overline{G} = \langle\phi(a)\rangle$.

Now suppose that $\overline{G} = \langle\phi(a)\rangle$. Clearly, $\langle a \rangle \subseteq G$. For any element b in G, we have $\phi(b) \in \langle\phi(a)\rangle$. So, for some integer k we have $\phi(b) = (\phi(a))^k = \phi(a^k)$. Because ϕ is one-to-one, $b = a^k$. This proves that $\langle a \rangle = G$.   ■ ■

Note that when the group operation is addition, property 2 of Theorem 6.2 is $\phi(na) = n\phi(a)$ and that property 4 says that an isomorphism between two cyclic groups takes a generator to a generator.

Property 5 is quite useful for showing that two groups are *not* isomorphic. Often b is picked to be the identity. For example, consider $\mathbf{C}^*$ and $\mathbf{R}^*$. Because the equation $x^4 = 1$ has four solutions in $\mathbf{C}^*$ but only two in $\mathbf{R}^*$, no matter how one attempts to define an isomorphism from $\mathbf{C}^*$ to $\mathbf{R}^*$, property 5 cannot hold.

Theorem 6.3   **Properties of Isomorphisms Acting on Groups**

> *Suppose that ϕ is an isomorphism from a group G onto a group $\overline{G}$. Then*
>
> *1. G is Abelian if and only if $\overline{G}$ is Abelian.*
> *2. G is cyclic if and only if $\overline{G}$ is cyclic.*
> *3. ϕ^{-1} is an isomorphism from $\overline{G}$ onto G.*
> *4. If K is a subgroup of G, then $\phi(K) = \{\phi(k) \mid k \in K\}$ is a subgroup of $\overline{G}$.*

PROOF.   Properties 1 and 2 are direct consequences of properties 3 and 5 of Theorem 6.2, respectively. Properties 3 and 4 are left as exercises (Exercises 19 and 20). ■ ■

Theorems 6.2 and 6.3 show that isomorphic groups have many properties in common. Actually, the definition is precisely formulated so that isomorphic groups have *all* group-theoretic properties in common. By this we mean that if two groups are isomorphic, then any property that can be expressed in the language of group theory is true for one if and only if it is true for the other. This is why algebraists speak of isomorphic groups as "equal" or "the same." Admittedly, calling such groups equivalent, rather than the same, might be more appropriate, but we bow to long-standing tradition.

AUTOMORPHISMS

Certain kinds of isomorphisms are referred to so often that they have been given special names.

DEFINITION   *Automorphism*

> An isomorphism from a group G onto itself is called an *automorphism* of G.

The isomorphism in Example 7 is an automorphism of $SL(2, \mathbf{R})$. Two more examples follow.

EXAMPLE 9   The function ϕ from $\mathbf{C}$ to $\mathbf{C}$ given by $\phi(a + bi) = a - bi$ is an automorphism of the group of complex numbers under addition. The restriction of ϕ to $\mathbf{C}^*$ is also an automorphism of the group of nonzero complex numbers under multiplication. (See Exercise 23.) ◆

EXAMPLE 10   Let $\mathbf{R}^2 = \{(a, b) \mid a, b \in \mathbf{R}\}$. Then $\phi(a, b) = (b, a)$ is an automorphism of the group $\mathbf{R}^2$ under componentwise addition. Geometrically, ϕ reflects each point in the plane across the line $y = x$. More generally, any reflection across a line passing through the origin or any rotation of the plane about the origin is an automorphism of $\mathbf{R}^2$. ◆

The isomorphism in Example 7 is a particular instance of an automorphism that arises often enough to warrant a name and notation of its own.

DEFINITION *Inner Automorphism Induced by a*

Let G be a group, and let $a \in G$. The function ϕ_a defined by $\phi_a(x) = axa^{-1}$ for all x in G is called the *inner automorphism of G induced by a.*

We leave it for the reader to show that ϕ_a is actually an automorphism of G. (Use Example 7 as a model.)

EXAMPLE 11 The action of the inner automorphism of D_4 induced by R_{90} is given in the following table.

$$x \quad \overset{\phi_{R_{90}}}{\to} \quad R_{90}\, x\, R_{90}{}^{-1}$$

$$R_0 \;\to\; R_{90}R_0R_{90}{}^{-1} = R_0$$
$$R_{90} \;\to\; R_{90}R_{90}R_{90}{}^{-1} = R_{90}$$
$$R_{180} \;\to\; R_{90}R_{180}R_{90}{}^{-1} = R_{180}$$
$$R_{270} \;\to\; R_{90}R_{270}R_{90}{}^{-1} = R_{270}$$
$$H \;\to\; R_{90}HR_{90}{}^{-1} = V$$
$$V \;\to\; R_{90}VR_{90}{}^{-1} = H$$
$$D \;\to\; R_{90}DR_{90}{}^{-1} = D'$$
$$D' \;\to\; R_{90}D'R_{90}{}^{-1} = D$$

◆

When G is a group, we use Aut(G) to denote the set of all automorphisms of G and Inn(G) to denote the set of all inner automorphisms of G. The reason these sets are noteworthy is demonstrated by the next theorem.

Theorem 6.4 **Aut(G) and Inn(G) Are Groups**[†]

> *The set of automorphisms of a group and the set of inner automorphisms of a group are both groups under the operation of function composition.*

PROOF. The proof of Theorem 6.4 is left as an exercise (Exercise 15). ■■

The determination of Inn(G) is routine. If $G = \{e, a, b, c. \ldots\}$, then Inn($G$) = $\{\phi_e, \phi_a, \phi_b, \phi_c, \ldots\}$. This latter list may have duplications,

---

[†]The group Aut(G) was first studied by O. Hölder in 1893 and, independently, by E. H. Moore in 1894.

however, since ϕ_a may be equal to ϕ_b even though $a \neq b$ (see Exercise 31). Thus, the only work involved in determining Inn(G) is deciding which distinct elements give the distinct automorphisms. On the other hand, the determination of Aut(G) is, in general, quite involved.

EXAMPLE 12   Inn(D_4)

To determine Inn(D_4), we first observe that the complete list of inner automorphisms is ϕ_{R_0}, $\phi_{R_{90}}$, $\phi_{R_{180}}$, $\phi_{R_{270}}$, ϕ_H, ϕ_V, ϕ_D, and $\phi_{D'}$. Our job is to determine the repetitions in this list. Since $R_{180} \in Z(D_4)$, we have $\phi_{R_{180}}(x) = R_{180}xR_{180}^{-1} = x$, so that $\phi_{R_{180}} = \phi_{R_0}$. Also, $\phi_{R_{270}}(x) = R_{270}xR_{270}^{-1} = R_{90}R_{180}xR_{180}^{-1}R_{90}^{-1} = R_{90}xR_{90}^{-1} = \phi_{R_{90}}(x)$. Similarly, since $H = R_{180}V$ and $D' = R_{180}D$, we have $\phi_H = \phi_V$ and $\phi_D = \phi_{D'}$. This proves that the previous list can be pared down to ϕ_{R_0}, $\phi_{R_{90}}$, ϕ_H, and ϕ_D. We leave it to the reader to show that these are distinct (Exercise 13).   ◆

EXAMPLE 13   Aut(Z_{10})

To compute Aut(Z_{10}), we try to discover enough information about an element α of Aut(Z_{10}) to determine how α must be defined. Because Z_{10} is so simple, this is not difficult to do. To begin with, observe that once we know $\alpha(1)$, we know $\alpha(k)$ for any k, because

$$\alpha(k) = \alpha(\underbrace{1 + 1 + \cdots + 1}_{k \text{ terms}})$$

$$= \underbrace{\alpha(1) + \alpha(1) + \cdots + \alpha(1)}_{k \text{ terms}} = k\alpha(1).$$

So, we need only determine the choices for $\alpha(1)$ that make α an automorphism of Z_{10}. Since property 5 of Theorem 6.2 tells us that $|\alpha(1)| = 10$, there are four candidates for $\alpha(1)$:

$$\alpha(1) = 1; \qquad \alpha(1) = 3; \qquad \alpha(1) = 7; \qquad \alpha(1) = 9.$$

To distinguish among the four possibilities, we refine our notation by denoting the mapping that sends 1 to 1 by α_1, 1 to 3 by α_3, 1 to 7 by α_7, and 1 to 9 by α_9. So the only possibilities for Aut(Z_{10}) are α_1, α_3, α_7, and α_9. But are all these automorphisms? Clearly, α_1 is the identity. Let us check α_3. Since $\alpha_3(1) = 3$ is a generator of Z_{10}, it follows that α_3 is onto (and, by Exercise 10 in Chapter 5, it is also one-to-one). Finally, since $\alpha_3(a + b) = 3(a + b) = 3a + 3b = \alpha_3(a) + \alpha_3(b)$, we see that α_3 is operation-preserving as well. Thus,

$\alpha_3 \in \text{Aut}(Z_{10})$. The same argument shows that α_7 and α_9 are also automorphisms.

This gives us the elements of $\text{Aut}(Z_{10})$ but not the structure. For instance, what is $\alpha_3\alpha_3$? Well, $(\alpha_3\alpha_3)(1) = \alpha_3(3) = 3 \cdot 3 = 9 = \alpha_9(1)$, so $\alpha_3\alpha_3 = \alpha_9$. Similar calculations show that $\alpha_3^3 = \alpha_7$ and $\alpha_3^4 = \alpha_1$, so that $|\alpha_3| = 4$. Thus, $\text{Aut}(Z_{10})$ is cyclic. Actually, the following Cayley tables reveal that $\text{Aut}(Z_{10})$ is isomorphic to $U(10)$.

$U(10)$	1	3	7	9
1	1	3	7	9
3	3	9	1	7
7	7	1	9	3
9	9	7	3	1

$\text{Aut}(Z_{10})$	α_1	α_3	α_7	α_9
α_1	α_1	α_3	α_7	α_9
α_3	α_3	α_9	α_1	α_7
α_7	α_7	α_1	α_9	α_3
α_9	α_9	α_7	α_3	α_1

◆

With Example 13 as a guide, we are now ready to tackle the group $\text{Aut}(Z_n)$. The result is particularly nice, since it relates the two kinds of groups we have most frequently encountered thus far—the cyclic groups Z_n and the U-groups $U(n)$.

Theorem 6.5   Aut(Z_n) ≈ U(n)

> *For every positive integer n, $\text{Aut}(Z_n)$ is isomorphic to $U(n)$.*

PROOF.   As in Example 13, any automorphism α is determined by the value of $\alpha(1)$, and $\alpha(1) \in U(n)$. Now consider the correspondence from $\text{Aut}(Z_n)$ to $U(n)$ given by $T: \alpha \rightarrow \alpha(1)$. The fact that $\alpha(k) = k\alpha(1)$ (see Example 13) implies that T is a one-to-one mapping. For if α and β belong to $\text{Aut}(Z_n)$ and $\alpha(1) = \beta(1)$, then $\alpha(k) = k\alpha(1) = k\beta(1) = \beta(k)$ for all k in Z_n, and therefore $\alpha = \beta$.

To prove that T is onto, let $r \in U(n)$ and consider the mapping α from Z_n to Z_n defined by $\alpha(s) = sr \pmod{n}$ for all s in Z_n. We leave it as an exercise to verify that α is an automorphism of Z_n (see Exercise 17). Then, since $T(\alpha) = \alpha(1) = r$, T is onto $U(n)$.

Finally, we establish the fact that T is operation-preserving. Let α, $\beta \in \text{Aut}(Z_n)$. We then have

$$T(\alpha\beta) = (\alpha\beta)(1) = \alpha(\beta(1)) = \alpha(\underbrace{1 + 1 + \cdots + 1}_{\beta(1)})$$

$$= \underbrace{\alpha(1) + \alpha(1) + \cdots + \alpha(1)}_{\beta(1)} = \alpha(1)\beta(1)$$

$$= T(\alpha)T(\beta).$$

This completes the proof. ■ ■

EXERCISES ▬▬

Being a mathematician is a bit like being a manic depressive: you spend your life alternating between giddy elation and black despair.

STEVEN G. KRANTZ, A Primer of Mathematical Writing

1. Find an isomorphism from the group of integers under addition to the group of even integers under addition.

2. Find $\text{Aut}(Z)$.

3. Let $\mathbf{R}^+$ be the group of positive real numbers under multiplication. Show that the mapping $\phi(x) = \sqrt{x}$ is an automorphism of $\mathbf{R}^+$.

4. Show that $U(8)$ is not isomorphic to $U(10)$.

5. Show that $U(8)$ is isomorphic to $U(12)$.

6. Prove that the notion of group isomorphism is transitive. That is, if G, H, and K are groups and $G \approx H$ and $H \approx K$, then $G \approx K$.

7. Prove that S_4 is not isomorphic to D_{12}.

8. Explain why the three parts of Exercise 1 of the Supplementary Exercises, Chapters 1–4, follow immediately from Theorem 6.3.

9. In the notation of Theorem 6.1, prove that T_e is the identity and that $(T_g)^{-1} = T_{g^{-1}}$.

10. Let G be a group. Prove that the mapping $\alpha(g) = g^{-1}$ for all g in G is an automorphism if and only if G is Abelian.

11. If g and h are elements from a group, prove that $\phi_g \phi_h = \phi_{gh}$.

12. Find two groups G and H such that $G \not\approx H$, but $\text{Aut}(G) \approx \text{Aut}(H)$.

13. Prove the assertion in Example 12 that the inner automorphisms ϕ_{R_0}, $\phi_{R_{90}}$, ϕ_H, and ϕ_D of D_4 are distinct.

14. Find $\text{Aut}(Z_6)$.

15. If G is a group, prove that $\text{Aut}(G)$ and $\text{Inn}(G)$ are groups.

16. Prove that the mapping from $U(16)$ to itself given by $x \to x^3$ is an automorphism. What about $x \to x^5$ and $x \to x^7$? Generalize.

17. Let $r \in U(n)$. Prove that the mapping $\alpha: Z_n \to Z_n$ defined by $\alpha(s) = sr$ mod n for all s in Z_n is an automorphism of Z_n. (This exercise is referred to in this chapter.)

18. Suppose that $\phi: Z_{50} \to Z_{50}$ is a group isomorphism with $\phi(7) = 13$. Determine $\phi(x)$.

19. Prove Property 3 of Theorem 6.3.

20. Prove Property 4 of Theorem 6.3.

21. Referring to Theorem 6.1, prove that T_g is indeed a permutation on the set G.

22. Prove or disprove that $U(20)$ and $U(24)$ are isomorphic.

23. Show that the mapping $\phi(a + bi) = a - bi$ is an automorphism of the group of complex numbers under addition. Show that ϕ preserves complex multiplication as well—that is, $\phi(xy) = \phi(x)\phi(y)$ for all x and y in **C**. (This exercise is referred to in Chapter 15.)

24. Let

$$G = \{a + b\sqrt{2} \mid a, b \text{ rational}\}$$

and

$$H = \left\{ \begin{bmatrix} a & 2b \\ b & a \end{bmatrix} \,\middle|\, a, b \text{ rational} \right\}.$$

Show that G and H are isomorphic under addition. Prove that G and H are closed under multiplication. Does your isomorphism preserve multiplication as well as addition? (G and H are examples of rings—a topic we will take up in Part 3.)

25. Prove that Z under addition is not isomorphic to Q under addition.

26. Prove that the quaternion group (see Exercise 4, Supplementary Exercises for Chapters 1–4) is not isomorphic to the dihedral group D_4.

27. Let **C** be the complex numbers and

$$M = \left\{ \begin{bmatrix} a & -b \\ b & a \end{bmatrix} \,\middle|\, a, b \in \mathbf{R} \right\}.$$

Prove that **C** and M are isomorphic under addition and that $\mathbf{C}^*$ and M^*, the nonzero elements of M, are isomorphic under multiplication.

28. Let $\mathbf{R}^n = \{(a_1, a_2, \ldots, a_n) \mid a_i \in \mathbf{R}\}$. Show that the mapping $\phi: (a_1, a_2, \ldots, a_n) \to (-a_1, -a_2, \ldots, -a_n)$ is an automorphism of the group $\mathbf{R}^n$ under componentwise addition. This automorphism is called *inversion*. Describe the action of ϕ geometrically.

29. Consider the following statement: The order of a subgroup divides the order of the group. Suppose you could prove this for finite permutation groups. Would the statement then be true for all finite groups? Explain.

30. Suppose that G is a finite Abelian group and G has no element of order 2. Show that the mapping $g \rightarrow g^2$ is an automorphism of G. Show, by example, that if G is infinite the mapping need not be an automorphism.

31. Let G be a group and let $g \in G$. If $z \in Z(G)$, show that the inner automorphism induced by g is the same as the inner automorphism induced by zg (that is, that the mappings ϕ_g and ϕ_{zg} are equal).

32. Show that the mapping $a \rightarrow \log_{10} a$ is an isomorphism from $\mathbf{R}^+$ under multiplication to $\mathbf{R}$ under addition.

33. Suppose that g and h induce the same inner automorphism of a group G. Prove that $h^{-1}g \in Z(G)$.

34. Combine the results of Exercises 31 and 33 into a single "if and only if" theorem.

35. Let a belong to a group G and let $|a|$ be finite. Let ϕ_a be the automorphism of G given by $\phi_a(x) = axa^{-1}$. Show that $|\phi_a|$ divides $|a|$. Exhibit an element a from a group for which $1 < |\phi_a| < |a|$.

36. Let $G = \{0, \pm2, \pm4, \pm6, \ldots\}$ and $H = \{0, \pm3, \pm6, \pm9, \ldots\}$. Show that G and H are isomorphic groups under addition. Does your isomorphism preserve multiplication?

37. Prove that Q under addition is not isomorphic to $\mathbf{R}^+$ under multiplication.

38. In Aut(Z_9), let α_i denote the automorphism that sends 1 to i where gcd(i, 9) $= 1$. Write α_5 and α_8 as permutations of $\{0, 1, \ldots, 8\}$ in disjoint cycle form. [For example, $\alpha_2 = (0)(124875)(36)$.]

39. Write the permutation corresponding to R_{90} in the left regular representation of D_4 in cycle form.

40. Show that every automorphism ϕ of the rational numbers Q under addition to itself has the form $\phi(x) = x\phi(1)$.

41. Prove that Q^+, the group of positive rational numbers under multiplication, is isomorphic to a proper subgroup of itself.

42. Prove that Q, the group of rational numbers under addition, is not isomorphic to a proper subgroup of itself.

43. Prove that every automorphism of $\mathbf{R}^*$, the group of nonzero real numbers under multiplication, maps positive numbers to positive numbers and negative numbers to negative numbers.

Suggested Videos

Cayley's Theorem, Films for the Humanities & Sciences, FFH 6354, 25 minutes, Box 2053, Princeton, NJ 08543-2053, 1-800-257-5126.

This video shows how the symmetry group of a regular octahedron can be represented as a group of permutations in several ways. It also shows how any group can be represented as a group of permutations.

Isomorphisms, Films for the Humanities & Sciences, FFH 6364, 24 minutes, Box 2053, Princeton, NJ 08543-2053, 1-800-257-5126.

This video uses colors to show that the set of motions in a traditional dance involving four couples forms a group isomorphic to Z_4 and $U(5)$, but that the group is not isomorphic to $U(8)$.

Arthur Cayley

Cayley is forging the weapons for future generations of physicists.

PETER TAIT

ARTHUR CAYLEY was born on August 16, 1821, in England. His genius showed itself at an early age. He published his first research paper while an undergraduate of 20, and in the next year he published eight papers. While still in his early twenties, he originated the concept of *n*-dimensional geometry.

After graduating from Trinity College, Cambridge, Cayley stayed on for three years as a tutor. At the age of 25, he began a 14-year career as a lawyer. During this period, he published approximately 200 mathematical papers, many of which are now classics.

In 1863, Cayley accepted the newly established Sadlerian professorship of mathematics at Cambridge University. He spent the rest of his life in that position. One of his notable accomplishments was his role in the successful effort to have women admitted to Cambridge.

Among Cayley's many innovations in mathematics were the notions of an abstract group and a group algebra, and the matrix concept. He made major contributions to geometry and linear algebra. Cayley and his lifelong friend and collaborator J. J. Sylvester were the founders of the theory of invariants, which was later to play an important role in the theory of relativity.

Cayley's collected works comprise 13 volumes, each about 600 pages in length. He died on January 26, 1895.

To find more information about Cayley, visit:

http://www-groups.dcs.st-and.ac.uk/~history/

7

Cosets and Lagrange's Theorem

It might be difficult, at this point, for the student to see the extreme importance of this result [Lagrange's Theorem]. As we penetrate the subject more deeply he will become more and more aware of its basic character.

I. N. HERSTEIN, *Topics in Algebra*

PROPERTIES OF COSETS

In this chapter, we will prove the single most important theorem in finite group theory—Lagrange's Theorem. But first, we introduce a new and powerful tool for analyzing a group—the notion of a coset. This notion was invented by Galois in 1830, although the term was coined by G. A. Miller in 1910.

DEFINITION Coset of H in G

Let G be a group and let H be a subset of G. For any $a \in G$, the set $\{ah \mid h \in H\}$ is denoted by aH. Analogously, $Ha = \{ha \mid h \in H\}$ and $aHa^{-1} = \{aha^{-1} \mid h \in H\}$. When H is a subgroup of G, the set aH is called the *left coset of H in G containing a*, whereas Ha is called the *right coset of H in G containing a*. In this case, the element a is called the *coset representative of aH (or Ha)*. We use $|aH|$ to denote the number of elements in the set aH, and $|Ha|$ to denote the number of elements in Ha.

EXAMPLE 1 Let $G = S_3$ and $H = \{(1), (13)\}$. Then the left cosets of H in G are

$$(1)H = H,$$
$$(12)H = \{(12), (12)(13)\} = \{(12), (132)\} = (132)H,$$
$$(13)H = \{(13), (1)\} = H,$$
$$(23)H = \{(23), (23)(13)\} = \{(23), (123)\} = (123)H. \qquad \blacklozenge$$

EXAMPLE 2   Let $\mathcal{H} = \{R_0, R_{180}\}$ in D_4, the dihedral group of order 8. Then,

$$R_0\mathcal{H} = \mathcal{H},$$
$$R_{90}\mathcal{H} = \{R_{90}, R_{270}\} = R_{270}\mathcal{H},$$
$$R_{180}\mathcal{H} = \{R_{180}, R_0\} = \mathcal{H},$$
$$V\mathcal{H} = \{V, H\} = H\mathcal{H},$$
$$D\mathcal{H} = \{D, D'\} = D'\mathcal{H}.$$

◆

EXAMPLE 3   Let $H = \{0, 3, 6\}$ in Z_9 under addition. In the case that the group operation is addition, we use the notation $a + H$ instead of aH. Then the cosets of H in Z_9 are

$$0 + H = \{0, 3, 6\} = 3 + H = 6 + H,$$
$$1 + H = \{1, 4, 7\} = 4 + H = 7 + H,$$
$$2 + H = \{2, 5, 8\} = 5 + H = 8 + H.$$

◆

The three preceding examples illustrate a few facts about cosets that are worthy of our attention. First, cosets are usually not subgroups. Second, aH may be the same as bH, even though a is not the same as b. Third, since in Example 1 $(12)H = \{(12), (132)\}$ whereas $H(12) = \{(12), (123)\}$, aH need not be the same as Ha.

These examples and observations raise many questions. When does $aH = bH$? Do aH and bH have any elements in common? When does $aH = Ha$? Which cosets are subgroups? Why are cosets important? The next lemma and theorem answer these questions. (Analogous results hold for right cosets.)

Lemma   **Properties of Cosets**

> Let H be a subgroup of G, and let a and b belong to G. Then,
>
> **1.** $a \in aH$,
> **2.** $aH = H$ if and only if $a \in H$,
> **3.** $aH = bH$ or $aH \cap bH = \varnothing$,
> **4.** $aH = bH$ if and only if $a^{-1}b \in H$,
> **5.** $|aH| = |bH|$,
> **6.** $aH = Ha$ if and only if $H = aHa^{-1}$,
> **7.** aH is a subgroup of G if and only if $a \in H$.

PROOF.
   1. $a = ae \in aH$.
   2. To verify property 2, we first suppose that $aH = H$. Then $a = ae \in aH = H$. Next, we assume that $a \in H$ and show that $aH \subseteq H$ and $H \subseteq aH$. The first inclusion follows directly from the closure of

H. To show that $H \subseteq aH$, let $h \in H$. Then, since $a \in H$ and $h \in H$, we know that $a^{-1}h \in H$. Thus, $h = eh = (aa^{-1})h = a(a^{-1}h) \in aH$.

3. To prove property 3, we suppose that $aH \cap bH \neq \emptyset$ and prove that $aH = bH$. Let $x \in aH \cap bH$. Then there exist h_1, h_2 in H such that $x = ah_1$ and $x = bh_2$. Thus, $a = xh_1^{-1} = bh_2h_1^{-1}$ and $aH = bh_2h_1^{-1}H = b(h_2h_1^{-1}H) = bH$, by property 2. [Here we use the fact that $(xy)H = x(yH)$.]

4. Observe that $aH = bH$ if and only if $H = a^{-1}bH$. The result now follows from property 2.

5. To prove that $|aH| = |bH|$, it suffices to define a one-to-one mapping from aH onto bH. Obviously, the correspondence $ah \rightarrow bh$ maps aH onto bH. That it is one-to-one follows directly from the cancellation property.

6. Note that $aH = Ha$ if and only if $(aH)a^{-1} = (Ha)a^{-1} = H(aa^{-1}) = H$—that is, if and only if $aHa^{-1} = H$.

7. If aH is a subgroup, then it contains the identity e. Thus, $aH \cap eH \neq \emptyset$; and, by property 3, we have $aH = eH = H$. Thus, from property 2, we have $a \in H$. Conversely, if $a \in H$, then, again by property 2, $aH = H$. ■ ■

Although most mathematical theorems are written in symbolic form, one should also know what they say *in words*. In the preceding lemma, property 1 says simply that the left coset of H containing a *does* contain a. Property 2 says that the H "absorbs" an element if and only if the element belongs to H. Property 3 says—and this is very important—that two left cosets of H are either identical or disjoint. Thus, a left coset of H is uniquely determined by any one of its elements. In particular, any element of a left coset can be used to represent the coset. Property 4 shows how we may transfer a question about equality of left cosets of H to a question about H itself and vice versa. Property 5 says that all left cosets of H have the same size. Property 6 is analogous to property 4 in that it shows how a question about the equality of the left and right cosets of H containing a is equivalent to a question about the equality of two subgroups of G. The last property of the lemma says that H itself is the only coset of H that is a subgroup of G.

Note that properties 1, 3, and 5 of the lemma guarantee that the left cosets of a subgroup H of G partition G into blocks of equal size. Indeed, we may view the cosets of H as a partitioning of G into equivalence classes under the equivalence relation defined by $a \sim b$ if $aH = bH$ (see Theorem 0.6).

In practice, the subgroup H is often chosen so that the cosets partition the group in some highly desirable fashion. For example, if G is 3-space $\mathbf{R}^3$ and H is a plane through the origin, then the coset $(a, b, c) + H$ (addition is done componentwise) is the plane passing through the point (a, b, c) and parallel to H. Thus, the cosets of H constitute a partition of 3-space into planes parallel to H. If $G = GL(2, \mathbf{R})$ and $H = SL(2, \mathbf{R})$, then for any matrix A in G, the coset AH is the set of *all* 2×2 matrices with the same determinant as A. Thus,

$$\begin{bmatrix} 2 & 0 \\ 0 & 1 \end{bmatrix} H \quad \text{is the set of all } 2 \times 2 \text{ matrices of determinant } 2$$

and

$$\begin{bmatrix} 1 & 2 \\ 2 & 1 \end{bmatrix} H \quad \text{is the set of all } 2 \times 2 \text{ matrices of determinant } -3.$$

Property 3 of the lemma is useful for actually finding the distinct cosets of a subgroup. We illustrate this in the next example.

EXAMPLE 4   To find the cosets of $H = \{1, 15\}$ in $G = U(32) = \{1, 3, 5, 7, 9, 11, 13, 15, 17, 19, 21, 23, 25, 27, 29, 31\}$ we begin with $H = \{1, 15\}$. We can find a second coset by choosing any element not in H, say 3, as a coset representative. This gives the coset $3H = \{3, 13\}$. We find our next coset by choosing a representative not already appearing in the two previously chosen cosets, say 5. This gives us the coset $5H = \{5, 11\}$. We continue to form cosets by picking elements from $U(32)$ that have not yet appeared in the previous cosets as representatives of the cosets until we have accounted for every element of $U(32)$. We then have the complete list of all distinct cosets of H.   ◆

LAGRANGE'S THEOREM AND CONSEQUENCES

We are now ready to prove a theorem that has been around for more than 200 years—longer than group theory itself! (This theorem was not originally stated in group theoretic terms.) At this stage, it should come as no surprise.

Theorem 7.1   **Lagrange's Theorem[†]: |H| Divides |G|**

> *If G is a finite group and H is a subgroup of G, then |H| divides |G|. Moreover, the number of distinct left (right) cosets of H in G is |G|/|H|.*

PROOF.   Let $a_1H, a_2H, \ldots, a_rH$ denote the distinct left cosets of H in G. Then, for each a in G, we have $aH = a_iH$ for some i. Also, by

---

[†]Lagrange stated his version of this theorem in 1770, but the first complete proof was given by Pietro Abbati some 30 years later.

property 1 of the lemma, $a \in aH$. Thus, each member of G belongs to one of the cosets a_iH. In symbols,

$$G = a_1H \cup \cdots \cup a_rH.$$

Now, property 3 of the lemma shows that this union is disjoint, so that

$$|G| = |a_1H| + |a_2H| + \cdots + |a_rH|.$$

Finally, since $|a_iH| = |H|$ for each i, we have $|G| = r|H|$. ■■

We pause to emphasize that Lagrange's Theorem is a subgroup candidate criterion; that is, it provides a list of candidates for the orders of the subgroups of a group. Thus, a group of order 12 may have subgroups of order 12, 6, 4, 3, 2, 1, but no others. *Warning!* The converse of Lagrange's Theorem is false. For example, a group of order 12 need not have a subgroup of order 6. We prove this in Example 5.

A special name and notation have been adopted for the number of left (or right) cosets of a subgroup in a group. The *index* of a subgroup H in G is the number of distinct left cosets of H in G. This number is denoted by $|G:H|$. As an immediate consequence of the proof of Lagrange's Theorem, we have the following useful formula for the number of distinct left (or right) cosets of H in G.

Corollary 1 $|G:H| = |G|/|H|$

> *If G is a finite group and H is a subgroup of G, then $|G:H| = |G|/|H|$.*

Corollary 2 $|a|$ **Divides** $|G|$

> *In a finite group, the order of each element of the group divides the order of the group.*

PROOF.   Recall that the order of an element is the order of the subgroup generated by that element. ■■

Corollary 3   **Groups of Prime Order Are Cyclic**

> *A group of prime order is cyclic.*

PROOF.   Suppose that G has prime order. Let $a \in G$ and $a \neq e$. Then, $|\langle a \rangle|$ divides $|G|$ and $|\langle a \rangle| \neq 1$. Thus, $|\langle a \rangle| = |G|$ and the corollary follows. ■■

Corollary 4 $a^{|G|} = e$

> Let G be a finite group, and let $a \in G$. Then, $a^{|G|} = e$.

PROOF.   By Corollary 2, $|G| = |a|k$ for some positive integer k. Thus, $a^{|G|} = a^{|a|k} = e^k = e$.   ■ ■

Corollary 5   **Fermat's Little Theorem**

> For every integer a and every prime p, $a^p \bmod p = a \bmod p$.

PROOF.   By the division algorithm, $a = pm + r$, where $0 \leq r < p$. Thus, $a \bmod p = r$, and it suffices to prove that $r^p \bmod p = r$. If $r = 0$, the result is trivial, so we may assume that $r \in U(p)$. [Recall that $U(p) = \{1, 2, \ldots, p - 1\}$ under multiplication modulo p.] Then, by the preceding corollary, $r^{p-1} \bmod p = 1$ and, therefore, $r^p \bmod p = r$.   ■ ■

Fermat's Little Theorem has been used in conjunction with computers to test for primality of certain numbers. One case concerned the number $p = 2^{257} - 1$. If p is prime, then we know from Fermat's Little Theorem that $10^p \bmod p = 10 \bmod p$ and, therefore, $10^{p+1} \bmod p = 100 \bmod p$. Using multiple precision and a simple loop, a computer was able to calculate $10^{p+1} \bmod p = 10^{2^{257}} \bmod p$ in a few seconds. The result was not 100, and so p is not prime.

EXAMPLE 5   The Converse of Lagrange's Theorem is False.[†]
The group A_4 of order 12 has no subgroups of order 6. To verify this, recall that A_4 has eight elements of order 3 (α_5 through a_{12} in the notation of Table 5.1) and suppose that H is a subgroup of order 6. Let a be any element of order 3 in A_4. Since H has index 2 in A_4, at most two of the cosets H, aH, and a^2H are distinct. But equality of any pair of these three implies that $aH = H$, so that $a \in H$. (For example, if $H = a^2H$, multiply on the left by a.) Thus, a subgroup of A_4 of order 6 would have to contain all eight elements of order 3, which is absurd. ◆

For any prime $p > 2$, we know that Z_{2p} and D_p are nonisomorphic groups of order $2p$. This naturally raises the question of whether there could be other possible groups of these orders. Remarkably, with just the simple machinery available to us at this point, we can answer this question.

[†]The first counterexample to the converse of Lagrange's Theorem was given by Paolo Ruffini in 1799.

Theorem 7.2 **Classification of Groups of Order 2*p***

> Let G be a group of order $2p$, where p is a prime greater than 2. Then G is isomorphic to Z_{2p} or D_p.

PROOF. We assume that G does not have an element of order $2p$ and show that $G \approx D_p$. We begin by first showing that G must have an element of order p. By our assumption and Lagrange's Theorem, any nonidentity element of G must have order 2 or p. Thus, to verify our assertion, we may assume that every nonidentity element of G has order 2. In this case, we have for all a and b in the group $ab = (ab)^{-1} = b^{-1}a^{-1} = ba$, so that G is Abelian. Then, for any nonidentity elements $a, b \in G$ with $a \neq b$, the set $\{e, a, b, ab\}$ is closed and therefore is a subgroup of G of order 4. Since this contradicts Lagrange's Theorem, we have proved that G must have an element of order p; call it a.

Now let b be any element not in $\langle a \rangle$. Then $b\langle a \rangle \neq \langle a \rangle$ and $G = \langle a \rangle \cup b\langle a \rangle$. We next claim that $|b| = 2$. To see this, observe that since $\langle a \rangle$ and $b\langle a \rangle$ are the only two distinct cosets of $\langle a \rangle$ in G, we must have $b^2\langle a \rangle = \langle a \rangle$ or $b^2\langle a \rangle = b\langle a \rangle$. We may rule out $b^2\langle a \rangle = b\langle a \rangle$, for then $b\langle a \rangle = \langle a \rangle$. On the other hand, $b^2\langle a \rangle = \langle a \rangle$ implies that $b^2 \in \langle a \rangle$ and, therefore, $|b^2| = 1$ or $|b^2| = p$. But $|b^2| = p$ and $|b| \neq 2p$ imply that $|b| = p$. Then $\langle b \rangle = \langle b^2 \rangle$ and therefore $b \in \langle a \rangle$, which is a contradiction. Thus, any element of G not in $\langle a \rangle$ has order 2.

Next consider ab. Since $ab \notin \langle a \rangle$, our argument above shows that $|ab| = 2$. Then $ab = (ab)^{-1} = b^{-1}a^{-1} = ba^{-1}$. Moreover, this relation completely determines the multiplication table for G. [For example, $a^3(ba^4) = a^2(ab)a^4 = a^2(ba^{-1})a^4 = a(ab)a^3 = a(ba^{-1})a^3 = (ab)a^2 = (ba^{-1})a^2 = ba$.] Since the multiplication table for all noncyclic groups of order $2p$ is uniquely determined by the relation $ab = ba^{-1}$, all noncyclic groups of order $2p$ must be isomorphic to each other. But of course, D_p, the dihedral group of order $2p$, is one such group. ∎

As an immediate corollary, we have that S_3, the symmetric group of degree 3, is isomorphic to D_3.

AN APPLICATION OF COSETS TO PERMUTATION GROUPS

Lagrange's Theorem and its corollaries dramatically demonstrate the fruitfulness of the coset concept. We next consider an application of cosets to permutation groups.

DEFINITION   *Stabilizer of a Point*

Let G be a group of permutations of a set S. For each i in S, let $\text{stab}_G(i) = \{\phi \in G \mid \phi(i) = i\}$. We call $\text{stab}_G(i)$ the *stabilizer of i in G*.

The student should verify that $\text{stab}_G(i)$ is a subgroup of G. (See Exercise 31 in Chapter 5.)

DEFINITION   *Orbit of a Point*

Let G be a group of permutations of a set S. For each s in S, let $\text{orb}_G(s) = \{\phi(s) \mid \phi \in G\}$. The set $\text{orb}_G(s)$ is a subset of S called the *orbit of s under G*. We use $|\text{orb}_G(s)|$ to denote the number of elements in $\text{orb}_G(s)$.

Example 6 should clarify these two definitions.

EXAMPLE 6   Let

$$G = \{(1), (132)(465)(78), (132)(465), (123)(456),$$
$$(123)(456)(78), (78)\}.$$

Then,

$$\text{orb}_G(1) = \{1, 3, 2\}, \qquad \text{stab}_G(1) = \{(1), (78)\},$$
$$\text{orb}_G(2) = \{2, 1, 3\}, \qquad \text{stab}_G(2) = \{(1), (78)\},$$
$$\text{orb}_G(4) = \{4, 6, 5\}, \qquad \text{stab}_G(4) = \{(1), (78)\},$$
$$\text{orb}_G(7) = \{7, 8\}, \qquad \text{stab}_G(7) = \{(1), (132)(465), (123)(456)\}. \; \blacklozenge$$

EXAMPLE 7   We may view D_4 as a group of permutations of a square region. Figure 7.1(a) illustrates the orbit of the point p under D_4, and Figure 7.1(b) illustrates the orbit of the point q under D_4. Observe that $\text{stab}_{D_4}(p) = \{R_0, D\}$, whereas $\text{stab}_{D_4}(q) = \{R_0\}$. $\qquad \blacklozenge$

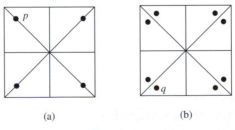

(a)   (b)

Figure 7.1

The preceding two examples also illustrate the following theorem.

Theorem 7.3 **Orbit-Stabilizer Theorem**

> Let G be a finite group of permutations of a set S. Then, for
> any i from S, $|G| = |orb_G(i)| \, |stab_G(i)|$.

PROOF. By Lagrange's Theorem, $|G|/|stab_G(i)|$ is the number of distinct left cosets of $stab_G(i)$ in G. Thus, it suffices to establish a one-to-one correspondence between the left cosets of $stab_G(i)$ and the elements in the orbit of i. To do this, we define a correspondence T by mapping the coset $\phi stab_G(i)$ to $\phi(i)$ under T. To show that T is a well-defined function, we must show that $\alpha stab_G(i) = \beta stab_G(i)$ implies $\alpha(i) = \beta(i)$. But $\alpha stab_G(i) = \beta stab_G(i)$ implies $\alpha^{-1}\beta \in stab_G(i)$, so that $(\alpha^{-1}\beta)(i) = i$ and, therefore, $\beta(i) = \alpha(i)$. Reversing the argument from the last step to the first step shows that T is also one-to-one. We conclude the proof by showing that T is onto $orb_G(i)$. Let $j \in orb_G(i)$. Then $\alpha(i) = j$ for some $\alpha \in G$ and clearly $T(\alpha stab_G(i)) = \alpha(i) = j$, so that T is onto. ■ ■

We leave as an exercise the proof of the important fact that the orbits of the elements of a set S under a group partition S (Exercise 29).

THE ROTATION GROUP OF A CUBE AND A SOCCER BALL

It cannot be overemphasized that Theorem 7.3 and Lagrange's Theorem (Theorem 7.1) are *counting* theorems.[†] They enable us to determine the numbers of elements in various sets. To see how Theorem 7.3 works, we will determine the order of the rotation group of a cube and a soccer ball. That is, we wish to find the number of essentially different ways in which we can take a cube or a soccer ball in a certain location in space, physically rotate it, and then return it to its original position in space.

EXAMPLE 8 Let G be the rotation group of a cube. Label the six faces of the cube 1 through 6. Since any rotation of the cube must carry each face of the cube to exactly one other face of the cube and different rotations induce different permutations of the faces, G can be viewed as a group of permutations on the set $\{1, 2, 3, 4, 5, 6\}$. Clearly, there is some rotation about a central horizontal or vertical axis that carries face number 1 to any other face, so that $|orb_G(1)| = 6$. Next, we consider $stab_G(1)$. Here, we are asking for all rotations of a cube that leave face number 1 where it is. Surely, there are only four such motions— rotations of 0°, 90°, 180°, and 270°—about the line perpendicular to

---

[†]*People who don't count won't count* (Anatole France).

the face and passing through its center (see Figure 7.2). Thus, by Theorem 7.3, $|G| = |\text{orb}_G(1)|\,|\text{stab}_G(1)| = 6 \cdot 4 = 24$.   ◆

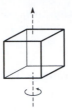

Figure 7.2   Axis of rotation of a cube.

Now that we know how many rotations a cube has, it is simple to determine the actual structure of the rotation group of a cube. Recall that S_4 is the symmetric group of degree 4.

Theorem 7.4   The Rotation Group of a Cube

> *The group of rotations of a cube is isomorphic to S_4.*

PROOF.   Since the group of rotations of a cube has the same order as S_4, we need only prove that the group of rotations is isomorphic to a subgroup of S_4. To this end, observe that a cube has four diagonals and that the rotation group induces a group of permutations on the four diagonals. But we must be careful not to assume that different rotations correspond to different permutations. To see that this is so, all we need do is show that all 24 permutations of the diagonals arise from rotations. Labeling the consecutive diagonals 1, 2, 3, and 4, it is obvious that there is a 90° rotation that yields the permutation $\alpha = (1234)$; another 90° rotation about an axis perpendicular to our first axis yields the permutation $\beta = (1423)$. See Figure 7.3. So, the group of permutations induced by the rotations contains the eight-element subgroup $\{\varepsilon, \alpha, \alpha^2, \alpha^3, \beta^2, \beta^2\alpha, \beta^2\alpha^2, \beta^2\alpha^3\}$ and $\alpha\beta$, which has order 3. Clearly, then, the rotations yield all 24 permutations since the order of the rotation group must be divisible by both 8 and 3.   ■■

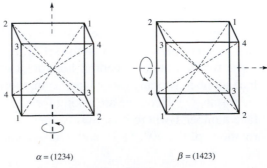

$\alpha = (1234)$ $\beta = (1423)$

Figure 7.3

EXAMPLE 9   A soccer ball has 20 faces that are regular hexagons and 12 faces that are regular pentagons. (The technical term for this solid is *truncated icosahedron*.) To determine the number of rotational symmetries of a soccer ball using Theorem 7.3, we may choose our set S to be the 20 hexagons or the 12 pentagons. Let us say that S is the set of 12 pentagons. Since any pentagon can be carried to any other pentagon by some rotation, the orbit of any pentagon is S. Also, there are five rotations that fix (stabilize) any particular pentagon. Thus, by the Orbit-Stabilizer Theorem, there are $12 \cdot 5 = 60$ rotational symmetries. (In case you are interested, the rotation group of a soccer ball is isomorphic to A_5.)   ◆

In 1985, chemists Robert Curl, Richard Smalley, and Harold Kroto caused tremendous excitement in the scientific community when they created a new form of carbon by using a laser beam to vaporize graphite. The structure of the new molecule is composed of 60 carbon atoms arranged in the shape of a soccer ball! Because the shape of the new molecule reminded them of the dome structures built by the architect R. Buckminster Fuller, Curl, Smalley, and Kroto named their discovery "buckyballs." Buckyballs are the roundest, most symmetrical large molecules known. Group theory has been particularly useful in illuminating the properties of buckyballs, since the absorption spectrum of a molecule depends on its symmetries and chemists classify various molecular states according to their symmetry properties. The buckyball discovery spurred a revolution in carbon chemistry. In 1996, Curl, Smalley, and Kroto received the Nobel Prize in chemistry for their discovery.

EXERCISES

I don't know, Marge. Trying is the first step towards failure.

HOMER SIMPSON

1. Let $H = \{(1), (12)(34), (13)(24), (14)(23)\}$. Find the left cosets of H in A_4 (see Table 5.1 on page 105).

2. Let H be as in Exercise 1. How many left cosets of H in S_4 are there? (Determine this without listing them.)

3. Let $H = \{0, \pm 3, \pm 6, \pm 9, \ldots\}$. Find all the left cosets of H in Z.

4. Rewrite the condition $a^{-1}b \in H$ given in property 4 of the lemma on page 138 in additive notation. Assume that the group is Abelian.

5. Let H be as in Exercise 3. Use Exercise 4 to decide whether or not the following cosets of H are the same.
   a. $11 + H$ and $17 + H$
   b. $-1 + H$ and $5 + H$
   c. $7 + H$ and $23 + H$

6. Let n be a positive integer. Let $H = \{0, \pm n, \pm 2n, \pm 3n, \ldots\}$. Find all left cosets of H in Z. How many are there?

7. Find all of the left cosets of $\{1, 11\}$ in $U(30)$.

8. Suppose that a has order 15. Find all of the left cosets of $\langle a^5 \rangle$ in $\langle a \rangle$.

9. Let $|a| = 30$. How many left cosets of $\langle a^4 \rangle$ in $\langle a \rangle$ are there? List them.

10. Let a and b be nonidentity elements of different orders in a group G of order 155. Prove that the only subgroup of G that contains a and b is G itself.

11. Let H be a subgroup of $\mathbf{R}^*$, the group of nonzero real numbers under multiplication. If $\mathbf{R}^+ \subseteq H \subseteq \mathbf{R}^*$, prove that $H = \mathbf{R}^+$ or $H = \mathbf{R}^*$.

12. Let $\mathbf{C}^*$ be the group of nonzero complex numbers under multiplication and let $H = \{a + bi \in \mathbf{C}^* | a^2 + b^2 = 1\}$. Give a geometric description of the cosets of H.

13. Let G be a group of order 60. What are the possible orders for the subgroups of G?

14. Suppose that K is a proper subgroup of H and H is a proper subgroup of G. If $|K| = 42$ and $|G| = 420$, what are the possible orders of H?

15. Suppose that $|G| = pq$, where p and q are prime. Prove that every proper subgroup of G is cyclic.

16. Recall that, for any integer n greater than 1, $\phi(n)$ denotes the number of integers less than n and relatively prime to n. Prove that if a is any integer relatively prime to n, then $a^{\phi(n)} \bmod n = 1$.

17. Compute 5^{15} mod 7 and 7^{13} mod 11.

18. Use Corollary 2 of Lagrange's Theorem (Theorem 7.1) to prove that the order of $U(n)$ is even when $n > 2$.

19. Suppose G is a finite group of order n and m is relatively prime to n. If $g \in G$ and $g^m = e$, prove that $g = e$.

20. Suppose H and K are subgroups of a group G. If $|H| = 12$ and $|K| = 35$, find $|H \cap K|$.

21. Suppose that G is an Abelian group with an odd number of elements. Show that the product of all of the elements of G is the identity.

22. Suppose that G is a group with more than one element and G has no proper, nontrivial subgroups. Prove that $|G|$ is prime. (Do not assume at the outset that G is finite.)

23. Let $|G| = 15$. If G has only one subgroup of order 3 and only one of order 5, prove that G is cyclic. Generalize to $|G| = pq$, where p and q are prime.

24. Let G be a group of order 25. Prove that G is cyclic or $g^5 = e$ for all g in G.

25. Let $|G| = 33$. What are the possible orders for the elements of G? Show that G must have an element of order 3.

26. Let $|G| = 8$. Show that G must have an element of order 2.

27. Let H and K be subgroups of a finite group G with $H \subseteq K \subseteq G$. Prove that $|G:H| = |G:K| \, |K:H|$.

28. Show that Q, the group of rational numbers under addition, has no proper subgroup of finite index.

29. Let G be a group of permutations of a set S. Prove that the orbits of the members of S constitute a partition of S. (This exercise is referred to in this chapter and in Chapter 29.)

30. Prove that every subgroup of D_n of odd order is cyclic.

31. Let $G = \{(1), (12)(34), (1234)(56), (13)(24), (1432)(56), (56)(13), (14)(23), (24)(56)\}$.
   a. Find the stabilizer of 1 and the orbit of 1.
   b. Find the stabilizer of 3 and the orbit of 3.
   c. Find the stabilizer of 5 and the orbit of 5.

32. Prove that 3, 5, and 7 are the only three consecutive odd integers that are prime.

33. Let G be a group of order p^n where p is prime. Prove that the center of G cannot have order p^{n-1}.

34. Prove that a group of order 12 must have an element of order 2.

35. Suppose that a group contains elements of orders 1 through 10. What is the minimum possible order of the group?

36. Let G be a finite Abelian group and let n be a positive integer that is relatively prime to $|G|$. Show that the mapping $a \rightarrow a^n$ is an automorphism of G.

37. Show that in a group G of odd order, the equation $x^2 = a$ has a unique solution for all a in G.

38. Let G be a group of order pqr, where p, q, and r are distinct primes. If H and K are subgroups of G with $|H| = pq$ and $|K| = qr$, prove that $|H \cap K| = q$.

39. Let $G = GL(2, \mathbf{R})$ and $H = SL(2, \mathbf{R})$. Let $A \in G$ and suppose that det $A = 2$. Prove that AH is the set of all 2×2 matrices in G that have determinant 2.

40. Let G be the group of rotations of a plane about a point P in the plane. Thinking of G as a group of permutations of the plane, describe the orbit of a point Q in the plane. (This is the motivation for the name "orbit.")

41. Let G be the rotation group of a cube. Label the faces of the cube 1 through 6, and let H be the subgroup of elements of G that carry face 1 to itself. If σ is a rotation that carries face 2 to face 1, give a physical description of the coset $H\sigma$.

42. The group D_4 acts as a group of permutations of the square regions shown below. (The axes of symmetry are drawn for reference purposes.) For each square region, locate the points in the orbit of the indicated point under D_4. In each case, determine the stabilizer of the indicated point.

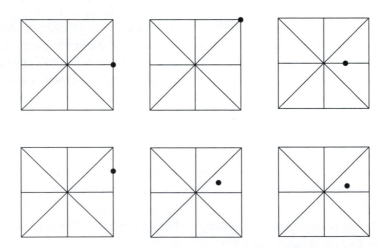

43. Let $G = GL(2, \mathbf{R})$, the group of 2×2 matrices over $\mathbf{R}$ with nonzero determinant. Let H be the subgroup of matrices of determinant ± 1. If a, b

$\in G$ and $aH = bH$, what can be said about det (a) and det (b)? Prove or disprove the converse. [Determinants have the property that det $(xy) =$ det (x)det (y).]

44. Calculate the orders of the following (refer to Figure 27.5 for illustrations):

    a. The group of rotations of a regular tetrahedron (a solid with four congruent equilateral triangles as faces)

    b. The group of rotations of a regular octahedron (a solid with eight congruent equilateral triangles as faces)

    c. The group of rotations of a regular dodecahedron (a solid with 12 congruent regular pentagons as faces)

    d. The group of rotations of a regular icosahedron (a solid with 20 congruent equilateral triangles as faces)

45. If G is a finite group with fewer than 100 elements and G has subgroups of orders 10 and 25, what is the order of G?

46. A soccer ball has 20 faces that are regular hexagons and 12 faces that are regular pentagons. Use Theorem 7.3 to explain why a soccer ball cannot have a 60° rotational symmetry about a line through the centers of two opposite hexagonal faces.

47. Determine all finite subgroups of $\mathbf{C}^*$, the group of nonzero complex numbers under multiplication.

COMPUTER EXERCISE ▪▪

In the fields of observation chance favors only the prepared mind.

LOUIS PASTEUR

Software for the computer exercise in this chapter is available at the website:

http://www.d.umn.edu/~jgallian

1. This software determines when Z_n is the only group of order n in the case that $n = pq$ where p and q are primes and $p < q$. Run the software for $n = 3 \cdot 5, 3 \cdot 7, 3 \cdot 11, 3 \cdot 13, 3 \cdot 17, 3 \cdot 31, 5 \cdot 7, 5 \cdot 11, 5 \cdot 13, 5 \cdot 17, 5 \cdot 31, 7 \cdot 11, 7 \cdot 13, 7 \cdot 17, 7 \cdot 19$, and $7 \cdot 43$. Conjecture a necessary and sufficient condition about p and q for Z_{pq} to be the only group of order pq, where p and q are primes and $p < q$.

Joseph Lagrange

Lagrange is the Lofty Pyramid of the Mathematical Sciences.

NAPOLEON BONAPARTE

This stamp was issued by France in Lagrange's honor in 1958.

JOSEPH LOUIS LAGRANGE was born in Italy of French ancestry on January 25, 1736. He became captivated by mathematics at an early age when he read an essay by Halley on Newton's calculus. At the age of 19, he became a professor of mathematics at the Royal Artillery School in Turin. Lagrange made significant contributions to many branches of mathematics and physics, among them the theory of numbers, the theory of equations, ordinary and partial differential equations, the calculus of variations, analytic geometry, fluid dynamics, and celestial mechanics. His methods for solving third- and fourth-degree polynomial equations by radicals laid the groundwork for the group-theoretic approach to solving polynomials taken by Galois. Lagrange was a very careful writer with a clear and elegant style.

At the age of 40, Lagrange was appointed Head of the Berlin Academy, succeeding Euler. In offering this appointment, Frederick the Great proclaimed that the "greatest king in Europe" ought to have the "greatest mathematician in Europe" at his court. In 1787, Lagrange was invited to Paris by Louis XVI and became a good friend of the king and his wife, Marie Antoinette. In 1793, Lagrange headed a commission, which included Laplace and Lavoisier, to devise a new system of weights and measures. Out of this came the metric system. Late in his life he was made a count by Napoleon. Lagrange died on April 10, 1813.

To find more information about Lagrange, visit:

http://www-groups.dcs.st-and.ac.uk/~history/

8 External Direct Products

The universe is an enormous direct product of representations of symmetry groups.

STEVEN WEINBERG[†]

DEFINITION AND EXAMPLES

In this chapter, we show how to piece together groups to make larger groups. In Chapter 9, we will show that we can often start with one large group and decompose it into a product of smaller groups in much the same way as a composite positive integer can be broken down into a product of primes. These methods will later be used to give us a simple way to construct all finite Abelian groups.

DEFINITION *External Direct Product*

Let $G_1, G_2, \ldots, G_n$ be a finite collection of groups. The *external direct product* of $G_1, G_2, \ldots, G_n$, written as $G_1 \oplus G_2 \oplus \cdots \oplus G_n$, is the set of all n-tuples for which the ith component is an element of G_i and the operation is componentwise.

In symbols,

$$G_1 \oplus G_2 \oplus \cdots \oplus G_n = \{(g_1, g_2, \ldots, g_n) \mid g_i \in G_i\},$$

where $(g_1, g_2, \ldots, g_n)(g_1', g_2', \ldots, g_n')$ is defined to be $(g_1 g_1', g_2 g_2', \ldots, g_n g_n')$. It is understood that each product $g_i g_i'$ is performed with the operation of G_i. We leave it to the reader to show that the external direct product of groups is itself a group (Exercise 1).

[†]Weinberg received the 1979 Nobel Prize in physics with Sheldon Glashow and Abdus Salam for their construction of a single theory incorporating weak and electromagnetic interactions.

This construction is not new to students who have had linear algebra or physics. Indeed, $\mathbf{R}^2 = \mathbf{R} \oplus \mathbf{R}$ and $\mathbf{R}^3 = \mathbf{R} \oplus \mathbf{R} \oplus \mathbf{R}$—the operation being componentwise addition. Of course, there is also scalar multiplication, but we ignore this for the time being, since we are interested only in the group structure at this point.

EXAMPLE 1

$$U(8) \oplus U(10) = \{(1, 1), (1, 3), (1, 7), (1, 9), (3, 1), (3, 3),$$
$$(3, 7), (3, 9), (5, 1), (5, 3), (5, 7), (5, 9), (7, 1), (7, 3),$$
$$(7, 7), (7, 9)\}.$$

The product $(3, 7)(7, 9) = (5, 3)$, since the first two components are combined by multiplication modulo 8, whereas the second two components are combined by multiplication modulo 10.   ◆

EXAMPLE 2

$$Z_2 \oplus Z_3 = \{(0, 0), (0, 1), (0, 2), (1, 0), (1, 1), (1, 2)\}.$$

Clearly, this is an Abelian group of order 6. Is this group related to another Abelian group of order 6 that we know, namely, Z_6? Consider the subgroup of $Z_2 \oplus Z_3$ generated by $(1, 1)$. Since the operation in each component is addition, we have $(1, 1) = (1, 1)$, $2(1, 1) = (0, 2)$, $3(1, 1) = (1, 0)$, $4(1, 1) = (0, 1)$, $5(1, 1) = (1, 2)$, and $6(1, 1) = (0, 0)$. Hence $Z_2 \oplus Z_3$ is cyclic. It follows that $Z_2 \oplus Z_3$ is isomorphic to Z_6.   ◆

In Theorem 7.2 we classified the groups of order $2p$ where p is an odd prime. Now that we have defined $Z_2 \oplus Z_2$, it is easy to classify the groups of order 4.

EXAMPLE 3   Classification of Groups of Order 4
A group of order 4 is isomorphic to Z_4 or $Z_2 \oplus Z_2$. To verify this, let $G = \{e, a, b, ab\}$. If G is not cyclic, then it follows from Lagrange's Theorem that $|a| = |b| = |ab| = 2$. Then the mapping $e \rightarrow (0, 0)$, $a \rightarrow (1, 0)$, $b \rightarrow (0, 1)$, and $ab \rightarrow (1, 1)$ is an isomorphism from G onto $Z_2 \oplus Z_2$.   ◆

On the basis of Example 2, one might be tempted to conjecture that $Z_m \oplus Z_n \approx Z_{mn}$ for all positive integers m and n. This is easily seen to be false by looking at $Z_2 \oplus Z_2$. Theorem 8.2 shows, however, that these groups are isomorphic in certain cases.

PROPERTIES OF EXTERNAL DIRECT PRODUCTS

Our first theorem gives a simple method for computing the order of an element in a direct product in terms of the orders of the component pieces.

Theorem 8.1   **Order of an Element in a Direct Product**

> *The order of an element in a direct product of a finite number of finite groups is the least common multiple of the orders of the components of the element. In symbols,*
>
> $$|(g_1, g_2, \ldots, g_n)| = \text{lcm}(|g_1|, |g_2|, \ldots, |g_n|).$$

PROOF.   Denote the identity of G_i by e_i. Suppose that t is a positive integer such that $(g_1, g_2, \ldots, g_n)^t = (g_1^t, g_2^t, \ldots, g_n^t) = (e_1, e_2, \ldots, e_n)$. It follows from Corollary 2 of Theorem 4.1 that t is a common multiple of $|g_1|, |g_2|, \ldots, |g_n|$. By definition, the smallest such positive integer is $\text{lcm}(|g_1|, |g_2|, \ldots, |g_n|)$.   ■ ■

The next two examples are applications of Theorem 8.1.

EXAMPLE 4   We determine the number of elements of order 5 in $Z_{25} \oplus Z_5$. By Theorem 8.1, we may count the number of elements (a, b) in $Z_{25} \oplus Z_5$ with the property that $5 = |(a, b)| = \text{lcm}(|a|, |b|)$. Clearly this requires that either $|a| = 5$ and $|b| = 1$ or 5, or $|b| = 5$ and $|a| = 1$ or 5. We consider two mutually exclusive cases.

CASE 1 $|a| = 5$ and $|b| = 1$ or 5. Here there are four choices for a (namely, 5, 10, 15, and 20) and five choices for b. This gives 20 elements of order 5.

CASE 2 $|a| = 1$ and $|b| = 5$. This time there is one choice for a and four choices for b, so we obtain four more elements of order 5.

Thus, $Z_{25} \oplus Z_5$ has 24 elements of order 5.   ◆

EXAMPLE 5   We determine the number of cyclic subgroups of order 10 in $Z_{100} \oplus Z_{25}$. We begin by counting the number of elements (a, b) of order 10.

CASE 1 $|a| = 10$ and $|b| = 1$ or 5. Since Z_{100} has a unique cyclic subgroup of order 10 and any cyclic group of order 10 has four generators (Theorem 4.4), there are four choices for a. Similarly, there are five choices for b. This gives 20 possibilities for (a, b).

CASE 2 $|a| = 2$ and $|b| = 5$. Since any finite cyclic group of even order has a unique subgroup of order 2 (Theorem 4.4), there is only one choice for a. Obviously, there are four choices for b. So, this case yields four more possibilities for (a, b).

Thus, $Z_{100} \oplus Z_{25}$ has 24 elements of order 10. Because each cyclic subgroup of order 10 has four elements of order 10 and no two of the cyclic subgroups can have an element of order 10 in common, there must be $24/4 = 6$ cyclic subgroups of order 10. (This method is analogous to determining the number of sheep in a flock by counting legs and dividing by 4.)   ◆

The direct product notation is convenient for specifying certain subgroups of a direct product.

EXAMPLE 6   Since 5 has order 6 in Z_{30} and 3 has order 4 in Z_{12}, $\langle 5 \rangle \oplus \langle 3 \rangle$ is a subgroup of order 24 in $Z_{30} \oplus Z_{12}$.   ◆

The next theorem and its first corollary characterize those direct products of cyclic groups that are themselves cyclic.

Theorem 8.2   **Criterion for $G \oplus H$ to Be Cyclic**

> *Let G and H be finite cyclic groups. Then $G \oplus H$ is cyclic if and only if $|G|$ and $|H|$ are relatively prime.*

PROOF.   Let $|G| = m$ and $|H| = n$, so that $|G \oplus H| = mn$. To prove the first half of the theorem, we assume $G \oplus H$ is cyclic and show that m and n are relatively prime. Suppose that $\gcd(m, n) = d$ and (g, h) is a generator of $G \oplus H$. Since $(g, h)^{mn/d} = ((g^m)^{n/d}, (h^n)^{m/d}) = (e, e)$, we have $mn = |(g, h)| \leq mn/d$. Thus, $d = 1$.

To prove the other half of the theorem, let $G = \langle g \rangle$ and $H = \langle h \rangle$ and suppose $\gcd(m, n) = 1$. Then, $|(g, h)| = \mathrm{lcm}(m, n) = mn = |G \oplus H|$, so that (g, h) is a generator of $G \oplus H$.   ■ ■

As a consequence of Theorem 8.2 and an induction argument, we obtain the following extension of Theorem 8.2.

Corollary 1   **Criterion for $G_1 \oplus G_2 \oplus \cdots \oplus G_n$ to Be Cyclic**

> *An external direct product $G_1 \oplus G_2 \oplus \cdots \oplus G_n$ of a finite number of finite cyclic groups is cyclic if and only if $|G_i|$ and $|G_j|$ are relatively prime when $i \neq j$.*

Corollary 2 **Criterion for $Z_{n_1 n_2 \cdots n_k} \approx Z_{n_1} \oplus Z_{n_2} \oplus \cdots \oplus Z_{n_k}$**

> *Let $m = n_1 n_2 \cdots n_k$. Then Z_m is isomorphic to $Z_{n_1} \oplus Z_{n_2} \oplus \cdots \oplus Z_{n_k}$ if and only if n_i and n_j are relatively prime when $i \neq j$.*

By using the results above in an iterative fashion, one can express the same group (up to isomorphism) in many different forms. For example, we have

$$Z_2 \oplus Z_2 \oplus Z_3 \oplus Z_5 \approx Z_2 \oplus Z_6 \oplus Z_5 \approx Z_2 \oplus Z_{30}.$$

Similarly,

$$Z_2 \oplus Z_2 \oplus Z_3 \oplus Z_5 \approx Z_2 \oplus Z_6 \oplus Z_5$$

$$\approx Z_2 \oplus Z_3 \oplus Z_2 \oplus Z_5 \approx Z_6 \oplus Z_{10}.$$

Thus, $Z_2 \oplus Z_{30} \approx Z_6 \oplus Z_{10}$. Note, however, that $Z_2 \oplus Z_{30} \not\approx Z_{60}$.

THE GROUP OF UNITS MODULO n AS AN EXTERNAL DIRECT PRODUCT

The *U*-groups provide a convenient way to illustrate the preceding ideas. We first introduce some notation. If k is a divisor of n, let

$$U_k(n) = \{x \in U(n) \mid x \bmod k = 1\}.$$

For example, $U_7(105) = \{1, 8, 22, 29, 43, 64, 71, 92\}$. It can be readily shown that $U_k(n)$ is indeed a subgroup of $U(n)$. (See Exercise 13 in Chapter 3.)

Theorem 8.3 **$U(n)$ as an External Direct Product**

> *Suppose s and t are relatively prime. Then $U(st)$ is isomorphic to the external direct product of $U(s)$ and $U(t)$. In short,*
>
> $$U(st) \approx U(s) \oplus U(t).$$
>
> *Moreover, $U_s(st)$ is isomorphic to $U(t)$ and $U_t(st)$ is isomorphic to $U(s)$.*

PROOF. An isomorphism from $U(st)$ to $U(s) \oplus U(t)$ is $x \rightarrow (x \bmod s, x \bmod t)$; an isomorphism from $U_s(st)$ to $U(t)$ is $x \rightarrow x \bmod t$; an isomorphism from $U_t(st)$ to $U(s)$ is $x \rightarrow x \bmod s$. We leave the verification that these mappings are operation-preserving, one-to-one, and onto to the reader. (See Exercises 11, 15, and 17 in Chapter 0; see also [1].) ▪ ■

As a consequence of Theorem 8.3, we have the following result.

Corollary

> *Let $m = n_1 n_2 \cdots n_k$, where $\gcd(n_i, n_j) = 1$ for $i \neq j$. Then,*
>
> $$U(m) \approx U(n_1) \oplus U(n_2) \oplus \cdots \oplus U(n_k).$$

To see how these results work, let's apply them to $U(105)$. We obtain

$$U(105) \approx U(7) \oplus U(15)$$
$$U(105) \approx U(21) \oplus U(5)$$
$$U(105) \approx U(3) \oplus U(5) \oplus U(7).$$

Moreover,

$U(7) \approx U_{15}(105) = \{1, 16, 31, 46, 61, 76\}$
$U(15) \approx U_7(105) = \{1, 8, 22, 29, 43, 64, 71, 92\}$
$U(21) \approx U_5(105) = \{1, 11, 16, 26, 31, 41, 46, 61, 71, 76, 86, 101\}$
$U(5) \approx U_{21}(105) = \{1, 22, 43, 64\}$
$U(3) \approx U_{35}(105) = \{1, 71\}.$

Among all groups, surely the cyclic groups Z_n have the simplest structures and, at the same time, are the easiest groups with which to compute. Direct products of groups of the form Z_n are only slightly more complicated in structure and computability. Because of this, algebraists endeavor to describe a finite Abelian group as such a direct product. Indeed, we shall soon see that every finite Abelian group can be so represented. With this goal in mind, let us reexamine the U-groups. Using the corollary to Theorem 8.3 and the fact (see [2, p. 93]), first proved by Carl Gauss in 1801, that

$$U(2) \approx \{0\}, \qquad U(4) \approx Z_2, \qquad U(2^n) \approx Z_2 \oplus Z_{2^{n-2}} \qquad \text{for } n \geq 3,$$

and

$$U(p^n) \approx Z_{p^n - p^{n-1}} \qquad \text{for } p \text{ an odd prime,}$$

we now can write any U-group as an external direct product of cyclic groups. For example,

$$U(105) = U(3 \cdot 5 \cdot 7) \approx U(3) \oplus U(5) \oplus U(7)$$
$$\approx Z_2 \oplus Z_4 \oplus Z_6$$

and

$$U(720) = U(16 \cdot 9 \cdot 5) \approx U(16) \oplus U(9) \oplus U(5)$$
$$\approx Z_2 \oplus Z_4 \oplus Z_6 \oplus Z_4.$$

What is the advantage of expressing a group in this form? Well, for one thing, we immediately see that the orders of the elements $U(720)$ can only be 1, 2, 3, 4, 6, and 12. This follows from the observations that an element from $Z_2 \oplus Z_4 \oplus Z_6 \oplus Z_4$ has the form (a, b, c, d), where $|a| = 1$ or 2, $|b| = 1, 2$, or 4, $|c| = 1, 2, 3$, or 6, and $|d| = 1, 2$, or 4, and that $|(a, b, c, d)| = \text{lcm}(|a|, |b|, |c|, |d|)$. For another thing, we can readily determine the number of elements of order 12, say, that $U(720)$ has. Because $U(720)$ is isomorphic to $Z_2 \oplus Z_4 \oplus Z_6 \oplus Z_4$, it suffices to calculate the number of elements of order 12 in $Z_2 \oplus Z_4 \oplus Z_6 \oplus Z_4$. But this is easy. By Theorem 8.1, an element (a, b, c, d) has order 12 if and only if $\text{lcm}(|a|, |b|, |c|, |d|) = 12$. Since $|a| = 1$ or 2, it does not matter how a is chosen. So, how can we have $\text{lcm}(|b|, |c|, |d|) = 12$? One way is to have $|b| = 4$, $|c| = 3$ or 6, and d arbitrary. By Theorem 4.4, there are two choices for b, four choices for c, and four choices for d. So, in this case, we have $2 \cdot 4 \cdot 4 = 32$ choices. The only other way to have $\text{lcm}(|b|, |c|, |d|) = 12$ is for $|d| = 4$, $|c| = 3$ or 6, and $|b| = 1$ or 2 (we exclude $|b| = 4$, since this was already accounted for). This gives $2 \cdot 4 \cdot 2 = 16$ new choices. Finally, since a can be either of the two elements in Z_2, we have a total of $2(32 + 16) = 96$ elements of order 12.

These calculations tell us more. Since $\text{Aut}(Z_{720})$ is isomorphic to $U(720)$, we also know that there are 96 automorphisms of Z_{720} of order 12. Imagine trying to deduce this information directly from $U(720)$ or, worse yet, from $\text{Aut}(Z_{720})$! These results beautifully illustrate the advantage of being able to represent a finite Abelian group as a direct product of cyclic groups. They also show the value of our theorems about $\text{Aut}(Z_n)$ and $U(n)$. After all, theorems are labor-saving devices. If you want to convince yourself of this, try to prove directly from the definitions that $\text{Aut}(Z_{720})$ has exactly 96 elements of order 12.

APPLICATIONS

We conclude this chapter with five applications of the material presented here—three to cryptography, the science of sending and deciphering secret messages, one to genetics, and one to electric circuits.

Data Security

Because computers are built from two-state electronic components, it is natural to represent information as strings of 0s and 1s called *binary strings*. A binary string of length n can naturally be thought of as an element of $Z_2 \oplus Z_2 \oplus \cdots \oplus Z_2$ (n copies) where the parentheses and the commas have been deleted. Thus the binary string 11000110 corresponds to the element (1, 1, 0, 0, 0, 1, 1, 0) in $Z_2 \oplus Z_2 \oplus Z_2 \oplus Z_2 \oplus Z_2 \oplus Z_2 \oplus Z_2 \oplus Z_2$. Similarly, two binary strings $a_1a_2 \cdots a_n$ and $b_1b_2 \cdots b_n$ are added componentwise modulo 2 just as their corresponding elements in $Z_2 \oplus Z_2 \oplus \cdots \oplus Z_2$ are. For example,

$$11000111 + 01110110 = 10110001$$

and

$$10011100 + 10011100 = 00000000.$$

The fact that the sum of two binary sequences $a_1a_2 \cdots a_n + b_1b_2 \cdots b_n = 00 \cdots 0$ if and only if the sequences are identical is the basis for a data security system used to protect internet transactions.

Suppose that you want to purchase a compact disc from www.Amazon.com. Need you be concerned that a hacker will intercept your credit card number during the transaction? As you might expect, your credit card number is sent to Amazon in a way that protects the data. We explain one way to send credit card numbers over the Web securely. When you place an order with Amazon the company sends your computer a randomly generated string of 0's and 1's called a *key*. This key has the same length as the binary string corresponding to your credit card number and the two strings are added (think of this process as "locking" the data.) The resulting sum is then transmitted to Amazon. Amazon in turn adds the same key to the received string which then produces the original string corresponding to your credit card number (adding the key a second time "unlocks" the data).

To illustrate the idea say you want to send an eight digit binary string such as $s = 10101100$ to Amazon (actual card card numbers have very long strings) and Amazon sends your computer the key $k = 00111101$. Your computer returns the string $s + k = 10101100 + 00111101 = 10010001$ to Amazon, and Amazon adds k to this string to get $10010001 + 00111101 = 10101100$, which is the string representing your credit card number. If someone intercepts the number $s + k = 10010001$ during transmission it is no value without knowing k.

The method is secure because the key sent by Amazon is randomly generated and used only one time. You can tell when you are using an encryption scheme on a Web transaction by looking to see if the Web address begins with "https" rather than the customary "http." You will also see a small padlock in the status bar at the bottom of the browser window.

Application to Public Key Cryptography

In the mid-1970s, Ronald Rivest, Adi Shamir, and Leonard Adleman devised an ingenious method that permits each person who is to receive a secret message to tell publicly how to scramble messages sent to him or her. And even though the method used to scramble the message is known publicly, only the person for whom it is intended will be able to unscramble the message. The idea is based on the fact that there exist efficient methods for finding very large prime numbers (say about 100 digits long) and for multiplying large numbers, but no one knows an efficient algorithm for factoring large integers (say about 200 digits long). So, the person who is to receive the message finds a pair of large primes p and q and chooses an integer r with $1 < r < m$, where $m = \text{lcm}(p - 1, q - 1)$, such that r is relatively prime to m (any such r will do). This person calculates $n = pq$ and announces that a message M is to be sent to him or her publicly as M^r mod n. Although r, n, and M^r are available to everyone, only the person who knows how to factor n as pq will be able to decipher the message.

To present a simple example that nevertheless illustrates the principal features of the method, say we wish to send the message "YES." We convert the message into a string of digits by replacing A by 01, B by 02, . . . , Z by 26, and a blank by 00. So, the message YES becomes 250519. To keep the numbers involved from becoming too unwieldy, we send the message in blocks of four digits and fill in with blanks when needed. Thus, the message YES is represented by the two blocks 2505 and 1900. The person to whom the message is to be sent has picked two primes p and q, say $p = 37$ and $q = 73$ (in actual practice, p and q would have 100 or so digits), and a number r that has no prime divisors in common with $\text{lcm}(p - 1, q - 1) = 72$, say $r = 5$, and has published $n = 37 \cdot 73 = 2701$ and $r = 5$ in a public directory. We will send the "scrambled" numbers $(2505)^5$ mod 2701 and $(1900)^5$ mod 2701 rather than 2505 and 1900, and the receiver will unscramble them. We show the work involved for us and the receiver only for the block 2505. The arithmetic involved in computing these numbers is simplified as follows:

$$2505 \bmod 2701 = 2505$$

$$(2505)^2 \bmod 2701 = 602$$

$$(2505)^4 \bmod 2701 = (602)(602) \bmod 2701 = 470.$$

So, $(2505)^5 \bmod 2701 = (2505)(470) \bmod 2701 = 2415$.

Thus, the number 2415 is sent to the receiver. Now the receiver must take the number he or she receives, 2415, and convert it back to 2505. To do so, the receiver takes the two factors of 2701, $p = 37$ and $q = 73$, and calculates the least common multiple of $p - 1 = 36$ and $q - 1 = 72$, which is 72. (This is where the knowledge of p and q is necessary.) Next, the receiver must find $s = r^{-1}$ in $U(72)$—that is, solve the equation $5 \cdot s = 1 \bmod 72$. This number is 29. (There is a simple algorithm for finding this number.) Then the receiver takes the number received, 2415, and calculates $(2415)^{29} \bmod 2701$. This calculation can be simplified as follows:

$$2415 \bmod 2701 = 2415$$
$$(2415)^2 \bmod 2701 = 766$$
$$(2415)^4 \bmod 2701 = (766)^2 \bmod 2701 = 639$$
$$(2415)^8 \bmod 2701 = (639)^2 \bmod 2701 = 470$$
$$(2415)^{16} \bmod 2701 = (470)^2 \bmod 2701 = 2119$$

So, $(2415)^{29} \bmod 2701 = (2415)^{16}(2415)^8(2415)^4(2415) \bmod 2701 = (2119)(470)(639)(2415) \bmod 2701 = (2119)(470) \bmod 2701 \times (639)(2415) \bmod 2701 = (1962)(914) \bmod 2701 = 2505$. [We compute the product $(2119)(470)(639)(2415)$ in two stages so that we may use a hand calculator.]

Thus the receiver correctly determines the code for "YE." On the other hand, without knowing how pq factors, one cannot find the modulus (in our case, 72) that is needed to determine the intended message.

The procedure just described is called the *RSA public key encryption scheme* in honor of the three people (Rivest, Shamir, and Adleman) who discovered the method. It is widely used in conjunction with web servers and browsers, e-mail programs, remote login sessions, and electronic financial transactions. The algorithm is summarized below.

Receiver

1. Pick very large primes p and q and compute $n = pq$.
2. Compute the least common multiple of $p - 1$ and $q - 1$; let us call it m.
3. Pick r relatively prime to m.
4. Find s such that $rs \bmod m = 1$.
5. Publicly announce n and r.

Sender

1. Convert the message to a string of digits. (In practice, the ASCII code is used.)
2. Break up the message into uniform blocks of digits; call them M_1, $M_2, \ldots, M_k$.
3. Check to see that the greatest common divisor of each M_i and n is 1. If not, n can be factored and our code is broken. (In practice, the primes p and q are so large that they exceed all M_i, so this step may be omitted.)
4. Calculate and send $R_i = M_i{}^r \bmod n$.

Receiver

1. For each received message R_i, calculate $R_i{}^s \bmod n$.
2. Convert the string of digits back to a string of characters.

Why does this method work? Well, we know that $U(n) \approx U(p) \oplus U(q) \approx Z_{p-1} \oplus Z_{q-1}$. Thus an element of the form x^m in $U(n)$ corresponds under an isomorphism to one of the form (mx_1, mx_2) in $Z_{p-1} \oplus Z_{q-1}$. Since m is the least common multiple of $p - 1$ and $q - 1$, we may write $m = u(p - 1)$ and $m = v(q - 1)$ for some u and v. Then $(mx_1, mx_2) = (u(p - 1)x_1, v(q - 1)x_2) = (0, 0)$ in $Z_{p-1} \oplus Z_{q-1}$, and it follows that $x^m = 1$ for all x in $U(n)$. So, because each message M_i is an element of $U(n)$ and r was chosen so that $rs = 1 + tm$ for some t, we have, modulo n,

$$R_i{}^s = (M_i{}^r)^s = M_i{}^{rs} = M_i{}^{1 + tm} = (M_i{}^m)^t M_i = 1^t M_i = M_i.$$

In 2002, Ronald Rivest, Adi Shamir, and Leonard Adleman received the Association for Computing Machinery A.M. Turing Award which is considered the "Nobel Prize of Computing" for their contribution to public key cryptography.

The software for Computer Exercise 5 in this chapter implements the RSA scheme for small primes.

Digital Signatures

With so many financial transactions now taking place electronically, the problem of authenticity is paramount. How is a stockbroker to know that an electronic message she receives that tells her to sell one stock and buy another actually came from her client? The technique used in public key cryptography allows for digital signatures as well. Let us say that person A wants to send a secret message to person B in such a way that only B can decode the message and B will know that only A could have sent it. Abstractly, let E_A and D_A denote the algorithms that A uses for encryption and decryption, respectively, and let E_B and D_B denote the algorithms that B uses for encryption and decryption, respectively. So, $D_B E_B$ and $E_A D_A$ applied to any message leave the message unchanged. Here we assume that E_A and E_B are available to the public, whereas D_A is known only to A and D_B is known only to B. Then A sends a message M to B as $E_B(D_A(M))$ and B decodes the received message by applying the function $E_A D_B$ to it to obtain

$$(E_A D_B)\,(E_B(D_A(M)) = E_A(D_B E_B)(D_A(M)) = E_A(D_A(M)) = M.$$

Notice that only A can execute the first step [i.e., create $D_A(M)$] and only B can implement the last step (i.e., apply $E_A D_B$ to the received message).

Transactions using digital signatures became legally binding in the United States in October 2000.

Application to Genetics[†]

The genetic code can be conveniently modeled using elements of $Z_4 \oplus Z_4 \oplus \cdots \oplus Z_4$ where we omit the parentheses and the commas and just use strings of 0s, 1s, 2s, and 3s and add componentwise modulo 4. A DNA molecule is composed of two long strands in the form of a double helix. Each strand is made up of strings of the four nitrogen bases adenine (A), thymine (T), guanine (G), and cytosine (C). Each base on one strand binds to a complementary base on the other strand. Adenine always is bound to thymine, and guanine always is bound to cytosine. To model this process, we identify A with 0, T with 2, G with 1, and C with 3. Thus, the DNA segment ACG-TAACAGGA and its complement segment TGCATTGTCCT are denoted by 03120030110 and 21302212332. Noting that in Z_4, $0+2 = 2$, $2+2 = 0$, $1+2 = 3$, and $3+2 = 1$, we see that adding 2 to elements of Z_4 interchanges 0 and 2 and 1 and 3. So, for any DNA segment $a_1 a_2 \cdots a_n$ represented by elements of $Z_4 \oplus Z_4 \oplus \cdots \oplus Z_4$, we see that its complementary segment is represented by $a_1 a_2 \cdots a_n + 22 \cdots 2$.

Application to Electric Circuits

Many homes have light fixtures that are operated by a pair of switches. They are wired so that when either switch is thrown the light

---

[†]This discussion is adapted from [3].

changes its status (from on to off or vice versa). Suppose the wiring is done so that the light is on when both switches are in the up position. We can conveniently think of the states of the two switches as being matched with the elements of $Z_2 \oplus Z_2$ with the two switches in the up position corresponding to $(0, 0)$ and the two switches in the down position corresponding to $(1, 1)$. Each time a switch is thrown, we add 1 to the corresponding component in the group $Z_2 \oplus Z_2$. We then see that the lights are on when the switches correspond to the elements of the subgroup $\langle(1, 1)\rangle$ and are off when the switches correspond to the elements in the coset $(1, 0) + \langle(1, 1)\rangle$. A similar analysis applies in the case of three switches with the subgroup $\{(0, 0, 0), (1, 1, 0), (0, 1, 1), (1, 0, 1)\}$ corresponding to the lights-on situation.

EXERCISES ■■

Happy is the man who can understand why things happen.

VIRGIL

1. Prove that the external direct product of any finite number of groups is a group. (This exercise is referred to in this chapter.)
2. Show that $Z_2 \oplus Z_2 \oplus Z_2$ has seven subgroups of order 2.
3. Let G be a group with identity e_G and let H be a group with identity e_H. Prove that G is isomorphic to $G \oplus \{e_H\}$ and that H is isomorphic to $\{e_G\} \oplus H$.
4. Show that $G \oplus H$ is Abelian if and only if G and H are Abelian.
5. Prove or disprove that $Z \oplus Z$ is a cyclic group.
6. Prove, by comparing orders of elements, that $Z_8 \oplus Z_2$ is not isomorphic to $Z_4 \oplus Z_4$.
7. Prove that $G_1 \oplus G_2$ is isomorphic to $G_2 \oplus G_1$.
8. Is $Z_3 \oplus Z_9$ isomorphic to Z_{27}? Why?
9. Is $Z_3 \oplus Z_5$ isomorphic to Z_{15}? Why?
10. How many elements of order 9 does $Z_3 \oplus Z_9$ have? (Do not do this exercise by brute force.)
11. How many elements of order 4 does $Z_4 \oplus Z_4$ have? (Do not do this by examining each element.) Explain why $Z_4 \oplus Z_4$ has the same number of elements of order 4 as does $Z_{8000000} \oplus Z_{400000}$. Generalize to the case $Z_m \oplus Z_n$.
12. The dihedral group D_n of order $2n$ $(n \geq 3)$ has a subgroup of n rotations and a subgroup of order 2. Explain why D_n cannot be isomorphic to the external direct product of two such groups.

13. Prove that the group of complex numbers under addition is isomorphic to $\mathbf{R} \oplus \mathbf{R}$.

14. Suppose that $G_1 \approx G_2$ and $H_1 \approx H_2$. Prove that $G_1 \oplus H_1 \approx G_2 \oplus H_2$.

15. If $G \oplus H$ is cyclic, prove that G and H are cyclic.

16. Determine the number of elements of order 15 and the number of cyclic subgroups of order 15 in $Z_{30} \oplus Z_{20}$.

17. What is the order of any nonidentity element of $Z_3 \oplus Z_3 \oplus Z_3$?

18. Let $m > 2$ be an even integer and let $n > 2$ be an odd integer. Find a formula for the number of elements of order 2 in $D_m \oplus D_n$.

19. Let M be the group of all real 2×2 matrices under addition. Let $N = \mathbf{R} \oplus \mathbf{R} \oplus \mathbf{R} \oplus \mathbf{R}$ under componentwise addition. Prove that M and N are isomorphic. What is the corresponding theorem for the group of $m \times n$ matrices under addition?

20. The group $S_3 \oplus Z_2$ is isomorphic to one of the following groups: Z_{12}, $Z_6 \oplus Z_2$, A_4, D_6. Determine which one by elimination.

21. Let G be a group, and let $H = \{(g, g) \mid g \in G\}$. Show that H is a subgroup of $G \oplus G$. (This subgroup is called the *diagonal* of $G \oplus G$.) When G is the set of real numbers under addition, describe $G \oplus G$ and H geometrically.

22. Find a subgroup of $Z_4 \oplus Z_2$ that is not of the form $H \oplus K$, where H is a subgroup of Z_4 and K is a subgroup of Z_2.

23. Find all subgroups of order 3 in $Z_9 \oplus Z_3$.

24. Find all subgroups of order 4 in $Z_4 \oplus Z_4$.

25. What is the largest order of any element in $Z_{30} \oplus Z_{20}$?

26. How many elements of order 2 are in $Z_{2000000} \oplus Z_{4000000}$?

27. Find a subgroup of $Z_{800} \oplus Z_{200}$ that is isomorphic to $Z_2 \oplus Z_4$.

28. Find a subgroup of $Z_{12} \oplus Z_4 \oplus Z_{15}$ that is of order 9.

29. Prove that $\mathbf{R}^* \oplus \mathbf{R}^*$ is not isomorphic to $\mathbf{C}^*$. (Compare this with Exercise 13.)

30. Let

$$H = \left\{ \begin{bmatrix} 1 & a & b \\ 0 & 1 & 0 \\ 0 & 0 & 1 \end{bmatrix} \middle| a, b \in Z_3 \right\}.$$

(See Exercise 34 in Chapter 2 for the definition of multiplication.) Show that H is an Abelian group of order 9. Is H isomorphic to Z_9 or to $Z_3 \oplus Z_3$?

31. Let $G = \{3^m 6^n \mid m, n \in Z\}$ under multiplication. Prove that G is isomorphic to $Z \oplus Z$.

32. Let $(a_1, a_2, \ldots, a_n) \in G_1 \oplus G_2 \oplus \cdots \oplus G_n$. Give a necessary and sufficient condition for $|(a_1, a_2, \ldots, a_n)| = \infty$.

33. Prove that $D_3 \oplus D_4 \neq D_{24}$.

34. Determine the number of cyclic subgroups of order 15 in $Z_{90} \oplus Z_{36}$.

35. If a group has exactly 24 elements of order 6, how many cyclic subgroups of order 6 does it have?

36. For any Abelian group G and any positive integer n, let $G^n = \{g^n \mid g \in G\}$ (see Exercise 15, Supplementary Exercises for Chapters 1–4). If H and K are Abelian, show that $(H \oplus K)^n = H^n \oplus K^n$.

37. Express Aut$(U(25))$ in the form $Z_m \oplus Z_n$.

38. Determine Aut$(Z_2 \oplus Z_2)$.

39. Suppose that $n_1, n_2, \ldots, n_k$ are positive even integers. How many elements of order 2 does $Z_{n_1} \oplus Z_{n_2} \oplus \cdots \oplus Z_{n_k}$ have ? How many are there if we drop the requirement that $n_1, n_2, \ldots, n_k$ must be even?

40. Is $Z_{10} \oplus Z_{12} \oplus Z_6 \approx Z_{60} \oplus Z_6 \oplus Z_2$?

41. Is $Z_{10} \oplus Z_{12} \oplus Z_6 \approx Z_{15} \oplus Z_4 \oplus Z_{12}$?

42. Find an isomorphism from Z_{12} to $Z_4 \oplus Z_3$.

43. How many isomorphisms are there from Z_{12} to $Z_4 \oplus Z_3$?

44. Suppose that ϕ is an isomorphism from $Z_3 \oplus Z_5$ to Z_{15} and $\phi(2, 3) = 2$. Find the element in $Z_3 \oplus Z_5$ that maps to 1.

45. Let (a, b) belong to $Z_m \oplus Z_n$. Prove that $|(a, b)|$ divides lcm(m, n).

46. Let $G = \{ax^2 + bx + c \mid a, b, c \in Z_3\}$. Add elements of G as you would polynomials with integer coefficients, except use modulo 3 addition. Prove that G is isomorphic to $Z_3 \oplus Z_3 \oplus Z_3$. Generalize.

47. Determine all cyclic groups that have exactly two generators.

48. Explain a way that a string of length n of the four nitrogen bases A, T, G, and C could be modeled with the external direct product of n copies of $Z_2 \oplus Z_2$.

49. Let p be a prime. Prove that $Z_p \oplus Z_p$ has exactly $p + 1$ subgroups of order p.

50. Express $U(165)$ as an external direct product of cyclic groups of the form Z_n.

51. Express $U(165)$ as an external direct product of U-groups in four different ways.

52. Without doing any calculations in Aut(Z_{20}), determine how many elements of Aut(Z_{20}) have order 4. How many have order 2?

53. Without doing any calculations in Aut(Z_{720}), determine how many elements of Aut(Z_{720}) have order 6.

54. Without doing any calculations in $U(27)$, decide how many subgroups $U(27)$ has.

55. What is the largest order of any element in $U(900)$?

56. Let p and q be odd primes and let m and n be positive integers. Explain why $U(p^m) \oplus U(q^n)$ is not cyclic.

57. Use the results presented in this chapter to prove that $U(55)$ is isomorphic to $U(75)$.

58. Use the results presented in this chapter to prove that $U(144)$ is isomorphic to $U(140)$.

59. For every $n > 2$, prove that $U(n)^2 = \{x^2 \mid x \in U(n)\}$ is a proper subgroup of $U(n)$.

60. Show that $U(55)^3 = \{x^3 \mid x \in U(55)\}$ is $U(55)$.

61. Find an integer n such that $U(n)$ contains a subgroup isomorphic to $Z_5 \oplus Z_5$.

62. Find a subgroup of order 6 in $U(700)$.

63. Show that there is a U-group containing a subgroup isomorphic to $Z_3 \oplus Z_3$.

64. Show that no U-group has order 14.

65. Show that there is a U-group containing a subgroup isomorphic to Z_{14}.

66. Show that no U-group is isomorphic to $Z_4 \oplus Z_4$.

67. Show that there is a U-group containing a subgroup isomorphic to $Z_4 \oplus Z_4$.

68. Using the RSA scheme with $p = 37$, $q = 73$, and $r = 5$, what number would be sent for the message "RM"?

69. Assuming that a message has been sent via the RSA scheme with $p = 37$, $q = 73$, and $r = 5$, decode the received message "34."

COMPUTER EXERCISES ▪▪

In a few minutes, a computer can make a mistake so great that it would take many men many months to equal it.

MERLE L. MEACHAM

Software for the computer exercises in this chapter is available at the website:

http://www.d.umn.edu/~jgallian

1. This software lists the elements of $U_s(st)$, where s and t are relatively prime. Run the program for $(s, t) = (5, 16), (16, 5), (8, 25), (5, 9), (9, 5), (9, 10), (10, 9)$, and $(10, 25)$.

2. This software computes the elements of the subgroup $U(n)^k = \{x^k \mid x \in U(n)\}$ of $U(n)$ and its order. Run the program for $(n, k) = (27, 3)$, (27,5), (27,7), and (27,11). Do you see a relationship connecting $|U(n)|$ and $|U(n)^k|$, $\phi(n)$, and k? Make a conjecture. Run the program for $(n, k) = (25, 3), (25, 5), (25, 7)$, and (25, 11). Do you see a relationship connecting $|U(n)|$ and $|U(n)^k|$, $\phi(n)$, and k? Make a conjecture. Run the program for $(n, k) = (32, 2), (32, 4)$, and (32, 8). Do you see a relationship connecting $|U(n)|$ and $|U(n)^k|$, $\phi(n)$, and k? Make a conjecture. Is your conjecture valid for (32, 16)? If not, restrict your conjecture. Run the program for $(n, k) = (77, 2), (77, 3), (77, 5), (77, 6), (77, 10)$, and (77, 15)? Do you see a relationship among $U(77, 6), U(77, 2)$, and $U(77, 3)$? What about $U(77, 10)$ $U(77, 2)$, and $U(77, 5)$? What about $U(77, 15), U(77, 3)$, and $U(77, 5)$? Make a conjecture. Use the theory developed in this chapter about expressing $U(n)$ as external direct products of cyclic groups of the form Z_n to analyze these groups to verify your conjectures.

3. This software implements the algorithm given on page 158 to express $U(n)$ as an external direct product of groups of the form Z_k. Run the program for $n = 3 \cdot 5 \cdot 7$, $16 \cdot 9 \cdot 5$, $8 \cdot 3 \cdot 25$, $9 \cdot 5 \cdot 11$, and $2 \cdot 27 \cdot 125$.

4. This software allows you to input positive integers $n_1, n_2, n_3, \ldots, n_k$, where $k \le 5$, and compute the number of elements in $Z_{n_1} \oplus Z_{n_2} \oplus \cdots \oplus Z_{n_k}$ of any specified order m. Use this software to verify the values obtained in Examples 4 and 5 and in Exercise 16. Run the software for $n_1 = 6, n_2 = 10, n_3 = 12$, and $m = 6$.

5. This program implements the RSA public key cryptography scheme. The user enters two primes p and q, an r that is relatively prime to m = least common multiple of $p - 1$ and $q - 1$, and the message M to be sent. Then the program computes s, which is the inverse of r mod m, and the value of M^r mod pq. Also, the user can input those numbers and have the computer raise the numbers to the s power to obtain the original input.

6. This software determines the order of $\text{Aut}(Z_p \oplus Z_p)$, where p is a prime. Run the software for $p = 3, 5$, and 7. Is the result always divisible by p? Is the result always divisible by $p - 1$? Is the result always divisible by $p + 1$? Make a conjecture about the order of $\text{Aut}(Z_p \oplus Z_p)$ for all primes p.

7. This software determines the order of $\text{Aut}(Z_p \oplus Z_p \oplus Z_p)$, where p is a prime. Run the software for $p = 3, 5$, and 7. What is the highest power of p that divides the order? What is the highest power of $p - 1$ that

divides the order? What is the highest power of $p + 1$ that divides the order? Make a conjecture about the order of Aut($Z_p \oplus Z_p \oplus Z_p$) for all primes p.

8. This software determines the order of Aut($Z_p \oplus Z_{p^2}$), where p is a prime. Run the software for $p = 3, 5$, and 7. What is the highest power of p that divides the order? What is the highest power of $p - 1$ that divides the order? Make a conjecture about the order of Aut($Z_p \oplus Z_{p^2}$) for all primes p.

9. This software determines the order of Aut($Z_p \oplus Z_{p^2} \oplus Z_{p^3}$), where p is a prime. Run the software for $p = 3, 5$, and 7. What is the highest power of p that divides the order? What is the highest power of $p - 1$ that divides the order? Make a conjecture about the order of Aut($Z_p \oplus Z_{p^2} \oplus Z_{p^3}$) for all primes p.

10. This software determines the order of Aut($Z_p \oplus Z_{p^2} \oplus Z_{p^3} \oplus Z_{p^4}$), where p is a prime. Run the software for $p = 3, 5$, and 7. What is the highest power of p that divides the order? What is the highest power of $p - 1$ that divides the order? Make a conjecture about the order of Aut($Z_p \oplus Z_{p^2} \oplus Z_{p^3} \oplus Z_{p^4}$) for all primes p.

References

1. J. A. Gallian and D. Rusin, "Factoring Groups of Integers Modulo *n*," *Mathematics Magazine* 53 (1980): 33–36.

2. D. Shanks, *Solved and Unsolved Problems in Number Theory,* 2nd ed., New York: Chelsea, 1978.

3. S. Washburn, T. Marlowe, and C. Ryan, *Discrete Mathematics,* Reading, MA: Addison-Wesley, 1999.

Suggested Readings

Y. Cheng, "Decomposition of *U*-groups," *Mathematics Magazine* 62 (1989): 271–273.

This article explores the decomposition of $U(st)$, where s and t are relatively prime, in greater detail than we have provided.

David J. Devries, "The Group of Units in Z_n," *Mathematics Magazine* 62 (1989): 340.

This article provides a simple proof that $U(n)$ is not cyclic when n is not of the form 1, 2, 4, p^k, or $2p^k$, where p is an odd prime.

David R. Guichard, "When is $U(n)$ Cyclic? An Algebra Approach," *Mathematics Magazine* 72 (1999): 139–142.

> The author provides a group-theoretic proof of the fact that $U(n)$ is cyclic if and only if n is 1, 2, 4, p^k, or $2p^k$, where p is an odd prime.

Leonard Adleman

". . . Dr. Adleman [has played] a central role in some of the most surprising, and provocative, discoveries in theoretical computer science."

GINA KOLATA, *The New York Times*, 13 December 1994.

LEONARD ADLEMAN grew up in San Francisco. He did not have any great ambitions for himself and, in fact, never even thought about becoming a mathematician. He enrolled at the University of California at Berkeley intending to be a chemist, then changed his mind and said he would be a doctor. Finally, he settled on a mathematics major. "I had gone through a zillion things and finally the only thing that was left where I could get out in a reasonable time was mathematics," he said.

Adleman graduated in five years, in 1968, "wondering what I wanted to do with my life." He took a job as a computer programmer at the Bank of America. Then he decided that maybe he should be a physicist, so he began taking classes at San Francisco State College while working at the bank. Once again, Adleman lost interest. "I didn't like doing experiments, I liked thinking about things," he said. Later, he returned to Berkeley with the aim of getting a Ph.D. in computer science. "I thought that getting a Ph.D. in computer science would at least further my career," he said.

But, while in graduate school, something else happened to Adleman. He finally understood the true nature and compelling beauty of mathematics. He discovered, he said, that mathematics "is less related to accounting than it is to philosophy."

"People think of mathematics as some kind of practical art," Adleman said. But, he added, "the point when you become a mathematician is where you somehow see through this and see the beauty and power of mathematics." Adleman got his Ph.D. in 1976 and immediately landed a job as an assistant professor of mathematics at the Massachusetts Institute of Technology. There he met Ronald Rivest and Adi Shamir, who were trying to invent an unbreakable public key system. They shared their excitement about the idea with Adleman, who greeted it with a polite yawn, thinking it impractical and not very interesting. Nevertheless, Adleman agreed to try to break the codes Rivest and Shamir proposed. Rivest and Shamir invented 42 coding systems, and each time Adleman broke the code. Finally, on their 43rd attempt, they hit upon what is now called the RSA scheme.

Adleman's mode of working is to find something that intrigues him and to dig in. He does not read mathematics journals, he says, because he does not want to be influenced by other people's ideas.

Asked what it is like to simply sit and think for six months, Adleman responded, "That's what a mathematician always does. Mathematicians are trained and inclined to sit and think. A mathematician can sit and think intensely about a problem for 12 hours a day, six months straight, with perhaps just a pencil and paper." The only prop he needs, he said, is a blackboard to stare at.

Adapted from an article by Gina Kolata, *The New York Times,* 13 December 1994.

SUPPLEMENTARY EXERCISES FOR CHAPTERS 5–8 ▬ ▪▪

My mind rebels at stagnation. Give me problems, give me work, give me the most obtruse cryptogram, or the most intricate analysis, and I am in my own proper atmosphere.

SHERLOCK HOLMES, The Sign of Four

True/False questions for Chapters 5–8 are available on the web at:

www.d.umn.edu/~jgallian/TF

1. A subgroup N of a group G is called a *characteristic subgroup* if $\phi(N) = N$ for all automorphisms ϕ of G. (The term *characteristic* was first applied by G. Frobenius in 1895.) Prove that every subgroup of a cyclic group is characteristic.

2. Prove that the center of a group is characteristic.

3. The *commutator subgroup* G' of a group G is the subgroup generated by the set $\{x^{-1}y^{-1}xy \mid x, y \in G\}$. (That is, every element of G' has the form $a_1^{i_1}a_2^{i_2} \cdots a_k^{i_k}$, where each a_j has the form $x^{-1}y^{-1}xy$, each $i_j = \pm 1$, and k is any positive integer.) Prove that G' is a characteristic subgroup of G. (This subgroup was first introduced by G. A. Miller in 1898.)

4. Prove that the property of being a characteristic subgroup is transitive. That is, if N is a characteristic subgroup of K and K is a characteristic subgroup of G, then N is a characteristic subgroup of G.

5. Let $G = Z_3 \oplus Z_3 \oplus Z_3$ and let H be the subgroup of $SL(3, Z_3)$, consisting of

$$H = \left\{ \begin{bmatrix} 1 & a & b \\ 0 & 1 & c \\ 0 & 0 & 1 \end{bmatrix} \;\middle|\; a, b, c \in Z_3 \right\}.$$

(See Exercise 34 in Chapter 2 for the definition of multiplication.) Determine the number of elements of each order in G and H. Are G and H isomorphic? (This exercise shows that two groups with the same number of elements of each order need not be isomorphic.)

6. Let H and K be subgroups of a group G and let $HK = \{hk \mid h \in H, k \in K\}$ and $KH = \{kh \mid k \in K, h \in H\}$. Prove that HK is a group if and only if $HK = KH$.

7. Let H and K be subgroups of a finite group G. Prove that

$$|HK| = \frac{|H|\,|K|}{|H \cap K|}.$$

(This exercise is referred to in Chapters 10, 11, and 24.)

8. The *exponent* of a group is the smallest positive integer n such that $x^n = e$ for all x in the group. Prove that every finite group has an exponent that divides the order of the group.

9. Determine all U-groups of exponent 2.

10. Can a group have more subgroups than it has elements?

11. Let $\mathbf{R}^+$ denote the multiplicative group of positive real numbers and let $T = \{a + bi \in \mathbf{C}^* | a^2 + b^2 = 1\}$ be the multiplicative group of complex numbers of norm 1. Show that every element of $\mathbf{C}^*$ can be uniquely expressed in the form rz, where $r \in \mathbf{R}^+$ and $z \in T$.

12. Prove that Q^* under multiplication is not isomorphic to $\mathbf{R}^*$ under multiplication.

13. Prove that Q under addition is not isomorphic to $\mathbf{R}$ under addition.

14. Prove that $\mathbf{R}$ under addition is not isomorphic to $\mathbf{R}^*$ under multiplication.

15. Show that Q^+ (the set of positive rational numbers) under multiplication is not isomorphic to Q under addition.

16. Suppose that $G = \{e, x, x^2, y, yx, yx^2\}$ is a non-Abelian group with $|x| = 3$ and $|y| = 2$. Show that $xy = yx^2$.

17. Let p be an odd prime. Show that 1 is the only solution of $x^{p-2} = 1$ in $U(p)$.

18. Let G be an Abelian group under addition. Let n be a fixed positive integer and let $H = \{(g, ng) \mid g \in G\}$. Show that H is a subgroup of $G \oplus G$. When G is the set of real numbers under addition, describe H geometrically.

19. Find a subgroup of $Z_{12} \oplus Z_{20}$ isomorphic to $Z_4 \oplus Z_5$.

20. Suppose that $G = G_1 \oplus G_2 \oplus \cdots \oplus G_n$. Prove that $Z(G) = Z(G_1) \oplus Z(G_2) \oplus \cdots \oplus Z(G_n)$.

21. Exhibit four nonisomorphic groups of order 18.

22. What is the order of the largest cyclic subgroup in $\text{Aut}(Z_{720})$? (*Hint:* It is not necessary to consider automorphisms of Z_{720}.)

23. Let G be the group of all permutations of the positive integers. Let H be the subset of elements of G that can be expressed as a product of a finite number of cycles. Prove that H is a subgroup of G.

24. Let G be a group and let $g \in G$. Show that $Z(G)\langle g \rangle$ is a subgroup of G.

25. Show that $D_{11} \oplus Z_3 \not\approx D_3 \oplus Z_{11}$. (This exercise is referred to in Chapter 24.)

26. Show that $D_{33} \not\approx D_{11} \oplus Z_3$. (This exercise is referred to in Chapter 24.)

27. Show that $D_{33} \not\approx D_3 \oplus Z_{11}$. (This exercise is referred to in Chapter 24.)

28. Exhibit four nonisomorphic groups of order 66. (This exercise is referred to in Chapter 24.)

29. Prove that $|\text{Inn}(G)| = 1$ if and only if G is Abelian.

30. Prove that $x^{100} = 1$ for all x in $U(1000)$.

31. Find a subgroup of order 6 in $U(450)$.

32. List four elements of $Z_{20} \oplus Z_5 \oplus Z_{60}$ that form a noncyclic subgroup.

33. In S_{10}, let $\beta = (13)(17)(265)(289)$. Find an element in S_{10} that commutes with β but is not a power of β.

34. Prove or disprove that $Z_4 \oplus Z_{15} \approx Z_6 \oplus Z_{10}$.

35. Prove or disprove that $D_{12} \approx Z_3 \oplus D_4$.

36. Describe a three-dimensional solid whose symmetry group is isomorphic to D_5.

37. Let $G = U(15) \oplus Z_{10} \oplus S_5$. Find the order of $(2, 3, (123)(15))$. Find the inverse of $(2, 3, (123)(15))$.

38. Let $G = Z \oplus Z_{10}$ and let $H = \{g \in G | \ |g| = \infty \text{ or } |g| = 1\}$. Prove or disprove that H is a subgroup of G.

39. Give five examples of groups of order 8, no two of which are isomorphic.

40. Find a noncyclic subgroup of order 4 in $Z_4 \oplus Z_{10}$.

41. Find an element of order 10 in A_9.

42. In the left regular representation for D_4, write $T_{R_{90}}$ and T_H in matrix form and in cycle form.

43. Find an Abelian subgroup of S_5 of order 6.

44. Prove that $S_3 \oplus S_4$ is not isomorphic to a subgroup of S_6.

45. Find a permutation β such that $\beta^2 = (13579)(268)$.

46. In $\mathbf{R} \oplus \mathbf{R}$ under componentwise addition, let $H = \{(x, 3x) \ | \ x \in \mathbf{R}\}$. (Note that H is the subgroup of all points on the line $y = 3x$.) Show that $(2, 5) + H$ is a straight line passing through the point $(2, 5)$ and parallel to the line $y = 3x$.

47. In $\mathbf{R} \oplus \mathbf{R}$, suppose that H is the subgroup of all points lying on a line through the origin. Show that any left coset of H is a line parallel to H.

48. Let G be a group of permutations on the set $\{1, 2, \ldots, n\}$. Recall that $\text{stab}_G(1) = \{\alpha \in G \ | \ \alpha(1) = 1\}$. If γ sends 1 to k, prove that $\gamma \ \text{stab}_G(1) = \{\beta \in G \ | \ \beta(1) = k\}$.

49. Let H be a subgroup of G and let $a, b \in G$. Show that $aH = bH$ if and only if $Ha^{-1} = Hb^{-1}$.

50. Suppose that H and K are subgroups of a group and that $|H|$ and $|K|$ are relatively prime. Show that $H \cap K = \{e\}$.

51. Suppose that n is odd and σ is an n-cycle in S_n. Prove that σ does not commute with any element of order 2.

52. Suppose that n is even and σ is an $(n - 1)$-cycle in S_n. Show that σ docs not commute with any element of order 2.

Normal Subgroups and Factor Groups

It is tribute to the genius of Galois that he recognized that those subgroups for which the left and right cosets coincide are distinguished ones. Very often in mathematics the crucial problem is to recognize and to discover what are the relevant concepts; once this is accomplished the job may be more than half done.

I. N. HERSTEIN, *Topics in Algebra*

NORMAL SUBGROUPS

As we saw in Chapter 7, if G is a group and H is a subgroup of G, it is not always true that $aH = Ha$ for all a in G. There are certain situations where this does hold, however, and these cases turn out to be of critical importance in the theory of groups. It was Galois, about 175 years ago, who first recognized that such subgroups were worthy of special attention.

DEFINITION   Normal Subgroup

A subgroup H of a group G is called a *normal* subgroup of G if $aH = Ha$ for all a in G. We denote this by $H \triangleleft G$.

Many students make the mistake of thinking that "H is normal in G" means $ah = ha$ for $a \in G$ and $h \in H$. This is not what normality of H means; rather, it means that if $a \in G$ and $h \in H$, then there exist elements h' and h'' in H such that $ah = h'a$ and $ha = ah''$. Think of it this way: You can switch the order of a product of an element from the group and an element from the normal subgroup, but you must "fudge" a bit on the element from the normal subgroup by using h' or h'' rather than h. (It is possible that $h' = h$ or $h'' = h$, but we may not assume this.)

There are several equivalent formulations of the definition of normality. We have chosen the one that is the easiest to use in applications. However, to *verify* that a subgroup is normal, it is usually better to use Theorem 9.1, which is a weaker version of property 6 of the lemma in Chapter 7. It allows us to substitute a condition about two subgroups of G for a condition about two cosets of G.

Theorem 9.1 **Normal Subgroup Test**

> *A subgroup H of G is normal in G if and only if $xHx^{-1} \subseteq H$ for all x in G.*

PROOF. If H is normal in G, then for any $x \in G$ and $h \in H$ there is an h' in H such that $xh = h'x$. Thus, $xhx^{-1} = h'$, and therefore $xHx^{-1} \subseteq H$.

Conversely, if $xHx^{-1} \subseteq H$ for all x, then, letting $x = a$, we have $aHa^{-1} \subseteq H$ or $aH \subseteq Ha$. On the other hand, letting $x = a^{-1}$, we have $a^{-1}H(a^{-1})^{-1} = a^{-1}Ha \subseteq H$ or $Ha \subseteq aH$. ∎ ■

EXAMPLE 1 Every subgroup of an Abelian group is normal. (In this case, $ah = ha$ for a in the group and h in the subgroup.) ◆

EXAMPLE 2 The center $Z(G)$ of a group is always normal. [Again, $ah = ha$ for any $a \in G$ and any $h \in Z(G)$.] ◆

EXAMPLE 3 The alternating group A_n of even permutations is a normal subgroup of S_n. [Note, for example, that for $(12) \in S_n$ and $(123) \in A_n$, we have $(12)(123) \neq (123)(12)$ but $(12)(123) = (132)(12)$ and $(132) \in A_n$.] ◆

EXAMPLE 4 The subgroup of rotations in D_n is normal in D_n. (For any rotation r and any reflection f, we have $fr = r^{-1}f$, whereas for any rotations r and r', we have $rr' = r'r$.) ◆

EXAMPLE 5 The group $SL(2, \mathbf{R})$ of 2×2 matrices with determinant 1 is a normal subgroup of $GL(2, \mathbf{R})$, the group of 2×2 matrices with nonzero determinant. To verify this, we use the normal subgroup test given in Theorem 9.1. Let $x \in GL(2, \mathbf{R}) = G$, $h \in SL(2, \mathbf{R}) = H$ and note that $\det xhx^{-1} = (\det x)(\det h)(\det x)^{-1} = (\det x)(\det x)^{-1} = 1$. So, $xhx^{-1} \in H$, and, therefore, $xHx^{-1} \subseteq H$. ◆

EXAMPLE 6 Referring to the group table for A_4 given in Table 5.1 on page 105, we may observe that $H = \{\alpha_1, \alpha_2, \alpha_3, \alpha_4\}$ is a normal subgroup of A_4, whereas $K = \{\alpha_1, \alpha_5, \alpha_9\}$ is *not* a normal subgroup of A_4. To see that H is normal, simply note that for any β in A_4, $\beta H \beta^{-1}$ is a subgroup of order 4 and H is the only subgroup of A_4 of order 4 (see

Table 5.1). Thus, $\beta H \beta^{-1} = H$. In contrast, $\alpha_2 \alpha_5 \alpha_2^{-1} = \alpha_7$, so that $\alpha_2 K \alpha_2^{-1} \not\subseteq K$. ◆

FACTOR GROUPS

We have yet to explain why normal subgroups are of special significance. The reason is simple. When the subgroup H of G is normal, then the set of left (or right) cosets of H in G is itself a group—called the *factor group of G by H* (or the *quotient group of G by H*). Quite often, one can obtain information about a group by studying one of its factor groups. This method will be illustrated in the next section of this chapter.

Theorem 9.2   Factor Groups (O. Hölder, 1889)

> *Let G be a group and let H be a normal subgroup of G. The set $G/H = \{aH \mid a \in G\}$ is a group under the operation $(aH)(bH) = abH$.*[†]

PROOF.   Our first task is to show that the operation is well defined; that is, we must show that the correspondence defined above from $G/H \times G/H$ into G/H is actually a function. To do this, assume that $aH = a'H$ and $bH = b'H$. Then $a' = ah_1$ and $b' = bh_2$ for some h_1, h_2 in H, and therefore $a'b'H = ah_1bh_2H = ah_1bH = ah_1Hb = aHb = abH$. Here we have made multiple use of associativity, property 2 of the lemma in Chapter 7, and the fact that $H \triangleleft G$. The rest is easy: $eH = H$ is the identity; $a^{-1}H$ is the inverse of aH; and $(aHbH)cH = (ab)HcH = (ab)cH = a(bc)H = aH(bc)H = aH(bHcH)$. This proves that G/H is a group.   ■ ■

Although it is merely a curiosity, we point out that the converse of Theorem 9.2 is also true; that is, if the correspondence $aHbH = abH$ defines a group operation on the set of left cosets of H in G, then H is normal in G.

The next few examples illustrate the factor group concept.

EXAMPLE 7   Let $4Z = \{0, \pm 4, \pm 8, \ldots\}$. To construct $Z/4Z$, we first must determine the left cosets of $4Z$ in Z. Consider the following four cosets:

$$0 + 4Z = 4Z = \{0, \pm 4, \pm 8, \ldots\},$$
$$1 + 4Z = \{1, 5, 9, \ldots; -3, -7, -11, \ldots\},$$

[†]The notation G/H was first used by C. Jordan.

$$2 + 4Z = \{2, 6, 10, \ldots; -2, -6, -10, \ldots\},$$
$$3 + 4Z = \{3, 7, 11, \ldots; -1, -5, -9, \ldots\}.$$

We claim that there are no others. For if $k \in Z$, then $k = 4q + r$, where $0 \le r < 4$; and, therefore, $k + 4Z = r + 4q + 4Z = r + 4Z$. Now that we know the elements of the factor group, our next job is to determine the structure of $Z/4Z$. Its Cayley table is

	$0 + 4Z$	$1 + 4Z$	$2 + 4Z$	$3 + 4Z$
$0 + 4Z$	$0 + 4Z$	$1 + 4Z$	$2 + 4Z$	$3 + 4Z$
$1 + 4Z$	$1 + 4Z$	$2 + 4Z$	$3 + 4Z$	$0 + 4Z$
$2 + 4Z$	$2 + 4Z$	$3 + 4Z$	$0 + 4Z$	$1 + 4Z$
$3 + 4Z$	$3 + 4Z$	$0 + 4Z$	$1 + 4Z$	$2 + 4Z$

Clearly, then, $Z/4Z \approx Z_4$. More generally, if for any $n > 0$ we let $nZ = \{0, \pm n, \pm 2n, \pm 3n, \ldots\}$, then Z/nZ is isomorphic to Z_n. ◆

EXAMPLE 8 Let $G = Z_{18}$ and let $H = \langle 6 \rangle = \{0, 6, 12\}$. Then $G/H = \{0 + H, 1 + H, 2 + H, 3 + H, 4 + H, 5 + H\}$. To illustrate how the group elements are combined, consider $(5 + H) + (4 + H)$. This should be one of the six elements listed in the set G/H. Well, $(5 + H) + (4 + H) = 5 + 4 + H = 9 + H = 3 + 6 + H = 3 + H$, since H absorbs all multiples of 6. ◆

A few words of caution about notation are warranted here. When H is a normal subgroup of G, the expression $|aH|$ has two possible interpretations. One could be thinking of aH as a *set* of elements and $|aH|$ as the size of the set; or, as is more often the case, one could be thinking of aH as a group element of the factor group G/H and $|aH|$ as the order of the *element aH* in G/H. In Example 8, for instance, the *set* $3 + H$ has size 3, since $3 + H = \{3, 9, 15\}$. But the *group element* $3 + H$ has order 2, since $(3 + H) + (3 + H) = 6 + H = 0 + H$. As is usually the case when one notation has more than one meaning, the appropriate interpretation will be clear from the context.

EXAMPLE 9 Let $\mathcal{H} = \{R_0, R_{180}\}$, and consider the factor group of the dihedral group D_4 (see page 33 for the multiplication table for D_4 by $\mathcal{H}$)

$$D_4/\mathcal{H} = \{\mathcal{H}, R_{90}\mathcal{H}, H\mathcal{H}, D\mathcal{H}\}.$$

The multiplication table for $D_4/\mathcal{H}$ is given in Table 9.1. (Notice that even though $R_{90}H = D'$, we have used $D\mathcal{H}$ in Table 9.1 for $H\mathcal{H}R_{90}\mathcal{H}$ because $D'\mathcal{H} = D\mathcal{H}$.)

TABLE 9.1

	$\mathcal{H}$	$R_{90}\mathcal{H}$	$H\mathcal{H}$	$D\mathcal{H}$
$\mathcal{H}$	$\mathcal{H}$	$R_{90}\mathcal{H}$	$H\mathcal{H}$	$D\mathcal{H}$
$R_{90}\mathcal{H}$	$R_{90}\mathcal{H}$	$\mathcal{H}$	$D\mathcal{H}$	$H\mathcal{H}$
$H\mathcal{H}$	$H\mathcal{H}$	$D\mathcal{H}$	$\mathcal{H}$	$R_{90}\mathcal{H}$
$D\mathcal{H}$	$D\mathcal{H}$	$H\mathcal{H}$	$R_{90}\mathcal{H}$	$\mathcal{H}$

$D_4/\mathcal{H}$ provides a good opportunity to demonstrate how a factor group of G is related to G itself. Suppose we arrange the heading of the Cayley table for D_4 in such a way that elements from the same coset of $\mathcal{H}$ are in adjacent columns (Table 9.2). Then, the multiplication table for D_4 can be blocked off into boxes that are cosets of $\mathcal{H}$, and the substitution that replaces a box containing the element x with the coset $x\mathcal{H}$ yields the Cayley table for $D_4/\mathcal{H}$.

Thus, when we pass from D_4 to $D_4/\mathcal{H}$, the box

H	V
V	H

in Table 9.2 becomes the element $H\mathcal{H}$ in Table 9.1. Similarly, the box

D	D'
D'	D

becomes the element $D\mathcal{H}$, and so on.   ◆

TABLE 9.2

	R_0	R_{180}	R_{90}	R_{270}	H	V	D	D'
R_0	R_0	R_{180}	R_{90}	R_{270}	H	V	D	D'
R_{180}	R_{180}	R_0	R_{270}	R_{90}	V	H	D'	D
R_{90}	R_{90}	R_{270}	R_{180}	R_0	D'	D	H	V
R_{270}	R_{270}	R_{90}	R_0	R_{180}	D	D'	V	H
H	H	V	D	D'	R_0	R_{180}	R_{90}	R_{270}
V	V	H	D'	D	R_{180}	R_0	R_{270}	R_{90}
D	D	D'	V	H	R_{270}	R_{90}	R_0	R_{180}
D'	D'	D	H	V	R_{90}	R_{270}	R_{180}	R_0

TABLE 9.3

	1	2	3	4	5	6	7	8	9	10	11	12
1	1	2	3	4	5	6	7	8	9	10	11	12
2	2	1	4	3	6	5	8	7	10	9	12	11
3	3	4	1	2	7	8	5	6	11	12	9	10
4	4	3	2	1	8	7	6	5	12	11	10	9
5	5	8	6	7	9	12	10	11	1	4	2	3
6	6	7	5	8	10	11	9	12	2	3	1	4
7	7	6	8	5	11	10	12	9	3	2	4	1
8	8	5	7	6	12	9	11	10	4	1	3	2
9	9	11	12	10	1	3	4	2	5	7	8	6
10	10	12	11	9	2	4	3	1	6	8	7	5
11	11	9	10	12	3	1	2	4	7	5	6	8
12	12	10	9	11	4	2	1	3	8	6	5	7

In this way, one can see that the formation of a factor group G/H causes a systematic collapse of the elements of G. In particular, all the elements in the coset of H containing a collapse to the single group element aH in G/H.

EXAMPLE 10   Consider the group A_4 as represented by Table 5.1 on page 105. (Here i denotes the permutation α_i.) Let $H = \{1, 2, 3, 4\}$. Then the three cosets of H are H, $5H = \{5, 6, 7, 8\}$, and $9H = \{9, 10, 11, 12\}$. (In this case, rearrangement of the headings is unnecessary.) Blocking off the table for A_4 into boxes that are cosets of H and replacing the boxes containing 1, 5, and 9 (see Table 9.3) with the cosets $1H$, $5H$, and $9H$, we obtain the Cayley table for G/H given in Table 9.4.

This procedure can be illustrated more vividly with colors. Let's say we had printed the elements of H in green, the elements of $5H$ in red, and the elements of $9H$ in blue. Then, in Table 9.3, each box would consist of elements of a uniform color. We could then think of

TABLE 9.4

	$1H$	$5H$	$9H$
$1H$	$1H$	$5H$	$9H$
$5H$	$5H$	$9H$	$1H$
$9H$	$9H$	$1H$	$5H$

the factor group as consisting of the three colors that define a group table isomorphic to G/H.

	Green	Red	Blue
Green	Green	Red	Blue
Red	Red	Blue	Green
Blue	Blue	Green	Red

It is instructive to see what happens if we attempt the same procedure with a group G and a subgroup H that is not normal in G—that is, if we arrange the headings of the Cayley table so that the elements from the same coset of H are in adjacent columns and attempt to block off the table into boxes that are also cosets of H to produce a Cayley table for the set of cosets. Say, for instance, we were to take G to be A_4 and $H = \{1, 5, 9\}$. The cosets of H would be H, $2H = \{2, 6, 10\}$, $3H = \{3, 7, 11\}$, and $4H = \{4, 8, 12\}$. Then the first three rows of the rearranged Cayley table for A_4 would be

	1	5	9	2	6	10	3	7	11	4	8	12
1	1	5	9	2	6	10	3	7	11	4	8	12
5	5	9	1	8	12	4	6	10	2	7	11	3
9	9	1	5	11	3	7	12	4	8	10	2	6

But already we are in trouble, for blocking these off into 3×3 boxes yields boxes that contain elements of different cosets. Hence, it is impossible to represent an entire box by a single element of the box in the same way we could for boxes made from the cosets of a normal subgroup. Had we printed the rearranged table in four colors with all members of the same coset having the same color, we would see multicolored boxes rather than the uniformly colored boxes produced by a normal subgroup. ◆

In Chapter 11, we will prove that every finite Abelian group is isomorphic to a direct product of cyclic groups. In particular, an Abelian group of order 8 is isomorphic to one of Z_8, $Z_4 \oplus Z_2$, or $Z_2 \oplus Z_2 \oplus Z_2$. In the next two examples, we examine Abelian factor groups of order 8 and determine the isomorphism type of each.

EXAMPLE 11   Let

$$G = U(32) = \{1, 3, 5, 7, 9, 11, 13, 15, 17, 19, 21, 23, 25, 27, 29, 31\}$$

and $H = U_{16}(32) = \{1, 17\}$. Then G/H is an Abelian group of order $16/2 = 8$. Which of the three Abelian groups of order 8 is it—Z_8,

$Z_4 \oplus Z_2$, or $Z_2 \oplus Z_2 \oplus Z_2$? To answer this question, we need only determine the elements of G/H and their orders. Observe that the eight cosets

$$1H = \{1, 17\}, \quad 3H = \{3, 19\}, \quad 5H = \{5, 21\}, \quad 7H = \{7, 23\},$$

$$9H = \{9, 25\}, \quad 11H = \{11, 27\}, \quad 13H = \{13, 29\}, \quad 15H = \{15, 31\}$$

are all distinct, so that they form the factor group G/H. Clearly, $(3H)^2 = 9H \neq H$, and so $3H$ has order at least 4. Thus, G/H is not $Z_2 \oplus Z_2 \oplus Z_2$. On the other hand, direct computations show that both $7H$ and $9H$ have order 2, so that G/H cannot be Z_8 either, since a cyclic group of even order has exactly one element of order 2 (Theorem 4.4). This proves that $U(32)/U_{16}(32) \approx Z_4 \oplus Z_2$, which (not so incidentally!) is isomorphic to $U(16)$. ◆

EXAMPLE 12 Let $G = U(32)$ and $K = \{1, 15\}$. Then $|G/K| = 8$, and we ask, which of the three Abelian groups of order 8 is G/K? Since $(3K)^4 = 81K = 17K \neq K$, $|3K| = 8$. Thus, $G/K \approx Z_8$. ◆

It is crucial to understand that when we factor out by a normal subgroup H, what we are essentially doing is defining every element in H to be the *identity*. Thus, in Example 9, we are making $R_{180}\mathcal{H} = \mathcal{H}$ the identity. Likewise, $R_{270}\mathcal{H} = R_{90}R_{180}\mathcal{H} = R_{90}\mathcal{H}$. Similarly, in Example 7, we are declaring any multiple of 4 to be 0 in the factor group $Z/4Z$. This is why $5 + 4Z = 1 + 4 + 4Z = 1 + 4Z$, and so on. In Example 11, we have $3H = 19H$, since $19 = 3 \cdot 17$ in $U(32)$ and going to the factor group makes 17 the identity. Algebraists often refer to the process of creating the factor group G/H as "killing" H.

APPLICATIONS OF FACTOR GROUPS

Why are factor groups important? Well, when G is finite and $H \neq \{e\}$, G/H is smaller than G, and its structure is usually less complicated than that of G. At the same time, G/H simulates G in many ways. In fact, we may think of a factor group of G as a less complicated approximation of G (similar to using the rational number 3.14 for the irrational number π). What makes factor groups important is that one can often deduce properties of G by examining the less complicated group G/H instead. We illustrate this by giving another proof that A_4 has no subgroup of order 6.

EXAMPLE 13 A_4 has no subgroup of order 6.
The group A_4 of even permutations on the set $\{1, 2, 3, 4\}$ has no subgroup H of order 6. To see this, suppose that A_4 does have a subgroup

H of order 6. By Exercise 7 in this chapter, we know that $H \triangleleft A_4$. Thus, the factor group A_4/H exists and has order 2. Since the order of an element divides the order of the group, we have for all $\alpha \in A_4$ that $\alpha^2 H = (\alpha H)^2 = H$. Thus, $\alpha^2 \in H$ for all α in A_4. Referring to the main diagonal of the group table for A_4 given in Table 5.1 on page 105, however, we observe that A_4 has nine different elements of the form α^2, all of which must belong to *H*, a subgroup of order 6. This is clearly impossible, so a subgroup of order 6 cannot exist in A_4.[†]   ◆

The next three theorems illustrate how knowledge of a factor group of *G* reveals information about *G* itself.

Theorem 9.3   The G/Z Theorem

> *Let G be a group and let Z(G) be the center of G. If G/Z(G)*
> *is cyclic, then G is Abelian.*

PROOF.   Let $gZ(G)$ be a generator of the factor group $G/Z(G)$, and let $a, b \in G$. Then there exist integers *i* and *j* such that

$$aZ(G) = (gZ(G))^i = g^i Z(G)$$

and

$$bZ(G) = (gZ(G))^j = g^j Z(G).$$

Thus, $a = g^i x$ for some *x* in $Z(G)$ and $b = g^j y$ for some *y* in $Z(G)$. It follows then that

$$ab = (g^i x)(g^j y) = g^i(xg^j)y = g^i(g^j x)y$$

$$= (g^i g^j)(xy) = (g^j g^i)(yx) = (g^j y)(g^i x) = ba.   ∎$$

A few remarks about Theorem 9.3 are in order. First, our proof shows that a better result is possible: If *G/H* is cyclic, where *H* is a subgroup of $Z(G)$, then *G* is Abelian. Second, in practice, it is the contrapositive of the theorem that is most often used—that is, if *G* is non-Abelian, then $G/Z(G)$ is not cyclic. For example, it follows immediately from this statement and Lagrange's Theorem that a non-Abelian group of order *pq*, where *p* and *q* are primes, must have a trivial center. Third, if $G/Z(G)$ is cyclic, it must be trivial.

[†]*How often have I said to you that when you have eliminated the impossible, whatever remains, however improbable, must be the truth.* Sherlock Holmes, *The Sign of Four*

Theorem 9.4 $G/Z(G) \approx \text{Inn}(G)$

> *For any group G, G/Z(G) is isomorphic to Inn(G).*

PROOF.   Consider the correspondence from $G/Z(G)$ to $\text{Inn}(G)$ given by $T: gZ(G) \rightarrow \phi_g$ [where, recall, $\phi_g(x) = gxg^{-1}$ for all x in G]. First, we show that T is a function. Suppose $gZ(G) = hZ(G)$, so that $h^{-1}g \in Z(G)$. Then, for all x in G, $h^{-1}gx = xh^{-1}g$. Thus, $gxg^{-1} = hxh^{-1}$ for all x in G, and, therefore, $\phi_g = \phi_h$. Reversing this argument shows that T is one-to-one, as well. Clearly, T is onto.

That T is operation-preserving follows directly from the fact that $\phi_g \phi_h = \phi_{gh}$ for all g and h in G.   ∎∎

As an application of Theorems 9.3 and 9.4, we may easily determine $\text{Inn}(D_6)$ without looking at $\text{Inn}(D_6)$!

EXAMPLE 14   We know from Example 11 in Chapter 3 that $|Z(D_6)| = 2$. Thus, $|D_6/Z(D_6)| = 6$. So, by our classification of groups of order 6 (Theorem 7.2), we know that $\text{Inn}(D_6)$ is isomorphic to D_3 or Z_6. Now, if $\text{Inn}(D_6)$ were cyclic, then, by Theorem 9.4, $D_6/Z(D_6)$ would be also. But then, Theorem 9.3 would tell us that D_6 is Abelian. So, $\text{Inn}(D_6)$ is isomorphic to D_3.   ◆

The next theorem demonstrates one of the most powerful proof techniques available in the theory of finite groups—the combined use of factor groups and induction.

Theorem 9.5   **Cauchy's Theorem for Abelian Groups**

> *Let G be a finite Abelian group and let p be a prime that divides the order of G. Then G has an element of order p.*

PROOF.   Clearly, this statement is true for the case in which G has order 2. We prove the theorem by using the Second Principle of Mathematical Induction on $|G|$. That is, we assume that the statement is true for all Abelian groups with fewer elements than G and use this assumption to show that the statement is true for G as well. Certainly, G has elements of prime order, for if $|x| = m$ and $m = qn$, where q is prime, then $|x^n| = q$. So let x be an element of G of some prime order q, say. If $q = p$, we are finished; so assume that $q \neq p$. Since every subgroup of an Abelian group is normal, we may construct the factor group $\overline{G} = G/\langle x \rangle$. Then $\overline{G}$ is Abelian and p divides $|\overline{G}|$, since $|\overline{G}| = |G|/q$. By induction, then, $\overline{G}$ has an element—call it $y\langle x \rangle$—of order p. The conclusion now follows from Exercise 61. ◼ ◼

INTERNAL DIRECT PRODUCTS

As we have seen, the external direct product provides a way of putting groups together into a larger group. It would be quite useful to be able to reverse this process—that is, to be able to start with a large group and break it down into a product of smaller groups. It is occasionally possible to do this. To this end, suppose that H and K are subgroups of some group G. We define the set $HK = \{hk \mid h \in H, k \in K\}$.

EXAMPLE 15 In $U(24) = \{1, 5, 7, 11, 13, 17, 19, 23\}$, let $H = \{1, 17\}$ and $K = \{1, 13\}$. Then, $HK = \{1, 13, 17, 5\}$, since $5 = 17 \cdot 13 \bmod 24$. ◆

EXAMPLE 16 In S_3, let $H = \{(1), (12)\}$ and $K = \{(1), (13)\}$. Then, $HK = \{(1), (13), (12), (12)(13)\} = \{(1), (13), (12), (132)\}$. ◆

The student should be careful not to assume that the set HK is a subgroup of G; in Example 15 it is, but in Example 16 it is not.

DEFINITION *Internal Direct Product of H and K*

We say that G is the *internal direct product* of H and K and write $G = H \times K$ if H and K are normal subgroups of G and

$$G = HK \quad \text{and} \quad H \cap K = \{e\}.$$

The wording of the phrase "internal direct product" is easy to justify. We want to call G the internal direct product of H and K if H and K are subgroups of G, and if G is naturally isomorphic to the external direct product of H and K. One forms the internal direct product by *starting* with a group G and then proceeding to find two subgroups H and K within G such that G is *isomorphic* to the external direct product of H and K. (The definition ensures that this is the case—see Theorem 9.6.) On the other hand, one forms an external direct product by *starting* with any two groups H and K, related or not, and proceeding to produce the larger group $H \oplus K$. The difference between the two products is that the internal direct product can be formed within G itself, using subgroups of G and the operation of G, whereas the external direct product can be formed with totally unrelated groups by creating a new set and a new operation. (See Figures 9.1 and 9.2.)

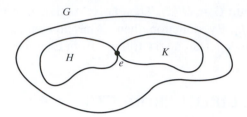

Figure 9.1   For the internal direct product, H and K must be subgroups of the same group.

Perhaps the following analogy with integers will be useful in clarifying the distinction between the two products of groups discussed in the preceding paragraph. Just as we may take any (finite) collection of integers and form their product, we may also take any collection of groups and form their external direct product. Conversely, just as we may start with a particular integer and express it as a product of certain of its divisors, we may be able to start with a particular group and factor it as an internal direct product of certain of its subgroups.

EXAMPLE 17   In D_6, the dihedral group of order 12, let F denote some reflection and let R_k denote a rotation of k degrees. Then,

$$D_6 = \{R_0, R_{120}, R_{240}, F, R_{120}F, R_{240}F\} \times \{R_0, R_{180}\}. \qquad \blacklozenge$$

Students should be cautioned about the necessity of having all conditions of the definition of internal direct product satisfied to ensure that $HK \approx H \oplus K$. For example, if we take

$$G = S_3, \qquad H = \langle(123)\rangle, \qquad \text{and} \qquad K = \langle(12)\rangle,$$

then $G = HK$, and $H \cap K = \{(1)\}$. But G is *not* isomorphic to $H \oplus K$, since, by Theorem 8.2, $H \oplus K$ is cyclic, whereas S_3 is not. Note that K is not normal.

A group G can also be the internal direct product of a collection of subgroups.

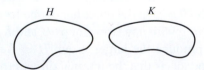

Figure 9.2   For the external direct product, H and K can be any groups.

DEFINITION *Internal Direct Product $H_1 \times H_2 \times \cdots \times H_n$*

> Let $H_1, H_2, \ldots, H_n$ be a finite collection of normal subgroups of G. We say that G is the *internal direct product of $H_1, H_2, \ldots, H_n$* and write $G = H_1 \times H_2 \times \cdots \times H_n$, if
>
> 1. $G = H_1H_2 \cdots H_n = \{h_1h_2 \cdots h_n \mid h_i \in H_i\}$
> 2. $(H_1H_2 \cdots H_i) \cap H_{i+1} = \{e\}$ for $i = 1, 2, \ldots, n - 1$.

This definition is somewhat more complicated than the one given for two subgroups. The student may wonder about the motivation for it—that is, why should we want the subgroups to be normal and why is it desirable for each subgroup to be disjoint from the product of all previous ones? The reason is quite simple. We want the internal direct product to be isomorphic to the external direct product. As the next theorem shows, the conditions in the definition of internal direct product were chosen to ensure that the two products are isomorphic.

Theorem 9.6 $H_1 \times H_2 \times \cdots \times H_n \approx H_1 \oplus H_2 \oplus \cdots \oplus H_n$

> *If a group G is the internal direct product of a finite number of subgroups $H_1, H_2, \ldots, H_n$, then G is isomorphic to the external direct product of $H_1, H_2, \ldots, H_n$.*

PROOF.   We first show that the normality of the H's together with the second condition of the definition guarantees that h's from different H_i's commute. For if $h_i \in H_i$ and $h_j \in H_j$ with $i \neq j$, then

$$(h_i h_j h_i^{-1})h_j^{-1} \in H_j h_j^{-1} = H_j$$

and

$$h_i(h_j h_i^{-1} h_j^{-1}) \in h_i H_i = H_i.$$

Thus, $h_i h_j h_i^{-1} h_j^{-1} \in H_i \cap H_j = \{e\}$ (see Exercise 3), and, therefore, $h_i h_j = h_j h_i$. We next claim that each member of G can be expressed uniquely in the form $h_1 h_2 \cdots h_n$, where $h_i \in H_i$. That there is at least one such representation is the content of condition 1 of the definition. To prove uniqueness, suppose that $g = h_1 h_2 \cdots h_n$ and $g = h_1' h_2' \cdots h_n'$, where h_i and h_i' belong to H_i for $i = 1, \ldots, n$. Then, using the fact that the h's from different H_i's commute, we can solve the equation

$$h_1 h_2 \cdots h_n = h_1' h_2' \cdots h_n' \tag{1}$$

for $h_n' h_n^{-1}$ to obtain

$$h_n' h_n^{-1} = (h_1')^{-1} h_1 (h_2')^{-1} h_2 \cdots (h_{n-1}')^{-1} h_{n-1}.$$

But then

$$h'_n h_n^{-1} \in H_1 H_2 \cdots H_{n-1} \cap H_n = \{e\},$$

so that $h'_n h_n^{-1} = e$ and, therefore, $h'_n = h_n$. At this point, we can cancel h_n and h'_n from opposite sides of the equal sign in Equation (1) and repeat the preceding argument to obtain $h_{n-1} = h'_{n-1}$. Continuing in this fashion, we eventually have $h_i = h'_i$ for $i = 1, \ldots, n$. With our claim established, we may now define a function ϕ from G to $H_1 \oplus H_2 \oplus \cdots \oplus H_n$ by $\phi(h_1 h_2 \cdots h_n) = (h_1, h_2, \ldots, h_n)$. We leave to the reader the easy verification that ϕ is an isomorphism. ■ ■

We mention in passing that if $G = H_1 \oplus H_2 \oplus \cdots \oplus H_n$, then G can be expressed as the internal direct product of subgroups isomorphic to $H_1, H_2, \ldots H_n$. For example, if $G = H_1 \oplus H_2$, then $G = \overline{H_1} \times \overline{H_2}$, where $\overline{H_1} = H_1 \oplus \{e\}$ and $\overline{H_2} = \{e\} \oplus H_2$.

The topic of direct products is one in which notation and terminology vary widely. Many authors use $H \times K$ to denote both the internal direct product and the external direct product of H and K, making no notational distinction between the two products. A few authors define only the external direct product. Many people reserve the notation $H \oplus K$ for the situation where H and K are Abelian groups under addition and call it the *direct sum* of H and K. In fact, we will adopt this terminology in the section on rings (Part 3), since rings are always Abelian groups under addition.

The U-groups provide a convenient way to illustrate the preceding ideas and to clarify the distinction between internal and external direct products. It follows directly from Theorem 8.3 and its corollary and Theorem 9.6 that if $m = n_1 n_2 \cdots n_k$, where $\gcd(n_i, n_j) = 1$ for $i \neq j$, then

$$U(m) = U_{m/n_1}(m) \times U_{m/n_2}(m) \times \cdots \times U_{m/n_k}(m)$$

$$\approx U(n_1) \oplus U(n_2) \oplus \cdots \oplus U(n_k).$$

Let us return to the examples given following Theorem 8.3.

$U(105) = U(15 \cdot 7) = U_{15}(105) \times U_7(105)$
$\qquad = \{1, 16, 31, 46, 61, 76\} \times \{1, 8, 22, 29, 43, 64, 71, 92\}$
$\qquad \approx U(7) \oplus U(15),$

$U(105) = U(5 \cdot 21) = U_5(105) \times U_{21}(105)$
$\qquad = \{1, 11, 16, 26, 31, 41, 46, 61, 71, 76, 86, 101\}$
$\qquad\quad \times \{1, 22, 43, 64\} \approx U(21) \oplus U(5),$

$U(105) = U(3 \cdot 5 \cdot 7) = U_{35}(105) \times U_{21}(105) \times U_{15}(105)$
$\qquad = \{1, 71\} \times \{1, 22, 43, 64\} \times \{1, 16, 31, 46, 61, 76\}$
$\qquad \approx U(3) \oplus U(5) \oplus U(7).$

EXERCISES

If there's no struggle, there's no progress.

FREDERICK DOUGLASS

1. Let $H = \{(1), (12)\}$. Is H normal in S_3?

2. Prove that A_n is normal in S_n.

3. Show that if G is the internal direct product of $H_1, H_2, \ldots, H_n$ and $i \neq j$ with $1 \leq i \leq n$, $1 \leq j \leq n$, then $H_i \cap H_j = \{e\}$. (This exercise is referred to in this chapter.)

4. Let $H = \left\{ \begin{bmatrix} a & b \\ 0 & d \end{bmatrix} \middle| a, b, d \in \mathbf{R}, \ ad \neq 0 \right\}$. Is H a normal subgroup of $GL(2, \mathbf{R})$?

5. Let $G = GL(2, \mathbf{R})$ and let K be a subgroup of $\mathbf{R}^*$. Prove that $H = \{A \in G| \det A \in K\}$ is a normal subgroup of G.

6. Viewing $\langle 3 \rangle$ and $\langle 12 \rangle$ as subgroups of Z, prove that $\langle 3 \rangle / \langle 12 \rangle$ is isomorphic to Z_4. Similarly, prove that $\langle 8 \rangle / \langle 48 \rangle$ is isomorphic to Z_6. Generalize to arbitrary integers k and n.

7. Prove that if H has index 2 in G, then H is normal in G. (This exercise is referred to in Chapters 24 and 25 and this chapter.)

8. Let $H = \{(1), (12)(34)\}$ in A_4.
   a. Show that H is not normal in A_4.
   b. Referring to the multiplication table for A_4 in Table 5.1 on page 105, show that, although $\alpha_6 H = \alpha_7 H$ and $\alpha_9 H = \alpha_{11} H$, it is not true that $\alpha_6 \alpha_9 H = \alpha_7 \alpha_{11} H$. Explain why this proves that the left cosets of H do not form a group under coset multiplication.

9. Let $G = Z_4 \oplus U(4)$, $H = \langle (2, 3) \rangle$, and $K = \langle (2, 1) \rangle$. Show that G/H is not isomorphic to G/K. (This shows that $H \approx K$ does not imply that $G/H \approx G/K$.)

10. Prove that a factor group of a cyclic group is cyclic.

11. Let H be a normal subgroup of G. If H and G/H are Abelian, must G be Abelian?

12. Prove that a factor group of an Abelian group is Abelian.

13. If H is a subgroup of G and $a, b \in G$, prove that $(ab)H = a(bH)$.

14. What is the order of the element $14 + \langle 8 \rangle$ in the factor group $Z_{24}/\langle 8 \rangle$?

15. What is the order of the element $4U_5(105)$ in the factor group $U(105)/U_5(105)$?

16. Recall that $Z(D_6) = \{R_0, R_{180}\}$. What is the order of the element $R_{60} Z(D_6)$ in the factor group $D_6/Z(D_6)$?

17. Let $G = Z/\langle 20 \rangle$ and $H = \langle 4 \rangle/\langle 20 \rangle$. List the elements of H and G/H.

18. What is the order of the factor group $Z_{60}/\langle 15 \rangle$?

19. What is the order of the factor group $(Z_{10} \oplus U(10))/\langle (2, 9) \rangle$?

20. Construct the Cayley table for $U(20)/U_5(20)$.

21. Prove that an Abelian group of order 33 is cyclic.

22. Determine the order of $(Z \oplus Z)/\langle (2, 2) \rangle$. Is the group cyclic?

23. Determine the order of $(Z \oplus Z)/\langle (4, 2) \rangle$. Is the group cyclic?

24. The group $(Z_4 \oplus Z_{12})/\langle (2, 2) \rangle$ is isomorphic to one of Z_8, $Z_4 \oplus Z_2$, or $Z_2 \oplus Z_2 \oplus Z_2$. Determine which one by elimination.

25. Let $G = U(32)$ and $H = \{1, 31\}$. The group G/H is isomorphic to one of Z_8, $Z_4 \oplus Z_2$, or $Z_2 \oplus Z_2 \oplus Z_2$. Determine which one by elimination.

26. Let G be the group of quarternions given by the table in Exercise 4 of the Supplementary Exercises for Chapters 1–4 on page 90, and let H be the subgroup $\{e, a^2\}$. Is G/H isomorphic to Z_4 or $Z_2 \oplus Z_2$?

27. Let $G = U(16)$, $H = \{1, 15\}$, and $K = \{1, 9\}$. Are H and K isomorphic? Are G/H and G/K isomorphic?

28. Let $G = Z_4 \oplus Z_4$, $H = \{(0, 0), (2, 0), (0, 2), (2, 2)\}$, and $K = \langle (1, 2) \rangle$. Is G/H isomorphic to Z_4 or $Z_2 \oplus Z_2$? Is G/K isomorphic to Z_4 or $Z_2 \oplus Z_2$?

29. Let $G = GL(2, \mathbf{R})$ and $H = \{A \in G \mid \det A = 3^k, k \in Z\}$. Prove that H is a normal subgroup of G.

30. Express $U(165)$ as an internal direct product of proper subgroups in four different ways.

31. Let $\mathbf{R}^*$ denote the group of all nonzero real numbers under multiplication. Let $\mathbf{R}^+$ denote the group of positive real numbers under multiplication. Prove that $\mathbf{R}^*$ is the internal direct product of $\mathbf{R}^+$ and the subgroup $\{1, -1\}$.

32. Prove that D_4 cannot be expressed as an internal direct product of two proper subgroups.

33. Let H and K be subgroups of a group G. If $G = HK$ and $g = hk$, where $h \in H$ and $k \in K$, is there any relationship among $|g|$, $|h|$, and $|k|$? What if $G = H \times K$?

34. In Z, let $H = \langle 5 \rangle$ and $K = \langle 7 \rangle$. Prove that $Z = HK$. Does $Z = H \times K$?

35. Let $G = \{3^a 6^b 10^c \mid a, b, c \in Z\}$ under multiplication and $H = \{3^a 6^b 12^c \mid a, b, c \in Z\}$ under multiplication. Prove that $G = \langle 3 \rangle \times \langle 6 \rangle \times \langle 10 \rangle$, whereas $H \neq \langle 3 \rangle \times \langle 6 \rangle \times \langle 12 \rangle$.

36. Determine all subgroups of $\mathbf{R}^*$ (nonzero reals under multiplication) of index 2.

37. Let G be a finite group and let H be a normal subgroup of G. Prove that the order of the element gH in G/H must divide the order of g in G.

38. Let H be a normal subgroup of G and let a belong to G. If the element aH has order 3 in the group G/H and $|H| = 10$, what are the possibilities for the order of a?

39. If H is a normal subgroup of a group G, prove that $C(H)$, the centralizer of H in G, is a normal subgroup of G.

40. An element is called a *square* if it can be expressed in the form b^2 for some b. Suppose that G is an Abelian group and H is a subgroup of G. If every element of H is a square and every element of G/H is a square, prove that every element of G is a square.

41. Show, by example, that in a factor group G/H it can happen that $aH = bH$ but $|a| \neq |b|$.

42. Observe from the table for A_4 given in Table 5.1 on page 105 that the subgroup given in Example 6 of this chapter is the only subgroup of A_4 of order 4. Why does this imply that this subgroup must be normal in A_4? Generalize this to arbitrary finite groups.

43. Suppose that G is a non-Abelian group of order p^3 (where p is a prime) and $Z(G) \neq \{e\}$. Prove that $|Z(G)| = p$.

44. If $|G| = pq$, where p and q are primes that are not necessarily distinct, prove that $|Z(G)| = 1$ or pq.

45. Let N be a normal subgroup of G and let H be a subgroup of G. If N is a subgroup of H, prove that H/N is a normal subgroup of G/N if and only if H is a normal subgroup of G.

46. Let G be an Abelian group and let H be the subgroup consisting of all elements of G that have finite order (See Exercise 18 in the Supplementary Exercises for Chapters 1–4). Prove that every nonidentity element in G/H has infinite order.

47. Determine all subgroups of $\mathbf{R}^*$ that have finite index.

48. Let $G = \{\pm 1, \pm i, \pm j, \pm k\}$, where $i^2 = j^2 = k^2 = -1$, $-i = (-1)i$, $1^2 = (-1)^2 = 1$, $ij = -ji = k$, $jk = -kj = i$, and $ki = -ik = j$.
  a. Construct the Cayley table for G.
  b. Show that $H = \{1, -1\} \triangleleft G$.
  c. Construct the Cayley table for G/H.

(The rules involving i, j, and k can be remembered by using the circle below.

Going clockwise, the product of two consecutive elements is the third one. The same is true for going counterclockwise, except that we obtain the negative of the third element.) This is the group of quaternions that was given in another form in Exercise 4 in the Supplementary Exercises for Chapters 1–4. It was invented by William Hamilton in 1843. The quaternions are used to describe rotations in three-dimensional space, and they are used in physics. The quaternions can be used to extend the complex numbers in a natural way.

49. In D_4, let $K = \{R_0, D\}$ and let $L = \{R_0, D, D', R_{180}\}$. Show that $K \lhd L \lhd D_4$, but that K is not normal in D_4. (Normality is not transitive. Compare Exercise 4, Supplementary Exercises for Chapters 5–8.)

50. Show that the intersection of two normal subgroups of G is a normal subgroup of G.

51. Let N be a normal subgroup of G and let H be any subgroup of G. Prove that NH is a subgroup of G. Give an example to show that NH need not be a subgroup of G if neither N nor H is normal. (This exercise is referred to in Chapter 24.)

52. If N and M are normal subgroups of G, prove that NM is also a normal subgroup of G.

53. Let N be a normal subgroup of a group G. If N is cyclic, prove that every subgroup of N is also normal in G. (This exercise is referred to in Chapter 24.)

54. Without looking at inner automorphisms of D_n, determine the number of such automorphisms.

55. Let H be a normal subgroup of a finite group G and let $x \in G$. If $\gcd(|x|, |G/H|) = 1$, show that $x \in H$. (This exercise is referred to in Chapter 25.)

56. Let G be a group and let G' be the subgroup of G generated by the set $S = \{x^{-1}y^{-1}xy \mid x, y \in G\}$. (See Exercise 3, Supplementary Exercises for Chapters 5–8, for a more complete description of G'.)

  a. Prove that G' is normal in G.

  b. Prove that G/G' is Abelian.

  c. If G/N is Abelian, prove that $G' \leq N$.

  d. Prove that if H is a subgroup of G and $G' \leq H$, then H is normal in G.

57. If N is a normal subgroup of G and $|G/N| = m$, show that $x^m \in N$ for all x in G.

58. Suppose that a group G has a subgroup of order n. Prove that the intersection of all subgroups of G of order n is a normal subgroup of G.

59. If G is non-Abelian, show that $\text{Aut}(G)$ is not cyclic.

60. Let $|G| = p^n m$, where p is prime and $\gcd(p, m) = 1$. Suppose that H is a normal subgroup of G of order p^n. If K is a subgroup of G of order p^k, show that $K \subseteq H$.

61. Suppose that H is a normal subgroup of a finite group G. If G/H has an element of order n, show that G has an element of order n. Show, by example, that the assumption that G is finite is necessary. (This exercise is referred to in this chapter.)

62. Recall that a subgroup N of a group G is called characteristic if $\phi(N) = N$ for all automorphisms ϕ of G. If N is a characteristic subgroup of G, show that N is a normal subgroup of G.

63. In D_4, let $\mathcal{K} = \{R_0, H\}$. Form an operation table for the cosets $\mathcal{K}$, $D\mathcal{K}$, $V\mathcal{K}$, and $D'\mathcal{K}$. Is the result a group table? Does your answer contradict Theorem 9.2?

64. Show that S_4 has a unique subgroup of order 12.

65. If $|G| = 30$ and $|Z(G)| = 5$, what is the structure of $G/Z(G)$?

66. If H is a normal subgroup of G and $|H| = 2$, prove that H is contained in the center of G.

67. Prove that A_5 cannot have a normal subgroup of order 2.

68. Let G be a finite group and let H be an odd-order subgroup of G of index 2. Show that the product of all the elements of G (taken in any order) cannot belong to H.

69. Let G be a group. If $H = \{g^2 \mid g \in G\}$ is a subgroup of G, prove that it is a normal subgroup of G.

70. Suppose that H is a normal subgroup of G. If $|H| = 4$ and gH has order 3 in G/H, find a subgroup of order 12 in G.

71. Prove that $A_4 \oplus Z_3$ has no subgroup of order 18.

Suggested Readings

Michael Brennan and Des MacHale, "Variations on a Theme: A_4 Definitely Has No Subgroup of Order Six!," *Mathematics Magazine*, 73 (2000): 36–40.

> The authors offer 11 proofs that A_4 has no subgroup of order 6. These proofs provide a review of many of the ideas covered thus far in this text.

J. A. Gallian, R. S. Johnson, and S. Peng. "On the Quotient Structure of Z^n," *Pi Mu Epsilon Journal*, 9 (1993): 524–526.

> The authors determine the structure of the group $(Z \oplus Z)/\langle(a, b)\rangle$ and related groups.

Tony Rothman, "Genius and Biographers: The Fictionalization of Évariste Galois," *The American Mathematical Monthly* 89 (1982): 84–106.

Achilleas Sinefakopoulos, "On Groups of Order p^2," *Mathematics Magazine*, 70 (1997): 212–213.

> This easy-to-read article gives a proof that a group of order p^2, where p is prime, must be Abelian.

Paul F. Zweifel, "Generalized Diatonic and Pentatonic Scales: A Group-theoretic Approach," *Perspectives of New Music*, 34 (1996): 140–161.

> The author discusses how group-theoretic notions such as subgroups, cosets, factor groups, and isomorphisms of Z_{12} and Z_{20} relate to musical scales, tuning, temperament, and structure.

Suggested Video

Cosets, Films for the Humanities & Sciences, FFH 6355, 25 minutes, Box 2053, Princeton, NJ 08543-2053, 1-800-257-5126.

> In this video, models are used to illustrate cosets, normal subgroups, and factor groups.

Évariste Galois

Galois at seventeen was making discoveries of epochal significance in the theory of equations, discoveries whose consequences are not yet exhausted after more than a century.

E. T. BELL, *Men of Mathematics*

This French stamp was issued as part of the 1984 "Celebrity Series" in support of the Red Cross Fund.

ÉVARISTE GALOIS (pronounced GAL-wah) was born on October 25, 1811, near Paris. Although he had mastered the works of Legendre and Lagrange at age 15, Galois twice failed his entrance examination to l'Ecole Polytechnique. He did not know some basic mathematics, and he did mathematics almost entirely in his head, to the annoyance of the examiner.

At 18, Galois wrote his important research on the theory of equations and submitted it to the French Academy of Sciences for publication. The paper was given to Cauchy for refereeing. Cauchy, impressed by the paper, agreed to present it to the academy, but never did. At the age of 19, Galois entered a paper of the highest quality in the competition for the Grand Prize in Mathematics, given by the French Academy of Sciences. The paper was given to Fourier, who died shortly thereafter. Galois's paper was never seen again.

Galois spent most of the last year and a half of his life in prison for revolutionary political offenses. While in prison, he attempted suicide and prophesied that he would die in a duel. On May 30, 1832, Galois was shot in a duel and died the next day at the age of 20.

Among the many concepts introduced by Galois are normal subgroups, isomorphisms, simple groups, finite fields, and Galois theory. His work provided a method for disposing of several famous constructability problems, such as trisecting an arbitrary angle and doubling a cube. Galois's entire collected works fill only 60 pages.

To find more information about Galois, visit:

http://www-groups.dcs.st-and.ac.uk/~history/

10 Group Homomorphisms

All modern theories of nuclear and electromagnetic interactions are based on group theory.

ANDREW WATSON, *New Scientist*

DEFINITION AND EXAMPLES

In this chapter, we consider one of the most fundamental ideas of algebra—homomorphisms. The term *homomorphism* comes from the Greek words *homo*, "like," and *morphe*, "form." We will see that a homomorphism is a natural generalization of an isomorphism and that there is an intimate connection between factor groups of a group and homomorphisms of a group. The concept of group homomorphisms was introduced by Camille Jordan in 1870, in his influential book *Traité des Substitutions*.

DEFINITION *Group Homomorphism*

A *homomorphism* ϕ from a group G to a group $\overline{G}$ is a mapping from G into $\overline{G}$ that preserves the group operation; that is, $\phi(ab) = \phi(a)\phi(b)$ for all a, b in G.

Before giving examples and stating numerous properties of homomorphisms, it is convenient to introduce an important subgroup that is intimately related to the image of a homomorphism.

DEFINITION *Kernel of a Homomorphism*

The *kernel* of a homomorphism ϕ from a group G to a group with identity e is the set $\{x \in G \mid \phi(x) = e\}$. The kernel of ϕ is denoted by Ker ϕ.

199

EXAMPLE 1 Any isomorphism is a homomorphism that is also onto and one-to-one. The kernel of an isomorphism is the trivial subgroup. ◆

EXAMPLE 2 Let $\mathbf{R}^*$ be the group of nonzero real numbers under multiplication. Then the determinant mapping $A \rightarrow \det A$ is a homomorphism from $GL(2, \mathbf{R})$ to $\mathbf{R}^*$. The kernel of the determinant mapping is $SL(2, \mathbf{R})$. ◆

EXAMPLE 3 The mapping ϕ from $\mathbf{R}^*$ to $\mathbf{R}^*$, defined by $\phi(x) = |x|$, is a homomorphism with Ker $\phi = \{1, -1\}$. ◆

EXAMPLE 4 Let $\mathbf{R}[x]$ denote the group of all polynomials with real coefficients under addition. For any f in $\mathbf{R}[x]$, let f' denote the derivative of f. Then the mapping $f \rightarrow f'$ is a homomorphism from $\mathbf{R}[x]$ to itself. The kernel of the derivative mapping is the set of all constant polynomials. ◆

EXAMPLE 5 The mapping ϕ from Z to Z_n, defined by $\phi(m) = m$ mod n, is a homomorphism (see Exercise 11 in Chapter 0). The kernel of this mapping is $\langle n \rangle$. ◆

EXAMPLE 6 The mapping $\phi(x) = x^2$ from $\mathbf{R}^*$, the nonzero real numbers under multiplication, to itself is a homomorphism, since $\phi(ab) = (ab)^2 = a^2b^2 = \phi(a)\phi(b)$ for all a and b in R^*. (See Exercise 5.) The kernel is $\{1, -1\}$. ◆

EXAMPLE 7 The mapping $\phi(x) = x^2$ from $\mathbf{R}$, the real numbers under addition, to itself is not a homomorphism, since $\phi(a + b) = (a + b)^2 = a^2 + 2ab + b^2$, whereas $\phi(a) + \phi(b) = a^2 + b^2$. ◆

When defining a homomorphism from a group in which there are several ways to represent the elements, caution must be exercised to ensure that the correspondence is a function. (The term *well-defined* is often used in this context.) For example, since $3(x + y) = 3x + 3y$ in Z_6, one might believe that the correspondence $x + \langle 3 \rangle \rightarrow 3x$ from $Z/\langle 3 \rangle$ to Z_6 is a homomorphism. But it is not a function, since $0 + \langle 3 \rangle = 3 + \langle 3 \rangle$ in $Z/\langle 3 \rangle$ but $3 \cdot 0 \neq 3 \cdot 3$ in Z_6.

For students who have had linear algebra, we remark that every linear transformation is a group homomorphism and the nullspace is the same as the kernel. An invertible linear transformation is a group isomorphism.

PROPERTIES OF HOMOMORPHISMS

Theorem 10.1   Properties of Elements Under Homomorphisms

> *Let ϕ be a homomorphism from a group G to a group $\overline{G}$ and let g be an element of G. Then*
>
> **1.** *ϕ carries the identity of G to the identity of $\overline{G}$.*
> **2.** *$\phi(g^n) = (\phi(g))^n$ for all n in Z.*
> **3.** *If $|g|$ is finite, then $|\phi(g)|$ divides $|g|$.*
> **4.** *Ker ϕ is a subgroup of G.*
> **5.** *$\phi(a) = \phi(b)$ if and only if aKer $\phi = b$Ker ϕ.*
> **6.** *If $\phi(g) = g'$, then $\phi^{-1}(g') = \{x \in G \mid \phi(x) = g'\} = g$Ker ϕ.*

PROOF.   The proofs of properties 1 and 2 are identical to the proofs of properties 1 and 2 of isomorphisms in Theorem 6.2. To prove property 3, notice that properties 1 and 2 together with $g^n = e$ imply that $e = \phi(e) = \phi(g^n) = (\phi(g))^n$. So, by Corollary 2 to Theorem 4.1, we have $|\phi(g)|$ divides n.

By property 1 we know that Ker ϕ is not empty. So, to prove property 4, we assume that $a, b \in$ Ker ϕ and show that $ab^{-1} \in$ Ker ϕ. Since $\phi(a) = e$ and $\phi(b) = e$, we have $\phi(ab^{-1}) = \phi(a)\phi(b^{-1}) = \phi(a)(\phi(b))^{-1} = ee^{-1} = e$. So, $ab^{-1} \in$ Ker ϕ.

To prove property 5, first assume that $\phi(a) = \phi(b)$. Then $e = (\phi(b))^{-1}\phi(a) = \phi(b^{-1})\phi(a) = \phi(b^{-1}a)$, so that $b^{-1}a \in$ Ker ϕ. It now follows from property 4 of the lemma in Chapter 7 that bKer $\phi = a$Ker ϕ. Reversing this argument completes the proof.

To prove property 6, we must show that $\phi^{-1}(g') \subseteq g$Ker ϕ and that gKer $\phi \subseteq \phi^{-1}(g')$. For the first inclusion, let $x \in \phi^{-1}(g')$, so that $\phi(x) = g'$. Then $\phi(g) = \phi(x)$ and by property 5 we have gKer $\phi = x$Ker ϕ and therefore $x \in g$Ker ϕ. This completes the proof that $\phi^{-1}(g') \subseteq g$Ker ϕ. To prove that gKer $\phi \subseteq \phi^{-1}(g')$, suppose that $k \in$ Ker ϕ. Then $\phi(gk) = \phi(g)\phi(k) = g'e = g'$. Thus, by definition, $gk \in \phi^{-1}(g')$. ∎

Theorem 10.2 **Properties of Subgroups Under Homomorphisms**

> *Let ϕ be a homomorphism from a group G to a group $\overline{G}$ and let H be a subgroup of G. Then*
>
> **1.** $\phi(H) = \{\phi(h) \mid h \in H\}$ *is a subgroup of $\overline{G}$.*
> **2.** *If H is cyclic, then $\phi(H)$ is cyclic.*
> **3.** *If H is Abelian, then $\phi(H)$ is Abelian.*
> **4.** *If H is normal in G, then $\phi(H)$ is normal in $\phi(G)$.*
> **5.** *If $|Ker\ \phi| = n$, then ϕ is an n-to-1 mapping from G onto $\phi(G)$.*
> **6.** *If $|H| = n$, then $|\phi(H)|$ divides n.*
> **7.** *If $\overline{K}$ is a subgroup of $\overline{G}$, then $\phi^{-1}(\overline{K}) = \{k \in G \mid \phi(k) \in \overline{K}\}$ is a subgroup of G.*
> **8.** *If $\overline{K}$ is a normal subgroup of $\overline{G}$, then $\phi^{-1}(\overline{K}) = \{k \in G \mid \phi(k) \in \overline{K}\}$ is a normal subgroup of G.*
> **9.** *If ϕ is onto and Ker $\phi = \{e\}$, then ϕ is an isomorphism from G to $\overline{G}$.*

PROOF. First note that the proofs of properties 1, 2, and 3 are identical to the proofs of properties 4, 2, and 1, respectively, of Theorem 6.3, since those proofs use only the fact that an isomorphism is an operation-preserving mapping.

To prove property 4, let $\phi(h) \in \phi(H)$ and $\phi(g) \in \phi(G)$. Then $\phi(g)\phi(h)\phi(g)^{-1} = \phi(ghg^{-1}) \in \phi(H)$, since H is normal in G.

Property 5 follows directly from property 6 of Theorem 10.1 and the fact that all cosets of Ker $\phi = \phi^{-1}\{e\}$ have the same number of elements.

To prove property 6, let ϕ_H denote the restriction of ϕ to the elements of H. Then ϕ_H is a homomorphism from H onto $\phi(H)$. Suppose $|Ker\ \phi_H| = t$. Then, by property 5, ϕ_H is a t-to-1 mapping. So, $|\phi(H)|t = |H|$.

To prove property 7, we use the One-Step Subgroup Test. Clearly, $e \in \phi^{-1}(\overline{K})$, so that $\phi^{-1}(\overline{K})$ is not empty. Let $k_1, k_2 \in \phi^{-1}(\overline{K})$. Then, by the definition of $\phi^{-1}(\overline{K})$, we know that $\phi(k_1), \phi(k_2) \in \overline{K}$. Thus, $\phi(k_2)^{-1} \in \overline{K}$ as well and $\phi(k_1 k_2^{-1}) = \phi(k_1)\phi(k_2)^{-1} \in \overline{K}$. So, by definition of $\phi^{-1}(\overline{K})$, we have $k_1 k_2^{-1} \in \phi^{-1}(\overline{K})$.

To prove property 8, we use the normality test given in Theorem 9.1. Note that every element in $x\phi^{-1}(\overline{K})x^{-1}$ has the form xkx^{-1}, where $\phi(k) \in \overline{K}$. Thus, since $\overline{K}$ is normal in $\overline{G}$, $\phi(xkx^{-1}) = \phi(x)\phi(k)(\phi(x))^{-1} \in \overline{K}$, and, therefore, $xkx^{-1} \in \phi^{-1}(\overline{K})$.

Finally, property 9 follows directly from property 5. ∎

A few remarks about Theorems 10.1 and 10.2 are in order. Students should remember the various properties of these theorems in words. For example, property 2 of Theorem 10.2 says that the homomorphic image of a cyclic group is cyclic. Property 4 of Theorem 10.2 says that the homomorphic image of a normal subgroup of G is normal in the image of G. Property 5 of Theorem 10.2 says that if ϕ is a homomorphism from G to $\overline{G}$, then every element of $\overline{G}$ that gets "hit" by ϕ gets hit the same number of times as does the identity. The set $\phi^{-1}(g')$ defined in property 6 of Theorem 10.1 is called the *inverse image of g'* (or the *pullback of g'*). Note that the inverse image of an element is a coset of the kernel and that every element in that coset has the same image. Similarly, the set $\phi^{-1}(\overline{K})$ defined in property 7 of Theorem 10.2 is called the *inverse image of $\overline{K}$* (or the *pullback of $\overline{K}$*).

Property 6 of Theorem 10.1 is reminiscent of something from linear algebra and differential equations. Recall that if x is a particular solution to a system of linear equations and S is the entire solution set of the corresponding homogeneous system of linear equations, then $x + S$ is the entire solution set of the nonhomogeneous system. In reality, this statement is just a special case of property 6. Properties 1 and 6 of Theorem 10.1 and property 5 of Theorem 10.2 are pictorially represented in Figure 10.1.

The special case of property 8 of Theorem 10.2, where $\overline{K} = \{e\}$, is of such importance that we single it out.

Corollary   **Kernels Are Normal**

> *Let ϕ be a group homomorphism from G to $\overline{G}$. Then Ker ϕ is a normal subgroup of G.*

The next two examples illustrate several properties of Theorems 10.1 and 10.2.

EXAMPLE 8   Consider the mapping ϕ from $\mathbf{C}^*$ to $\mathbf{C}^*$ given by $\phi(x) = x^4$. Since $(xy)^4 = x^4y^4$, ϕ is a homomorphism. Clearly, Ker $\phi = \{x \mid x^4 = 1\} = \{1, -1, i, -i\}$. So, by property 5 of Theorem 10.2, we know that ϕ is a 4-to-1 mapping. Now let's find all elements that map to, say, 2. Certainly, $\phi(\sqrt[4]{2}) = 2$. Then, by property 6 of Theorem 10.1, the set of all elements that map to 2 is $\sqrt[4]{2}$ Ker $\phi = \{\sqrt[4]{2}, -\sqrt[4]{2}, \sqrt[4]{2}i, -\sqrt[4]{2}i\}$.

Finally, we verify a specific instance of property 3 of Theorem 10.1 and of property 2 and property 6 of Theorem 10.2. Let $H =$

G

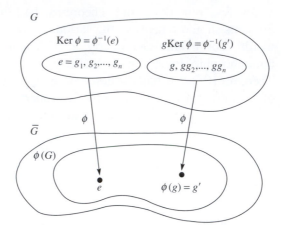

Figure 10.1

$\langle\cos 30° + i \sin 30°\rangle$. It follows from DeMoivre's Theorem (Example 8 in Chapter 0) that $|H| = 12$, $\phi(H) = \langle\cos 120° + i \sin 120°\rangle$, and $|\phi(H)| = 3$. ◆

EXAMPLE 9   Define $\phi:Z_{12} \rightarrow Z_{12}$ by $\phi(x) = 3x$. To verify that ϕ is a homomorphism, we observe that in Z_{12}, $3(a + b) = 3a + 3b$ (since the group operation is addition). Direct calculations show that Ker $\phi = \{0, 4, 8\}$. Thus, we know from property 5 of Theorem 10.2 that ϕ is a 3-to-1 mapping. Since $\phi(2) = 6$, we have by property 6 of Theorem 10.1 that $\phi^{-1}(6) = 2 + $ Ker $\phi = \{2, 6, 10\}$. Notice also that $\langle 2 \rangle$ is cyclic and $\phi(\langle 2 \rangle) = \{0, 6\}$ is cyclic. Moreover, $|2| = 6$ and $|\phi(2)| = |6| = 2$, as required by property 3 of Theorem 10.1. Letting $\overline{K} = \{0, 6\}$, we see that the subgroup $\phi^{-1}(\overline{K})$ = $\{0, 2, 4, 6, 8, 10\}$. This verifies property 7 of Theorem 10.2 in this particular case. ◆

The next example illustrates how one can easily determine all homomorphisms from a cyclic group to a cyclic group.

EXAMPLE 10   We determine all homomorphisms from Z_{12} to Z_{30}. By property 2 of Theorem 10.1, such a homomorphism is completely specified by the image of 1. That is, if 1 maps to a, then x maps to xa. Lagrange's Theorem and property 3 of Theorem 10.1 require that $|a|$ divide both 12 and 30. So, $|a| = 1, 2, 3,$ or 6. Thus, $a = 0, 15, 10, 20, 5,$ or 25. This gives us a list of candidates for the homomorphisms. That each of these six possibilities yields an operation-preserving, well-defined function can now be verified by direct calculations. [Note that gcd(12, 30) = 6. This is not a coincidence!] ◆

EXAMPLE 11 The mapping from S_n to Z_2 that takes an even permutation to 0 and an odd permutation to 1 is a homomorphism. Figure 10.2 illustrates the telescoping nature of the mapping. ◆

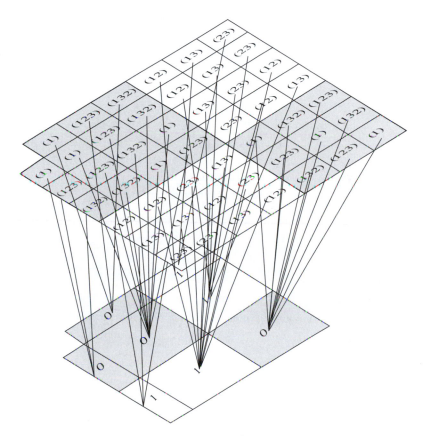

Figure 10.2 Homomorphism from S_3 to Z_2.

THE FIRST ISOMORPHISM THEOREM

In Chapter 9, we showed that for a group G and a normal subgroup H, we could arrange the Cayley table of G into boxes that represented the cosets of H in G, and that these boxes then became a Cayley table for G/H. The next theorem shows that for any homomorphism ϕ of G and the normal subgroup Ker ϕ, the same process produces a Cayley table isomorphic to the homomorphic image of G. Thus, homomorphisms, like factor groups, cause a *systematic* collapse of a group to a simpler but closely related group. This can be likened to viewing a group through the reverse end of a telescope—the general features of the group are present, but the apparent size is diminished. The important relationship between homomorphisms and factor groups given below is often called the Fundamental Theorem of Group Homomorphisms.

Theorem 10.3   **First Isomorphism Theorem (Jordan, 1870)**

> *Let ϕ be a group homomorphism from G to $\overline{G}$. Then the mapping from $G/\text{Ker }\phi$ to $\phi(G)$, given by $g\text{Ker }\phi \to \phi(g)$, is an isomorphism. In symbols, $G/\text{Ker }\phi \approx \phi(G)$.*

PROOF.   Let us use ψ to denote the correspondence $g\text{Ker}\phi \to \phi(g)$. That ψ is well defined (that is, the correspondence is independent of the particular coset representative chosen) and one-to-one follows directly from property 5 of Theorem 10.1. To show that ψ is operation-preserving, observe that $\psi(x\text{Ker }\phi \ y\text{Ker }\phi) = \psi(xy\text{Ker }\phi) = \phi(xy) = \phi(x) \ \phi(y) = \psi(x\text{Ker }\phi)\psi(y\text{Ker }\phi)$.   ∎

The next corollary follows directly from Theorem 10.3, property 1 of Theorem 10.2, and Lagrange's Theorem.

Corollary

> *If ϕ is a homomorphism from a finite group G to $\overline{G}$, then $|\phi(G)|$ divides $|G|$ and $|\overline{G}|$.*

EXAMPLE 12   To illustrate Theorem 10.3 and its proof, consider the homomorphism ϕ from D_4 to itself given by

$$
\begin{array}{cccccccc}
R_0 & R_{180} & R_{90} & R_{270} & H & V & D & D' \\
\searrow & \swarrow & \searrow & \swarrow & \searrow & \swarrow & \searrow & \swarrow \\
R_0 & & H & & R_{180} & & V
\end{array}
$$

Then Ker $\phi = \{R_0, R_{180}\}$, and the mapping ψ in Theorem 10.3 is $R_0\text{Ker }\phi \to R_0$, $R_{90}\text{Ker }\phi \to H$, $H\text{Ker }\phi \to R_{180}$, $D\text{Ker }\phi \to V$. It is straightforward to verify that the mapping ψ is an isomorphism.   ◆

Mathematicians often give a pictorial representation of Theorem 10.3, as follows:

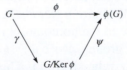

where $\gamma:G \to G/\text{Ker } \phi$ is defined as $\gamma(g) = g\text{Ker } \phi$. The mapping γ is called the *natural mapping* from G to $G/\text{Ker } \phi$. Our proof of Theorem 10.3 shows that $\psi\gamma = \phi$. In this case, one says that the preceding diagram is *commutative*.

As a consequence of Theorem 10.3, we see that all homomorphic images of G can be determined using G. We may simply consider the various factor groups of G. For example, we know that the homomorphic image of an Abelian group is Abelian because the factor group of an Abelian group is Abelian. We know that the number of homomorphic images of a cyclic group G of order n is the number of divisors of n, since there is exactly one subgroup of G (and therefore one factor group of G) for each divisor of n. (Be careful: The number of homomorphisms of a cyclic group of order n need not be the same as the number of divisors of n, since different homomorphisms can have the same image.)

An appreciation for Theorem 10.3 can be gained by looking at a few examples.

EXAMPLE 13 $Z/\langle n\rangle \approx Z_n$

Consider the mapping from Z to Z_n defined in Example 5. Clearly, its kernel is $\langle n\rangle$. So, by Theorem 10.3, $Z/\langle n\rangle \approx Z_n$.   ◆

EXAMPLE 14   The Wrapping Function

Recall the wrapping function W from trigonometry. The real number line is wrapped around a unit circle in the plane centered at $(0, 0)$ with the number 0 on the number line at the point $(1, 0)$, the positive reals in the counterclockwise direction and the negative reals in the clockwise direction (see Figure 10.3). The function W assigns to each real number a the point a radians from $(1, 0)$ on the circle. This mapping is a homomorphism from the group **R** under addition onto the circle group (the group of complex numbers of magnitude 1 under multiplication). Indeed, it follows from elementary facts of trigonometry that $W(x) = \cos x + i \sin x$ and $W(x + y) = W(x)W(y)$. Since W is periodic of period 2π, $\text{Ker } W = \langle 2\pi\rangle$. So, from the First Isomorphism Theorem, we see that $\mathbf{R}/\langle 2\pi\rangle$ is isomorphic to the circle group.   ◆

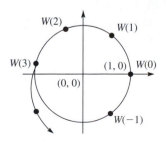

Figure 10.3

Our next example is a theorem that is used repeatedly in Chapters 24 and 25.

EXAMPLE 15   The *N/C* Theorem

Let H be a subgroup of a group G. Recall that the normalizer of H in G is $N(H) = \{x \in G \mid xHx^{-1} = H\}$ and the centralizer of H in G is $C(H) = \{x \in G \mid xhx^{-1} = h$ for all h in $H\}$. Consider the mapping from $N(H)$ to Aut(H) given by $g \rightarrow \phi_g$, where ϕ_g is the inner automorphism of H induced by g [that is, $\phi_g(h) = ghg^{-1}$ for all h in H]. This mapping is a homomorphism with kernel $C(H)$. So, by Theorem 10.3, $N(H)/C(H)$ is isomorphic to a subgroup of Aut(H). ◆

As an application of the *N/C* Theorem, we will show that every group of order 35 is cyclic.

EXAMPLE 16   Let G be a group of order 35. By Lagrange's Theorem, every nonidentity element of G has order 5, 7, or 35. If some element has order 35, G is cyclic. So we may assume that all nonidentity elements have order 5 or 7. However, not all such elements can have order 5, since elements of order 5 come 4 at a time (if $|x| = 5$, then $|x^2| = |x^3| = |x^4| = 5$) and 4 does not divide 34. Similarly, since 6 does not divide 34, not all nonidentity elements can have order 7. So, G has elements of order 7 and order 5. Since G has an element of order 7, it has a subgroup of order 7. Let us call it H. In fact, H is the only subgroup of G of order 7, for if K is another subgroup of G of order 7, we have by Exercise 7 of the Supplementary Exercises for Chapters 5–8 that $|HK| = |H||K|/|H \cap K| = 7 \cdot 7/1 = 49$. But, of course, this is impossible in a group of order 35. Since for every a in G, aHa^{-1} is also a subgroup of G of order 7 (see Exercise 1 of the Supplementary Exercises for Chapters 1–4), we must have $aHa^{-1} = H$. So, $N(H) = G$. Since H has prime order, it is cyclic and therefore Abelian. In particular, $C(H)$ contains H. So, 7 divides $|C(H)|$ and $|C(H)|$ divides 35. It follows, then, that $C(H) = G$ or $C(H) = H$. If $C(H) = G$, then we may obtain an element x of order 35 by letting $x = hk$, where h is a nonidentity element of H and k has order 5. On the other hand, if $C(H) = H$, then $|C(H)| = 7$ and $|N(H)/C(H)| = 35/7 = 5$. However, 5 does not divide $|\text{Aut}(H)| = |\text{Aut}(Z_7)| = 6$. This contradiction shows that G is cyclic. ◆

The corollary of Theorem 10.2 says that the kernel of every homomorphism of a group is a normal subgroup of the group. We conclude this chapter by verifying that the converse of this statement is also true.

Theorem 10.4 **Normal Subgroups Are Kernels**

> *Every normal subgroup of a group G is the kernel of a*
> *homomorphism of G. In particular, a normal subgroup N is*
> *the kernel of the mapping g → gN from G to G/N.*

PROOF.   Define $\gamma{:}G \to G/N$ by $\gamma(g) = gN$. (This mapping is called the *natural homomorphism* from G to G/N.) Then, $\gamma(xy) = (xy)N = xNyN = \gamma(x)\gamma(y)$. Moreover, $g \in \text{Ker } \gamma$ if and only if $gN = \gamma(g) = N$, which is true if and only if $g \in N$ (see Part 2 of the lemma in Chapter 7).   ∎

  Examples 13, 14, and 15 illustrate the utility of the First Isomorphism Theorem. But what about homomorphisms in general? Why would one care to study a homomorphism of a group? The answer is that, just as was the case with factor groups of a group, homomorphic images of a group tell us *some* of the properties of the original group. One measure of the likeness of a group and its homomorphic image is the size of the kernel. If the kernel of the homomorphism of group G is the identity, then the image of G tells us everything (group theoretically) about G (the two being isomorphic). On the other hand, if the kernel of the homomorphism is G itself, then the image tells us nothing about G. Between these two extremes, some information about G is preserved and some is lost. The utility of a particular homomorphism lies in its ability to preserve the group properties we want, while losing some inessential ones. In this way, we have replaced G by a group less complicated (and therefore easier to study) than G; but, in the process, we have saved enough information to answer questions that we have about G itself. For example, if G is a group of order 60 and G has a homomorphic image of order 12 that is cyclic, then we know from properties 5, 7, and 8 of Theorem 10.2 that G has normal subgroups of orders 5, 10, 15, 20, 30, and 60. To illustrate further, suppose we are asked to find an infinite group that is the union of three proper subgroups. Instead of attempting to do this directly, we first make the problem easier by finding a finite group that is the union of three proper subgroups. Observing that $Z_2 \oplus Z_2$ is the union of $H_1 = \langle 1, 0 \rangle$, $H_2 = \langle 0, 1 \rangle$, and $H_3 = \langle 1, 1 \rangle$, we have found our finite group. Now all we need do is think of an infinite group that has $Z_2 \oplus Z_2$ as a homomorphic image and pull back H_1, H_2, and H_3, and our original problem is solved. Clearly, the mapping from $Z_2 \oplus Z_2 \oplus Z$ onto $Z_2 \oplus Z_2$ given by $\phi(a, b, c) = (a, b)$ is such a mapping, and therefore $Z_2 \oplus$

$Z_2 \oplus Z$ is the union of $\phi^{-1}(H_1) = \{(a, 0, c,) \mid a \in Z_2, c \in Z\}$, $\phi^{-1}(H_2)$ $= \{(0, b, c,) \mid b \in Z_2, c \in Z\}$, and $\phi^{-1}(H_3) = \{(a, a, c,) \mid a \in Z_2, c \in Z\}$.

Although an isomorphism is a special case of a homomorphism, the two concepts have entirely different roles. Whereas isomorphisms allow us to look at a group in an alternative way, homomorphisms act as investigative tools. The following analogy between homomorphisms and photography may be instructive.[†] A photograph of a person cannot tell us the person's exact height, weight, or age. Nevertheless, we *may* be able to decide from a photograph whether the person is tall or short, heavy or thin, old or young, male or female. In the same way, a homomorphic image of a group gives us *some* information about the group.

In certain branches of group theory, and especially in physics and chemistry, one often wants to know all homomorphic images of a group that are matrix groups over the complex numbers (these are called *group representations*). Here, we may carry our analogy with photography one step further by saying that this is like wanting photographs of a person from many different angles (front view, profile, head-to-toe view, close-up, etc.), as well as x-rays! Just as this composite information from the photographs reveals much about the person, several homomorphic images of a group reveal much about the group.

EXERCISES ▪▪

The greater the difficulty, the more glory in surmounting it. Skillful pilots gain their reputation from storms and tempests.

EPICURUS

1. Prove that the mapping given in Example 2 is a homomorphism.
2. Prove that the mapping given in Example 3 is a homomorphism.
3. Prove that the mapping given in Example 4 is a homomorphism.
4. Prove that the mapping given in Example 11 is a homomorphism.
5. Let $\mathbf{R}^*$ be the group of nonzero real numbers under multiplication, and let r be a positive integer. Show that the mapping that takes x to x^r is a homomorphism from $\mathbf{R}^*$ to $\mathbf{R}^*$. Which values of r yield an isomorphism?
6. Let G be the group of all polynomials with real coefficients under addition. For each f in G, let $\int f$ denote the antiderivative of f that passes through the

---

[†]*All perception of truth is the detection of an analogy.* Henry David Thoreau, *Journal.*

point $(0, 0)$. Show that the mapping $f \rightarrow \int f$ from G to G is a homomorphism. What is the kernel of this mapping? Is this mapping a homomorphism if $\int f$ denotes the antiderivative of f that passes through $(0, 1)$?

7. If ϕ is a homomorphism from G to H and σ is a homomorphism from H to K, show that $\sigma\phi$ is a homomorphism from G to K.

8. Let G be a group of permutations. For each σ in G, define

$$\text{sgn}(\sigma) = \begin{cases} +1 & \text{if } \sigma \text{ is an even permutation,} \\ -1 & \text{if } \sigma \text{ is an odd permutation.} \end{cases}$$

Prove that sgn is a homomorphism from G to the multiplicative group $\{+1, -1\}$. What is the kernel?

9. Prove that the mapping from $G \oplus H$ to G given by $(g, h) \rightarrow g$ is a homomorphism. What is the kernel? This mapping is called the *projection* of $G \oplus H$ onto G.

10. Let G be a subgroup of some dihedral group. For each x in G, define

$$\phi(x) = \begin{cases} +1 & \text{if } x \text{ is a rotation,} \\ -1 & \text{if } x \text{ is a reflection.} \end{cases}$$

Prove that ϕ is a homomorphism from G to the multiplicative group $\{+1, -1\}$. What is the kernel of ϕ?

11. Prove that $(Z \oplus Z)/(\langle(a, 0)\rangle \times \langle(0, b)\rangle)$ is isomorphic to $Z_a \oplus Z_b$.

12. Suppose that k is a divisor of n. Prove that $Z_n/\langle k \rangle \approx Z_k$.

13. Prove that $(A \oplus B)/(A \oplus \{e\}) \approx B$.

14. Explain why the correspondence $x \rightarrow 3x$ from Z_{12} to Z_{10} is not a homomorphism.

15. Suppose that ϕ is a homomorphism from Z_{30} to Z_{30} and Ker $\phi = \{0, 10, 20\}$. If $\phi(23) = 9$, determine all elements that map to 9.

16. Prove that there is no homomorphism from $Z_8 \oplus Z_2$ onto $Z_4 \oplus Z_4$.

17. Prove that there is no homomorphism from $Z_{16} \oplus Z_2$ onto $Z_4 \oplus Z_4$.

18. Can there be a homomorphism from $Z_4 \oplus Z_4$ onto Z_8? Can there be a homomorphism from Z_{16} onto $Z_2 \oplus Z_2$? Explain your answers.

19. Suppose that there is a homomorphism ϕ from Z_{17} to some group and that ϕ is not one-to-one. Determine ϕ.

20. How many homomorphisms are there from Z_{20} onto Z_8? How many are there to Z_8?

21. If ϕ is a homomorphism from Z_{30} onto a group of order 5, determine the kernel of ϕ.

22. Suppose that ϕ is a homomorphism from a finite group G onto $\overline{G}$ and that $\overline{G}$ has an element of order 8. Prove that G has an element of order 8. Generalize.

23. Suppose that ϕ is a homomorphism from Z_{36} to a group of order 24.

    a. Determine the possible homomorphic images.

    b. For each image in part a, determine the corresponding kernel of ϕ.

24. Suppose that $\phi\colon Z_{50} \to Z_{15}$ is a group homomorphism with $\phi(7) = 6$.

    a. Determine $\phi(x)$.

    b. Determine the image of ϕ.

    c. Determine the kernel of ϕ.

    d. Determine $\phi^{-1}(3)$.

25. How many homomorphisms are there from Z_{20} onto Z_{10}? How many are there to Z_{10}?

26. Determine all homomorphisms from Z_4 to $Z_2 \oplus Z_2$.

27. Determine all homomorphisms from Z_n to itself.

28. Suppose that ϕ is a homomorphism from S_4 onto Z_2. Determine Ker ϕ. Determine all homomorphisms from S_4 to Z_2.

29. Suppose that there is a homomorphism from a finite group G onto Z_{10}. Prove that G has normal subgroups of indexes 2 and 5.

30. Suppose that ϕ is a homomorphism from a group G onto $Z_6 \oplus Z_2$ and that the kernel of ϕ has order 5. Explain why G must have normal subgroups of orders 5, 10, 15, 20, 30, and 60.

31. Suppose that ϕ is a homomorphism from $U(30)$ to $U(30)$ and that Ker $\phi = \{1, 11\}$. If $\phi(7) = 7$, find all elements of $U(30)$ that map to 7.

32. Find a homomorphism ϕ from $U(30)$ to $U(30)$ with kernel $\{1, 11\}$ and $\phi(7) = 7$.

33. Suppose that ϕ is a homomorphism from $U(40)$ to $U(40)$ and that Ker $\phi = \{1, 9, 17, 33\}$. If $\phi(11) = 11$, find all elements of $U(40)$ that map to 11.

34. Find a homomorphism ϕ from $U(40)$ to $U(40)$ with kernel $\{1, 9, 17, 33\}$ and $\phi(11) = 11$.

35. Prove that the mapping $\phi\colon Z \oplus Z \to Z$ given by $(a, b) \to a - b$ is a homomorphism. What is the kernel of ϕ? Describe the set $\phi^{-1}(3)$ (that is, all elements that map to 3).

36. Suppose that there is a homomorphism ϕ from $Z \oplus Z$ to a group G such that $\phi((3, 2)) = a$ and $\phi((2, 1)) = b$. Determine $\phi((4, 4))$ in terms of a and b. Assume that the operation of G is addition.

37. Prove that the mapping $x \to x^6$ from $\mathbf{C}^*$ to $\mathbf{C}^*$ is a homomorphism. What is the kernel?

38. For each pair of positive integers m and n, we can define a homomorphism from Z to $Z_m \oplus Z_n$ by $x \to (x \bmod m, x \bmod n)$. What is the kernel when $(m, n) = (3, 4)$? What is the kernel when $(m, n) = (6, 4)$? Generalize.

39. (Second Isomorphism Theorem) If K is a subgroup of G and N is a normal subgroup of G, prove that $K/(K \cap N)$ is isomorphic to KN/N.

40. (Third Isomorphism Theorem) If M and N are normal subgroups of G and $N \leq M$, prove that $(G/N)/(M/N) \approx G/M$.

41. Let $\phi(d)$ denote the Euler phi function of d (see page 80). Show that the number of homomorphisms from Z_n to Z_k is $\Sigma\phi(d)$, where the sum runs over all common divisors d of n and k. [It follows from number theory that this sum is actually $\gcd(n, k)$.]

42. Let k be a divisor of n. Consider the homomorphism from $U(n)$ to $U(k)$ given by $x \rightarrow x \bmod k$. What is the relationship between this homomorphism and the subgroup $U_k(n)$ of $U(n)$?

43. Determine all homomorphic images of D_4 (up to isomorphism).

44. Let N be a normal subgroup of a finite group G. Use the theorems of this chapter to prove that the order of the group element gN in G/N divides the order of g.

45. Suppose that G is a finite group and that Z_{10} is a homomorphic image of G. What can we say about $|G|$?

46. Suppose that Z_{10} and Z_{15} are both homomorphic images of a finite group G. What can be said about $|G|$?

47. Suppose that for each prime p, Z_p is the homomorphic image of a group G. What can we say about $|G|$? Give an example of such a group.

48. (For students who have had linear algebra.) Suppose that x is a particular solution to a system of linear equations and that S is the entire solution set of the corresponding homogeneous system of linear equations. Explain why property 6 of Theorem 10.1 guarantees that $x + S$ is the entire solution set of the nonhomogeneous system. In particular, describe the relevant groups and the homomorphism between them.

49. Let N be a normal subgroup of a group G. Use property 7 of Theorem 10.2 to prove that every subgroup of G/N has the form H/N, where H is a subgroup of G. (This exercise is referred to in Chapter 24.)

50. Show that a homomorphism defined on a cyclic group is completely determined by its action on a generator of the group.

51. Use the First Isomorphism Theorem to prove Theorem 9.4.

52. Let α and β be group homomorphisms from G to $\overline{G}$ and let $H = \{g \in G \mid \alpha(g) = \beta(g)\}$. Prove or disprove that H is a subgroup of G.

53. Let $Z[x]$ be the group of polynomials in x with integer coefficients under addition. Prove that the mapping from $Z[x]$ into Z given by $f(x) \rightarrow f(3)$ is a homomorphism. Give a geometric description of the kernel of this homomorphism. Generalize.

54. If H and K are normal subgroups of G and $H \cap K = \{e\}$, prove that G is isomorphic to a subgroup of $G/H \oplus G/K$.

55. Suppose that H and K are distinct subgroups of G of index 2. Prove that $H \cap K$ is a normal subgroup of G of index 4 and that $G/(H \cap K)$ is not cyclic.

56. Suppose that the number of homomorphisms from G to H is n. How many homomorphisms are there from G to $H \oplus H \oplus \cdots \oplus H$ (s terms)? When H is Abelian, how many homomorphisms are there from $G \oplus G \oplus \cdots \oplus G$ (s terms) to H?

57. Prove that every group of order 77 is cyclic.

58. Determine all homomorphisms from Z onto S_3. Determine all homomorphisms from Z to S_3.

59. Suppose G is an Abelian group under addition with the property that for every positive integer n the set $nG = \{ng \mid g \in G\} = G$. Show that every proper subgroup of G is properly contained in a proper subgroup of G. Name two familiar groups that satisfy the hypothesis.

COMPUTER EXERCISE ▬▬▬▬▬▬▬▬▬▬▬▬▬▬▬▬▬▬▬▬▬ ■■

A computer lets you make more mistakes faster than any invention in human history—with the possible exceptions of handguns and tequila.

MITCH RATLIFFE

Software for the computer exercise in this chapter is available at the website:

http://www.d.umn.edu/~jgallian

1. This software determines the homomorphisms from Z_m to Z_n. (Recall that a homomorphism from Z_m is completely determined by the image of 1.) Run the program for $m = 20$ with various choices for n. Run the program for $m = 15$ with various choices for n. What relationship do you see between m and n and the number of homomorphisms from Z_m to Z_n? For each choice of m and n, observe the smallest positive image of 1. Try to see the relationship between this image and the values of m and n. What relationship do you see between the smallest positive image of 1 and the other images of 1? Test your conclusions with other choices of m and n.

Suggested Video

Clumping in Groups, Films for the Humanities & Sciences, FFH 6357, 25 minutes, Box 2053, Princeton, NJ 08543-2053, 1-800-257-5126.

This video uses the group of symmetries of the regular tetrahedron and the group of nonzero complex numbers to illustrate the Fundamental Theorem of Group Homomorphisms.

Camille Jordan

Although these contributions [to analysis and topology] would have been enough to rank Jordan very high among his mathematical contemporaries, it is chiefly as an algebraist that he reached celebrity when he was barely thirty; and during the next forty years he was universally regarded as the undisputed master of group theory.

J. DIEUDONNÉ, *Dictionary of Scientific Biography*

CAMILLE JORDAN was born into a well-to-do family on January 5, 1838, in Lyons, France. Like his father, he graduated from the École Polytechnique and became an engineer. Nearly all of his 120 research papers in mathematics were written before his retirement from engineering in 1885. From 1873 until 1912, Jordan taught simultaneously at the École Polytechnique and at the College of France.

In the great French tradition, Jordan was a universal mathematician who published in nearly every branch of mathematics. Among the concepts named after him are the Jordan canonical form in matrix theory, the Jordan curve theorem from topology, and the Jordan-Hölder theorem from group theory. His classic book *Traité des Substitutions,* published in 1870, was the first to be devoted solely to group theory and its applications to other branches of mathematics.

Another book that had great influence and set a new standard for rigor was his *Cours d'analyse.* This book gave the first clear definitions of the notions of volume and multiple integral. Nearly 100 years after this book appeared, the distinguished mathematician and mathematical historian B. L. van der Waerden wrote, "For me, every single chapter of the *Cours d'analyse* is a pleasure to read." Jordan died in Paris on January 22, 1922.

To find more information about Jordan, visit (click on Jordan, Marie):

http://www-groups.dcs.st-and.ac.uk/~history/

11

Fundamental Theorem of Finite Abelian Groups

By a small sample we may judge of the whole piece.
MIGUEL DE CERVANTES, *Don Quixote*

THE FUNDAMENTAL THEOREM

In this chapter, we present a theorem that describes to an algebraist's eye (that is, up to isomorphism) all finite Abelian groups in a standardized way. Before giving the proof, which is long and difficult, we discuss some consequences of the theorem and its proof. The first proof of the theorem was given by Leopold Kronecker in 1858.

Theorem 11.1 Fundamental Theorem of Finite Abelian Groups

Every finite Abelian group is a direct product of cyclic groups of prime-power order. Moreover, the number of terms in the product and the orders of the cyclic groups are uniquely determined by the group.

Since a cyclic group of order n is isomorphic to Z_n, Theorem 11.1 shows that every finite Abelian group G is isomorphic to a group of the form

$$Z_{p_1^{n_1}} \oplus Z_{p_2^{n_2}} \oplus \cdots \oplus Z_{p_k^{n_k}},$$

where the p_i's are not necessarily distinct primes and the prime-powers $p_1^{n_1}, p_2^{n_2}, \ldots, p_k^{n_k}$ are uniquely determined by G. Writing a group in this form is called *determining the isomorphism class of G.*

THE ISOMORPHISM CLASSES
OF ABELIAN GROUPS

The Fundamental Theorem is extremely powerful. As an application, we can use it as an algorithm for constructing all Abelian groups of any order. Let's look at groups whose orders have the form p^k, where p is prime and $k \leq 4$. In general, there is one group of order p^k for each set of positive integers whose sum is k (such a set is called a *partition* of k); that is, if k can be written as

$$k = n_1 + n_2 + \cdots + n_t,$$

where each n_i is a positive integer, then

$$Z_{p^{n_1}} \oplus Z_{p^{n_2}} \oplus \cdots \oplus Z_{p^{n_t}}$$

is an Abelian group of order p^k.

Order of G	Partitions of k	Possible direct products for G
p	1	Z_p
p^2	2	Z_{p^2}
	$1 + 1$	$Z_p \oplus Z_p$
p^3	3	Z_{p^3}
	$2 + 1$	$Z_{p^2} \oplus Z_p$
	$1 + 1 + 1$	$Z_p \oplus Z_p \oplus Z_p$
p^4	4	Z_{p^4}
	$3 + 1$	$Z_{p^3} \oplus Z_p$
	$2 + 2$	$Z_{p^2} \oplus Z_{p^2}$
	$2 + 1 + 1$	$Z_{p^2} \oplus Z_p \oplus Z_p$
	$1 + 1 + 1 + 1$	$Z_p \oplus Z_p \oplus Z_p \oplus Z_p$

Furthermore, the uniqueness portion of the Fundamental Theorem guarantees that distinct partitions of k yield distinct isomorphism classes. Thus, for example, $Z_9 \oplus Z_3$ is not isomorphic to $Z_3 \oplus Z_3 \oplus Z_3$. A reliable mnemonic for comparing external direct products is the cancellation property: If A is *finite*, then

$$A \oplus B \approx A \oplus C \qquad \text{if and only if} \qquad B \approx C \quad \text{(see [1])}.$$

Thus $Z_4 \oplus Z_4$ is not isomorphic to $Z_4 \oplus Z_2 \oplus Z_2$ because Z_4 is not isomorphic to $Z_2 \oplus Z_2$.

To appreciate fully the potency of the Fundamental Theorem, contrast the ease with which the Abelian groups of order p^k, $k \leq 4$, were determined with the corresponding problem for non-Abelian groups.

Even a description of the two non-Abelian groups of order 8 is a challenge (see Chapter 26), and a description of the nine non-Abelian groups of order 16 is well beyond the scope of this text.

Now that we know how to construct all the Abelian groups of prime-power order, we move to the problem of constructing all Abelian groups of a certain order n, where n has two or more distinct prime divisors. We begin by writing n in prime-power decomposition form $n = p_1{}^{n_1}p_2{}^{n_2} \cdots p_k{}^{n_k}$. Next, we individually form all Abelian groups of order $p_1{}^{n_1}$, then $p_2{}^{n_2}$, and so on, as described earlier. Finally, we form all possible external direct products of these groups. For example, let $n = 1176 = 2^3 \cdot 3 \cdot 7^2$. Then, the complete list of the distinct isomorphism classes of Abelian groups of order 1176 is

$$Z_8 \oplus Z_3 \oplus Z_{49},$$
$$Z_4 \oplus Z_2 \oplus Z_3 \oplus Z_{49},$$
$$Z_2 \oplus Z_2 \oplus Z_2 \oplus Z_3 \oplus Z_{49},$$
$$Z_8 \oplus Z_3 \oplus Z_7 \oplus Z_7,$$
$$Z_4 \oplus Z_2 \oplus Z_3 \oplus Z_7 \oplus Z_7,$$
$$Z_2 \oplus Z_2 \oplus Z_2 \oplus Z_3 \oplus Z_7 \oplus Z_7.$$

If we are given any particular Abelian group G of order 1176, the question we want to answer about G is: Which of the preceding six isomorphism classes represents the structure of G? We can answer this question by comparing the orders of the elements of G with the orders of the elements in the six direct products, since it can be shown that two finite Abelian groups are isomorphic if and only if they have the same number of elements of each order. For instance, we could determine whether G has any elements of order 8. If so, then G must be isomorphic to the first or fourth group above, since these are the only ones with elements of order 8. To narrow G down to a single choice, we now need only check whether or not G has an element of order 49, since the first product above has such an element, whereas the fourth one does not.

What if we have some specific Abelian group G of order $p_1{}^{n_1}p_2{}^{n_2} \cdots p_k{}^{n_k}$, where the p_i's are distinct primes? How can G be expressed as an *internal* direct product of cyclic groups of prime-power order? For simplicity, let us say that the group has 2^n elements. First, we must compute the orders of the elements. After this is done, pick an element of maximum order 2^r, call it a_1. Then $\langle a_1 \rangle$ is one of the factors in the desired internal direct product. If $G \neq \langle a_1 \rangle$, choose an element a_2 of maximum order 2^s such that $s \leq n - r$ and none of $a_2, a_2{}^2, a_2{}^4, \ldots,$ $a_2{}^{2^{s-1}}$ is in $\langle a_1 \rangle$. Then $\langle a_2 \rangle$ is a second direct factor. If $n \neq r + s$, select an element a_3 of maximum order 2^t such that $t \leq n - r - s$ and none of $a_3, a_3^2, a_3^4, \ldots, a_3^{2^{t-1}}$ is in $\langle a_1 \rangle \times \langle a_2 \rangle = \{a_1^i a_2^j \mid 0 \leq i < 2^r, 0 \leq j < 2^s\}$. Then $\langle a_3 \rangle$ is another direct factor. We continue in this fashion until our direct product has the same order as G.

A formal presentation of this algorithm for any Abelian group G of prime-power order p^n is as follows.

Greedy Algorithm for an Abelian Group of Order p^n

1. Compute the orders of the elements of the group G.
2. Select an element a_1 of maximum order and define $G_1 = \langle a_1 \rangle$.
   Set $i = 1$.
3. If $|G| = |G_i|$, stop. Otherwise, replace i by $i + 1$.
4. Select an element a_i of maximum order p^k such that $p^k \leq |G|/|G_{i-1}|$ and none of $a_i, a_i^p, a_i^{p^2}, \ldots, a_i^{p^{k-1}}$ is in G_{i-1}, and define $G_i = G_{i-1} \times \langle a_i \rangle$.
5. Return to step 3.

In the general case where $|G| = p_1^{n_1} p_2^{n_2} \cdots p_k^{n_k}$, we simply use the algorithm to build up a direct product of order $p_1^{n_1}$, then another of order $p_2^{n_2}$, and so on. The direct product of all of these pieces is the desired factorization of G. The following example is small enough that we can compute the appropriate internal and external direct products by hand.

EXAMPLE 1   Let $G = \{1, 8, 12, 14, 18, 21, 27, 31, 34, 38, 44, 47, 51, 53, 57, 64\}$ under multiplication modulo 65. Since G has order 16, we know it is isomorphic to one of

$$Z_{16},$$
$$Z_8 \oplus Z_2,$$
$$Z_4 \oplus Z_4,$$
$$Z_4 \oplus Z_2 \oplus Z_2,$$
$$Z_2 \oplus Z_2 \oplus Z_2 \oplus Z_2.$$

To decide which one, we dirty our hands to calculate the orders of the elements of G.

Element	1	8	12	14	18	21	27	31	34	38	44	47	51	53	57	64
Order	1	4	4	2	4	4	4	4	4	4	4	4	2	4	4	2

From the table of orders, we can instantly rule out all but $Z_4 \oplus Z_4$ and $Z_4 \oplus Z_2 \oplus Z_2$ as possibilities. Finally, we observe that since this latter group has a subgroup isomorphic to $Z_2 \oplus Z_2 \oplus Z_2$, it has more than three elements of order 2, and therefore we must have $G \approx Z_4 \oplus Z_4$.

Expressing G as an internal direct product is even easier. Pick an element of maximum order, say the element 8. Then $\langle 8 \rangle$ is a factor in the product. Next, choose a second element, say a, so that a has order 4 and a and a^2 are not in $\langle 8 \rangle = \{1, 8, 64, 57\}$. Since 12 has this property, we have $G = \langle 8 \rangle \times \langle 12 \rangle$.   ◆

Example 1 illustrates how quickly and easily one can write an Abelian group as a direct product given the orders of the elements of the group. But calculating all those orders is certainly not an appealing prospect! The good news is that, in practice, a combination of theory and calculation of the orders of a few elements will usually suffice.

EXAMPLE 2   Let $G = \{1, 8, 17, 19, 26, 28, 37, 44, 46, 53, 62, 64, 71, 73, 82, 89, 91, 98, 107, 109, 116, 118, 127, 134\}$ under multiplication modulo 135. Since G has order 24, it is isomorphic to one of

$$Z_8 \oplus Z_3 \approx Z_{24},$$
$$Z_4 \oplus Z_2 \oplus Z_3 \approx Z_{12} \oplus Z_2,$$
$$Z_2 \oplus Z_2 \oplus Z_2 \oplus Z_3 \approx Z_6 \oplus Z_2 \oplus Z_2.$$

Consider the element 8. Direct calculations show that $8^6 = 109$ and $8^{12} = 1$. (Be sure to mod as you go. For example, $8^3 \bmod 135 = 512 \bmod 135 = 107$, so compute 8^4 as $8 \cdot 107$ rather than $8 \cdot 512$.) But now we know G. Why? Clearly, $|8| = 12$ rules out the third group in the list. At the same time, $|109| = 2 = |134|$ (remember, $134 = -1 \bmod 135$) implies that G is not Z_{24} (see Theorem 4.4). Thus, $G \approx Z_{12} \oplus Z_2$, and $G = \langle 8 \rangle \times \langle 134 \rangle$.   ◆

Rather than express an Abelian group as a direct product of cyclic groups of prime-power orders, it is often more convenient to combine the cyclic factors of relatively prime order, as we did in Example 2, to obtain a direct product of the form $Z_{n_1} \oplus Z_{n_2} \oplus \cdots \oplus Z_{n_k}$, where n_i divides n_{i-1}. For example, $Z_4 \oplus Z_4 \oplus Z_2 \oplus Z_9 \oplus Z_3 \oplus Z_5$ would be written as $Z_{180} \oplus Z_{12} \oplus Z_2$ (see Exercise 11). The algorithm above is easily adapted to accomplish this by replacing step 4 by 4′: select an element a_i of maximum order m such that $m \le |G|/|G_{i-1}|$ and none of $a_i, a_i^2, \ldots, a_i^{m-1}$ is in G_{i-1}, and define $G_i = G_{i-1} \times \langle a_i \rangle$.

As a consequence of the Fundamental Theorem of Finite Abelian Groups, we have the following corollary.

Corollary   **Existence of Subgroups of Abelian Groups**

> *If m divides the order of a finite Abelian group G, then G has a subgroup of order m.*

It is instructive to verify this corollary for a specific case. Let us say that G is an Abelian group of order 72 and we wish to produce a subgroup of order 12. According to the Fundamental Theorem, G is isomorphic to one of the following six groups:

$$Z_8 \oplus Z_9, \qquad\qquad Z_8 \oplus Z_3 \oplus Z_3,$$
$$Z_4 \oplus Z_2 \oplus Z_9, \qquad\qquad Z_4 \oplus Z_2 \oplus Z_3 \oplus Z_3,$$
$$Z_2 \oplus Z_2 \oplus Z_2 \oplus Z_9, \qquad Z_2 \oplus Z_2 \oplus Z_2 \oplus Z_3 \oplus Z_3.$$

Obviously, $Z_8 \oplus Z_9 \approx Z_{72}$ and $Z_4 \oplus Z_2 \oplus Z_3 \oplus Z_3 \approx Z_{12} \oplus Z_6$ both have a subgroup of order 12. To construct a subgroup of order 12 in $Z_4 \oplus Z_2 \oplus Z_9$, we simply piece together all of Z_4 and the subgroup of order 3 in Z_9; that is, $\{(a, 0, b) \mid a \in Z_4, b \in \{0, 3, 6\}\}$. A subgroup of order 12 in $Z_8 \oplus Z_3 \oplus Z_3$ is given by $\{(a, b, 0) \mid a \in \{0, 2, 4, 6\}, b \in Z_3\}$. An analogous procedure applies to the remaining cases and indeed to any finite Abelian group.

PROOF OF THE FUNDAMENTAL THEOREM

Because of the length and complexity of the proof of the Fundamental Theorem of Finite Abelian Groups, we will break it up into a series of lemmas.

Lemma 1

> *Let G be a finite Abelian group of order $p^n m$, where p is a prime that does not divide m. Then $G = H \times K$, where $H = \{x \in G \mid x^{p^n} = e\}$ and $K = \{x \in G \mid x^m = e\}$. Moreover, $|H| = p^n$.*

PROOF.   It is an easy exercise to prove that H and K are subgroups of G (see Exercise 23 in Chapter 3). Because G is Abelian, to prove that $G = H \times K$ we need only prove that $G = HK$ and $H \cap K = \{e\}$. Since we have $\gcd(m, p^n) = 1$, there are integers s and t such that $1 = sm + tp^n$. For any x in G, we have $x = x^1 = x^{sm + tp^n} = x^{sm} x^{tp^n}$ and, by

Corollary 4 of Lagrange's Theorem (Theorem 7.1), $x^{sm} \in H$ and $x^{tp^n} \in K$. Thus, $G = HK$. Now suppose that some $x \in H \cap K$. Then $x^{p^n} = e = x^m$ and, by Corollary 2 to Theorem 4.1, $|x|$ divides both p^n and m. Since p does not divide m, we have $|x| = 1$ and, therefore, $x = e$.

To prove the second assertion of the lemma, note that $p^n m = |HK| = |H||K|/|H \cap K| = |H||K|$ (see Exercise 7 in the Supplementary Exercises for Chapters 5–8). It follows from Theorem 9.5 and Corollary 2 to Theorem 4.1 that p does not divide $|K|$ and therefore $|H| = p^n$. ∎

Given an Abelian group G with $|G| = p_1^{n_1} p_2^{n_2} \cdots p_k^{n_k}$, where the p's are distinct primes, we let $G(p_i)$ denote the set $\{x \in G \mid x^{p_i^{n_i}} = e\}$. It then follows immediately from Lemma 1 and induction that $G = G(p_1) \times G(p_2) \times \cdots \times G(p_k)$ and $|G(p_i)| = p_i^{n_i}$. Hence, we turn our attention to groups of prime-power order.

Lemma 2

> Let G be an Abelian group of prime-power order and let a be an element of maximal order in G. Then G can be written in the form $\langle a \rangle \times K$.

PROOF. We denote $|G|$ by p^n and induct on n. If $n = 1$, then $G = \langle a \rangle \times \langle e \rangle$. Now assume that the statement is true for all Abelian groups of order p^k, where $k < n$. Among all the elements of G, choose a of maximal order p^m. Then $x^{p^m} = e$ for all x in G. We may assume that $G \neq \langle a \rangle$, for otherwise there is nothing to prove. Now, among all the elements of G, choose b of smallest order such that $b \notin \langle a \rangle$. We claim that $\langle a \rangle \cap \langle b \rangle = \{e\}$. Clearly, we may establish this claim by showing that $|b| = p$. Since $|b^p| = |b|/p$, we know that $b^p \in \langle a \rangle$ by the manner in which b was chosen. Say $b^p = a^i$. Notice that $e = b^{p^m} = (b^p)^{p^{m-1}} = (a^i)^{p^{m-1}}$, so $|a^i| \leq p^{m-1}$. Thus, a^i is not a generator of $\langle a \rangle$ and, therefore, by Corollary 2 to Theorem 4.2, $\gcd(p^m, i) \neq 1$. This proves that p divides i, so that we can write $i = pj$. Then $b^p = a^i = a^{pj}$. Consider the element $c = a^{-j}b$. Certainly, c is not in $\langle a \rangle$, for if it were, b would be, too. Also, $c^p = a^{-jp}b^p = a^{-i}b^p = b^{-p}b^p = e$. Thus, we have found an element c of order p such that $c \notin \langle a \rangle$. Since b was chosen to have smallest order such that $b \notin \langle a \rangle$, we conclude that b also has order p, and our claim is verified.

Now consider the factor group $\overline{G} = G/\langle b \rangle$. To simplify the notation, we let $\overline{x}$ denote the coset $x\langle b \rangle$ in $\overline{G}$. If $|\overline{a}| < |a| = p^m$, then $\bar{a}^{p^{m-1}} = \bar{e}$. This means that $(a\langle b\rangle)^{p^{m-1}} = a^{p^{m-1}}\langle b\rangle = \langle b\rangle$, so that $a^{p^{m-1}} \in \langle a\rangle \cap \langle b\rangle = \{e\}$, contradicting the fact that $|a| = p^m$. Thus, $|\bar{a}| = |a| = p^m$, and therefore $\bar{a}$ is an element of maximal order in $\bar{G}$. By induction, we know that $\bar{G}$ can be written in the form $\langle \bar{a}\rangle \times \bar{K}$ for some subgroup $\bar{K}$ of $\bar{G}$. Let K be the pullback of $\bar{K}$ under the natural homomorphism from G to $\bar{G}$ (that is, $K = \{x \in G \mid \bar{x} \in \bar{K}\}$). We claim that $\langle a\rangle \cap K = \{e\}$. For if $x \in \langle a\rangle \cap K$, then $\bar{x} \in \langle \bar{a}\rangle \cap \bar{K} = \{\bar{e}\} = \langle b\rangle$ and $x \in \langle a\rangle \cap \langle b\rangle = \{e\}$. It now follows from an order argument (see Exercise 33) that $G = \langle a\rangle K$, and therefore $G = \langle a\rangle \times K$. ■■

Lemma 2 and induction on the order of the group now give the following.

Lemma 3

> *A finite Abelian group of prime-power order is an internal direct product of cyclic groups.*

Let us pause to determine where we are in our effort to prove the Fundamental Theorem of Finite Abelian Groups. The remark following Lemma 1 shows that $G = G(p_1) \times G(p_2) \times \cdots \times G(p_n)$, where each $G(p_i)$ is a group of prime-power order, and Lemma 3 shows that each of these factors is an internal direct product of cyclic groups. Thus, we have proved that G is an internal direct product of cyclic groups of prime-power order. All that remains to be proved is the uniqueness of the factors. Certainly the groups $G(p_i)$ are uniquely determined by G, since they comprise the elements of G whose orders are powers of p_i. So we must prove that there is only one way (up to isomorphism and rearrangement of factors) to write each $G(p_i)$ as an internal direct product of cyclic groups.

Lemma 4

> *Suppose that G is a finite Abelian group of prime-power order. If $G = H_1 \times H_2 \times \cdots \times H_m$ and $G = K_1 \times K_2 \times \cdots \times K_n$, where the H's and K's are nontrivial cyclic subgroups with $|H_1| \geq |H_2| \geq \cdots \geq |H_m|$ and $|K_1| \geq |K_2| \geq \cdots \geq |K_n|$, then $m = n$ and $|H_i| = |K_i|$ for all i.*

PROOF. We proceed by induction on $|G|$. Clearly, the case where $|G| = p$ is true. Now suppose that the statement is true for all Abelian groups of order less than $|G|$. For any Abelian group L, the set $L^p = \{x^p \mid x \in L\}$ is a subgroup of L (see Exercise 15 in the Supplementary Exercises for Chapters 1–4) and is a proper subgroup if p divides $|L|$. It follows that $G^p = H_1{}^P \times H_2{}^P \times \cdots \times H_{m'}{}^P$, and $G^p = K_1{}^P \times K_2{}^P \times \cdots \times K_{n'}{}^P$, where m' is the largest integer i such that $|H_i| > p$, and n' is the largest integer j such that $|K_j| > p$. (This ensures that our two direct products for G^p do not have trivial factors.) Since $|G^p| < |G|$, we have, by induction, $m' = n'$ and $|H_i{}^P| = |K_i{}^P|$ for $i = 1, \ldots, m'$. Since $|H_i| = p|H_i{}^P|$, this proves that $|H_i| = |K_i|$ for all $i = 1, \ldots, m'$. All that remains to be proved is that the number of H_i of order p equals the number of K_i of order p; that is, we must prove that $m - m' = n - n'$ (since $n' = m'$). This follows directly from the facts that $|H_1||H_2| \cdots |H_{m'}|p^{m-m'} = |G| = |K_1||K_2| \cdots |K_{n'}|p^{n-n'}$, $|H_i| = |K_i|$, and $m' = n'$. ▪ ▪

EXERCISES

You know it ain't easy, you know how hard it can be.

JOHN LENNON AND PAUL MCCARTNEY,
"The Ballad of John and Yoko"

1. What is the smallest positive integer n such that there are two nonisomorphic groups of order n? Name the two groups.
2. What is the smallest positive integer n such that there are three nonisomorphic Abelian groups of order n? Name the three groups.
3. What is the smallest positive integer n such that there are exactly four nonisomorphic Abelian groups of order n? Name the four groups.
4. Calculate the number of elements of order 2 in each of Z_{16}, $Z_8 \oplus Z_2$, $Z_4 \oplus Z_4$, and $Z_4 \oplus Z_2 \oplus Z_2$. Do the same for the elements of order 4.
5. Prove that any Abelian group of order 45 has an element of order 15. Does every Abelian group of order 45 have an element of order 9?
6. Show that there are two Abelian groups of order 108 that have exactly one subgroup of order 3.
7. Show that there are two Abelian groups of order 108 that have exactly four subgroups of order 3.
8. Show that there are two Abelian groups of order 108 that have exactly 13 subgroups of order 3.

9. Suppose that G is an Abelian group of order 120 and that G has exactly three elements of order 2. Determine the isomorphism class of G.

10. Find all Abelian groups (up to isomorphism) of order 360.

11. Prove that every finite Abelian group can be expressed as the (external) direct product of cyclic groups of orders $n_1, n_2, \ldots, n_t$, where n_{i+1} divides n_i for $i = 1, 2, \ldots, t - 1$. (This exercise is referred to in this Chapter and in Chapter 22.)

12. Suppose that the order of some finite Abelian group is divisible by 10. Prove that the group has a cyclic subgroup of order 10.

13. Show, by example, that if the order of a finite Abelian group is divisible by 4, the group need not have a cyclic subgroup of order 4.

14. On the basis of Exercises 12 and 13, draw a general conclusion about the existence of cyclic subgroups of a finite Abelian group.

15. How many Abelian groups (up to isomorphism) are there
    a. of order 6?
    b. of order 15?
    c. of order 42?
    d. of order pq, where p and q are distinct primes?
    e. of order pqr, where p, q, and r are distinct primes?
    f. Generalize parts d and e.

16. How does the number (up to isomorphism) of Abelian groups of order n compare with the number (up to isomorphism) of Abelian groups of order m where
    a. $n = 3^2$ and $m = 5^2$?
    b. $n = 2^4$ and $m = 5^4$?
    c. $n = p^r$ and $m = q^r$, where p and q are prime?
    d. $n = p^r$ and $m = p^r q$, where p and q are distinct primes?
    e. $n = p^r$ and $m = p^r q^2$, where p and q are distinct primes?

17. The symmetry group of a nonsquare rectangle is an Abelian group of order 4. Is it isomorphic to Z_4 or $Z_2 \oplus Z_2$?

18. Verify the corollary to the Fundamental Theorem of Finite Abelian Groups in the case that the group has order 1080 and the divisor is 180.

19. The set $\{1, 9, 16, 22, 29, 53, 74, 79, 81\}$ is a group under multiplication modulo 91. Determine the isomorphism class of this group.

20. Suppose that G is a finite Abelian group that has exactly one subgroup for each divisor of $|G|$. Show that G is cyclic.

21. Characterize those integers n such that the only Abelian groups of order n are cyclic.

22. Characterize those integers n such that any Abelian group of order n belongs to one of exactly four isomorphism classes.

23. Refer to Example 1 in this chapter and explain why it is unnecessary to compute the orders of the last five elements listed to determine the isomorphism class of G.

24. Let $G = \{1, 7, 17, 23, 49, 55, 65, 71\}$ under multiplication modulo 96. Express G as an external and an internal direct product of cyclic groups.

25. Let $G = \{1, 7, 43, 49, 51, 57, 93, 99, 101, 107, 143, 149, 151, 157, 193, 199\}$ under multiplication modulo 200. Express G as an external and an internal direct product of cyclic groups.

26. The set $G = \{1, 4, 11, 14, 16, 19, 26, 29, 31, 34, 41, 44\}$ is a group under multiplication modulo 45. Write G as an external and an internal direct product of cyclic groups of prime-power order.

27. Suppose that G is an Abelian group of order 9. What is the maximum number of elements (excluding the identity) of which one needs to compute the order to determine the isomorphism class of G? What if G has order 18? What about 16?

28. Suppose that G is an Abelian group of order 16, and in computing the orders of its elements, you come across an element of order 8 and two elements of order 2. Explain why no further computations are needed to determine the isomorphism class of G.

29. Let G be an Abelian group of order 16. Suppose that there are elements a and b in G such that $|a| = |b| = 4$ and $a^2 \neq b^2$. Determine the isomorphism class of G.

30. Prove that an Abelian group of order $2^n (n \geq 1)$ must have an odd number of elements of order 2.

31. Without using Lagrange's Theorem, show that an Abelian group of odd order cannot have an element of even order.

32. Let G be the group of all $n \times n$ diagonal matrices with ± 1 diagonal entries. What is the isomorphism class of G?

33. Prove the assertion made in the proof of Lemma 2 that $G = \langle a \rangle K$.

34. Suppose that G is a finite Abelian group. Prove that G has order p^n, where p is prime, if and only if the order of every element of G is a power of p.

35. Dirichlet's Theorem says that, for every pair of relatively prime integers a and b, there are infinitely many primes of the form $at + b$. Use Dirichlet's Theorem to prove that every finite Abelian group is isomorphic to a subgroup of a U-group.

36. Determine the isomorphism class of $\text{Aut}(Z_2 \oplus Z_3 \oplus Z_5)$.

COMPUTER EXERCISES ▪▪

The purpose of computation is insight, not numbers.

RICHARD HAMMING

Software for the computer exercises in this chapter is available at the website:

http://www.d.umn.edu/~jgallian

1. This software lists the isomorphism classes of all finite Abelian groups of any particular order n. Run the program for $n = 16, 24, 512, 2048, 441000$, and 999999.

2. This software determines how many integers in a given interval are the order of exactly one Abelian group, of exactly two Abelian groups, and so on, up to exactly nine Abelian groups. Run your program for the integers up to 1000. Then from 10001 to 11000. Then choose your own interval of 1000 consecutive integers. Is there much difference in the results?

3. This software expresses a U-group as an internal direct product of subgroups $H_1 \times H_2 \times \cdots \times H_i$, where $|H_i|$ divides $|H_{i-1}|$. Run the program for the groups $U(32)$, $U(80)$, and $U(65)$.

Reference

1. R. Hirshon, "On Cancellation in Groups," *American Mathematical Monthly* 76 (1969): 1037–1039.

Suggested Readings

J. A. Gallian, "Computers in Group Theory," *Mathematics Magazine* 49 (1976): 69–73.

This paper discusses several computer-related projects in group theory done by undergraduate students.

J. Kane, "Distribution of Orders of Abelian Groups," *Mathematics Magazine* 49 (1976): 132–135.

In this article, the author determines the percentages of integers k between 1 and n, for sufficiently large n, that have exactly one isomorphism class of Abelian groups of order k, exactly two isomorphism classes of Abelian groups of order k, and so on, up to 13 isomorphism classes.

G. Mackiw, "Computing in Abstract Algebra," *The College Mathematics Journal* 27 (1996): 136–142.

This article explains how one can use computer software to implement the algorithm given in this chapter for expressing an Abelian group as an internal direct product.

Suggested Website

To find more information about the development of group theory, visit:

http://www-groups.dcs.st-and.ac.uk/~history/

SUPPLEMENTARY EXERCISES FOR CHAPTERS 9–11   ▬ ▪▪

Every prospector drills many a dry hole, pulls out his rig, and moves on.

JOHN L. HESS

True/false questions for Chapters 9–11 are available on the Web at:

http://www.d.umn.edu/~jgallian/TF

1. Suppose that H is a subgroup of G and that each left coset of H in G is some right coset of H in G. Prove that H is normal in G.

2. Use a factor group-induction argument to prove that a finite Abelian group of order n has a subgroup of order m for every positive divisor m of n.

3. Let $\text{diag}(G) = \{(g, g) \mid g \in G\}$. Prove that $\text{diag}(G) \lhd G \oplus G$ if and only if G is Abelian. When G is finite, what is the index of $\text{diag}(G)$ in $G \oplus G$?

4. Let H be any group of rotations in D_n. Prove that H is normal in D_n.

5. Prove that $\text{Inn}(G) \lhd \text{Aut}(G)$.

6. Let H be a subgroup of G. Prove that H is a normal subgroup if and only if for all a and b in $Gab \in H$ implies $ba \in H$.

7. The factor group $GL(2, \mathbf{R})/SL(2, \mathbf{R})$ is isomorphic to some very familiar group. What is the group?

8. Let k be a divisor of n. The factor group $(Z/\langle n \rangle)/(\langle k \rangle/\langle n \rangle)$ is isomorphic to some very familiar group. What is the group?

9. Let

$$H = \left\{ \begin{bmatrix} 1 & a & b \\ 0 & 1 & c \\ 0 & 0 & 1 \end{bmatrix} \;\middle|\; a, b, c \in Q \right\}$$

   under matrix multiplication.

   a. Prove that $Z(H)$ is isomorphic to Q under addition.
   b. Prove that $H/Z(H)$ is isomorphic to $Q \oplus Q$.
   c. Are your proofs for parts a and b valid when Q is replaced by $\mathbf{R}$? Are they valid when Q is replaced by Z_p, where p is prime?

10. Prove that $D_4/Z(D_4)$ is isomorphic to $Z_2 \oplus Z_2$.

11. Prove that Q/Z under addition is an infinite group in which every element has finite order.

12. Show that the intersection of any collection of normal subgroups of a group is a normal subgroup.

13. Let $n > 1$ be a fixed integer and let G be a group. If the set $H = \{x \in G \mid |x| = n\}$ together with the identity forms a subgroup of G, prove that it is a normal subgroup of G. In the case where such a subgroup exists, what can be said about n? Give an example of a non-Abelian group that has such a subgroup. Give an example of a group G and a prime n for which the set H together with the identity is not a subgroup.

14. Show that Q/Z has a unique subgroup of order n for each positive integer n.

15. If H and K are normal Abelian subgroups of a group and $H \cap K = \{e\}$, prove that HK is Abelian.

16. Prove that the mapping $x \to x^2$ from a finite group to itself is one-to-one if the group has odd order.

17. Suppose that G is a group of permutations on some set. If $|G| = 60$ and $\mathrm{orb}_G(5) = \{1, 5\}$, prove that $\mathrm{stab}_G(5)$ is normal in G.

18. Suppose that $G = H \times K$ and that N is a normal subgroup of H. Prove that N is normal in G.

19. Show that there is no homomorphism from $Z_8 \oplus Z_2 \oplus Z_2$ onto $Z_4 \oplus Z_4$.

20. Show that there is no homomorphism from A_4 onto a group of order 2, 4, or 6, but that there is a homomorphism from A_4 onto a group of order 3.

21. Let H be a normal subgroup of S_4 of order 4. Prove that S_4/H is isomorphic to S_3.

22. Suppose that ϕ is a homomorphism of $U(36)$, $\mathrm{Ker}\ \phi = \{1, 13, 25\}$, and $\phi(5) = 17$. Determine all elements that map to 17.

23. Let $n = 2m$, where m is odd. How many elements of order 2 does $D_n/Z(D_n)$ have? How many elements are in the subgroup $\langle R_{360/n} \rangle / Z(D_n)$? How do these numbers compare with the number of elements of order 2 in D_m?

24. Suppose that H is a normal subgroup of a group G of odd order and that $|H| = 5$. Show that $H \subseteq Z(G)$.

25. Let G be an Abelian group and let n be a positive integer. Let $G_n = \{g \mid g^n = e\}$ and $G^n = \{g^n \mid g \in G\}$. Prove that G/G_n is isomorphic to G^n.

26. Let $\mathbf{R}^+$ denote the multiplicative group of positive reals and let $T = \{a + bi \in \mathbf{C} \mid a^2 + b^2 = 1\}$ be the multiplicative group of complex numbers of norm 1. Show that $\mathbf{C}^*$ is the internal direct product of $\mathbf{R}^+$ and T.

27. Let G be a finite group and let p be a prime. If $p^2 > |G|$, show that any subgroup of order p is normal in G.

28. Let $G = Z \oplus Z$ and $H = \{(x, y) \mid x \text{ and } y \text{ are even integers}\}$. Show that H is a subgroup of G. Determine the order of G/H. To which familiar group is G/H isomorphic?

29. Let n be a positive integer. Prove that every element of order n in Q/Z is contained in $\langle 1/n + Z \rangle$.

30. (1997 Putnam Competition) Let G be a group and let $\phi : G \rightarrow G$ be a function such that

$$\phi(g_1)\phi(g_2)\phi(g_3) = \phi(h_1)\phi(h_2)\phi(h_3)$$

whenever $g_1g_2g_3 = e = h_1h_2h_3$. Prove that there exists an element a in G such that $\psi(x) = a\phi(x)$ is a homomorphism.

31. Prove that every homomorphism from $Z \oplus Z$ into Z has the form $(x, y) \rightarrow ax + by$, where a and b are integers.

32. Prove that every homomorphism from $Z \oplus Z$ into $Z \oplus Z$ has the form $(x, y) \rightarrow (ax + by, cx + dy)$, where a, b, c, and d are integers.

33. Prove that Q/Z is not isomorphic to a proper subgroup of itself.

34. Prove that for each positive integer n, the group Q/Z has exactly $\phi(n)$ elements of order n (ϕ is the Euler phi function).

35. Show that any group with more than two elements has an automorphism other than the identity mapping.

36. A proper subgroup H of a group G is called *maximal* if there is no subgroup K such that $H \subset K \subset G$. Prove that Q under addition has no maximal subgroups.

PART 3

Rings

 For online student resources,
visit this textbook's website at
math.college.hmco.com/students.

12 Introduction to Rings

Example is the school of mankind, and they will learn at no other.

EDMUND BURKE, *On a Regicide Peace*

MOTIVATION AND DEFINITION

Many sets are naturally endowed with two binary operations: addition and multiplication. Examples that quickly come to mind are the integers, the integers modulo n, the real numbers, matrices, and polynomials. When considering these sets as groups, we simply used addition and ignored multiplication. In many instances, however, one wishes to take into account both addition and multiplication. One abstract concept that does this is the concept of a ring.[†] This notion was originated in the mid-nineteenth century by Richard Dedekind, although its first formal abstract definition was not given until Abraham Fraenkel presented it in 1914.

DEFINITION   *Ring*

A *ring* R is a nonempty set with two binary operations, addition (denoted by $a + b$) and multiplication (denoted by ab), such that for all a, b, c in R:

1. $a + b = b + a$.
2. $(a + b) + c = a + (b + c)$.
3. There is an additive identity 0. That is, there is an element 0 in R such that $a + 0 = a$ for all a in R.
4. There is an element $-a$ in R such that $a + (-a) = 0$.
5. $a(bc) = (ab)c$.
6. $a(b + c) = ab + ac$ and $(b + c)a = ba + ca$.

[†]The term *ring* was first applied in 1897 by the German mathematician David Hilbert (1862–1943).

So, a ring is an Abelian group under addition, also having an associative multiplication that is left and right distributive over addition. Note that multiplication need not be commutative. When it is, we say that the ring is *commutative*. Also, a ring need not have an identity under multiplication. A *unity* (or *identity*) in a ring is a nonzero element that is an identity under multiplication. A nonzero element of a commutative ring with unity need not have a multiplicative inverse. When it does, we say that it is a *unit* of the ring. Thus, a is a unit if a^{-1} exists.

The following terminology and notation are convenient. If a and b belong to a commutative ring R and a is nonzero, we say that a *divides* b (or that a is a *factor* of b) and write $a \mid b$, if there exists an element c in R such that $b = ac$. If a does not divide b, we write $a \nmid b$.

Recall that if a is an element from a group under the operation of addition and n is a positive integer, na means $a + a + \cdots + a$, where there are n summands. When dealing with rings, this notation can cause confusion, since we also use juxtaposition for the ring multiplication. When there is the potential for confusion, we will use $n \cdot a$ to mean $a + a + \cdots + a$ (n summands).

For an abstraction to be worthy of study, it must have many diverse concrete realizations. The following list of examples shows that the ring concept is pervasive.

EXAMPLES OF RINGS

EXAMPLE 1   The set Z of integers under ordinary addition and multiplication is a commutative ring with unity 1. The units of Z are 1 and -1.   ◆

EXAMPLE 2   The set $Z_n = \{0, 1, \ldots, n - 1\}$ under addition and multiplication modulo n is a commutative ring with unity 1. The set of units is $U(n)$.   ◆

EXAMPLE 3   The set $Z[x]$ of all polynomials in the variable x with integer coefficients under ordinary addition and multiplication is a commutative ring with unity $f(x) = 1$.   ◆

EXAMPLE 4   The set $M_2(Z)$ of 2×2 matrices with integer entries is a noncommutative ring with unity $\begin{bmatrix} 1 & 0 \\ 0 & 1 \end{bmatrix}$.   ◆

EXAMPLE 5   The set 2Z of even integers under ordinary addition and multiplication is a commutative ring without unity.   ◆

EXAMPLE 6   The set of all continuous real-valued functions of a real variable whose graphs pass through the point $(1, 0)$ is a commutative ring without unity under the operations of pointwise addition and multiplication [that is, the operations $(f + g)(a) = f(a) + g(a)$ and $(fg)(a) = f(a)g(a)$].   ◆

EXAMPLE 7   Let $R_1, R_2, \ldots, R_n$ be rings. We can use these to construct a new ring as follows. Let

$$R_1 \oplus R_2 \oplus \cdots \oplus R_n = \{(a_1, a_2, \ldots, a_n) \mid a_i \in R_i\}$$

and perform componentwise addition and multiplication; that is, define

$$(a_1, a_2, \ldots, a_n) + (b_1, b_2, \ldots, b_n) = (a_1 + b_1, a_2 + b_2, \ldots, a_n + b_n)$$

and

$$(a_1, a_2, \ldots, a_n)(b_1, b_2, \ldots, b_n) = (a_1 b_1, a_2 b_2, \ldots, a_n b_n).$$

This ring is called the *direct sum* of $R_1, R_2, \ldots, R_n$.   ◆

PROPERTIES OF RINGS

Our first theorem shows how the operations of addition and multiplication intertwine. We use $b - c$ to denote $b + (-c)$.

Theorem 12.1   **Rules of Multiplication**

> *Let a, b, and c belong to a ring R. Then*
>
> **1.** $a0 = 0a = 0$.
> **2.** $a(-b) = (-a)b = -(ab)$.
> **3.** $(-a)(-b) = ab$.[†]
> **4.** $a(b - c) = ab - ac$ and $(b - c)a = ba - ca$.
>
> *Furthermore, if R has a unity element 1, then*
>
> **5.** $(-1)a = -a$.
> **6.** $(-1)(-1) = 1$.

[†]*Minus times minus is plus.*
The reason for this we need not discuss.
W. H. Auden

PROOF.   We will prove rules 1 and 2 and leave the rest as easy exercises (see Exercise 11). To prove statements such as those in Theorem 12.1, we need only "play off" the distributive property against the fact that R is a group under addition with additive identity 0. Consider rule 1. Clearly,

$$0 + a0 = a0 = a(0 + 0) = a0 + a0.$$

So, by cancellation, $0 = a0$. Similarly, $0a = 0$.

To prove rule 2, we observe that $a(-b) + ab = a(-b + b) = a0 = 0$. So, adding $-(ab)$ to both sides yields $a(-b) = -(ab)$. The remainder of rule 2 is done analogously.   ■ ■

Recall that in the case of groups, the identity and inverses are unique. The same is true for rings, provided that these elements exist. The proofs are identical to the ones given for groups and therefore are omitted.

Theorem 12.2   Uniqueness of the Unity and Inverses

> *If a ring has a unity, it is unique. If a ring element has a multiplicative inverse, it is unique.*

Many students have the mistaken tendency to treat a ring as if it were a group under *multiplication.* It is not. The two most common errors are the assumptions that ring elements have multiplicative inverses—they need not—and that a ring has a multiplicative identity—it need not. For example, if a, b, and c belong to a ring, $a \neq 0$ and $ab = ac$, we *cannot* conclude that $b = c$. Similarly, if $a^2 = a$, we *cannot* conclude that $a = 0$ or 1 (as is the case with real numbers). In the first place, the ring need not have multiplicative cancellation, and in the second place, the ring need not have a multiplicative identity. There is an important class of rings wherein multiplicative identities exist and for which multiplicative cancellation holds. This class is taken up in the next chapter.

SUBRINGS

In our study of groups, subgroups played a crucial role. Subrings, the analogous structures in ring theory, play a much less prominent role than their counterparts in group theory. Nevertheless, subrings are important.

DEFINITION *Subring*

> A subset S of a ring R is a *subring of R* if S is itself a ring with the operations of R.

Just as was the case for subgroups, there is a simple test for subrings.

Theorem 12.3 Subring Test

> A nonempty subset S of a ring R is a subring if S is closed under subtraction and multiplication—that is, if $a - b$ and ab are in S whenever a and b are in S.

PROOF. Since addition in R is commutative and S is closed under subtraction, we know by the One-Step Subgroup Test (Theorem 3.1) that S is an Abelian group under addition. Also, since multiplication in R is associative as well as distributive over addition, the same is true for multiplication in S. Thus, the only condition remaining to be checked is that multiplication is a binary operation on S. But this is exactly what closure means. ∎

We leave it to the student to confirm that each of the following examples is a subring.

EXAMPLE 8 $\{0\}$ and R are subrings of any ring R. $\{0\}$ is called the *trivial* subring of R. ◆

EXAMPLE 9 $\{0, 2, 4\}$ is a subring of the ring Z_6, the integers modulo 6. Note that although 1 is the unity in Z_6, 4 is the unity in $\{0, 2, 4\}$. ◆

EXAMPLE 10 For each positive integer n, the set

$$nZ = \{0, \pm n, \pm 2n, \pm 3n, \ldots\}$$

is a subring of the integers Z. ◆

EXAMPLE 11 The set of Gaussian integers

$$Z[i] = \{a + bi \mid a, b \in Z\}$$

is a subring of the complex numbers **C**. ◆

EXAMPLE 12 Let R be the ring of all real-valued functions of a single real variable under pointwise addition and multiplication. The subset S of R of functions whose graphs pass through the origin forms a subring of R.   ◆

EXAMPLE 13   The set

$$\left\{ \begin{bmatrix} a & 0 \\ 0 & b \end{bmatrix} \middle| \, a, b \in Z \right\}$$

of diagonal matrices is a subring of the ring of all 2×2 matrices over Z.   ◆

We can picture the relationship between a ring and its various subrings by way of a subring lattice diagram. In such a diagram, any ring is a subring of all the rings that it is connected to by one or more upward lines. Figure 12.1 shows the relationship among some of the rings we have already discussed.

In the next several chapters, we will see that many of the fundamental concepts of group theory can be naturally extended to rings. In particular, we will introduce ring homomorphisms and factor rings.

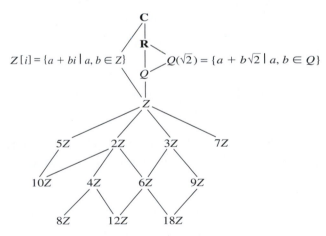

Figure 12.1   Partial subring lattice diagram of **C**.

EXERCISES

There is no substitute for hard work.
THOMAS ALVA EDISON, Life

1. Give an example of a finite noncommutative ring. Give an example of an infinite noncommutative ring that does not have a unity.

2. The ring $\{0, 2, 4, 6, 8\}$ under addition and multiplication modulo 10 has a unity. Find it.

3. Let $Z[\sqrt{2}] = \{a + b\sqrt{2} \mid a, b \in Z\}$. Prove that $Z[\sqrt{2}]$ is a ring under the ordinary addition and multiplication of real numbers. Is your proof valid if 2 is replaced by any positive integer n?

4. Show, by example, that for fixed nonzero elements a and b in a ring, the equation $ax = b$ can have more than one solution. How does this compare with groups?

5. Prove Theorem 12.2.

6. Find an integer n that shows that the rings Z_n need not have the following properties that the ring of integers has.
   a. $a^2 = a$ implies $a = 0$ or $a = 1$.
   b. $ab = 0$ implies $a = 0$ or $b = 0$.
   c. $ab = ac$ and $a \neq 0$ imply $b = c$.
   Is the n you found prime?

7. Show that the three properties listed in Exercise 6 are valid for Z_p, where p is prime.

8. Show that a ring is commutative if it has the property that $ab = ca$ implies $b = c$ when $a \neq 0$.

9. Prove that the intersection of any collection of subrings of a ring R is a subring of R.

10. Verify that Examples 8 through 13 in this chapter are as stated.

11. Prove parts 3 through 6 of Theorem 12.1.

12. Give an example of a noncommutative ring that has exactly 16 elements.

13. Describe all the subrings of the ring of integers.

14. Let a and b belong to a ring R and let m be an integer. Prove that $m \cdot (ab) = (m \cdot a)b = a(m \cdot b)$.

15. Show that if m and n are integers and a and b are elements from a ring, then $(m \cdot a)(n \cdot b) = (mn) \cdot (ab)$. (This exercise is referred to in Chapters 13 and 15.)

16. Show that if n is an integer and a is an element from a ring, then $n \cdot (-a) = -(n \cdot a)$.

17. Show that a ring that is cyclic under addition is commutative.

18. Let a belong to a ring R. Let $S = \{x \in R \mid ax = 0\}$. Show that S is a subring of R.

19. Let R be a ring. The *center of R* is the set $\{x \in R \mid ax = xa \text{ for all } a \text{ in } R\}$. Prove that the center of a ring is a subring.

20. Describe the elements of $M_2(Z)$ (see Example 4) that have multiplicative inverses.

21. Suppose that $R_1, R_2, \ldots, R_n$ are rings that contain nonzero elements. Show that $R_1 \oplus R_2 \oplus \cdots \oplus R_n$ has a unity if and only if each R_i has a unity.

22. Let R be a commutative ring with unity and let $U(R)$ denote the set of units of R. Prove that $U(R)$ is a group under the multiplication of R. (This group is called the *group of units of R*.)

23. Determine $U(Z[i])$ (see Example 11).

24. If $R_1, R_2, \ldots, R_n$ are commutative rings with unity, show that $U(R_1 \oplus R_2 \oplus \cdots \oplus R_n) = U(R_1) \oplus U(R_2) \oplus \cdots \oplus U(R_n)$.

25. Determine $U(Z[x])$. (This exercise is referred to in Chapter 17.)

26. Determine $U(\mathbf{R}[x])$.

27. Show that a unit of a ring divides every element of the ring.

28. In Z_6, show that $4 \mid 2$; in Z_8, show that $3 \mid 7$; in Z_{15}, show that $9 \mid 12$.

29. Suppose that a and b belong to a commutative ring R. If a is a unit of R and $b^2 = 0$, show that $a + b$ is a unit of R.

30. Suppose that there is an integer $n > 1$ such that $x^n = x$ for all elements x of some ring. If m is a positive integer and $a^m = 0$ for some a, show that $a = 0$.

31. Give an example of ring elements a and b with the properties that $ab = 0$ but $ba \neq 0$.

32. Let n be an integer greater than 1. In a ring in which $x^n = x$ for all x, show that $ab = 0$ implies $ba = 0$.

33. Suppose that R is a ring such that $x^3 = x$ for all x in R. Prove that $6x = 0$ for all x in R.

34. Suppose that a belongs to a ring and $a^4 = a^2$. Prove that $a^{2n} = a^2$ for all $n \geq 1$.

35. Find an integer $n > 1$ such that $a^n = a$ for all a in Z_6. Do the same for Z_{10}. Show that no such n exists for Z_m when m is divisible by the square of some prime.

36. Let m and n be positive integers and let k be the least common multiple of m and n. Show that $mZ \cap nZ = kZ$.

37. Explain why every subgroup of Z_n under addition is also a subring of Z_n.

38. Is Z_6 a subring of Z_{12}?

39. Suppose that R is a ring with unity 1 and a is an element of R such that $a^2 = 1$. Let $S = \{ara \mid r \in R\}$. Prove that S is a subring of R. Does S contain 1?

40. Let $M_2(Z)$ be the ring of all 2×2 matrices over the integers and let $R = \left\{ \begin{bmatrix} a & a + b \\ a + b & b \end{bmatrix} \mid a, b \in Z \right\}$. Prove or disprove that R is a subring of $M_2(Z)$.

41. Let $M_2(Z)$ be the ring of all 2×2 matrices over the integers and let $R = \left\{ \begin{bmatrix} a & a-b \\ a-b & b \end{bmatrix} \middle| a, b \in Z \right\}$. Prove or disprove that R is a subring of $M_2(Z)$.

42. Let $R = \left\{ \begin{bmatrix} a & a \\ b & b \end{bmatrix} \middle| a, b \in Z \right\}$. Prove or disprove that R is a subring of $M_2(Z)$.

43. Let $R = Z \oplus Z \oplus Z$ and $S = \{(a, b, c) \in R \mid a + b = c\}$. Prove or disprove that S is a subring of R.

44. Suppose that there is a positive even integer n such that $a^n = a$ for all elements a of some ring. Show that $-a = a$ for all a in the ring.

45. Let R be a ring with unity 1. Show that $S = \{n \cdot 1 \mid n \in Z\}$ is a subring of R.

46. Show that $2Z \cup 3Z$ is not a subring of Z.

47. Determine the smallest subring of Q that contains 1/2. (That is, find the subring S with the property that S contains 1/2 and, if T is any subring containing 1/2, then T contains S.)

48. Determine the smallest subring of Q that contains 2/3.

49. Let R be a ring. Prove that $a^2 - b^2 = (a + b)(a - b)$ for all a, b in R if and only if R is commutative.

50. Suppose that R is a ring and that $a^2 = a$ for all a in R. Show that R is commutative. [A ring in which $a^2 = a$ for all a is called a *Boolean* ring, in honor of the English mathematician George Boole (1815–1864).]

51. Give an example of a Boolean ring with four elements. Give an example of an infinite Boolean ring.

52. Give an example of a subset of a ring that is a subgroup under addition but not a subring.

COMPUTER EXERCISES

Theory is the general; experiments are the soldiers.

LEONARDO DA VINCI

Software for the computer exercises in this chapter is available at the website:

http://www.d.umn.edu/~jgallian

1. This software finds all solutions to the equation $x^2 + y^2 = 0$ in Z_p. Run the software for all odd primes up to 37. Make a conjecture about the existence of nontrivial solutions in Z_p (p a prime) and the form of p.

2. Let $Z_n[i] = \{a + bi \mid a, b$ belong to $Z_n, i^2 = -1\}$ (the Gaussian integers modulo n). This software finds the group of units of this ring and the order of each element of the group. Run the program for $n = 3, 7, 11$, and 23. Is the group of units cyclic for these cases? Try to guess a formula for the order of the group of units of $Z_n[i]$ as a function of n when n is a prime and $n \mod 4 = 3$. Run the program for $n = 9$ and 27. Are the groups cyclic? Try to guess a formula for the order when $n = 3^k$. Run the program for $n = 5, 13, 17$, and 29. Is the group cyclic for these cases? What is the largest order of any element in the group? Try to guess a formula for the order of the group of units of $Z_n[i]$ as a function of n when n is a prime and $n \mod 4 = 1$. Try to guess a formula for the largest order of any element in the group of units of $Z_n[i]$ as a function of n when n is a prime and $n \mod 4 = 1$. On the basis of the orders of the elements of the group of units, try to guess the isomorphism class of the group. Run the program for $n = 25$. Is this group cyclic? Based on the number of elements in this group and the orders of the elements, try to guess the isomorphism class of the group.

3. This software determines the isomorphism class of the group of units of $Z_n[i]$. Run the program for $n = 5, 13, 17, 29$, and 37. Make a conjecture. Run the program for $n = 3, 7, 11, 19, 23$, and 31. Make a conjecture. Run the program for $n = 5, 25$, and 125. Make a conjecture. Run the program for $n = 13$ and 169. Make a conjecture. Run the program for $n = 3, 9$, and 27. Make a conjecture. Run the program for $n = 7$ and 49. Make a conjecture. Run the program for $n = 11$ and 121. Make a conjecture. Make a conjecture about the case where $n = p^k$ where p is a prime and $p \mod 4 = 1$. Make a conjecture about the case where $n = p^k$ where p is a prime and $p \mod 4 = 3$.

4. This software determines the order of the group of units in the ring of 2×2 matrices over Z_n (that is, the group $GL(2, Z_n)$) and the subgroup $SL(2, Z_n)$. Run the program for $n = 2, 3, 5, 7, 11$, and 13. What relationship do you see between the order of $GL(2, Z_n)$ and the order of $SL(2, Z_n)$ in these cases? Run the program for $n = 16, 27, 25$, and 49. Make a conjecture about the relationship between the order of $GL(2, Z_n)$ and the order of $SL(2, Z_n)$ when n is a power of a prime. Run the program for $n = 32$. (Notice that when you run the program for $n = 32$, the table shows the orders for all divisors of 32 greater than 1.) How do the orders of the two groups change each time you increase the power of 2 by 1? Run the program for $n = 27$. How do the orders of the two groups change each time you increase the power of 3 by 1? Run the program for $n = 25$. How do

the orders of the two groups change when you increase the power of 5 by 1? Make a conjecture about the relationship between $|SL(2, Z_{p^i})|$ and $|SL(2, Z_{p^{i+1}})|$. Make a conjecture about the relationship between $|GL(2, Z_{p^i})|$ and $|GL(2, Z_{p^{i+1}})|$. Run the program for $n = 12, 15, 20, 21$, and 30. Make a conjecture about the order of $GL(2, Z_n)$ in terms of the orders of $GL(2, Z_s)$ and $GL(2, Z_t)$ where $n = st$ and s and t are relatively prime. (Notice that when you run the program for st, the table shows the values for st, s, and t.) For each value of n, is the order of $SL(2, Z_n)$ divisible by n? Is it divisible by $n + 1$? Is it divisible by $n - 1$?

5. In the ring Z_n, this software finds the number of solutions to the equation $x^2 = -1$. Run the program for all primes between 3 and 29. How does the answer depend on the prime? Make a conjecture about the number of solutions when n is a prime greater than 2. Run the program for the squares of all primes between 3 and 29. Make a conjecture about the number of solutions when n is the square of a prime greater than 2. Run the program for the cubes of primes between 3 and 29. Make a conjecture about the number of solutions when n is any power of an odd prime. Run the program for $n = 2, 4, 8, 16$, and 32. Make a conjecture about the number of solutions when n is a power of 2. Run the program for $n = 12$, 20, 24, 28, and 36. Make a conjecture about the number of solutions when n is a multiple of 4. Run the program for various cases where $n = pq$ and $n = 2pq$ where p and q are odd primes. Make a conjecture about the number of solutions when $n = pq$ or $n = 2pq$ where p and q are odd primes. What relationship do you see among the numbers of solutions for $n = p$, $n = q$, and $n = pq$? Run the program for various cases where $n = pqr$ and $n = 2pqr$ where p, q, and r are odd primes. Make a conjecture about the number of solutions when $n = pqr$ or $n = 2pqr$ where p, q, and r are odd primes. What relationship do you see among the numbers of solutions when $n = p$, $n = q$, and $n = r$ and the case that $n = pqr$?

6. This software determines the number of solutions to the equation $X^2 = -I$ where X is a 2×2 matrix with entries from Z_n and I is the identity. Run the program for $n = 32$. Make a conjecture about the number of solutions when $n = 2^k$ where $k > 1$. Run the program for $n = 3, 11, 19, 23$, and 31. Make a conjecture about the number of solutions when n is a prime of the form $4q + 3$. Run the program for $n = 27$ and 49. Make a conjecture about the number of solutions when n has the form p^i where p is a prime of the form $4q + 3$. Run the program for $n = 5, 13, 17, 29$, and 37. Make a conjecture about the number of solutions when n is a prime of the form $4q + 1$. Run the program for $n = 6, 10, 14, 22, 15$, 21, 33, 39, 30, 42. What seems to be the relationship between the number of solutions for a given n and the number of solutions for the prime power factors of n?

Suggested Reading

D. B. Erickson, "Orders for Finite Noncommutative Rings," *American Mathematical Monthly* 73 (1966): 376–377.

> In this elementary paper, it is shown that there exists a noncommutative ring of order $m > 1$ if and only if m is divisible by the square of a prime.

I. N. Herstein

A whole generation of textbooks and an entire generation of mathematicians, myself included, have been profoundly influenced by that text [Herstein's Topics in Algebra*].*
   GEORGIA BENKART

I. N. HERSTEIN was born on March 28, 1923, in Poland. His family moved to Canada when he was seven. He grew up in a poor and tough environment, on which he commented that in his neighborhood you became either a gangster or a college professor. During his school years he played football, hockey, golf, tennis, and pool. During this time he worked as a steeplejack and as a barber at a fair. Herstein received a B.S. degree from the University of Manitoba, an M.A. from the University of Toronto, and, in 1948, a Ph.D. degree from Indiana University under the supervision of Max Zorn. Before permanently settling at the University of Chicago in 1962, he held positions at the University of Kansas, Ohio State University, the University of Pennsylvania, and Cornell University.

Herstein wrote more than 100 research papers and a dozen books. Although his principal interest was noncommutative ring theory, he also wrote papers on finite groups, linear algebra, and mathematical economics. His textbook *Topics in Algebra,* first published in 1964, dominated the field for 20 years and has become a classic. Herstein had great influence through his teaching and his collaboration with colleagues. He had 30 Ph.D. students, and traveled and lectured widely. His nonmathematical interests included languages and art. He spoke Italian, Hebrew, Polish, and Portuguese. Herstein died on February 9, 1988, after a long battle with cancer.

To find more information about Herstein, visit:

http://www-groups.dcs.st-and.ac.uk/~history/

13 | **Integral Domains**

DEFINITION AND EXAMPLES

To a certain degree, the notion of a ring was invented in an attempt to
put the algebraic properties of the integers into an abstract setting. A
ring is not the appropriate abstraction of the integers, however, for
too much is lost in the process. Besides the two obvious properties of
commutativity and existence of a unity, there is one other essential
feature of the integers that rings in general do not enjoy—the cancellation property. In this chapter, we introduce integral domains—
a particular class of rings that have all three of these properties.
Integral domains play a prominent role in number theory and algebraic geometry.

DEFINITION *Zero-Divisors*

A *zero-divisor* is a nonzero element a of a commutative ring R
such that there is a nonzero element $b \in R$ with $ab = 0$.

DEFINITION *Integral Domain*

An *integral domain* is a commutative ring with unity and no zero-
divisors.

Thus, in an integral domain, a product is 0 only when one of the
factors is 0; that is, $ab = 0$ only when $a = 0$ or $b = 0$. The following

examples show that many familiar rings are integral domains and some familiar rings are not. For each example, the student should verify the assertion made.

EXAMPLE 1   The ring of integers is an integral domain.   ◆

EXAMPLE 2   The ring of Gaussian integers $Z[i] = \{a + bi \mid a, b \in Z\}$ is an integral domain.   ◆

EXAMPLE 3   The ring $Z[x]$ of polynomials with integer coefficients is an integral domain.   ◆

EXAMPLE 4   The ring $Z[\sqrt{2}] = \{a + b\sqrt{2} \mid a, b \in Z\}$ is an integral domain.   ◆

EXAMPLE 5   The ring Z_p of integers modulo a prime p is an integral domain.   ◆

EXAMPLE 6   The ring Z_n of integers modulo n is *not* an integral domain when n is not prime.   ◆

EXAMPLE 7   The ring $M_2(Z)$ of 2×2 matrices over the integers is *not* an integral domain.   ◆

EXAMPLE 8 $Z \oplus Z$ is *not* an integral domain.   ◆

What makes integral domains particularly appealing is that they have an important multiplicative group-theoretic property, in spite of the fact that the nonzero elements need not form a group under multiplication. This property is cancellation.

Theorem 13.1   **Cancellation**

> *Let a, b, and c belong to an integral domain. If $a \neq 0$ and $ab = ac$, then $b = c$.*

PROOF.   From $ab = ac$, we have $a(b - c) = 0$. Since $a \neq 0$, we must have $b - c = 0$.   ■ ■

Many authors prefer to define integral domains by the cancellation property—that is, as commutative rings with unity in which the cancellation property holds. This definition is equivalent to ours.

FIELDS

In many applications, a particular kind of integral domain called a *field* is necessary.

DEFINITION   Field

A *field* is a commutative ring with unity in which every nonzero element is a unit.

To verify that every field is an integral domain, observe that if a and b belong to a field with $a \neq 0$ and $ab = 0$, we can multiply both sides of the last expression by a^{-1} to obtain $b = 0$.

It is often helpful to think of ab^{-1} as a divided by b. With this in mind, a field can be thought of as simply an algebraic system that is closed under addition, subtraction, multiplication, and division (except by 0). We have had numerous examples of fields: the complex numbers, the real numbers, the rational numbers. The abstract theory of fields was initiated by Heinrich Weber in 1893. Groups, rings, and fields are the three main branches of abstract algebra. Theorem 13.2 says that, in the finite case, fields and integral domains are the same.

Theorem 13.2   Finite Integral Domains Are Fields

A finite integral domain is a field.

PROOF.   Let D be a finite integral domain with unity 1. Let a be any nonzero element of D. We must show that a is a unit. If $a = 1$, a is its own inverse, so we may assume that $a \neq 1$. Now consider the following sequence of elements of D: $a, a^2, a^3, \ldots$. Since D is finite, there must be two positive integers i and j such that $i > j$ and $a^i = a^j$. Then, by cancellation, $a^{i-j} = 1$. Since $a \neq 1$, we know that $i - j > 1$, and we have shown that a^{i-j-1} is the inverse of a.   ■■

Corollary Z_p Is a Field

For every prime p, Z_p, the ring of integers modulo p, is a field.

PROOF.   According to Theorem 13.2, we need only prove that Z_p has no zero-divisors. So, suppose that $a, b \in Z_p$ and $ab = 0$. Then $ab = pk$

for some integer k. But then, by Euclid's Lemma (see Chapter 0), p divides a or p divides b. Thus, in Z_p, $a = 0$ or $b = 0$. ■ ■

Putting the preceding corollary together with Example 6, we see that Z_n is a field if and only if n is prime. In Chapter 22, we will describe how all finite fields can be constructed. For now, we give one example of a finite field that is not of the form Z_p.

EXAMPLE 9 Field with Nine Elements
Let $\qquad Z_3[i] = \{a + bi \mid a, b \in Z_3\}$

$$= \{0, 1, 2, i, 1 + i, 2 + i, 2i, 1 + 2i, 2 + 2i\},$$

where $i^2 = -1$. This is the ring of Gaussian integers modulo 3. Elements are added and multiplied as in the complex numbers, except that the coefficients are reduced modulo 3. In particular, $-1 = 2$. Table 13.1 is the multiplication table for the nonzero elements of $Z_3[i]$. ◆

EXAMPLE 10
Let $Q[\sqrt{2}] = \{a + b\sqrt{2} \mid a, b \in Q\}$. It is easy to see that $Q[\sqrt{2}]$ is a ring. Viewed as an element of **R**, the multiplicative inverse of any nonzero element of the form $a + b\sqrt{2}$ is simply $1/(a + b\sqrt{2})$. To verify that $Q[\sqrt{2}]$ is a field, we must show that $1/(a + b\sqrt{2})$ can be written in the form $c + d\sqrt{2}$. In high school algebra, this process is called "rationalizing the denominator." Specifically,

$$\frac{1}{a + b\sqrt{2}} = \frac{1}{a + b\sqrt{2}} \frac{a - b\sqrt{2}}{a - b\sqrt{2}} = \frac{a}{a^2 - 2b^2} - \frac{b}{a^2 - 2b^2}\sqrt{2}.$$

(Note that $a + b\sqrt{2} \neq 0$ guarantees that $a - b\sqrt{2} \neq 0$.) ◆

TABLE 13.1 Multiplication Table for $Z_3[i]$

	1	2	i	$1 + i$	$2 + i$	$2i$	$1 + 2i$	$2 + 2i$
1	1	2	i	$1 + i$	$2 + i$	$2i$	$1 + 2i$	$2 + 2i$
2	2	1	$2i$	$2 + 2i$	$1 + 2i$	i	$2 + i$	$1 + i$
i	i	$2i$	2	$2 + i$	$2 + 2i$	1	$1 + i$	$1 + 2i$
$1 + i$	$1 + i$	$2 + 2i$	$2 + i$	$2i$	1	$1 + 2i$	2	i
$2 + i$	$2 + i$	$1 + 2i$	$2 + 2i$	1	i	$1 + i$	$2i$	2
$2i$	$2i$	i	1	$1 + 2i$	$1 + i$	2	$2 + 2i$	$2 + i$
$1 + 2i$	$1 + 2i$	$2 + i$	$1 + i$	2	$2i$	$2 + 2i$	i	1
$2 + 2i$	$2 + 2i$	$1 + i$	$1 + 2i$	i	2	$2 + i$	1	$2i$

CHARACTERISTIC OF A RING

Note that for any element x in $Z_3[i]$, we have $3x = x + x + x = 0$, since addition is done modulo 3. Similarly, in the subring $\{0, 3, 6, 9\}$ of Z_{12}, we have $4x = x + x + x + x = 0$ for all x. This observation motivates the following definition.

DEFINITION   *Characteristic of a Ring*

> The *characteristic* of a ring R is the least positive integer n such that $nx = 0$ for all x in R. If no such integer exists, we say that R has characteristic 0. The characteristic of R is denoted by char R.

Thus, the ring of integers has characteristic 0, and Z_n has characteristic n. An infinite ring can have a nonzero characteristic. Indeed, the ring $Z_2[x]$ of all polynomials with coefficients in Z_2 has characteristic 2. (Addition and multiplication are done as for polynomials with ordinary integer coefficients except that the coefficients are reduced modulo 2.) When a ring has a unity, the task of determining the characteristic is simplified by Theorem 13.3.

Theorem 13.3   Characteristic of a Ring with Unity

> *Let R be a ring with unity 1. If 1 has infinite order under addition, then the characteristic of R is 0. If 1 has order n under addition, then the characteristic of R is n.*

PROOF.   If 1 has infinite order, then there is no positive integer n such that $n \cdot 1 = 0$, so R has characteristic 0. Now suppose that 1 has additive order n. Then $n \cdot 1 = 0$, and n is the least positive integer with this property. So, for any x in R, we have

$$n \cdot x = x + x + \cdots + x \ (n \text{ summands})$$
$$= 1x + 1x + \cdots + 1x \ (n \text{ summands})$$
$$= (1 + 1 + \cdots + 1)x \ (n \text{ summands})$$
$$= (n \cdot 1)x = 0x = 0.$$

Thus, R has characteristic n.   ∎

In the case of an integral domain, the possibilities for the characteristic are severely limited.

Theorem 13.4 **Characteristic of an Integral Domain**

> *The characteristic of an integral domain is 0 or prime.*

PROOF. By Theorem 13.3, it suffices to show that if the additive order of 1 is finite, it must be prime. Suppose that 1 has order n and that $n = st$, where $1 \le s, t \le n$. Then, by Exercise 15 in Chapter 12,

$$0 = n \cdot 1 = (st) \cdot 1 = (s \cdot 1)(t \cdot 1).$$

So, $s \cdot 1 = 0$ or $t \cdot 1 = 0$. Since n is the least positive integer with the property that $n \cdot 1 = 0$, we must have $s = n$ or $t = n$. Thus, n is prime. ∎

We conclude this chapter with a brief discussion of polynomials with coefficients from a ring—a topic we will consider in detail in later chapters. The existence of zero-divisors in a ring causes unusual results when one is finding zeros of polynomials with coefficients in the ring. Consider, for example, the equation $x^2 - 4x + 3 = 0$. In the integers, we could find all solutions by factoring

$$x^2 - 4x + 3 = (x - 3)(x - 1) = 0$$

and setting each factor equal to 0. But notice that when we say we can find *all* solutions in this manner, we are using the fact that the only way for a product to equal 0 is for one of the factors to be 0—that is, we are using the fact that Z is an integral domain. In Z_{12}, there are many pairs of nonzero elements whose products are 0: $2 \cdot 6 = 0, 3 \cdot 4 = 0, 4 \cdot 6 = 0, 6 \cdot 8 = 0$, and so on. So, how do we find *all* solutions of $x^2 - 4x + 3 = 0$ in Z_{12}? The easiest way is simply to try every element! Upon doing so, we find four solutions: $x = 1, x = 3, x = 7$, and $x = 9$. Observe that we can find all solutions of $x^2 - 4x + 3 = 0$ over Z_{11} or Z_{13}, say, by setting the two factors $x - 3$ and $x - 1$ equal to 0. Of course, the reason this works for these rings is that they are integral domains. Perhaps this will convince you that integral domains are particularly advantageous rings. Table 13.2 gives a summary of some of the rings we have introduced and their properties.

TABLE 13.2 Summary of Rings and Their Properties

Ring	Form of Element	Unity	Commutative	Integral Domain	Field	Characteristic
Z	k	1	Yes	Yes	No	0
Z_n, n composite	k	1	Yes	No	No	n
Z_p, p prime	k	1	Yes	Yes	Yes	p
$Z[x]$	$a_nx^n + \cdots + a_1x + a_0$	$f(x) = 1$	Yes	Yes	No	0
nZ, $n > 1$	nk	None	Yes	No	No	0
$M_2(Z)$	$\begin{bmatrix} a & b \\ c & d \end{bmatrix}$	$\begin{bmatrix} 1 & 0 \\ 0 & 1 \end{bmatrix}$	No	No	No	0
$M_2(2Z)$	$\begin{bmatrix} 2a & 2b \\ 2c & 2d \end{bmatrix}$	None	No	No	No	0
$Z[i]$	$a + bi$	1	Yes	Yes	No	0
$Z_3[i]$	$a + bi; a, b \in Z_3$	1	Yes	Yes	Yes	3
$Z[\sqrt{2}]$	$a + b\sqrt{2}; a, b \in Z$	1	Yes	Yes	No	0
$Q[\sqrt{2}]$	$a + b\sqrt{2}; a, b \in Q$	1	Yes	Yes	Yes	0
$Z \oplus Z$	(a, b)	$(1, 1)$	Yes	No	No	0

EXERCISES ▪■

It looked absolutely impossible. But it so happens that you go on worrying away at a problem in science and it seems to get tired, and lies down and lets you catch it.

WILLIAM LAWRENCE BRAGG[†]

1. Verify that Examples 1 through 8 are as claimed.
2. Which of Examples 1 through 5 are fields?
3. Show that a commutative ring with the cancellation property (under multiplication) has no zero-divisors.
4. List all zero-divisors in Z_{20}. Can you see a relationship between the zero-divisors of Z_{20} and the units of Z_{20}?
5. Show that every nonzero element of Z_n is a unit or a zero-divisor.
6. Find a nonzero element in a ring that is neither a zero-divisor nor a unit.
7. Let R be a finite commutative ring with unity. Prove that every nonzero element of R is either a zero-divisor or a unit. What happens if we drop the "finite" condition on R?

[†]Bragg, at age 24, won the Nobel Prize for the invention of x-ray crystallography. He remains the youngest person ever to receive the Nobel Prize.

8. Describe all zero-divisors and units of $Z \oplus Q \oplus Z$.

9. Let d be an integer. Prove that $Z[\sqrt{d}] = \{a + b\sqrt{d} \mid a, b \in Z\}$ is an integral domain. (This exercise is referred to in Chapter 18.)

10. In Z_7, give a reasonable interpretation for the expressions $1/2$, $-2/3$, $\sqrt{-3}$, and $-1/6$.

11. Give an example of a commutative ring without zero-divisors that is not an integral domain.

12. Find two elements a and b in a ring such that both a and b are zero-divisors, $a + b \neq 0$, and $a + b$ is not a zero-divisor.

13. Let a belong to a ring R with unity and suppose that $a^n = 0$ for some positive integer n. (Such an element is called *nilpotent*.) Prove that $1 - a$ has a multiplicative inverse in R. [*Hint:* Consider $(1 - a)(1 + a + a^2 + \cdots + a^{n-1})$.]

14. Show that the nilpotent elements of a commutative ring form a subring.

15. Show that 0 is the only nilpotent element in an integral domain.

16. A ring element a is called an *idempotent* if $a^2 = a$. Prove that the only idempotents in an integral domain are 0 and 1.

17. Prove that the set of idempotents of a commutative ring is closed under multiplication.

18. Find a zero-divisor in $Z_5[i] = \{a + bi \mid a, b \in Z_5\}$.

19. Find an idempotent in $Z_5[i] = \{a + bi \mid a, b \in Z_5\}$.

20. Find all units, zero-divisors, idempotents, and nilpotent elements in $Z_3 \oplus Z_6$.

21. Determine all elements of a ring that are both units and idempotents.

22. Let R be the set of all real-valued functions defined for all real numbers under function addition and multiplication.
    a. Determine all zero-divisors of R.
    b. Determine all nilpotent elements of R.
    c. Show that every nonzero element is a zero-divisor or a unit.

23. (Subfield Test) Let F be a field and let K be a subset of F with at least two elements. Prove that K is a subfield of F if, for any a, b ($b \neq 0$) in K, $a - b$ and ab^{-1} belong to K.

24. Let d be a positive integer. Prove that $Q[\sqrt{d}] = \{a + b\sqrt{d} \mid a, b \in Q\}$ is a field.

25. Let R be a ring with unity 1. If the product of any pair of nonzero elements of R is nonzero, prove that $ab = 1$ implies $ba = 1$.

26. Let $R = \{0, 2, 4, 6, 8\}$ under addition and multiplication modulo 10. Prove that R is a field.

27. Formulate the appropriate definition of a subdomain (that is, a "sub" integral domain). Let D be an integral domain with unity 1. Show that

$P = \{n \cdot 1 \mid n \in Z\}$ (that is, all integral multiples of 1) is a subdomain of D. Show that P is contained in every subdomain of D. What can we say about the order of P?

28. Prove that there is no integral domain with exactly six elements. Can your argument be adapted to show that there is no integral domain with exactly four elements? What about 15 elements? Use these observations to guess a general result about the number of elements in a finite integral domain.

29. Let F be a field of order 2^n. Prove that char $F = 2$.

30. Determine all elements of an integral domain that are their own inverses under multiplication.

31. Suppose that a and b belong to an integral domain.
   a. If $a^5 = b^5$ and $a^3 = b^3$, prove that $a = b$.
   b. If $a^m = b^m$ and $a^n = b^n$, where m and n are positive integers that are relatively prime, prove that $a = b$.

32. Find an example of an integral domain and distinct positive integers m and n such that $a^m = b^m$ and $a^n = b^n$, but $a \neq b$.

33. If a is an idempotent in Z_n, show that $1 - a$ is also an idempotent.

34. Construct a multiplication table for $Z_2[i]$, the ring of Gaussian integers modulo 2. Is this ring a field? Is it an integral domain?

35. The nonzero elements of $Z_3[i]$ form an Abelian group of order 8 under multiplication. Is it isomorphic to Z_8, $Z_4 \oplus Z_2$, or $Z_2 \oplus Z_2 \oplus Z_2$?

36. Show that $Z_7[\sqrt{3}] = \{a + b\sqrt{3} \mid a, b \in Z_7\}$ is a field. For any positive integer k and any prime p, determine a necessary and sufficient condition for $Z_p[\sqrt{k}] = \{a + b\sqrt{k} \mid a, b \in Z_p\}$ to be a field.

37. Show that a finite commutative ring with no zero-divisors and at least two elements has a unity.

38. Suppose that a and b belong to a commutative ring and ab is a zero-divisor. Show that either a or b is a zero-divisor.

39. Suppose that R is a commutative ring without zero-divisors. Show that all the nonzero elements of R have the same additive order.

40. Suppose that R is a commutative ring without zero-divisors. Show that the characteristic of R is 0 or prime.

41. Let x and y belong to a commutative ring R with char $R = p \neq 0$.
   a. Show that $(x + y)^p = x^p + y^p$.
   b. Show that, for all positive integers n, $(x + y)^{p^n} = x^{p^n} + y^{p^n}$.
   c. Find elements x and y in a ring of characteristic 4 such that $(x + y)^4 \neq x^4 + y^4$. (This exercise is referred to in Chapter 20.)

42. Let R be a commutative ring with unity 1 and prime characteristic. If $a \in R$ is nilpotent, prove that there is a positive integer k such that $(1 + a)^k = 1$.

43. Show that any finite field has order p^n, where p is a prime. *Hint:* Use facts about finite Abelian groups. (This exercise is referred to in Chapter 22.)

44. Give an example of an infinite integral domain that has characteristic 3.

45. Let R be a ring and let $M_2(R)$ be the ring of 2×2 matrices with entries from R. Explain why these two rings have the same characteristic.

46. Let R be a ring with m elements. Show that the characteristic of R divides m.

47. Explain why a finite ring with at least two elements must have a nonzero characteristic.

48. Find all solutions of $x^2 - x + 2 = 0$ over $Z_3[i]$. (See Example 9.)

49. Consider the equation $x^2 - 5x + 6 = 0$.
   a. How many solutions does this equation have in Z_7?
   b. Find all solutions of this equation in Z_8.
   c. Find all solutions of this equation in Z_{12}.
   d. Find all solutions of this equation in Z_{14}.

50. Find the characteristic of $Z_4 \oplus 4Z$.

51. Suppose that R is an integral domain in which $20 \cdot 1 = 0$ and $12 \cdot 1 = 0$. (Recall that $n \cdot 1$ means the sum $1 + 1 + \cdots + 1$ with n terms.) What is the characteristic of R?

52. In a commutative ring of characteristic 2, prove that the idempotents form a subring.

53. Describe the smallest subfield of the field of real numbers that contains $\sqrt{2}$. (That is, describe the subfield K with the property that K contains $\sqrt{2}$ and if F is any subfield containing $\sqrt{2}$, then F contains K.)

54. Let F be a finite field with n elements. Prove that $x^{n-1} = 1$ for all nonzero x in F.

55. Let F be a field of prime characteristic p. Prove that $K = \{x \in F \mid x^p = x\}$ is a subfield of F.

56. Suppose that a and b belong to a field of order 8 and that $a^2 + ab + b^2 = 0$. Prove that $a = 0$ and $b = 0$. Do the same when the field has order 2^n with n odd.

57. Let F be a field of characteristic 2 with more than two elements. Show that $(x + y)^3 \neq x^3 + y^3$ for some x and y in F.

58. Suppose that F is a field with characteristic not 2, and that the nonzero elements of F form a cyclic group under multiplication. Prove that F is finite.

59. Suppose that D is an integral domain and that ϕ is a nonconstant function from D to the nonnegative integers such that $\phi(xy) = \phi(x)\phi(y)$. If x is a unit in D, show that $\phi(x) = 1$.

60. Let F be a field of order 32. Show that the only subfields of F are F itself and $\{0, 1\}$.

COMPUTER EXERCISES

The basic unit of mathematics is conjecture.

ARNOLD ROSS

Software for the computer exercises in this chapter is available at the website:

http://www.d.umn.edu/~jgallian

1. This software lists the idempotents (see Exercise 16 for the definition) in Z_n. Run the program for various values of n. Use these data to make conjectures about the number of idempotents in Z_n as a function of n. For example, how many idempotents are there when n is a prime-power? What about when n is divisible by exactly two distinct primes? In the case where n is of the form pq where p and q are primes, can you see a relationship between the two idempotents that are not 0 and 1? Can you see a relationship between the number of idempotents for a given n and the number of distinct prime divisors of n?

2. This software lists the nilpotent elements (see Exercise 13 for the definition) in Z_n. Run the program for various values of n. Use these data to make conjectures about the number of nilpotent elements in Z_n as a function of n.

3. This software determines which rings of the form $Z_p[i]$ are fields. Run the program for all primes up to 37. From these data, make a conjecture about the form of the primes that yield a field.

4. This software finds the idempotents in $Z_n[i] = \{a + bi \mid a, b \in Z_n\}$ (Gaussian integers modulo n). Run the software for $n = 4, 8, 16,$ and 32. Make a conjecture about the number of idempotents when $n = 2^k$. Run the software for $n = 13, 17, 29,$ and 37. What do these values of n have in common? Make a conjecture about the number of idempotents for these n. Run the software for $n = 7, 11, 19, 23, 31,$ and 43. What do these values of n have in common? Make a conjecture about the number of idempotents for these n.

5. This software finds the nilpotent elements in $Z_n[i] = \{a + bi \mid a, b \in Z_n\}$. Run the software for $n = 4, 8, 16,$ and 32. Make a conjecture about the number of nilpotent elements when $n = 2^k$. Run the software for $n = 3, 5, 7, 11, 13,$ and 17. What do these values of n have in common? Make a conjecture about the number of nilpotent elements for these n. Run the program for $n = 9$. Do you need to revise the conjecture you made based on $n = 3, 5, 7, 11, 13,$ and 17? Run the software for $n = 9, 25,$ and 49. What do these values of n have in common? Make a conjecture about the number of nilpotent elements for these n. Run the program for $n = 81$. Do you need to revise the conjecture you made based on $n = 9, 25,$ and 49? What do these values of n have in common? Make a conjecture about the number of nilpotent elements for these n. Run the program for $n = 27$. Do you need to revise the conjecture you made based on $n = 9$, 25, and 49? Run your program for $n = 125$ (this may take a few minutes). On the basis of all of your data for this exercise, make a single conjecture in the case that $n = p^k$ where p is any prime. Run the program for $n = 6$, 15, and 21. Make a conjecture. Run the program for 12, 20, 28, and 45. Make a conjecture. Run the program for 36 and 100 (this may take a few minutes). On the basis of all your data for this exercise, make a single conjecture that covers all integers $n > 1$.

6. This software determines the zero-divisors in $Z_n[i] = \{a + bi \mid a, b \in Z_n\}$. Use the software to formulate and test conjectures about the number of zero-divisors in $Z_n[i]$ based on various conditions of n.

Suggested Readings

Eric Berg, "A Family of Fields," *Pi Mu Epsilon* 9 (1990): 154–155.

> In this article, the author uses properties of logarithms and exponents to define recursively an infinite family of fields starting with the real numbers.

N. A. Khan, "The Characteristic of a Ring," *American Mathematical Monthly* 70 (1963): 736.

> Here it is shown that a ring has nonzero characteristic n if and only if n is the maximum of the orders of the elements of R.

K. Robin McLean, "Groups in Modular Arithmetic," *The Mathematical Gazette* 62 (1978): 94–104.

> This article explores the interplay between various groups of integers under multiplication modulo n and the ring Z_n. It shows how to construct groups of integers in which the identity is not obvious; for example, 1977 is the identity of the group $\{1977, 5931\}$ under multiplication modulo 7908.

Nathan Jacobson

Few mathematicians have been as productive over such a long career or have had as much influence on the profession as has Professor Jacobson.

CITATION FOR THE STEELE PRIZE
FOR LIFETIME ACHIEVEMENT

NATHAN JACOBSON was born on September 8, 1910, in Warsaw, Poland. After arriving in the United States in 1917, Jacobson grew up in Alabama, Mississippi, and Georgia, where his father owned small clothing stores. He received a B.A. degree from the University of Alabama in 1930 and a Ph.D. from Princeton in 1934. After brief periods as a professor at Bryn Mawr, the University of Chicago, the University of North Carolina, and Johns Hopkins, Jacobson accepted a position at Yale, where he remained until his retirement in 1981.

Jacobson's principal contributions to algebra were in the areas of rings, Lie algebras, and Jordan algebras. In particular, he developed structure theories for these systems. He was the author of nine books and numerous articles and he had 33 Ph.D. students.

Jacobson held visiting positions in France, India, Italy, Israel, China, Australia, and Switzerland. Among his many honors were the presidency of the American Mathematical Society, memberships in the National Academy of Sciences and the American Academy of Arts and Sciences, a Guggenheim Fellowship, and an honorary degree from the University of Chicago. Jacobson died on December 5, 1999, at the age of 89.

To find more information about Jacobson, visit:

http://www-groups.dcs.st-and.ac.uk/~history/

14 | Ideals and Factor Rings

The secret of science is to ask the right questions, and it is the choice of problem more than anything else that marks the man of genius in the scientific world.

SIR HENRY TIZARD IN C. P. SNOW,
A postscript to *Science and Government*

IDEALS

Normal subgroups play a special role in group theory—they permit us to construct factor groups. In this chapter, we introduce the analogous concepts for rings—ideals and factor rings.

DEFINITION Ideal

A subring A of a ring R is called a (two-sided) *ideal* of R if for every $r \in R$ and every $a \in A$ both ra and ar are in A.

So, a subring A of a ring R is an ideal of R if A "absorbs" elements from R—that is, if $rA = \{ra \mid a \in A\} \subseteq A$ and $Ar = \{ar \mid a \in A\} \subseteq A$ for all $r \in R$.

An ideal A of R is called a *proper* ideal of R if A is a proper subset of R. In practice, one identifies ideals with the following test, which is an immediate consequence of the definition of ideal and the subring test given in Theorem 12.3.

Theorem 14.1 Ideal Test

A nonempty subset A of a ring R is an ideal of R if

1. $a - b \in A$ whenever $a, b \in A$.
2. ra and ar are in A whenever $a \in A$ and $r \in R$.

EXAMPLE 1   For any ring R, $\{0\}$ and R are ideals of R. The ideal $\{0\}$ is called the *trivial* ideal.   ◆

EXAMPLE 2   For any positive integer n, the set $nZ = \{0, \pm n, \pm 2n, \ldots\}$ is an ideal of Z.   ◆

EXAMPLE 3   Let R be a commutative ring with unity and let $a \in R$. The set $\langle a \rangle = \{ra \mid r \in R\}$ is an ideal of R called the *principal ideal generated by* a. (Notice that $\langle a \rangle$ is also the notation we used for the cyclic subgroup generated by a. However, the intended meaning will always be clear from the context.) The assumption that R is commutative is necessary in this example (see Exercise 29 in the Supplementary Exercises for Chapters 12–14).   ◆

EXAMPLE 4   Let $\mathbf{R}[x]$ denote the set of all polynomials with real coefficients and let A denote the subset of all polynomials with constant term 0. Then A is an ideal of $\mathbf{R}[x]$ and $A = \langle x \rangle$.   ◆

EXAMPLE 5   Let R be a commutative ring with unity and let a_1, $a_2, \ldots, a_n$ belong to R. Then $I = \langle a_1, a_2, \ldots, a_n \rangle = \{r_1 a_1 + r_2 a_2 + \cdots + r_n a_n \mid r_i \in R\}$ is an ideal of R called the *ideal generated by* a_1, $a_2, \ldots, a_n$. The verification that I is an ideal is left as an easy exercise (Exercise 3).   ◆

EXAMPLE 6   Let $Z[x]$ denote the ring of all polynomials with integer coefficients and let I be the subset of $Z[x]$ of all polynomials with even constant terms. Then I is an ideal of $Z[x]$ and $I = \langle x, 2 \rangle$ (see Exercise 35).   ◆

EXAMPLE 7   Let R be the ring of all real-valued functions of a real variable. The subset S of all differentiable functions is a subring of R but not an ideal of R.   ◆

FACTOR RINGS

Let R be a ring and let A be an ideal of R. Since R is a group under addition and A is a normal subgroup of R, we may form the factor group $R/A = \{r + A \mid r \in R\}$. The natural question at this point is: How may we form a ring of this group of cosets? The addition is already taken care of, and, by analogy with groups of cosets, we define the product of two cosets of $s + A$ and $t + A$ as $st + A$. The next theorem shows that this definition works as long as A is an ideal, and not just a subring, of R.

Theorem 14.2 **Existence of Factor Rings**

> *Let R be a ring and let A be a subring of R. The set of cosets*
> *$\{r + A \mid r \in R\}$ is a ring under the operations $(s + A) +$*
> *$(t + A) = s + t + A$ and $(s + A)(t + A) = st + A$ if and only*
> *if A is an ideal of R.*

PROOF. We know that the set of cosets forms a group under addition. Once we know that multiplication is indeed a binary operation on the cosets, it is trivial to check that the multiplication is associative and that multiplication is distributive over addition. Hence, the proof boils down to showing that multiplication is well defined if and only if A is an ideal of R. To do this, let us suppose that A is an ideal and let $s + A = s' + A$ and $t + A = t' + A$. Then we must show that $st + A = s't' + A$. Well, by definition, $s = s' + a$ and $t = t' + b$, where a and b belong to A. Then

$$st = (s' + a)(t' + b) = s't' + at' + s'b + ab,$$

and so

$$st + A = s't' + at' + s'b + ab + A = s't' + A,$$

since A absorbs $at' + s'b + ab$. Thus, multiplication is well defined when A is an ideal.

On the other hand, suppose that A is a subring of R that is not an ideal of R. Then there exist elements $a \in A$ and $r \in R$ such that $ar \notin A$ or $ra \notin A$. For convenience, say $ar \notin A$. Consider the elements $a + A = 0 + A$ and $r + A$. Clearly, $(a + A)(r + A) = ar + A$ but $(0 + A)(r + A) = 0 \cdot r + A = A$. Since $ar + A \neq A$, the multiplication is not well defined and the set of cosets is not a ring. ■ ■

Let's look at a few factor rings.

EXAMPLE 8 $Z/4Z = \{0 + 4Z, 1 + 4Z, 2 + 4Z, 3 + 4Z\}$. To see how to add and multiply, consider $2 + 4Z$ and $3 + 4Z$.

$$(2 + 4Z) + (3 + 4Z) = 5 + 4Z = 1 + 4 + 4Z = 1 + 4Z,$$
$$(2 + 4Z)(3 + 4Z) = 6 + 4Z = 2 + 4 + 4Z = 2 + 4Z.$$

One can readily see that the two operations are essentially modulo 4 arithmetic. ◆

EXAMPLE 9 $2Z/6Z = \{0 + 6Z, 2 + 6Z, 4 + 6Z\}$. Here the operations are essentially modulo 6 arithmetic. For example, $(4 + 6Z) + (4 + 6Z) = 2 + 6Z$ and $(4 + 6Z)(4 + 6Z) = 4 + 6Z$. ◆

Here is a noncommutative example of an ideal and factor ring.

EXAMPLE 10   Let $R = \left\{ \begin{bmatrix} a_1 & a_2 \\ a_3 & a_4 \end{bmatrix} \middle| a_i \in Z \right\}$ and let I be the sub-

set of R consisting of matrices with even entries. It is easy to show that I is indeed an ideal of R (Exercise 19). Consider the factor ring R/I. The interesting question about this ring is: What is its size? We claim it is 16; in fact, $R/I = \left\{ \begin{bmatrix} r_1 & r_2 \\ r_3 & r_4 \end{bmatrix} + I \,\middle|\, r_i \in \{0, 1\} \right\}$. An ex-

ample illustrates the typical situation. Which of the 16 elements is $\begin{bmatrix} 7 & 8 \\ 5 & -3 \end{bmatrix} + I$? Well, observe that $\begin{bmatrix} 7 & 8 \\ 5 & -3 \end{bmatrix} + I = \begin{bmatrix} 1 & 0 \\ 1 & 1 \end{bmatrix} + \begin{bmatrix} 6 & 8 \\ 4 & -4 \end{bmatrix} + I = \begin{bmatrix} 1 & 0 \\ 1 & 1 \end{bmatrix} + I$, since an ideal absorbs its own elements. The general case is left to the reader (Exercise 21).   ◆

EXAMPLE 11   Consider the factor ring of the Gaussian integers $R = Z[i]/\langle 2 - i \rangle$. What does this ring look like? Of course, the elements of R have the form $a + bi + \langle 2 - i \rangle$, where a and b are integers, but the important question is: What do the *distinct* cosets look like? The fact that $2 - i + \langle 2 - i \rangle = 0 + \langle 2 - i \rangle$ means that *when dealing with coset representatives*, we may treat $2 - i$ as equivalent to 0, so that $2 = i$. For example, the coset $3 + 4i + \langle 2 - i \rangle = 3 + 8 + \langle 2 - i \rangle = 11 + \langle 2 - i \rangle$. Similarly, all the elements of R can be written in the form $a + \langle 2 - i \rangle$, where a is an integer. But we can further reduce the set of distinct coset representatives by observing that *when dealing with coset representatives*, $2 = i$ implies (by squaring both sides) that $4 = -1$ or $5 = 0$. Thus, the coset $3 + 4i + \langle 2 - i \rangle = 11 + \langle 2 - i \rangle = 1 + 5 + 5 + \langle 2 - i \rangle = 1 + \langle 2 - i \rangle$. In this way, we can show that every element of R is equal to one of the following cosets: $0 + \langle 2 - i \rangle$, $1 + \langle 2 - i \rangle$, $2 + \langle 2 - i \rangle$, $3 + \langle 2 - i \rangle$, $4 + \langle 2 - i \rangle$. Is any further reduction possible? To demonstrate that there is not, we will show that these five cosets are distinct. It suffices to show that $1 + \langle 2 - i \rangle$ has additive order 5. Since $5(1 + \langle 2 - i \rangle) = 5 + \langle 2 - i \rangle = 0 + \langle 2 - i \rangle$, $1 + \langle 2 - i \rangle$ has order 1 or 5. If the order is actually 1, then $1 + \langle 2 - i \rangle = 0 + \langle 2 - i \rangle$, so $1 \in \langle 2 - i \rangle$. Thus, $1 = (2 - i)(a + bi) = 2a + b + (-a + 2b)i$ for some integers a and b. But this equation implies that $1 = 2a + b$ and $0 = -a + 2b$, and solving these simultaneously yields $b = 1/5$, which is a contradiction. It should be clear that the ring R is essentially the same as the field Z_5.   ◆

EXAMPLE 12   Let $\mathbf{R}[x]$ denote the ring of polynomials with real coefficients and let $\langle x^2 + 1 \rangle$ denote the principal ideal generated by $x^2 + 1$; that is,

$$\langle x^2 + 1 \rangle = \{f(x)(x^2 + 1) \mid f(x) \in \mathbf{R}[x]\}.$$

Then

$$\mathbf{R}[x]/\langle x^2 + 1 \rangle = \{g(x) + \langle x^2 + 1 \rangle \mid g(x) \in \mathbf{R}[x]\}$$
$$= \{ax + b + \langle x^2 + 1 \rangle \mid a, b \in \mathbf{R}\}.$$

To see this last equality, note that if $g(x)$ is any member of $\mathbf{R}[x]$, then we may write $g(x)$ in the form $q(x)(x^2 + 1) + r(x)$, where $q(x)$ is the quotient and $r(x)$ is the remainder upon dividing $g(x)$ by $x^2 + 1$. In particular, $r(x) = 0$ or the degree of $r(x)$ is less than 2, so that $r(x) = ax + b$ for some a and b in $\mathbf{R}$. Thus,

$$g(x) + \langle x^2 + 1 \rangle = q(x)(x^2 + 1) + r(x) + \langle x^2 + 1 \rangle$$
$$= r(x) + \langle x^2 + 1 \rangle,$$

since the ideal $\langle x^2 + 1 \rangle$ absorbs the term $q(x)(x^2 + 1)$.

How is multiplication done? Since

$$x^2 + 1 + \langle x^2 + 1 \rangle = 0 + \langle x^2 + 1 \rangle,$$

one should think of $x^2 + 1$ as 0 or, equivalently, as $x^2 = -1$. So, for example,

$$(x + 3 + \langle x^2 + 1 \rangle) \cdot (2x + 5 + \langle x^2 + 1 \rangle)$$
$$= 2x^2 + 11x + 15 + \langle x^2 + 1 \rangle = 11x + 13 + \langle x^2 + 1 \rangle.$$

In view of the fact that the elements of this ring have the form $ax + b + \langle x^2 + 1 \rangle$, where $x^2 + \langle x^2 + 1 \rangle = -1 + \langle x^2 + 1 \rangle$, it is perhaps not surprising that this ring turns out to be algebraically the same ring as the ring of complex numbers. This observation was first made by Cauchy in 1847.   ◆

Examples 11 and 12 illustrate one of the most important applications of factor rings—the construction of rings with highly desirable properties. In particular, we shall show how one may use factor rings to construct integral domains and fields.

PRIME IDEALS AND MAXIMAL IDEALS

DEFINITION *Prime Ideal, Maximal Ideal*

A *prime ideal* A of a commutative ring R is a proper ideal of R such that $a, b \in R$ and $ab \in A$ imply $a \in A$ or $b \in A$. A proper ideal A of a commutative ring R is a *maximal ideal* of R if, whenever B is an ideal of R and $A \subseteq B \subseteq R$, then $B = A$ or $B = R$.

So, the only ideal that properly contains a maximal ideal is the entire ring. The motivation for the definition of a prime ideal comes from the integers.

EXAMPLE 13 Let n be an integer greater than 1. Then, in the ring of integers, the ideal nZ is prime if and only if n is prime (Exercise 9). ($\{0\}$ is also a prime ideal of Z.) ◆

EXAMPLE 14 The lattice of ideals of Z_{36} (Figure 14.1) shows that only $\langle 2 \rangle$ and $\langle 3 \rangle$ are maximal ideals. ◆

EXAMPLE 15 The ideal $\langle x^2 + 1 \rangle$ is maximal in $\mathbf{R}[x]$. To see this, assume that A is an ideal of $\mathbf{R}[x]$ that properly contains $\langle x^2 + 1 \rangle$. We will prove that $A = \mathbf{R}[x]$ by showing that A contains some nonzero real number c. [This is the constant polynomial $h(x) = c$ for all x.] Then $1 = (1/c)c \in A$ and therefore, by Exercise 15, $A = \mathbf{R}[x]$. To this end, let $f(x) \in A$, but $f(x) \notin \langle x^2 + 1 \rangle$. Then

$$f(x) = q(x)(x^2 + 1) + r(x),$$

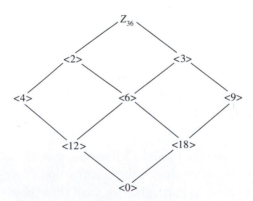

Figure 14.1

where $r(x) \neq 0$ and degree $r(x) < 2$. It follows that $r(x) = ax + b$, where a and b are not both 0, and

$$ax + b = r(x) = f(x) - q(x)(x^2 + 1) \in A.$$

Thus,

$$a^2x^2 - b^2 = (ax + b)(ax - b) \in A \qquad \text{and} \qquad a^2(x^2 + 1) \in A.$$

So,

$$0 \neq a^2 + b^2 = (a^2x^2 + a^2) - (a^2x^2 - b^2) \in A. \qquad \blacklozenge$$

EXAMPLE 16 The ideal $\langle x^2 + 1 \rangle$ is not prime in $Z_2[x]$, since it contains $(x + 1)^2 = x^2 + 2x + 1 = x^2 + 1$ but does not contain $x + 1$. $\qquad \blacklozenge$

Theorem 14.3 *R/A* Is an Integral Domain if and Only if *A* Is Prime

> Let R be a commutative ring with unity and let A be an ideal of R. Then R/A is an integral domain if and only if A is prime.

PROOF. Suppose that R/A is an integral domain and $ab \in A$. Then $(a + A)(b + A) = ab + A = A$, the zero element of the ring R/A. So, either $a + A = A$ or $b + A = A$; that is, either $a \in A$ or $b \in A$. Hence, A is prime.

To prove the other half of the theorem, we first observe that R/A is a commutative ring with unity for any proper ideal A. Thus, our task is simply to show that when A is prime, R/A has no zero-divisors. So, suppose that A is prime and $(a + A)(b + A) = 0 + A = A$. Then $ab \in A$ and, therefore, $a \in A$ or $b \in A$. Thus, one of $a + A$ or $b + A$ is the zero coset in R/A. ■ ■

For maximal ideals, we can do even better.

Theorem 14.4 *R/A* Is a Field if and Only if *A* Is Maximal

> Let R be a commutative ring with unity and let A be an ideal of R. Then R/A is a field if and only if A is maximal.

PROOF. Suppose that R/A is a field and B is an ideal of R that properly contains A. Let $b \in B$ but $b \notin A$. Then $b + A$ is a nonzero element of R/A and, therefore, there exists an element $c + A$ such that $(b + A)(c + A) = 1 + A$, the multiplicative identity of R/A. Since

$b \in B$, we have $bc \in B$. Because

$$1 + A = (b + A)(c + A) = bc + A,$$

we have $1 - bc \in A \subset B$. So, $1 = (1 - bc) + bc \in B$. By Exercise 15, $B = R$. This proves that A is maximal.

Now suppose that A is maximal and let $b \in R$ but $b \notin A$. It suffices to show that $b + A$ has a multiplicative inverse. (All other properties for a field follow trivially.) Consider $B = \{br + a \mid r \in R, a \in A\}$. This is an ideal of R that properly contains A (Exercise 23). Since A is maximal, we must have $B = R$. Thus, $1 \in B$, say, $1 = bc + a'$, where $a' \in A$. Then

$$1 + A = bc + a' + A = bc + A = (b + A)(c + A). \quad \blacksquare\blacksquare$$

When a commutative ring has a unity, it follows from Theorems 14.3 and 14.4 that a maximal ideal is a prime ideal. The next example shows that a prime ideal need not be maximal.

EXAMPLE 17   The ideal $\langle x \rangle$ is a prime ideal in $Z[x]$ but not a maximal ideal in $Z[x]$. To verify this, we begin with the observation that $\langle x \rangle = \{f(x) \in Z[x] \mid f(0) = 0\}$ (see Exercise 27). Thus, if $g(x)h(x) \in \langle x \rangle$, then $g(0)h(0) = 0$. And since $g(0)$ and $h(0)$ are integers, we have $g(0) = 0$ or $h(0) = 0$.

To see that $\langle x \rangle$ is not maximal, we simply note that $\langle x \rangle \subset \langle x, 2 \rangle \subset Z[x]$ (see Exercise 35). ◆

EXERCISES ▬▬▬▬▬▬▬▬▬▬▬▬▬▬▬▬▬▬▬▬▬▬▬▬▬▬▬▬▬ ■■

Problems worthy of attack
prove their worth by hitting back.
       PIET HEIN, "Problems,"
          Grooks (1966)

1. Verify that the set defined in Example 3 is an ideal.
2. Verify that the set A in Example 4 is an ideal and that $A = \langle x \rangle$.
3. Verify that the set I in Example 5 is an ideal and that if J is any ideal of R that contains $a_1, a_2, \ldots, a_n$, then $I \subseteq J$. (Hence, $\langle a_1, a_2, \ldots, a_n \rangle$ is the smallest ideal of R that contains $a_1, a_2, \ldots, a_n$.)
4. Find a subring of $Z \oplus Z$ that is not an ideal of $Z \oplus Z$.
5. Let $S = \{a + bi \mid a, b \in Z, b \text{ is even}\}$. Show that S is a subring of $Z[i]$, but not an ideal of $Z[i]$.
6. Find all maximal ideals in
    a. Z_8,   **b.** Z_{10},   **c.** Z_{12},   **d.** Z_n.

7. Let a belong to a commutative ring R. Show that $aR = \{ar \mid r \in R\}$ is an ideal of R. If R is the ring of even integers, list the elements of $4R$.

8. Prove that the intersection of any set of ideals of a ring is an ideal.

9. If n is an integer greater than 1, show that $\langle n \rangle = nZ$ is a prime ideal of Z if and only if n is prime. (This exercise is referred to in this chapter.)

10. If A and B are ideals of a ring, show that the *sum* of A and B, $A + B = \{a + b \mid a \in A, b \in B\}$, is an ideal.

11. In the ring of integers, find a positive integer a such that
    a. $\langle a \rangle = \langle 2 \rangle + \langle 3 \rangle$,
    b. $\langle a \rangle = \langle 3 \rangle + \langle 6 \rangle$,
    c. $\langle a \rangle = \langle m \rangle + \langle n \rangle$.

12. If A and B are ideals of a ring, show that the *product* of A and B, $AB = \{a_1 b_1 + a_2 b_2 + \cdots + a_n b_n \mid a_i \in A, b_i \in B, n \text{ a positive integer}\}$, is an ideal.

13. Find a positive integer a such that
    a. $\langle a \rangle = \langle 3 \rangle \langle 4 \rangle$,
    b. $\langle a \rangle = \langle 6 \rangle \langle 8 \rangle$,
    c. $\langle a \rangle = \langle m \rangle \langle n \rangle$.

14. Let A and B be ideals of a ring. Prove that $AB \subseteq A \cap B$.

15. If A is an ideal of a ring R and 1 belongs to A, prove that $A = R$. (This exercise is referred to in this chapter.)

16. If A and B are ideals of a commutative ring R with unity and $A + B = R$, show that $A \cap B = AB$.

17. If an ideal I of a ring R contains a unit, show that $I = R$.

18. Suppose that in the ring Z the ideal $\langle 35 \rangle$ is a proper ideal of J and J is a proper ideal of I. What are the possibilities for J? What are the possibilities for I?

19. Let R and I be as described in Example 10. Prove that I is an ideal of R.

20. Let $I = \langle 2 \rangle$. Prove that $I[x]$ is not a maximal ideal of $Z[x]$ even though I is a maximal ideal of Z.

21. Verify the claim made in Example 10 about the size of R/I.

22. Give an example of a commutative ring that has a maximal ideal that is not a prime ideal.

23. Show that the set B in the latter half of the proof of Theorem 14.4 is an ideal of R. (This exercise is referred to in this chapter.)

24. If R is a commutative ring with unity and A is a proper ideal of R, show that R/A is a commutative ring with unity.

25. Prove that the only ideals of a field F are $\{0\}$ and F itself.

26. Show that $\mathbf{R}[x]/\langle x^2 + 1 \rangle$ is a field.

27. In $Z[x]$, the ring of polynomials with integer coefficients, let $I = \{f(x) \in Z[x] \mid f(0) = 0\}$. Prove that $I = \langle x \rangle$. (This exercise is referred to in Chapter 15.)

28. Show that $A = \{(3x, y) \mid x, y \in Z\}$ is a maximal ideal of $Z \oplus Z$.

29. Let R be the ring of continuous functions from **R** to **R**. Show that $A = \{f \in R \mid f(0) = 0\}$ is a maximal ideal of R.

30. Let $R = Z_8 \oplus Z_{30}$. Find all maximal ideals of R, and for each maximal ideal I, identify the size of the field R/I.

31. How many elements are in $Z[i]/\langle 3 + i \rangle$? Give reasons for your answer.

32. In $Z[x]$, the ring of polynomials with integer coefficients, let $I = \{f(x) \in Z[x] \mid f(0) = 0\}$. Prove that I is not a maximal ideal.

33. In $Z \oplus Z$, let $I = \{(a, 0) \mid a \in Z\}$. Show that I is a prime ideal but not a maximal ideal.

34. Let R be a ring and let I be an ideal of R. Prove that the factor ring R/I is commutative if and only if $rs - sr \in I$ for all r and s in R.

35. In $Z[x]$, let $I = \{f(x) \in Z[x] \mid f(0)$ is an even integer$\}$. Prove that $I = \langle x, 2 \rangle$. Is I a prime ideal of $Z[x]$? Is I a maximal ideal? How many elements does $Z[x]/I$ have? (This exercise is referred to in this chapter.)

36. Prove that $I = \langle 2 + 2i \rangle$ is not a prime ideal of $Z[i]$. How many elements are in $Z[i]/I$? What is the characteristic of $Z[i]/I$?

37. In $Z_5[x]$, let $I = \langle x^2 + x + 2 \rangle$. Find the multiplicative inverse of $2x + 3 + I$ in $Z_5[x]/I$.

38. Let R be a ring and let p be a fixed prime. Show that $I_p = \{r \in R \mid$ additive order of r is a power of $p\}$ is an ideal of R.

39. An integral domain D is called a *principal ideal domain* if every ideal of D has the form $\langle a \rangle = \{ad \mid d \in D\}$ for some a in D. Show that Z is a principal ideal domain. (This exercise is referred to in Chapter 18.)

40. Let a and b belong to a commutative ring R. Prove that $\{x \in R \mid ax \in bR\}$ is an ideal.

41. Let R be a commutative ring and let A be any subset of R. Show that the *annihilator* of A, $\mathrm{Ann}(A) = \{r \in R \mid ra = 0$ for all a in $A\}$, is an ideal.

42. Let R be a commutative ring and let A be any ideal of R. Show that the *nil radical* of A, $N(A) = \{r \in R \mid r^n \in A$ for some positive integer n (n depends on r)$\}$, is an ideal of R. [$N(\langle 0 \rangle)$ is called the *nil radical* of R.]

43. Let $R = Z_{27}$. Find

   a. $N(\langle 0 \rangle)$,   **b.** $N(\langle 3 \rangle)$,   **c.** $N(\langle 9 \rangle)$.

44. Let $R = Z_{36}$. Find

   a. $N(\langle 0 \rangle)$,   **b.** $N(\langle 4 \rangle)$,   **c.** $N(\langle 6 \rangle)$.

45. Let R be a commutative ring. Show that $R/N(\langle 0 \rangle)$ has no nonzero nilpotent elements.

46. Let A be an ideal of a commutative ring. Prove that $N(N(A)) = N(A)$.

47. Let $Z_2[x]$ be the ring of all polynomials with coefficients in Z_2 (that is, coefficients are 0 or 1, and addition and multiplication of coefficients are done modulo 2). Show that $Z_2[x]/\langle x^2 + x + 1 \rangle$ is a field.

48. List the elements of the field given in Exercise 47, and make an addition and multiplication table for the field.

49. Show that $Z_3[x]/\langle x^2 + x + 1 \rangle$ is not a field.

50. Let R be a commutative ring without unity, and let $a \in R$. Describe the smallest ideal I of R that contains a (that is, if J is any ideal that contains a, then $I \subseteq J$).

51. Let R be the ring of continuous functions from $\mathbf{R}$ to $\mathbf{R}$. Let $A = \{f \in R \mid f(0)$ is an even integer$\}$. Show that A is a subring of R, but not an ideal of R.

52. Show that $Z[i]/\langle 1 - i \rangle$ is a field. How many elements does this field have?

53. If R is a principal ideal domain and I is an ideal of R, prove that every ideal of R/I is principal (see Exercise 39).

54. How many elements are in $Z_5[i]/\langle 1 + i \rangle$?

55. Let R be a commutative ring with unity that has the property that $a^2 = a$ for all a in R. Let I be a prime ideal in R. Show that $|R/I| = 2$.

56. Let R be a commutative ring with unity and let I be a proper ideal with the property that every element of R that is not in I is a unit of R. Prove that I is the unique maximal ideal of R.

57. Let $I_0 = \{f(x) \in Z[x] \mid f(0) = 0\}$. For any positive integer n, show that there exists a sequence of strictly increasing ideals such that $I_0 \subset I_1 \subset I_2 \subset \cdots \subset I_n \subset Z[x]$.

58. Let $R = \{(a_1, a_2, a_3, \ldots)\}$, where each $a_i \in Z$. Let $I = \{(a_1, a_2, a_3, \ldots)\}$, where only a finite number of terms are nonzero. Prove that I is not a principal ideal of R.

59. Let R be a commutative ring with unity and let $a, b \in R$. Show that $\langle a, b \rangle$, the smallest ideal of R containing a and b, is $I = \{ra + sb \mid r, s \in R\}$. That is, show that I contains a and b and that any ideal that contains a and b also contains I.

COMPUTER EXERCISES ▬▬▬▬▬▬▬▬▬▬▬▬▬▬▬▬▬ ■■

What is the common denominator of intellectual accomplishment? In math, science, economics, history, or any other subject, the answer is the same: great thinkers notice patterns.

DAVID NIVEN, PSYCHOLOGIST

Software for the computer exercises in this chapter is available at the website:

http://www.d.umn.edu/~jgallian

1. This software determines the number of elements in the ring $Z[i]/\langle a + bi \rangle$ (where $i^2 = -1$). Run the program for several cases and formulate a conjecture based on your data.

2. This software determines the characteristic of the ring $Z[i]/\langle a + bi \rangle$ (where $i^2 = -1$). Let $d = \gcd(a, b)$. Run the program for several cases with $d = 1$ and formulate a conjecture based on your data. Run the program for several cases with $d > 1$ and formulate a conjecture in terms of a, b, and d based on your data. Does the formula you found for $d > 1$ also work in the case that $d = 1$?

3. This software determines when the ring $Z[i]/\langle a + bi \rangle$ (where $i^2 = -1$) is isomorphic to the ring $Z_{a^2+b^2}$. Run the program for several cases and formulate a conjecture based on your data.

Richard Dedekind

*Richard Dedekind was not only
a mathematician, but one of the
wholly great in the history of
mathematics, now and in the
past, the last hero of a great
epoch, the last pupil of Gauss, for
four decades himself a classic,
from whose works not only we,
but our teachers and the teachers
of our teachers, have drawn.*

> EDMUND LANDAU,
> *Commemorative Address
> to the Royal Society of
> Göttingen*

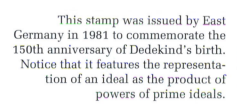

This stamp was issued by East
Germany in 1981 to commemorate the
150th anniversary of Dedekind's birth.
Notice that it features the representation of an ideal as the product of
powers of prime ideals.

RICHARD DEDEKIND was born on October 6, 1831, in Brunswick, Germany, the
birthplace of Gauss. Dedekind was the youngest of four children of a law professor. His early interests were in chemistry and physics, but he obtained a
doctor's degree in mathematics at the age of 21 under Gauss at the University
of Göttingen. Dedekind continued his studies at Göttingen for a few years,
and in 1854 he began to lecture there.

Dedekind spent the years 1858–1862 as a professor in Zürich. Then he
accepted a position at an institute in Brunswick where he had once been a
student. Although this school was less than university level, Dedekind
remained there for the next 50 years. He died in Brunswick in 1916.

During his career, Dedekind made numerous fundamental contributions
to mathematics. His treatment of irrational numbers, "Dedekind cuts," put
analysis on a firm, logical foundation. His work on unique factorization led to
the modern theory of algebraic numbers. He was a pioneer in the theory of
rings and fields. The notion of ideals as well as the term itself are attributed to
Dedekind. Mathematics historian Morris Kline has called him "the effective
founder of abstract algebra."

To find more information about Dedekind, visit:

http://www-groups.dcs.st-and.ac.uk/~history/

Emmy Noether

In the judgment of the most competent living mathematicians, Fräulein Noether was the most significant creative mathematical genius thus far produced since the higher education of women began. In the realm of algebra, in which the most gifted mathematicians have been busy for centuries, she discovered methods which have proved of enormous importance in the development of the present-day younger generation of mathematicians.

ALBERT EINSTEIN, *The New York Times*

EMMY NOETHER was born on March 23, 1882, in Germany. When she entered the University of Erlangen, she was one of only two women among the 1000 students. Noether completed her doctorate in 1907.

In 1916, Noether went to Göttingen and, under the influence of David Hilbert and Felix Klein, became interested in general relativity. While there, she made a major contribution to physics with her theorem that, whenever there is a symmetry in nature, there is also a conservation law, and vice versa. Hilbert tried unsuccessfully to obtain a faculty appointment at Göttingen for Noether, saying, "I do not see that the sex of the candidate is an argument against her admission as Privatdozent. After all, we are a university and not a bathing establishment."

It was not until she was 38 that Noether's true genius revealed itself. Over the next 13 years, she used an axiomatic method to develop a general theory of ideals and noncommutative algebras. With this abstract theory, Noether was able to weld together many important concepts. Her approach was even more important than the individual results. Hermann Weyl said of Noether, "She originated above all a new and epoch-making style of thinking in algebra."

With the rise of Hitler in 1933, Noether, a Jew, fled to the United States and took a position at Bryn Mawr College. She died suddenly on April 14, 1935, following an operation.

To find more information about Noether, visit:

http://www-groups.dcs.st-and.ac.uk/~history/

SUPPLEMENTARY EXERCISES FOR CHAPTERS 12–14 ▬ ▬▬

If at first you do succeed—try to hide your astonishment.

HARRY F. BANKS

True/false questions for Chapters 12–14 are available on the Web at:

http://www.d.umn.edu/~jgallian/TF

1. Find all idempotent elements in Z_{10}, Z_{20}, and Z_{30}. (Recall that a is idempotent if $a^2 = a$.)

2. If m and n are relatively prime integers greater than 1, prove that Z_{mn} has at least two idempotents besides 0 and 1.

3. Suppose that R is a ring in which $a^2 = 0$ implies $a = 0$. Show that R has no nonzero nilpotent elements. (Recall that b is nilpotent if $b^n = 0$ for some positive integer n.)

4. Let R be a commutative ring with more than one element. Prove that if for every nonzero element a of R we have $aR = R$, then R is a field.

5. Let A, B, and C be ideals of a ring R. If $AB \subseteq C$ and C is a prime ideal of R, show that $A \subseteq C$ or $B \subseteq C$. (Compare this with Euclid's Lemma in Chapter 0.)

6. Show, by example, that the intersection of two prime ideals need not be a prime ideal.

7. Let **R** denote the ring of real numbers. Determine all ideals of $\mathbf{R} \oplus \mathbf{R}$. What happens if **R** is replaced by any field F?

8. Determine all factor rings of Z.

9. Let p be a prime. Show that $A = \{(px, y) \mid x, y \in Z\}$ is a maximal ideal of $Z \oplus Z$.

10. Let R be a commutative ring with unity. Suppose that a is a unit and b is nilpotent. Show that $a + b$ is a unit. (*Hint:* See Exercise 29 in Chapter 12.)

11. Let A, B, and C be subrings of a ring R. If $A \subseteq B \cup C$, show that $A \subseteq B$ or $A \subseteq C$.

12. For any element a in a ring R, define $\langle a \rangle$ to be the smallest ideal of R that contains a. If R is a commutative ring with unity, show that $\langle a \rangle = aR = \{ar \mid r \in R\}$. Show, by example, that if R is commutative but does not have a unity, then $\langle a \rangle$ and aR may be different.

13. Let R be a ring with unity. Show that $\langle a \rangle = \{s_1 a t_1 + s_2 a t_2 + \cdots + s_n a t_n \mid s_i, t_i \in R$ and n is a positive integer$\}$.

14. Show that $Z_n[x]$ has characteristic n.

15. Let A and B be ideals of a ring R. If $A \cap B = \{0\}$, show that $ab = 0$ when $a \in A$ and $b \in B$.

16. Show that the direct sum of two integral domains is not an integral domain.

17. Consider the ring $R = \{0, 2, 4, 6, 8, 10\}$ under addition and multiplication modulo 12. What is the characteristic of R?

18. What is the characteristic of $Z_m \oplus Z_n$?

19. Let R be a commutative ring with unity. Suppose that the only ideals of R are $\{0\}$ and R. Show that R is a field.

20. Suppose that I is an ideal of J and that J is an ideal of R. Prove that if I has a unity, then I is an ideal of R. (Be careful not to assume that the unity of I is the unity of R. It need not be—see Exercise 2 in Chapter 12.)

21. Recall that an idempotent element b in a ring is one with the property that $b^2 = b$. Find a nontrivial idempotent (that is, not 0 and not 1) in $Q[x]/\langle x^4 + x^2 \rangle$.

22. In a principal ideal domain, show that every nontrivial prime ideal is a maximal ideal.

23. Find an example of a commutative ring R with unity such that $a, b \in R$, $a \neq b$, and $a^n = b^n$ and $a^m = b^m$, where n and m are positive integers that are relatively prime. (Compare with Exercise 31, part **b**, in Chapter 13.)

24. Let $Q(\sqrt[3]{2})$ denote the smallest subfield of **R** that contains Q and $\sqrt[3]{2}$. [That is, $Q(\sqrt[3]{2})$ is the subfield with the property that $Q(\sqrt[3]{2})$ contains Q and $\sqrt[3]{2}$ and if F is any subfield containing Q and $\sqrt[3]{2}$, then F contains $Q(\sqrt[3]{2})$.] Describe the elements of $Q(\sqrt[3]{2})$.

25. Let R be an integral domain with nonzero characteristic. If A is a proper ideal of R, show that R/A has the same characteristic as R.

26. Let F be a field of order p^n. Determine the group isomorphism class of F under the operation addition.

27. If R is a finite commutative ring with unity, prove that every prime ideal of R is a maximal ideal of R.

28. Let R be a noncommutative ring and let $C(R)$ be the center of R (see Exercise 19 in Chapter 12). Prove that the additive group of $R/C(R)$ is not cyclic.

29. Let

$$R = \left\{ \begin{bmatrix} a & b \\ c & d \end{bmatrix} \middle| a, b, c, d \in Z_2 \right\}$$

with ordinary matrix addition and multiplication modulo 2. Show that

$$A = \left\{ \begin{bmatrix} 1 & 0 \\ 0 & 0 \end{bmatrix} r \middle| r \in R \right\}$$

is not an ideal of R. (Hence, in Exercise 7 in Chapter 14, the commutativity assumption is necessary.)

30. If R is an integral domain and A is a proper ideal of R, must R/A be an integral domain?

31. Let $A = \{a + bi \mid a, b \in Z, a \bmod 2 = b \bmod 2\}$. Show that A is a maximal ideal of $Z[i]$. How many elements does $Z[i]/A$ have?

32. Suppose that R is a commutative ring with unity such that for each a in R there is a positive integer n greater than 1 (n depends on a) such that $a^n = a$. Prove that every prime ideal of R is a maximal ideal of R.

33. State a "finite subfield test"; that is, state conditions that guarantee that a finite subset of a field is a subfield.

34. Let F be a finite field with more than two elements. Prove that the sum of all of the elements of F is 0.

35. Show that if there are nonzero elements a and b in Z_n such that $a^2 + b^2 = 0$, then the ring $Z_n[i] = \{x + yi \mid x, y \in Z_n\}$ has zero-divisors. Use this fact to find a zero-divisor in $Z_{13}[i]$.

36. Suppose that R is a ring with no zero-divisors and that R contains a nonzero element b such that $b^2 = b$. Show that b is a unity for R.

37. Find the characteristic of $Z[i]/\langle 2 + i \rangle$.

38. Show that the characteristic of $Z[i]/\langle a + bi \rangle$ divides $a^2 + b^2$.

39. Show that $4x^2 + 6x + 3$ is a unit in $Z_8[x]$.

40. For any commutative ring R, $R[x, y]$ is the ring of polynomials in x and y with coefficients in R (that is, $R[x, y]$ consists of all finite sums of terms of the form ax^iy^j, where $a \in R$ and i and j are nonnegative integers). (For example, $x^4 - 3x^2y - y^3 \in Z[x, y]$.) Prove that $\langle x, y \rangle$ is a prime ideal in $Z[x, y]$ but not a maximal ideal in $Z[x, y]$.

41. Prove that $\langle x, y \rangle$ is a maximal ideal in $Z_5[x, y]$.

42. Prove that $\langle 2, x, y \rangle$ is a maximal ideal in $Z[x, y]$.

43. Let R and S be rings. Prove that (a, b) is nilpotent in $R \oplus S$ if and only if both a and b are nilpotent.

44. Let R and S be rings with more than one element. Prove that (a,b) is a zero-divisor in $R \oplus S$ if and only if a or b is a zero-divisor or exactly one of a or b is 0.

45. Determine all idempotents in Z_{p^k}, where p is a prime.

46. Let R be a commutative ring with unity 1. Show that a is an idempotent if and only if there exists an element b in R such that $ab = 0$ and $a + b = 1$.

47. Let $Z_n[\sqrt{2}] = \{a + b\sqrt{2} \mid a, b \in Z_n\}$. Define addition and multiplication as in $Z[\sqrt{2}]$, except that modulo n arithmetic is used to combine the coefficients. Show that $Z_3[\sqrt{2}]$ is a field but $Z_7[\sqrt{2}]$ is not.

Ring Homomorphisms

If there is one central idea which is common to all aspects of modern algebra it is the notion of homomorphism.

I. N. HERSTEIN, *Topics in Algebra*

DEFINITION AND EXAMPLES

In our work with groups, we saw that one way to discover information about a group is to examine its interaction with other groups by way of homomorphisms. It should not be surprising to learn that this concept extends to rings with equally profitable results.

Just as a group homomorphism preserves the group operation, a ring homomorphism preserves the ring operations.

DEFINITIONS   *Ring Homomorphism, Ring Isomorphism*

A *ring homomorphism* ϕ from a ring R to a ring S is a mapping from R to S that preserves the two ring operations; that is, for all a, b in R,

$$\phi(a + b) = \phi(a) + \phi(b) \qquad \text{and} \qquad \phi(ab) = \phi(a)\phi(b).$$

A ring homomorphism that is both one-to-one and onto is called a *ring isomorphism*.

As is the case for groups, in the preceding definition the operations on the left of the equal signs are those of R, whereas the operations on the right of the equal signs are those of S.

Again as with group theory, the roles of isomorphisms and homomorphisms are entirely distinct. An isomorphism is used to show that two rings are algebraically identical; a homomorphism is used to simplify a ring while retaining certain of its features.

A schematic representation of a ring homomorphism is given in Figure 15.1. The dashed arrows indicate the results of performing the ring operations.

The following examples illustrate ring homomorphisms. The reader should supply the missing details.

EXAMPLE 1   For any positive integer n, the mapping $k \rightarrow k \bmod n$ is a ring homomorphism from Z onto Z_n (see Exercise 11 in Chapter 0). This mapping is called the *natural homomorphism* from Z to Z_n. ◆

EXAMPLE 2   The mapping $a + bi \rightarrow a - bi$ is a ring isomorphism from the complex numbers onto the complex numbers (see Exercise 23 in Chapter 6). ◆

EXAMPLE 3   Let **R**[x] denote the ring of all polynomials with real coefficients. The mapping $f(x) \rightarrow f(1)$ is a ring homomorphism from **R**[x] onto **R**. ◆

EXAMPLE 4   The correspondence $\phi: x \rightarrow 5x$ from Z_4 to Z_{10} is a ring homomorphism. Although showing that $\phi(x + y) = \phi(x) + \phi(y)$ appears to be accomplished by the simple statement that $5(x + y) = 5x + 5y$, we must bear in mind that the addition on the left is done modulo 4, whereas the addition on the right and the multiplication on both sides are done modulo 10. An analogous difficulty arises in showing that ϕ preserves multiplication. So, to verify that ϕ preserves both operations, we write $x + y = 4q_1 + r_1$ and $xy = 4q_2 + r_2$, where $0 \leq r_1 < 4$ and $0 \leq r_2 < 4$. Then $\phi(x + y) = \phi(r_1) = 5r_1 = 5(x + y - 4q_1) = 5x + 5y - 20q_1 = 5x + 5y = \phi(x) + \phi(y)$ in Z_{10}. Similarly, using the fact that $5 \cdot 5 = 5$ in Z_{10}, we have $\phi(xy) = \phi(r_2) = 5r_2 = 5(xy - 4q_2) = 5xy - 20q_2 = (5 \cdot 5)xy = 5x5y = \phi(x)\phi(y)$ in Z_{10}. ◆

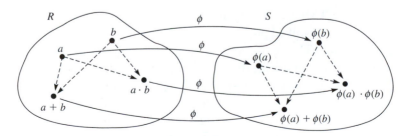

Figure 15.1

EXAMPLE 5   We determine all ring homomorphisms from Z_{12} to Z_{30}. By Example 10 in Chapter 10, the only group homomorphisms from Z_{12} to Z_{30} are $x \rightarrow ax$, where $a = 0, 15, 10, 20, 5,$ or 25. But, since $1 \cdot 1 = 1$ in Z_{12}, we must have $a \cdot a = a$ in Z_{30}. This requirement rules out 20 and 5 as possibilities for a. Finally, simple calculations show that each of the remaining four choices does yield a ring homomorphism.

EXAMPLE 6   Let R be a commutative ring of characteristic 2. Then the mapping $a \rightarrow a^2$ is a ring homomorphism from R to R.   ◆

EXAMPLE 7   Although $2Z$, the group of even integers under addition, is group-isomorphic to the group Z under addition, the ring $2Z$ is not ring-isomorphic to the ring Z. (Quick! What does Z have that $2Z$ doesn't?)   ◆

Our next two examples are applications of the natural homomorphism given in Example 1 to number theory.

EXAMPLE 8   Test for Divisibility by 9
An integer n with decimal representation $a_k a_{k-1} \cdots a_0$ is divisible by 9 if and only if $a_k + a_{k-1} + \cdots + a_0$ is divisible by 9. To verify this, observe that $n = a_k 10^k + a_{k-1} 10^{k-1} + \cdots + a_0$. Then, letting α denote the natural homomorphism from Z to Z_9 [in particular, $\alpha(10) = 1$], we note that n is divisible by 9 if and only if

$$0 = \alpha(n) = \alpha(a_k)(\alpha(10))^k + \alpha(a_{k-1})(\alpha(10))^{k-1} + \cdots + \alpha(a_0)$$

$$= \alpha(a_k) + \alpha(a_{k-1}) + \cdots + \alpha(a_0)$$

$$= \alpha(a_k + a_{k-1} + \cdots + a_0).$$

But $\alpha(a_k + a_{k-1} + \cdots + a_0) = 0$ is equivalent to $a_k + a_{k-1} + \cdots + a_0$ being divisible by 9.   ◆

EXAMPLE 9 (Theorem of Gersonides).   Among the most important unsolved problems in number theory is the so-called "abc conjecture." This conjecture is a natural generalization of a theorem first proved in the fourteenth century by the Rabbi Gersonides. Gersonides proved that the only pairs of positive integers that are powers of 2 and powers of 3 which differ by 1 are 1, 2; 2, 3; 3, 4; and 8, 9. That is, these four pairs are the only solutions to the equations $2^m = 3^n \pm 1$. To verify that this is so for $2^m = 3^n + 1$, observe that for all n we have 3^n

mod $8 = 3$ or 1. Thus, $3^n + 1$ mod $8 = 4$ or 2. On the other hand, for $m > 3$, we have 2^m mod $8 = 0$. To handle the case where $2^m = 3^n - 1$, we first note that for all n, 3^n mod $16 = 3, 9, 11$, or 1, depending on the value of n mod 4. Thus, $(3^n - 1)$ mod $16 = 2, 8, 10$, or 0. Since 2^m mod $16 = 0$ for $m \geq 4$, we have ruled out the cases where n mod $4 = 1, 2$, or 3. Because 3^{4k} mod $5 = (3^4)^k$ mod $5 = 1^k$ mod $5 = 1$, we know that $(3^{4k} - 1)$ mod $5 = 0$. But the only values for 2^m mod 5 are $2, 4, 3$, and 1. This contradiction completes the proof.   ◆

PROPERTIES OF RING HOMOMORPHISMS

Theorem 15.1   **Properties of Ring Homomorphisms**

> *Let ϕ be a ring homomorphism from a ring R to a ring S. Let A be a subring of R and let B be an ideal of S.*
>
> **1.** *For any $r \in R$ and any positive integer n, $\phi(nr) = n\phi(r)$ and $\phi(r^n) = (\phi(r))^n$.*
> **2.** *$\phi(A) = \{\phi(a) \mid a \in A\}$ is a subring of S.*
> **3.** *If A is an ideal and ϕ is onto S, then $\phi(A)$ is an ideal.*
> **4.** *$\phi^{-1}(B) = \{r \in R \mid \phi(r) \in B\}$ is an ideal of R.*
> **5.** *If R is commutative, then $\phi(R)$ is commutative.*
> **6.** *If R has a unity 1, $S \neq \{0\}$, and ϕ is onto, then $\phi(1)$ is the unity of S.*
> **7.** *ϕ is an isomorphism if and only if ϕ is onto and Ker $\phi = \{r \in R \mid \phi(r) = 0\} = \{0\}$.*
> **8.** *If ϕ is an isomorphism from R onto S, then ϕ^{-1} is an isomorphism from S onto R.*

PROOF.   The proofs of these properties are similar to those given in Theorems 10.1 and 10.2 and are left as exercises (Exercise 1).   ■ ■

The student should learn the various properties of Theorem 15.1 in words in addition to the symbols. Property 2 says that the homomorphic image of a subring is a subring. Property 4 says that the pullback of an ideal is an ideal, and so on.

The next three theorems parallel results we had for groups. The proofs are nearly identical to their group theory counterparts and are left as exercises (Exercises 2, 3, and 4).

Theorem 15.2 **Kernels Are Ideals**

> *Let ϕ be a homomorphism from a ring R to a ring S. Then $Ker\ \phi = \{r \in R \mid \phi(r) = 0\}$ is an ideal of R.*

Theorem 15.3 **First Isomorphism Theorem for Rings**

> *Let ϕ be a ring homomorphism from R to S. Then the mapping from $R/Ker\ \phi$ to $\phi(R)$, given by $r + Ker\ \phi \rightarrow \phi(r)$, is an isomorphism. In symbols, $R/Ker\ \phi \approx \phi(R)$.*

Theorem 15.4 **Ideals Are Kernels**

> *Every ideal of a ring R is the kernel of a ring homomorphism of R. In particular, an ideal A is the kernel of the mapping $r \rightarrow r + A$ from R to R/A.*

The homomorphism from R to R/A given in Theorem 15.4 is called the *natural homomorphism* from R to R/A. Theorem 15.3 is often referred to as the Fundamental Theorem of Ring Homomorphisms.

In Example 17 in Chapter 14 we gave a direct proof that $\langle x \rangle$ is a prime ideal of $Z[x]$ but not a maximal ideal. In the following example we illustrate a better way to do this kind of problem.

EXAMPLE 10   Since the mapping ϕ from $Z[x]$ onto Z given by $\phi(f(x)) = f(0)$ is a ring homomorphism with Ker $\phi = \langle x \rangle$ (see Exercise 27 in Chapter 14), we have, by Theorem 15.3, $Z[x]/\langle x \rangle \approx Z$. And because Z is an integral domain but not a field, we know by Theorems 14.3 and 14.4 that the ideal $\langle x \rangle$ is prime but not maximal in $Z[x]$.   ◆

Theorem 15.5 **Homomorphism from Z to a Ring with Unity**

> *Let R be a ring with unity 1. The mapping $\phi: Z \rightarrow R$ given by $n \rightarrow n \cdot 1$ is a ring homomorphism.*

PROOF.   Let $m, n \in Z$. To show that addition is preserved, we consider three cases. First suppose that both m and n are nonnegative. Then

$$\phi(m + n) = (m + n) \cdot 1 = \underbrace{1 + 1 + \cdots + 1}_{(m + n) \text{ summands}}$$

$$= \underbrace{(1 + 1 + \cdots + 1)}_{m \text{ summands}} + \underbrace{(1 + 1 + \cdots + 1)}_{n \text{ summands}}$$

$$= m \cdot 1 + n \cdot 1 = \phi(m) + \phi(n).$$

Next, suppose that both m and n are negative. Then,

$$\phi(m + n) = (m + n) \cdot 1 = (-m - n) \cdot (-1)$$
$$= (-m) \cdot (-1) + (-n) \cdot (-1)$$
$$= m \cdot 1 + n \cdot 1 = \phi(m) + \phi(n).$$

Third, suppose that one of m and n is nonnegative and the other is negative, say $m \geq 0, n < 0$.

Then,

$$\phi(m + n) = (m + n) \cdot 1$$

$$= \underbrace{(1 + 1 + \cdots + 1)}_{m \text{ summands}} - \underbrace{(1 + 1 + \cdots + 1)}_{|n| \text{ summands}}$$

$$= m \cdot 1 + (-n) \cdot (-1) = m \cdot 1 + n \cdot 1 = \phi(m) + \phi(n).$$

So, ϕ preserves addition.

The multiplication can be handled in a single case with the aid of Exercise 15 in Chapter 12, which says that $(m \cdot a)(n \cdot b) = (mn) \cdot (ab)$ for all integers m and n. Thus, $\phi(mn) = (mn) \cdot 1 = (mn) \cdot ((1)(1)) = (m \cdot 1)(n \cdot 1) = \phi(m)\phi(n)$. So, ϕ preserves multiplication as well. ■ ■

Corollary 1   A Ring with Unity Contains Z_n or Z

> *If R is a ring with unity and the characteristic of R is*
> *$n > 0$, then R contains a subring isomorphic to Z_n. If the*
> *characteristic of R is 0, then R contains a subring*
> *isomorphic to Z.*

PROOF.   Let 1 be the unity of R and let $S = \{k \cdot 1 \mid k \in Z\}$. Theorem 15.5 shows that the mapping ϕ from Z to S given by $\phi(k) = k \cdot 1$ is a homomorphism, and by the First Isomorphism Theorem for rings, we have $Z/\text{Ker } \phi \approx S$. But, clearly, Ker $\phi = \langle n \rangle$, where n is the additive order of 1 and, by Theorem 13.3, n is also the characteristic of R. So, when R has characteristic n, $S \approx Z/\langle n \rangle \approx Z_n$. When R has characteristic 0, $S \approx Z/\langle 0 \rangle \approx Z$. ■ ■

Corollary 2 Z_m Is a Homomorphic Image of Z

> *For any positive integer m, the mapping of ϕ: $Z \rightarrow Z_m$ given by $x \rightarrow x$ mod m is a ring homomorphism.*

PROOF.   This follows directly from the statement of Theorem 15.5, since in the ring Z_m, the integer x mod m is $x \cdot 1$. (For example, in Z_3, if $x = 5$, we have $5 \cdot 1 = 1 + 1 + 1 + 1 + 1 = 2$.)   ■ ■

Corollary 3   A Field Contains Z_p or Q (Steinitz, 1910)

> *If F is a field of characteristic p, then F contains a subfield isomorphic to Z_p. If F is a field of characteristic 0, then F contains a subfield isomorphic to the rational numbers.*

PROOF.   By Corollary 1, F contains a subring isomorphic to Z_p if F has characteristic p, and F has a subring S isomorphic to Z if F has characteristic 0. In the latter case, let

$$T = \{ab^{-1} \mid a, b \in S, b \neq 0\}.$$

Then T is isomorphic to the rationals (Exercise 57).   ■ ■

Since the intersection of all subfields of a field is itself a subfield (Exercise 9), every field has a smallest subfield (that is, a subfield that is contained in every subfield). This subfield is called the *prime subfield* of the field. It follows from Corollary 3 that the prime subfield of a field of characteristic p is isomorphic to Z_p, whereas the prime subfield of a field of characteristic 0 is isomorphic to Q. (See Exercise 61.)

THE FIELD OF QUOTIENTS

Although the integral domain Z is not a field, it is at least contained in a field—the field of rational numbers. And notice that the field of rational numbers is nothing more than quotients of integers. Can we mimic the construction of the rationals from the integers for other integral domains? Yes. The field constructed in Theorem 15.6 is called the *field of quotients of D*. Throughout the proof of Theorem 15.6, you should keep in mind that we are using the construction of the rationals from the integers as a model for our construction of the field of quotients of D.

Theorem 15.6   **Field of Quotients**

> *Let D be an integral domain. Then there exists a field F*
> *(called the field of quotients of D) that contains a subring*
> *isomorphic to D.*

PROOF.   Let $S = \{(a, b) \mid a, b \in D, b \neq 0\}$. We define an equivalence relation on S by $(a, b) \equiv (c, d)$ if $ad = bc$ (compare with
Example 15 in Chapter 0). Now, let F be the set of equivalence classes
of S under the relation $\equiv$ and denote the equivalence class that contains (x, y) by x/y. We define addition and multiplication on F by

$$a/b + c/d = (ad + bc)/(bd) \qquad \text{and} \qquad a/b \cdot c/d = (ac)/(bd).$$

(Notice that here we need the fact that D is an integral domain to ensure
that multiplication is closed; that is, $bd \neq 0$ whenever $b \neq 0$ and $d \neq 0$.)

Since there are many representations of any particular element of
F (just as in the rationals, we have $1/2 = 3/6 = 4/8$), we must show
that these two operations are well defined. To do this, suppose that a/b
$= a'/b'$ and $c/d = c'/d'$, so that $ab' = a'b$ and $cd' = c'd$. It then
follows that

$$\begin{aligned}
(ad + bc)b'd' &= adb'd' + bcb'd' = (ab')dd' + (cd')bb' \\
&= (a'b)dd' + (c'd)bb' = a'd'bd + b'c'bd \\
&= (a'd' + b'c')bd.
\end{aligned}$$

Thus, by definition, we have

$$(ad + bc)/(bd) = (a'd' + b'c')/(b'd'),$$

and, therefore, addition is well defined. We leave the verification that
multiplication is well defined as an exercise (Exercise 49). That F is a
field is straightforward. Let 1 denote the unity of D. Then $0/1$ is the
additive identity of F. The additive inverse of a/b is $-a/b$; the multiplicative inverse of a nonzero element a/b is b/a. The remaining field
properties can be checked easily.

Finally, the mapping $\phi: D \rightarrow F$ given by $x \rightarrow x/1$ is a ring isomorphism from D to $\phi(D)$ (see Exercise 63).   ∎

EXAMPLE 11   Let $D = Z[x]$. Then the field of quotients of D is
$\{f(x)/g(x) \mid f(x), g(x) \in D$, where $g(x)$ is not the zero polynomial$\}$.   ◆

When F is a field, the field of quotients of $F[x]$ is traditionally
denoted by $F(x)$.

EXAMPLE 12   Let p be a prime. Then $Z_p(x) = \{f(x)/g(x) \mid f(x), g(x)$
$\in Z_p[x], g(x) \neq 0\}$ is an infinite field of characteristic p.   ◆

EXERCISES ▪▪

We can work it out.

TITLE OF SONG BY JOHN LENNON AND
PAUL MCCARTNEY, December 1965

1. Prove Theorem 15.1.

2. Prove Theorem 15.2.

3. Prove Theorem 15.3.

4. Prove Theorem 15.4.

5. Show that the correspondence $x \to 5x$ from Z_5 to Z_{10} does not preserve addition.

6. Show that the correspondence $x \to 3x$ from Z_4 to Z_{12} does not preserve multiplication.

7. **a.** Is the ring $2Z$ isomorphic to the ring $3Z$?

   b. Is the ring $2Z$ isomorphic to the ring $4Z$?

8. Prove that every ring homomorphism ϕ from Z_n to itself has the form $\phi(x) = ax$, where $a^2 = a$.

9. Prove that the intersection of any collection of subfields of a field F is a subfield of F. (This exercise is referred to in this chapter.)

10. Let $Z_3[i] = \{a + bi \mid a, b \in Z_3\}$ (see Example 9 in Chapter 13). Show that the field $Z_3[i]$ is ring-isomorphic to the field $Z_3[x]/\langle x^2 + 1 \rangle$.

11. Let

$$S = \left\{ \begin{bmatrix} a & b \\ -b & a \end{bmatrix} \middle| a, b \in \mathbf{R} \right\}.$$

Show that $\phi: \mathbf{C} \to S$ given by

$$\phi(a + bi) = \begin{bmatrix} a & b \\ -b & a \end{bmatrix}$$

is a ring isomorphism.

12. Let $Z[\sqrt{2}] = \{a + b\sqrt{2} \mid a, b \in Z\}$. Let

$$H = \left\{ \begin{bmatrix} a & 2b \\ b & a \end{bmatrix} \middle| a, b \in Z \right\}.$$

Show that $Z[\sqrt{2}]$ and H are isomorphic as rings.

13. Consider the mapping from $M_2(Z)$ into Z given by $\begin{bmatrix} a & b \\ c & d \end{bmatrix} \to a$. Prove or disprove that this is a ring homomorphism.

14. Let $R = \left\{ \begin{bmatrix} a & b \\ 0 & c \end{bmatrix} \mid a, b, c \in Z \right\}$. Prove or disprove that the mapping

$\begin{bmatrix} a & b \\ 0 & c \end{bmatrix} \to a$ is a ring homomorphism.

15. Is the mapping from Z_5 to Z_{30} given by $x \to 6x$ a ring homomorphism? Note that the image of the unity is the unity of the image but not the unity of Z_{30}.

16. Is the mapping from Z_{10} to Z_{10} given by $x \to 2x$ a ring homomorphism?

17. Describe the kernel of the homomorphism given in Example 3.

18. Determine all ring isomorphisms from Z_n to itself.

19. In Z, let $A = \langle 2 \rangle$ and $B = \langle 8 \rangle$. Show that the group A/B is isomorphic to the group Z_4 but that the ring A/B is not ring-isomorphic to the ring Z_4.

20. Determine all ring homomorphisms from Z_6 to Z_6. Determine all ring homomorphisms from Z_{20} to Z_{30}.

21. Determine all ring homomorphisms from Z to Z.

22. Suppose ϕ is a ring homomorphism from $Z \oplus Z$ into $Z \oplus Z$. What are the possibilities for $\phi((1, 0))$?

23. Determine all ring homomorphisms from $Z \oplus Z$ into $Z \oplus Z$.

24. Recall that a ring element a is called an idempotent if $a^2 = a$. Prove that a ring homomorphism carries an idempotent to an idempotent.

25. Let R be a ring with unity and let ϕ be a ring homomorphism from R onto S. Where S has more than one element. Prove that S has a unity.

26. Show that $(Z \oplus Z)/(\langle a \rangle \oplus \langle b \rangle)$ is ring-isomorphic to $Z_a \oplus Z_b$.

27. Determine all ring homomorphisms from $Z \oplus Z$ to Z.

28. Prove that the sum of the squares of three consecutive integers cannot be a square.

29. Let m be a positive integer and let n be an integer obtained from m by rearranging the digits of m in some way. (For example, 72345 is a rearrangement of 35274.) Show that $m - n$ is divisible by 9.

30. (Test for divisibility by 11) Let n be an integer with decimal representation $a_k a_{k-1} \cdots a_1 a_0$. Prove that n is divisible by 11 if and only if $a_0 - a_1 + a_2 - \cdots (-1)^k a_k$ is divisible by 11.

31. Show that the number 7,176,825,942,116,027,211 is divisible by 9 but not divisible by 11.

32. Show that the number 9,897,654,527,609,805 is divisible by 99.

33. (Test for divisibility by 3) Let n be an integer with decimal representation $a_k a_{k-1} \cdots a_1 a_0$. Prove that n is divisible by 3 if and only if $a_k + a_{k-1} + \cdots + a_1 + a_0$ is divisible by 3.

34. (Test for divisibility by 4) Let n be an integer with decimal representation $a_k a_{k-1} \cdots a_1 a_0$. Prove that n is divisible by 4 if and only if $a_1 a_0$ is divisible by 4.

35. In your head, determine $(2 \cdot 10^{75} + 2)^{100}$ mod 3 and $(10^{100} + 1)^{99}$ mod 3.

36. Determine all ring homomorphisms from Q to Q.

37. Let R and S be commutative rings with unity. If ϕ is a homomorphism from R onto S and the characteristic of R is nonzero, prove that the characteristic of S divides the characteristic of R.

38. Let R be a commutative ring of prime characteristic p. Show that the *Frobenius* map $x \rightarrow x^p$ is a ring homomorphism from R to R.

39. Is there a ring homomorphism from the reals to some ring whose kernel is the integers?

40. Show that a homomorphism from a field onto a ring with more than one element must be an isomorphism.

41. Suppose that R and S are commutative rings with unities. Let ϕ be a ring homomorphism from R onto S and let A be an ideal of S.
    a. If A is prime in S, show that $\phi^{-1}(A) = \{x \in R \mid \phi(x) \in A\}$ is prime in R.
    b. If A is maximal in S, show that $\phi^{-1}(A)$ is maximal in R.

42. A *principle ideal ring* is a ring with the property that every ideal has the form (a). Show that the homomorphic image of a principal ideal ring is a principal ideal ring.

43. Let R and S be rings.
    a. Show that the mapping from $R \oplus S$ onto R given by $(a, b) \rightarrow a$ is a ring homomorphism.
    b. Show that the mapping from R to $R \oplus S$ given by $a \rightarrow (a, 0)$ is a one-to-one ring homomorphism.
    c. Show that $R \oplus S$ is ring-isomorphic to $S \oplus R$.

44. Show that if m and n are distinct positive integers, then mZ is not ring-isomorphic to nZ.

45. Prove or disprove that the field of real numbers is ring-isomorphic to the field of complex numbers.

46. Show that the only ring automorphism of the real numbers is the identity mapping.

47. Determine all ring homomorphisms from $\mathbf{R}$ to $\mathbf{R}$.

48. Suppose that n divides m and that a is an idempotent of Z_n (that is, $a^2 = a$). Show that the mapping $x \rightarrow ax$ is a ring homomorphism from Z_m to Z_n. Show that the same correspondence need not yield a ring homomorphism if n does not divide m.

49. Show that the operation of multiplication defined in the proof of Theorem 15.6 is well defined.

50. Let $Q[\sqrt{2}] = \{a + b\sqrt{2} \mid a, b \in Q\}$ and $Q[\sqrt{5}] = \{a + b\sqrt{5} \mid a, b \in Q\}$. Show that these two rings are not ring-isomorphic.

51. Let $Z[i] = \{a + bi \mid a, b \in Z\}$. Show that the field of quotients of $Z[i]$ is ring-isomorphic to $Q[i] = \{r + si \mid r, s \in Q\}$. (This exercise is referred to in Chapter 18.)

52. Let F be a field. Show that the field of quotients of F is ring-isomorphic to F.

53. Let D be an integral domain and let F be the field of quotients of D. Show that if E is any field that contains D, then E contains a subfield that is ring-isomorphic to F. (Thus, the field of quotients of an integral domain D is the smallest field containing D.)

54. Explain why a commutative ring with unity that is not an integral domain cannot be contained in a field. (Compare with Theorem 15.6.)

55. Show that the relation $\equiv$ defined in the proof of Theorem 15.6 is an equivalence relation.

56. Give an example of a ring without unity that is contained in a field.

57. Prove that the set T in the proof of Corollary 3 to Theorem 15.5 is ring-isomorphic to the field of rational numbers.

58. Suppose that $\phi: R \rightarrow S$ is a ring homomorphism and that the image of ϕ is not $\{0\}$. If R has a unity and S is an integral domain, show that ϕ carries the unity of R to the unity of S. Give an example to show that the preceding statement need not be true if S is not an integral domain.

59. Let $f(x) \in \mathbf{R}[x]$. If $a + bi$ is a complex zero of $f(x)$ (here $i = \sqrt{-1}$), show that $a - bi$ is a zero of $f(x)$. (This exercise is referred to in Chapter 32.)

60. Let $R = \left\{ \begin{bmatrix} a & b \\ b & a \end{bmatrix} \mid a, b \in Z \right\}$, and let ϕ be the mapping that takes $\begin{bmatrix} a & b \\ b & a \end{bmatrix}$ to $a - b$.

   a. Show that ϕ is a homomorphism.
   b. Determine the kernel of ϕ.
   c. Show that $R/\text{Ker } \phi$ is isomorphic to Z.
   d. Is $\text{Ker } \phi$ a prime ideal?
   e. Is $\text{Ker } \phi$ a maximal ideal?

61. Show that the prime subfield of a field of characteristic p is ring-isomorphic to Z_p and that the prime subfield of a field of characteristic

0 is ring-isomorphic to Q. (This exercise is referred to in this chapter.)

62. Let n be a positive integer. Show that there is a ring isomorphism from Z_2 to a subring of Z_{2n} if and only if n is odd.

63. Show that the mapping $\phi\colon D \to F$ in the proof of Theorem 15.6 is a ring homomorphism.

Suggested Readings

J. A. Gallian and J. Van Buskirk, "The Number of Homomorphisms from Z_m into Z_n," *American Mathematical Monthly* 91 (1984): 196–197.

    In this article, formulas are given for the number of group homomorphisms from Z_m into Z_n and the number of ring homomorphisms from Z_m into Z_n.

Lillian Kinkade and Joyce Wagner, "When Polynomial Rings Are Principal Ideal Rings," *Journal of Undergraduate Mathematics* 23 (1991): 59–62.

    In this article written by undergraduates, it is shown that $R[x]$ is a principal ideal ring if and only if $R \approx R_1 \oplus R_2 \oplus \cdots \oplus R_n$, where each R_i is a field.

Mohammad Saleh and Hasan Yousef, "The Number of Ring Homomorphisms from $Z_{m_1} \oplus \cdots \oplus Z_{m_r}$ into $Z_{k_1} \oplus \cdots \oplus Z_{k_s}$," *American Mathematical Monthly* 105 (1998): 259–260.

    This article gives a formula for the number described in the title.

Suggested Website

http://www.d.umn.edu/~jgallian/puzzle
This site has a math puzzle that is based on the ideas presented in this chapter. The user selects an integer and then proceeds through a series of steps to produce a new integer. Finally, another integer is created by using all but one of the digits of the previous integer in any order. The software then determines the digit not used.

16 Polynomial Rings

Wit lies in recognising the resemblance among things which differ and the difference between things which are alike.

MADAME DE STAËL

NOTATION AND TERMINOLOGY

One of the mathematical concepts that students are most familiar with and most comfortable with is that of a polynomial. In high school, students study polynomials with integer coefficients, rational coefficients, real coefficients, and perhaps even complex coefficients. In earlier chapters of this book, we introduced something that was probably new—polynomials with coefficients from Z_n. Notice that all of these sets of polynomials are rings, and, in each case, the set of coefficients is also a ring. In this chapter, we abstract all of these examples into one.

DEFINITION *Ring of Polynomials over R*

Let R be a commutative ring. The set of formal symbols

$$R[x] = \{a_nx^n + a_{n-1}x^{n-1} + \cdots + a_1x + a_0 \mid a_i \in R,$$
$$n \text{ is a nonnegative integer}\}$$

is called the *ring of polynomials over R in the indeterminate x.*

Two elements

$$a_nx^n + a_{n-1}x^{n-1} + \cdots + a_1x + a_0$$

and

$$b_mx^m + b_{m-1}x^{m-1} + \cdots + b_1x + b_0$$

of $R[x]$ are considered equal if and only if $a_i = b_i$ for all nonnegative integers i. (Define $a_i = 0$ when $i > n$ and $b_i = 0$ when $i > m$.)

291

In this definition, the symbols x, x^2, $\ldots$, x^n do not represent "unknown" elements or variables from the ring R. Rather, their purpose is to serve as convenient placeholders that separate the ring elements a_n, a_{n-1}, $\ldots$, a_0. We could have avoided the x's by defining a polynomial as an infinite sequence a_0, a_1, a_2, $\ldots$, a_n, $0, 0, 0, \ldots$, but our method takes advantage of the student's experience in manipulating polynomials where x does represent a variable. The disadvantage of our method is that one must be careful not to confuse a polynomial with the function determined by a polynomial. For example, in $Z_3[x]$, the polynomials $f(x) = x^3$ and $g(x) = x^5$ determine the same function from Z_3 to Z_3, since $f(a) = g(a)$ for all a in Z_3.[†] But $f(x)$ and $g(x)$ are different elements of $Z_3[x]$. Also, in the ring $Z_n[x]$, be careful to reduce only the coefficients and not the exponents modulo n. For example, in $Z_3[x]$, $5x = 2x$, but $x^5 \neq x^2$.

To make $R[x]$ into a ring, we define addition and multiplication in the usual way.

DEFINITION   *Addition and Multiplication in R[x]*

Let R be a commutative ring and let

$$f(x) = a_n x^n + a_{n-1} x^{n-1} + \cdots + a_1 x + a_0$$

and

$$g(x) = b_m x^m + b_{m-1} x^{m-1} + \cdots + b_1 x + b_0$$

belong to $R[x]$. Then

$$f(x) + g(x) = (a_s + b_s)x^s + (a_{s-1} + b_{s-1})x^{s-1}$$
$$+ \cdots + (a_1 + b_1)x + a_0 + b_0,$$

where s is the maximum of m and n, $a_i = 0$ for $i > n$, and $b_i = 0$ for $i > m$. Also,

$$f(x)g(x) = c_{m+n}x^{m+n} + c_{m+n-1}x^{m+n-1} + \cdots + c_1 x + c_0,$$

where

$$c_k = a_k b_0 + a_{k-1} b_1 + \cdots + a_1 b_{k-1} + a_0 b_k$$

for $k = 0, \ldots, m + n$.

[†]In general, given $f(x)$ in $R[x]$ and a in R, $f(a)$ means substitute a for x in the formula for $f(x)$. This substitution is a homomorphism from $R[x]$ to R.

Although the definition of multiplication might appear complicated, it is just a formalization of the familiar process of using the distributive property and collecting like terms. So, just multiply polynomials over a commutative ring R in the same way that polynomials are always multiplied. Here is an example.

Consider $f(x) = 2x^3 + x^2 + 2x + 2$ and $g(x) = 2x^2 + 2x + 1$ in $Z_3[x]$. Then, in our preceding notation, $a_5 = 0$, $a_4 = 0$, $a_3 = 2$, $a_2 = 1$, $a_1 = 2$, $a_0 = 2$, and $b_5 = 0$, $b_4 = 0$, $b_3 = 0$, $b_2 = 2$, $b_1 = 2$, $b_0 = 1$. Now, using the definitions and remembering that addition and multiplication of the coefficients are done modulo 3, we have

$$f(x) + g(x) = (2 + 0)x^3 + (1 + 2)x^2 + (2 + 2)x + (2 + 1)$$
$$= 2x^3 + 0x^2 + 1x + 0$$

and

$$\begin{aligned} f(x) \cdot g(x) &= (0 \cdot 1 + 0 \cdot 2 + 2 \cdot 2 + 1 \cdot 0 + 2 \cdot 0 + 2 \cdot 0)x^5 \\ &+ (0 \cdot 1 + 2 \cdot 2 + 1 \cdot 2 + 2 \cdot 0 + 2 \cdot 0)x^4 \\ &+ (2 \cdot 1 + 1 \cdot 2 + 2 \cdot 2 + 2 \cdot 0)x^3 \\ &+ (1 \cdot 1 + 2 \cdot 2 + 2 \cdot 2)x^2 + (2 \cdot 1 + 2 \cdot 2)x + 2 \cdot 1 \\ &= x^5 + 0x^4 + 2x^3 + 0x^2 + 0x + 2. \end{aligned}$$

Our definitions for addition and multiplication of polynomials were formulated so that they are commutative and associative, and so that multiplication is distributive over addition. We leave the verification that $R[x]$ is a ring to the reader.

It is time to introduce some terminology for polynomials. If

$$f(x) = a_n x^n + a_{n-1} x^{n-1} + \cdots + a_1 x + a_0,$$

where $a_n \neq 0$, we say that $f(x)$ has *degree n*; the term a_n is called the *leading coefficient* of $f(x)$, and if the leading coefficient is the multiplicative identity element of R, we say that $f(x)$ is a *monic* polynomial. The polynomial $f(x) = 0$ has no degree. Polynomials of the form $f(x) = a_0$ are called *constant*. We often write $\deg f(x) = n$ to indicate that $f(x)$ has degree n. In keeping with our experience with polynomials with real coefficients, we adopt the following notational conventions: We may insert or delete terms of the form $0x^k$; $1x^k$ will be denoted by x^k; $+ (-a_k)x^k$ will be denoted by $-a_k x^k$.

Very often properties of R carry over to $R[x]$. Our first theorem is a case in point.

Theorem 16.1   **D an Integral Domain Implies D[x] an Integral Domain**

> *If D is an integral domain, then D[x] is an integral domain.*

PROOF.   Since we already know that $D[x]$ is a ring, all we need to show is that $D[x]$ is commutative with a unity and has no zero-divisors. Clearly, $D[x]$ is commutative whenever D is. If 1 is the unity element of D, it is easy to check that $f(x) = 1$ is the unity element of $D[x]$. Finally, suppose that

$$f(x) = a_n x^n + a_{n-1} x^{n-1} + \cdots + a_0$$

and

$$g(x) = b_m x^m + b_{m-1} x^{m-1} + \cdots + b_0,$$

where $a_n \neq 0$ and $b_m \neq 0$. Then, by definition, $f(x)g(x)$ has leading coefficient $a_n b_m$ and, since D is an integral domain, $a_n b_m \neq 0$.   ■ ■

THE DIVISION ALGORITHM AND CONSEQUENCES

One of the properties of integers that we have used repeatedly is the division algorithm: If a and b are integers and $b \neq 0$, then there exist unique integers q and r such that $a = bq + r$, where $0 \leq r < |b|$. The next theorem is the analogous statement for polynomials over a field.

Theorem 16.2   **Division Algorithm for F[x]**

> *Let F be a field and let f(x) and g(x) $\in$ F[x] with g(x) $\neq$ 0. Then there exist unique polynomials q(x) and r(x) in F[x] such that f(x) = g(x)q(x) + r(x) and either r(x) = 0 or deg r(x) < deg g(x).*

PROOF.   We begin by showing the existence of $q(x)$ and $r(x)$. If $f(x) = 0$ or deg $f(x)$ < deg $g(x)$, we simply set $q(x) = 0$ and $r(x) = f(x)$. So, we may assume that $n =$ deg $f(x) \geq$ deg $g(x) = m$ and let $f(x) = a_n x^n + \cdots + a_0$ and $g(x) = b_m x^m + \cdots + b_0$. The idea behind this proof is to begin just as if you were going to "long divide" $g(x)$ into $f(x)$, then use the Second Principle of Mathematical Induction on deg $f(x)$ to finish up. Thus, resorting to long division, we let $f_1(x) = f(x) -$

$a_n b_m^{-1} x^{n-m} g(x)$.[†] Then, $f_1(x) = 0$ or deg $f_1(x) <$ deg $f(x)$; so, by our induction hypothesis, there exist $q_1(x)$ and $r_1(x)$ in $F[x]$ such that $f_1(x) = g(x)q_1(x) + r_1(x)$, where $r_1(x) = 0$ or deg $r_1(x) <$ deg $g(x)$. [Technically, we should get the induction started by proving the case in which deg $f(x) = 0$, but this is trivial.] Thus,

$$f(x) = a_n b_m^{-1} x^{n-m} g(x) + f_1(x)$$
$$= a_n b_m^{-1} x^{n-m} g(x) + q_1(x)g(x) + r_1(x)$$
$$= [a_n b_m^{-1} x^{n-m} + q_1(x)]g(x) + r_1(x).$$

So, the polynomials $q(x) = a_n b_m^{-1} x^{n-m} + q_1(x)$ and $r(x) = r_1(x)$ have the desired properties.

To prove uniqueness, suppose that $f(x) = g(x)q(x) + r(x)$ and $f(x) = g(x)\bar{q}(x) + \bar{r}(x)$, where $r(x) = 0$ or deg $r(x) <$ deg $g(x)$ and $\bar{r}(x) = 0$ or deg $\bar{r}(x) <$ deg $g(x)$. Then, subtracting these two equations, we obtain

$$0 = g(x)[q(x) - \bar{q}(x)] + [r(x) - \bar{r}(x)]$$

or

$$\bar{r}(x) - r(x) = g(x)[q(x) - \bar{q}(x)].$$

Thus, $\bar{r}(x) - r(x)$ is 0, or the degree of $\bar{r}(x) - r(x)$ is at least that of $g(x)$. Since the latter is clearly impossible, we have $\bar{r}(x) = r(x)$ and $q(x) = \bar{q}(x)$ as well. ■ ■

The polynomials $q(x)$ and $r(x)$ in the division algorithm are called the *quotient* and *remainder* in the division of $f(x)$ by $g(x)$. When the ring of coefficients of a polynomial ring is a field, we can use the long division process to determine the quotient and remainder.

[†]For example,

$$
\begin{array}{r}
(3/2)x^2 \qquad\qquad \\
2x^2 + 2 \overline{)3x^4 \qquad\quad + x + 1} \\
3x^4 + 3x^2 \qquad\qquad \\
\overline{- 3x^2 + x + 1}
\end{array}
$$

So,

$$-3x^2 + x + 1 = 3x^4 + x + 1 - (3/2)x^2(2x^2 + 2)$$

In general,

$$
\begin{array}{r}
a_n b_m^{-1} x^{n-m} \qquad\qquad \\
b_m x^m + \cdots \overline{)a_n x^n + \cdots} \\
a_n x^n + \cdots \qquad \\
\overline{f_1(x)}
\end{array}
$$

So,

$$f_1(x) = (a_n x^n + \cdots) - a_n b_m^{-1} x^{n-m}(b_m x^m + \cdots)$$

EXAMPLE 1 To find the quotient and remainder upon dividing $f(x) = 3x^4 + x^3 + 2x^2 + 1$ by $g(x) = x^2 + 4x + 2$, where $f(x)$ and $g(x)$ belong to $Z_5[x]$, we may proceed by long division, provided we keep in mind that addition and multiplication are done modulo 5. Thus,

$$
\begin{array}{r}
3x^2 + 4x \\
x^2 + 4x + 2 \overline{)3x^4 + x^3 + 2x^2 \qquad +1} \\
\underline{3x^4 + 2x^3 + x^2} \\
4x^3 + x^2 \qquad +1 \\
\underline{4x^3 + x^2 + 3x} \\
2x + 1
\end{array}
$$

So, $3x^2 + 4x$ is the quotient and $2x + 1$ is the remainder. Therefore,

$$3x^4 + x^3 + 2x^2 + 1 = (x^2 + 4x + 2)(3x^2 + 4x) + 2x + 1. \qquad \blacklozenge$$

Let D be an integral domain. If $f(x)$ and $g(x) \in D[x]$, we say that $g(x)$ *divides* $f(x)$ in $D[x]$ [and write $g(x) \mid f(x)$] if there exists an $h(x) \in D[x]$ such that $f(x) = g(x)h(x)$. In this case, we also call $g(x)$ a *factor* of $f(x)$. An element a is a *zero* (or a *root*) of a polynomial $f(x)$ if $f(a) = 0$. [Recall that $f(a)$ means substitute a for x in the expression for $f(x)$.] When F is a field, $a \in F$, and $f(x) \in F[x]$, we say that a is a *zero of multiplicity* k $(k \geq 1)$ if $(x - a)^k$ is a factor of $f(x)$ but $(x - a)^{k+1}$ is not a factor of $f(x)$. With these definitions, we may now give several important corollaries of the division algorithm. No doubt you have seen these for the special case where F is the field of real numbers.

Corollary 1 **The Remainder Theorem**

> *Let F be a field, $a \in F$, and $f(x) \in F[x]$. Then $f(a)$ is the remainder in the division of $f(x)$ by $x - a$.*

PROOF. The proof of Corollary 1 is left as an exercise (Exercise 5). ■■

Corollary 2 **The Factor Theorem**

> *Let F be a field, $a \in F$, and $f(x) \in F[x]$. Then a is a zero of $f(x)$ if and only if $x - a$ is a factor of $f(x)$.*

PROOF. The proof of Corollary 2 is left as an exercise (Exercise 7). ■■

Corollary 3   **Polynomials of Degree *n* Have at Most *n* Zeros**

> *A polynomial of degree n over a field has at most n zeros counting multiplicity.*

PROOF.   We proceed by induction on n. Clearly, a polynomial of degree 0 over a field has no zeros. Now suppose that $f(x)$ is a polynomial of degree n over a field and a is a zero of $f(x)$ of multiplicity k. Then, $f(x) = (x - a)^k q(x)$ and $q(a) \neq 0$; and, since $n = \deg f(x) = \deg (x - a)^k q(x) = k + \deg q(x)$, we have $k \leq n$ (see Exercise 17). If $f(x)$ has no zeros other than a, we are done. On the other hand, if $b \neq a$ and b is a zero of $f(x)$, then $0 = f(b) = (b - a)^k q(b)$, so that b is also a zero of $q(x)$. By the Second Principle of Mathematical Induction, we know that $q(x)$ has at most $\deg q(x) = n - k$ zeros, counting multiplicity. Thus, $f(x)$ has at most $k + n - k = n$ zeros, counting multiplicity.   ■ ■

We remark that Corollary 3 is not true for arbitrary polynomial rings. For example, the polynomial $x^2 + 3x + 2$ has four zeros in Z_6. (See Exercise 3.) Lagrange was the first to prove Corollary 3 for polynomials in $Z_p[x]$.

EXAMPLE 2   The Complex Zeros of $x^n - 1$
We find all complex zeros of $x^n - 1$. Let $\omega = \cos(360°/n) + i \sin(360°/n)$. It follows from DeMoivre's Theorem (see Example 8 in Chapter 0) that $\omega^n = 1$ and $\omega^k \neq 1$ for $1 \leq k < n$. Thus, each of 1, $\omega, \omega^2, \ldots, \omega^{n-1}$ is a zero of $x^n - 1$ and, by Corollary 3, there are no others.   ◆

The complex number ω in Example 2 is called a *primitive nth root of unity.*
We conclude this chapter with an important theoretical application of the division algorithm, but first an important definition.

DEFINITION   *Principal Ideal Domain (PID)*

> A *principal ideal domain* is an integral domain R in which every ideal has the form $\langle a \rangle = \{ra \mid r \in R\}$ for some a in R.

Theorem 16.3   **F[x] Is a PID**

> *Let F be a field. Then F[x] is a principal ideal domain.*

PROOF. By Theorem 16.1, we know that $F[x]$ is an integral domain. Now, let I be an ideal in $F[x]$. If $I = \{0\}$, then $I = \langle 0 \rangle$. If $I \neq \{0\}$, then among all the elements of I, let $g(x)$ be one of minimum degree. We will show that $I = \langle g(x) \rangle$. Since $g(x) \in I$, we have $\langle g(x) \rangle \subseteq I$. Now let $f(x) \in I$. Then, by the division algorithm, we may write $f(x) = g(x)q(x) + r(x)$, where $r(x) = 0$ or deg $r(x) <$ deg $g(x)$. Since $r(x) = f(x) - g(x)q(x) \in I$, the minimality of deg $g(x)$ implies that the latter condition cannot hold. So, $r(x) = 0$ and, therefore, $f(x) \in \langle g(x) \rangle$. This shows that $I \subseteq \langle g(x) \rangle$.   ■ ■

The proof of Theorem 16.3 also establishes the following.

Theorem 16.4   **Criterion for $I = \langle g(x) \rangle$**

> *Let F be a field, I a nonzero ideal in $F[x]$, and $g(x)$ an element of $F[x]$. Then, $I = \langle g(x) \rangle$ if and only if $g(x)$ is a nonzero polynomial of minimum degree in I.*

As an application of the First Isomorphism Theorem for Rings (Theorem 15.3) and Theorem 16.4, we verify the remark we made in Example 12 in Chapter 14 that the ring $\mathbf{R}[x]/\langle x^2 + 1 \rangle$ is isomorphic to the ring of complex numbers.

EXAMPLE 3   Consider the homomorphism ϕ from $\mathbf{R}[x]$ onto $\mathbf{C}$ given by $f(x) \to f(i)$ (that is, evaluate a polynomial in $\mathbf{R}[x]$ at i). Then $x^2 + 1 \in \text{Ker } \phi$ and is clearly a polynomial of minimum degree in Ker ϕ. Thus, Ker $\phi = \langle x^2 + 1 \rangle$ and $\mathbf{R}[x]/\langle x^2 + 1 \rangle$ is isomorphic to $\mathbf{C}$.   ◆

EXERCISES   ■ ■

If I feel unhappy, I do mathematics to become happy. If I am happy, I do mathematics to keep happy.

PAUL TURÁN

1. Let $f(x) = 4x^3 + 2x^2 + x + 3$ and $g(x) = 3x^4 + 3x^3 + 3x^2 + x + 4$, where $f(x), g(x) \in Z_5[x]$. Compute $f(x) + g(x)$ and $f(x) \cdot g(x)$.

2. In $Z_3[x]$, show that $x^4 + x$ and $x^2 + x$ determine the same function from Z_3 to Z_3.

3. Show that $x^2 + 3x + 2$ has four zeros in Z_6. (This exercise is referred to in this chapter.)

4. If R is a commutative ring, show that the characteristic of $R[x]$ is the same as the characteristic of R.

5. Prove Corollary 1 of Theorem 16.2.

6. List all the polynomials of degree 2 in $Z_2[x]$.

7. Prove Corollary 2 of Theorem 16.2.

8. Let R be a commutative ring. Show that $R[x]$ has a subring isomorphic to R.

9. If $\phi: R \rightarrow S$ is a ring homomorphism, define $\bar{\phi}:R[x] \rightarrow S[x]$ by $(a_nx^n + \cdots + a_0) \rightarrow \phi(a_n)x^n + \cdots + \phi(a_0)$. Show that $\bar{\phi}$ is a ring homomorphism. (This exercise is referred to in Chapter 33.)

10. If the rings R and S are isomorphic, show that $R[x]$ and $S[x]$ are isomorphic.

11. Let $f(x) = x^3 + 2x + 4$ and $g(x) = 3x + 2$ in $Z_5[x]$. Determine the quotient and remainder upon dividing $f(x)$ by $g(x)$.

12. Let $f(x) = 5x^4 + 3x^3 + 1$ and $g(x) = 3x^2 + 2x + 1$ in $Z_7[x]$. Determine the quotient and remainder upon dividing $f(x)$ by $g(x)$.

13. Show that the polynomial $2x + 1$ in $Z_4[x]$ has a multiplicative inverse in $Z_4[x]$.

14. Are there any nonconstant polynomials in $Z[x]$ that have multiplicative inverses? Explain your answer.

15. Let p be a prime. Are there any nonconstant polynomials in $Z_p[x]$ that have multiplicative inverses? Explain your answer.

16. Show that Corollary 3 of Theorem 16.2 is false for any commutative ring that has a zero divisor.

17. (Degree Rule) Let D be an integral domain and $f(x), g(x) \in D[x]$. Prove that $\deg (f(x) \cdot g(x)) = \deg f(x) + \deg g(x)$. (This exercise is referred to in this chapter, Chapter 17, and Chapter 18.)

18. Prove that the ideal $\langle x \rangle$ in $Q[x]$ is maximal.

19. Let F be an infinite field and let $f(x) \in F[x]$. If $f(a) = 0$ for infinitely many elements a of F, show that $f(x) = 0$.

20. Let F be an infinite field and let $f(x), g(x) \in F[x]$. If $f(a) = g(a)$ for infinitely many elements a of F, show that $f(x) = g(x)$.

21. Let F be a field and let $p(x) \in F[x]$. If $f(x), g(x) \in F[x]$ and $\deg f(x) < \deg p(x)$ and $\deg g(x) < \deg p(x)$, show that $f(x) + \langle p(x) \rangle = g(x) + \langle p(x) \rangle$ implies $f(x) = g(x)$. (This exercise is referred to in Chapter 20.)

22. Prove that $Z[x]$ is not a principal ideal domain. (Compare this with Theorem 16.3.)

23. Find a polynomial with integer coefficients that has $1/2$ and $-1/3$ as zeros.

24. Let $f(x) \in \mathbf{R}[x]$. Suppose that $f(a) = 0$ but $f'(a) \neq 0$, where $f'(x)$ is the derivative of $f(x)$. Show that a is a zero of $f(x)$ of multiplicity 1.

25. Show that Corollary 2 of Theorem 16.2 is true over any commutative ring with unity.

26. Show that Corollary 3 of Theorem 16.2 is true for polynomials over integral domains.

27. Let F be a field and let

$$I = \{a_n x^n + a_{n-1} x^{n-1} + \cdots + a_0 \mid a_n, a_{n-1}, \ldots, a_0 \in F \quad \text{and}$$
$$a_n + a_{n-1} + \cdots + a_0 = 0\}.$$

Show that I is an ideal of $F[x]$ and find a generator for I.

28. Let F be a field and let $f(x) = a_n x^n + a_{n-1} x^{n-1} + \cdots + a_0 \in F[x]$. Prove that $x - 1$ is a factor of $f(x)$ if and only if $a_n + a_{n-1} + \cdots + a_0 = 0$.

29. Let m be a fixed positive integer. For any integer a, let $\bar{a}$ denote a mod m. Show that the mapping of $\phi: Z[x] \rightarrow Z_m[x]$ given by

$$\phi(a_n x^n + a_{n-1} x^{n-1} + \cdots + a_0) = \bar{a}_n x^n + \bar{a}_{n-1} x^{n-1} + \cdots + \bar{a}_0$$

is a ring homomorphism. (This exercise is referred to in Chapters 17 and 33.)

30. Find infinitely many polynomials $f(x)$ in $Z_3[x]$ such that $f(a) = 0$ for all a in Z_3.

31. For every prime p, show that

$$x^{p-1} - 1 = (x - 1)(x - 2) \cdots [x - (p - 1)]$$

in $Z_p[x]$.

32. (Wilson's Theorem) For every integer $n > 1$, prove that $(n - 1)!$ mod n $= n - 1$ if and only if n is prime.

33. For every prime p, show that $(p - 2)!$ mod $p = 1$.

34. Find the remainder upon dividing 98! by 101.

35. Prove that $(50!)^2$ mod $101 = -1$ mod 101.

36. If I is an ideal of a ring R, prove that $I[x]$ is an ideal of $R[x]$.

37. Give an example of a commutative ring R with unity and a maximal ideal I of R such that $I[x]$ is not a maximal ideal of $R[x]$.

38. Let R be a commutative ring with unity. If I is a prime ideal of R, prove that $I[x]$ is a prime ideal of $R[x]$.

39. Let F be a field, and let $f(x)$ and $g(x)$ belong to $F[x]$. If there is no polynomial of positive degree in $F[x]$ that divides both $f(x)$ and $g(x)$ [in this case, $f(x)$ and $g(x)$ are said to be *relatively prime*], prove that there exist polynomials $h(x)$ and $k(x)$ in $F[x]$ with the property that $f(x)h(x) + g(x)k(x) = 1$. (This exercise is referred to in Chapter 20.)

40. Prove that $Q[x]/\langle x^2 - 2 \rangle$ is ring-isomorphic to $Q[\sqrt{2}] = \{a + b\sqrt{2} \mid a, b \in Q\}$.

41. Let $f(x) \in \mathbf{R}[x]$. If $f(a) = 0$ and $f'(a) = 0$ [$f'(a)$ is the derivative of $f(x)$ at a], show that $(x - a)^2$ divides $f(x)$.

42. Let F be a field and let $I = \{f(x) \in F[x] \mid f(a) = 0 \text{ for all } a \text{ in } F\}$. Prove that I is an ideal in $F[x]$. Prove that I is infinite when F is finite and $I = \{0\}$ when F is infinite.

43. Let $g(x)$ and $h(x)$ belong to $Z[x]$ and let $h(x)$ be monic. If $h(x)$ divides $g(x)$ in $Q[x]$, show that $h(x)$ divides $g(x)$ in $Z[x]$. (This exercise is referred to in Chapter 33.)

44. For any field F, recall that $F(x)$ denotes the field of quotients of the ring $F[x]$. Prove that there is no element in $F(x)$ whose square is x.

45. Let F be a field. Show that there exist $a, b \in F$ with the property that $x^2 + x + 1$ divides $x^{43} + ax + b$.

46. Let $f(x) = a_m x^m + a_{m-1} x^{m-1} + \cdots + a_0$ and $g(x) = b_n + b_{n-1} x^{n-1} + \cdots + b_0$ belong to $Q[x]$ and suppose that fg belongs to $Z[x]$. Prove that $a_i b_j$ is an integer for every i and j.

47. Let $f(x)$ belong to $Z[x]$. If $a \bmod m = b \bmod m$, prove that $f(a) \bmod m = f(b) \bmod m$. Prove that if both $f(0)$ and $f(1)$ are odd then f has no zero in Z.

48. Find the remainder when x^{51} is divided by $x + 4$ in $Z_7[x]$.

49. Show that 1 is the only solution of $x^{25} - 1 = 0$ in Z_{37}.

Saunders Mac Lane

The 1986 Steele Prize for cumulative influence is awarded to Saunders Mac Lane for his many contributions to algebra and algebraic topology, and in particular for his pioneering work in homological and categorical algebra.

CITATION FOR THE STEELE PRIZE

SAUNDERS MAC LANE ranks among the most influential mathematicians in the twentieth century. He was born on August 4, 1909, in Norwich, Connecticut. In 1933, at the height of the Depression, he was newly married; despite having degrees from Yale, the University of Chicago, and the University of Göttingen, he had no prospects for a position at a college or university. After applying for employment as a master at a private preparatory school for boys, Mac Lane received a two-year instructorship at Harvard in 1934. He then spent a year at Cornell and a year at the University of Chicago before returning to Harvard in 1938. In 1947, he went back to Chicago permanently.

Much of Mac Lane's work focuses on the interconnections among algebra, topology, and geometry. His book, *Survey of Modern Algebra,* coauthored with Garrett Birkhoff, influenced generations of mathematicians and is now a classic. Mac Lane has served as president of the Mathematical Association of America and the American Mathematical Society. He was elected to the National Academy of Sciences and received the Presidents Medal of Science at the White House in 1989.

To find more information about Mac Lane, visit:

http://www-groups.dcs.st-and.ac.uk/~history/

17 Factorization of Polynomials

> *The value of a principle is the number of things it will explain.*
> RALPH WALDO EMERSON

REDUCIBILITY TESTS

In high school, students spend much time factoring polynomials and finding their zeros. In this chapter, we consider the same problems in a more abstract setting.

To discuss factorization of polynomials, we must first introduce the polynomial analog of a prime integer.

DEFINITION Irreducible Polynomial, Reducible Polynomial

Let D be an integral domain. A polynomial $f(x)$ from $D[x]$ that is neither the zero polynomial nor a unit in $D[x]$ is said to be *irreducible over D* if, whenever $f(x)$ is expressed as a product $f(x) = g(x)h(x)$, with $g(x)$ and $h(x)$ from $D[x]$, then $g(x)$ or $h(x)$ is a unit in $D[x]$. A nonzero, nonunit element of $D[x]$ that is not irreducible over D is called *reducible over D*.

In the case that an integral domain is a field F, it is equivalent and more convenient to define a nonconstant $f(x) \in F[x]$ to be irreducible if $f(x)$ cannot be expressed as a product of two polynomials of lower degree.

EXAMPLE 1 The polynomial $f(x) = 2x^2 + 4$ is irreducible over Q but reducible over Z, since $2x^2 + 4 = 2(x^2 + 2)$ and neither 2 nor $x^2 + 2$ is a unit in $Z[x]$. ◆

EXAMPLE 2   The polynomial $f(x) = 2x^2 + 4$ is irreducible over **R** but reducible over **C**.   ◆

EXAMPLE 3   The polynomial $x^2 - 2$ is irreducible over Q but reducible over **R**.   ◆

EXAMPLE 4   The polynomial $x^2 + 1$ is irreducible over Z_3 but reducible over Z_5.   ◆

In general, it is a difficult problem to decide whether or not a particular polynomial is reducible over an integral domain, but there are special cases when it is easy. Our first theorem is a case in point. It applies to the three preceding examples.

Theorem 17.1   **Reducibility Test for Degrees 2 and 3**

> *Let F be a field. If $f(x) \in F[x]$ and deg $f(x) = 2$ or 3, then $f(x)$ is reducible over F if and only if $f(x)$ has a zero in F*

PROOF.   Suppose that $f(x) = g(x)h(x)$, where both $g(x)$ and $h(x)$ belong to $F[x]$ and have degrees less than that of $f(x)$. Since deg $f(x) = $ deg $g(x) + $ deg $h(x)$ (Exercise 17 in Chapter 16) and deg $f(x) = 2$ or 3, at least one of $g(x)$ and $h(x)$ has degree 1. Say $g(x) = ax + b$. Then, clearly, $-a^{-1}b$ is a zero of $g(x)$ and therefore a zero of $f(x)$ as well.

Conversely, suppose that $f(a) = 0$, where $a \in F$. Then, by the Factor Theorem, we know that $x - a$ is a factor of $f(x)$ and, therefore, $f(x)$ is reducible over F.   ■ ■

Theorem 17.1 is particularly easy to use when the field is Z_p, because, in this case, we can check for reducibility of $f(x)$ by simply testing to see if $f(a) = 0$ for $a = 0, 1, \ldots, p - 1$. For example, since 2 is a zero of $x^2 + 1$ over Z_5, $x^2 + 1$ is reducible over Z_5. On the other hand, because neither 0, 1, nor 2 is a zero of $x^2 + 1$ over Z_3, $x^2 + 1$ is irreducible over Z_3.

Note that polynomials of degree larger than 3 may be reducible over a field, even though they do not have zeros in the field. For example, in $Q[x]$, the polynomial $x^4 + 2x^2 + 1$ is equal to $(x^2 + 1)^2$, but has no zeros in Q.

Our next three tests deal with polynomials with integer coefficients. To simplify the proof of the first of these, we introduce some terminology and isolate a portion of the argument in the form of a lemma.

DEFINITION   *Content of Polynomial, Primitive Polynomial*

The *content* of a nonzero polynomial $a_nx^n + a_{n-1}x^{n-1} + \cdots + a_0$, where the a's are integers, is the greatest common divisor of the integers $a_n, a_{n-1}, \ldots, a_0$. A *primitive polynomial* is an element of $Z[x]$ with content 1.

Gauss's Lemma

The product of two primitive polynomials is primitive.

PROOF.   Let $f(x)$ and $g(x)$ be primitive polynomials, and suppose that $f(x)g(x)$ is not primitive. Let p be a prime divisor of the content of $f(x)g(x)$, and let $\overline{f}(x)$, $\overline{g}(x)$, and $\overline{f(x)g(x)}$ be the polynomials obtained from $f(x)$, $g(x)$, and $f(x)g(x)$ by reducing the coefficients modulo p. Then, $\overline{f}(x)$ and $\overline{g}(x)$ belong to the integral domain $Z_p[x]$ and $\overline{f}(x)\overline{g}(x) = \overline{f(x)g(x)} = 0$, the zero element of $Z_p[x]$ (see Exercise 29 in Chapter 16). Thus, $\overline{f}(x) = 0$ or $\overline{g}(x) = 0$. This means that either p divides every coefficient of $f(x)$ or p divides every coefficient of $g(x)$. Hence, either $f(x)$ is not primitive or $g(x)$ is not primitive. This contradiction completes the proof.   ∎∎

Remember that the question of reducibility depends on which ring of coefficients one permits. Thus, $x^2 - 2$ is irreducible over Z but reducible over $Q[\sqrt{2}]$. In Chapter 20, we will prove that every polynomial of degree greater than 1 with coefficients from an integral domain is reducible over some field. Theorem 17.2 shows that in the case of polynomials irreducible over Z, this field must be larger than the field of rational numbers.

Theorem 17.2   Over *Q* Implies over *Z*

Let $f(x) \in Z[x]$. If $f(x)$ is reducible over Q, then it is reducible over Z.

PROOF.   Suppose that $f(x) = g(x)h(x)$, where $g(x)$ and $h(x) \in Q[x]$. Clearly, we may assume that $f(x)$ is primitive because we can divide both $f(x)$ and $g(x)h(x)$ by the content of $f(x)$. Let a be the least common multiple of the denominators of the coefficients of $g(x)$, and b the least common multiple of the denominators of the coefficients of $h(x)$. Then $abf(x) = ag(x) \cdot bh(x)$, where $ag(x)$ and $bh(x) \in Z[x]$. Let c_1 be the content of $ag(x)$ and let c_2 be the content of $bh(x)$. Then $ag(x) = c_1 g_1(x)$ and $bh(x) = c_2 h_1(x)$, where both $g_1(x)$ and $h_1(x)$ are primitive and $abf(x) = c_1 c_2 g_1(x) h_1(x)$. Since $f(x)$ is primitive, the content of $abf(x)$ is ab. Also, since the product of two primitive polynomials is primitive, it follows that the content of $c_1 c_2 g_1(x) h_1(x)$ is $c_1 c_2$. Thus, $ab = c_1 c_2$ and $f(x) = g_1(x) h_1(x)$, where $g_1(x)$ and $h_1(x) \in Z[x]$ and deg $g_1(x) = \deg g(x)$ and deg $h_1(x) = \deg h(x)$. ■ ■

EXAMPLE 5   We illustrate the proof of Theorem 17.2 by tracing through it for the polynomial $f(x) = 6x^2 + x - 2 = (3x - 3/2)(2x + 4/3) = g(x)h(x)$. In this case we have $a = 2$, $b = 3$, $c_1 = 3$, $c_2 = 2$, $g_1(x) = 2x - 1$, and $h_1(x) = 3x + 2$, so that $2 \cdot 3(6x^2 + x - 2) = 3 \cdot 2(2x - 1)(3x + 2)$ or $6x^2 + x - 2 = (2x - 1)(3x + 2)$. ◆

IRREDUCIBILITY TESTS

Theorem 17.1 reduces the question of irreducibility of a polynomial of degree 2 or 3 to one of finding a zero. The next theorem often allows us to simplify the problem even further.

Theorem 17.3   **Mod p Irreducibility Test**

> *Let p be a prime and suppose that $f(x) \in Z[x]$ with deg $f(x) \geq 1$. Let $\overline{f}(x)$ be the polynomial in $Z_p[x]$ obtained from $f(x)$ by reducing all the coefficients of $f(x)$ modulo p. If $\overline{f}(x)$ is irreducible over Z_p and deg $\overline{f}(x) = \deg f(x)$, then $f(x)$ is irreducible over Q.*

PROOF.   It follows from the proof of Theorem 17.2 that if $f(x)$ is reducible over Q, then $f(x) = g(x)h(x)$ with $g(x), h(x) \in Z[x]$, and both $g(x)$ and $h(x)$ have degree less than that of $f(x)$. Let $\overline{f}(x)$, $\overline{g}(x)$, and $\overline{h}(x)$ be the polynomials obtained from $f(x)$, $g(x)$, and $h(x)$ by reducing all the coefficients modulo p. Since deg $f(x) = \deg \overline{f}(x)$, we have deg $\overline{g}(x) \leq \deg g(x) < \deg \overline{f}(x)$ and deg $\overline{h}(x) \leq \deg h(x) < \deg \overline{f}(x)$. But, $\overline{f}(x) = \overline{g}(x)\overline{h}(x)$, and this contradicts our assumption that $\overline{f}(x)$ is irreducible over Z_p. ■ ■

EXAMPLE 6 Let $f(x) = 21x^3 - 3x^2 + 2x + 9$. Then, over Z_2, we have $\overline{f}(x) = x^3 + x^2 + 1$ and, since $\overline{f}(0) = 1$ and $\overline{f}(1) = 1$, we see that $\overline{f}(x)$ is irreducible over Z_2. Thus, $f(x)$ is irreducible over Q. Notice that, over Z_3, $\overline{f}(x) = 2x$ is irreducible, but we may *not* apply Theorem 17.3 to conclude that $f(x)$ is irreducible over Q. ◆

Be careful not to use the converse of Theorem 17.3. If $f(x) \in Z[x]$ and $\overline{f}(x)$ is reducible over Z_p for some p, $f(x)$ may still be irreducible over Q. For example, consider $f(x) = 21x^3 - 3x^2 + 2x + 8$. Then, over Z_2, $\overline{f}(x) = x^3 + x^2 = x^2(x + 1)$. But over Z_5, $\overline{f}(x)$ has no zeros and therefore is irreducible over Z_5. So, $f(x)$ is irreducible over Q. Note that this example shows that the Mod p Irreducibility Test may fail for some p and work for others. To conclude that a particular $f(x)$ in $Z[x]$ is irreducible over Q, all we need to do is find a single p for which the corresponding polynomial $\overline{f}(x)$ in Z_p is irreducible. However, this is not always possible, since $f(x) = x^4 + 1$ is irreducible over Q but reducible over Z_p for *every* prime p. (See Exercise 29.)

The Mod p Irreducibility Test can also be helpful in checking for irreducibility of polynomials of degree greater than 3 and polynomials with rational coefficients.

EXAMPLE 7 Let $f(x) = (3/7)x^4 - (2/7)x^2 + (9/35)x + 3/5$. We will show that $f(x)$ is irreducible over Q. First, let $h(x) = 35f(x) = 15x^4 - 10x^2 + 9x + 21$. Then $f(x)$ is irreducible over Q if $h(x)$ is irreducible over Z. Next, applying the Mod 2 Irreducibility Test to $h(x)$, we get $\overline{h}(x) = x^4 + x + 1$. Clearly, $\overline{h}(x)$ has no zeros in Z_2. Furthermore, $\overline{h}(x)$ has no quadratic factor in $Z_2[x]$ either. [For if so, the factor would have to be either $x^2 + x + 1$ or $x^2 + 1$. Long division shows that $x^2 + x + 1$ is not a factor, and $x^2 + 1$ cannot be a factor because it has a zero whereas $\overline{h}(x)$ does not.] Thus $\overline{h}(x)$ is irreducible over $Z_2[x]$. This guarantees that $h(x)$ is irreducible over Q. ◆

EXAMPLE 8 Let $f(x) = x^5 + 2x + 4$. Obviously, neither Theorem 17.1 nor the Mod 2 Irreducibility Test helps here. Let's try mod 3. Substitution of 0, 1, and 2 into $\overline{f}(x)$ does not yield 0, so there are no linear factors. But $\overline{f}(x)$ may have a quadratic factor. If so, we may assume it has the form $x^2 + ax + b$ (see Exercise 5). This gives nine possibilities to check. We can immediately rule out each of the nine that has a zero over

Z_3, since $\bar{f}(x)$ does not have one. This leaves only $x^2 + 1$, $x^2 + x + 2$, and $x^2 + 2x + 2$ to check. These are eliminated by long division. So, since $\bar{f}(x)$ is irreducible over Z_3, $f(x)$ is irreducible over Q. (Why is it unnecessary to check for cubic or fourth-degree factors?)   ◆

Another important irreducibility test is the following one, credited to Ferdinand Eisenstein (1823–1852), a student of Gauss. The corollary was first proved by Gauss by a different method.

Theorem 17.4   Eisenstein's Criterion (1850)

Let
$$f(x) = a_n x^n + a_{n-1}x^{n-1} + \cdots + a_0 \in Z[x].$$
If there is a prime p such that $p \nmid a_n$, $p \mid a_{n-1}, \ldots, p \mid a_0$ and $p^2 \nmid a_0$, then $f(x)$ is irreducible over Q.

PROOF.   If $f(x)$ is reducible over Q, we know by Theorem 17.2 that there exist elements $g(x)$ and $h(x)$ in $Z[x]$ such that $f(x) = g(x)h(x)$ and $1 \leq \deg g(x)$, $\deg h(x) < n$. Say $g(x) = b_r x^r + \cdots + b_0$ and $h(x) = c_s x^s + \cdots + c_0$. Then, since $p \mid a_0$, $p^2 \nmid a_0$, and $a_0 = b_0 c_0$, it follows that p divides one of b_0 and c_0 but not the other. Let us say $p \mid b_0$ and $p \nmid c_0$. Also, since $p \nmid a_n = b_r c_s$, we know that $p \nmid b_r$. So, there is a least integer t such that $p \nmid b_t$. Now, consider $a_t = b_t c_0 + b_{t-1} c_1 + \cdots + b_0 c_t$. By assumption, p divides a_t and, by choice of t, every summand on the right after the first one is divisible by p. Clearly, this forces p to divide $b_t c_0$ as well. This is impossible, however, since p is prime and p divides neither b_t nor c_0.   ■ ■

Corollary   Irreducibility of *p*th Cyclotomic Polynomial

For any prime p, the pth cyclotomic polynomial
$$\Phi_p(x) = \frac{x^p - 1}{x - 1} = x^{p-1} + x^{p-2} + \cdots + x + 1$$
is irreducible over Q.

PROOF.   Let

$$f(x) = \Phi_p(x+1) = \frac{(x+1)^p - 1}{(x+1) - 1} = x^{p-1} + \binom{p}{1}x^{p-2} + \binom{p}{2}x^{p-3} + \cdots + \binom{p}{1}.$$

Then, since every coefficient except that of x^{p-1} is divisible by p, and the constant term is not divisible by p^2, by Eisenstein's Criterion, $f(x)$ is irreducible over Q. So, if $\Phi_p(x) = g(x)h(x)$ were a nontrivial factorization of $\Phi_p(x)$ over Q, then $f(x) = \Phi_p(x + 1) = g(x + 1)h(x + 1)$ would be a nontrivial factorization of $f(x)$ over Q. Since this is impossible, we conclude that $\Phi_p(x)$ is irreducible over Q. ■ ■

EXAMPLE 9 The polynomial $3x^5 + 15x^4 - 20x^3 + 10x + 20$ is irreducible over Q because $5 \nmid 3$ and $25 \nmid 20$ but 5 does divide 15, -20, 10, and 20. ◆

The principal reason for our interest in irreducible polynomials stems from the fact that there is an intimate connection among them, maximal ideals, and fields. This connection is revealed in the next theorem and its first corollary.

Theorem 17.5 **$\langle p(x) \rangle$ Is Maximal if and Only if $p(x)$ Is Irreducible**

> *Let F be a field and let $p(x) \in F[x]$. Then $\langle p(x) \rangle$ is a maximal ideal in $F[x]$ if and only if $p(x)$ is irreducible over F.*

PROOF. Suppose first that $\langle p(x) \rangle$ is a maximal ideal in $F[x]$. Clearly, $p(x)$ is neither the zero polynomial nor a unit in $F[x]$, because neither $\{0\}$ nor $F[x]$ is a maximal ideal in $F[x]$. If $p(x) = g(x)h(x)$ is a factorization of $p(x)$ over F, then $\langle p(x) \rangle \subseteq \langle g(x) \rangle \subseteq F[x]$. Thus, $\langle p(x) \rangle = \langle g(x) \rangle$ or $F[x] = \langle g(x) \rangle$. In the first case, we must have deg $p(x) =$ deg $g(x)$. In the second case, it follows that deg $g(x) = 0$ and, consequently, deg $h(x) =$ deg $p(x)$. Thus, $p(x)$ cannot be written as a product of two polynomials in $F[x]$ of lower degree.

Now, suppose that $p(x)$ is irreducible over F. Let I be any ideal of $F[x]$ such that $\langle p(x) \rangle \subseteq I \subseteq F[x]$. Because $F[x]$ is a principal ideal domain, we know that $I = \langle g(x) \rangle$ for some $g(x)$ in $F[x]$. So, $p(x) \in \langle g(x) \rangle$ and, therefore, $p(x) = g(x)h(x)$, where $h(x) \in F[x]$. Since $p(x)$ is irreducible over F, it follows that either $g(x)$ is a constant or $h(x)$ is a constant. In the first case, we have $I = F[x]$; in the second case, we have $\langle p(x) \rangle = \langle g(x) \rangle = I$. So, $\langle p(x) \rangle$ is maximal in $F[x]$. ■ ■

Corollary 1 **$F[x]/\langle p(x) \rangle$ Is a Field**

> *Let F be a field and $p(x)$ an irreducible polynomial over F. Then $F[x]/\langle p(x) \rangle$ is a field.*

PROOF.   This follows directly from Theorems 17.5 and 14.4.   ■ ■

The next corollary is a polynomial analog of Euclid's Lemma for primes (see Chapter 0).

Corollary 2 $p(x) \mid a(x)b(x)$ **Implies** $p(x) \mid a(x)$ **or** $p(x) \mid b(x)$

> *Let F be a field and let $p(x)$, $a(x)$, $b(x) \in F[x]$. If $p(x)$ is irreducible over F and $p(x) \mid a(x)b(x)$, then $p(x) \mid a(x)$ or $p(x) \mid b(x)$.*

PROOF.   Since $p(x)$ is irreducible, $F[x]/\langle p(x)\rangle$ is a field and, therefore, an integral domain. From Theorem 14.3, we know that $\langle p(x)\rangle$ is a prime ideal, and since $p(x)$ divides $a(x)b(x)$, we have $a(x)b(x) \in \langle p(x)\rangle$. Thus, $a(x) \in \langle p(x)\rangle$ or $b(x) \in \langle p(x)\rangle$. This means that $p(x) \mid a(x)$ or $p(x) \mid b(x)$.   ■ ■

The next two examples put the theory to work.

EXAMPLE 10   We construct a field with eight elements. By Theorem 17.1 and Corollary 1 of Theorem 17.5, it suffices to find a cubic polynomial over Z_2 that has no zero in Z_2. By inspection, $x^3 + x + 1$ fills the bill. Thus, $Z_2[x]/\langle x^3 + x + 1\rangle = \{ax^2 + bx + c + \langle x^3 + x + 1\rangle \mid a, b, c \in Z_2\}$ is a field with eight elements. For practice, let us do a few calculations in this field. Since the sum of two polynomials of the form $ax^2 + bx + c$ is another one of the same form, addition is easy. For example,

$$(x^2 + x + 1 + \langle x^3 + x + 1\rangle) + (x^2 + 1 + \langle x^3 + x + 1\rangle)$$

$$= x + \langle x^3 + x + 1\rangle.$$

On the other hand, multiplication of two coset representatives need not yield one of the original eight coset representatives:

$$(x^2 + x + 1 + \langle x^3 + x + 1\rangle) \cdot (x^2 + 1 + \langle x^3 + x + 1\rangle)$$

$$= x^4 + x^3 + x + 1 + \langle x^3 + x + 1\rangle = x^4 + \langle x^3 + x + 1\rangle$$

(since the ideal absorbs the last three terms). How do we express this in the form $ax^2 + bx + c + \langle x^3 + x + 1\rangle$? One way is to long divide x^4 by $x^3 + x + 1$ to obtain the remainder of $x^2 + x$ (just as one reduces $12 + \langle 5\rangle$ to $2 + \langle 5\rangle$ by dividing 12 by 5 to obtain the remainder 2). Another way is to observe that $x^3 + x + 1 + \langle x^3 + x + 1\rangle = 0 +$

TABLE 17.1 A Partial Multiplication Table for Example 10

	1	x	$x + 1$	x^2	$x^2 + 1$	$x^2 + x$	$x^2 + x + 1$
1	1	x	$x + 1$	x^2	$x^2 + 1$	$x^2 + x$	$x^2 + x + 1$
x	x	x^2	$x^2 + x$	$x + 1$	1	$x^2 + x + 1$	$x^2 + 1$
$x + 1$	$x + 1$	$x^2 + x$	$x^2 + 1$	$x^2 + x + 1$	x^2	1	x
x^2	x^2	$x + 1$	$x^2 + x + 1$	$x^2 + x$	x	$x^2 + 1$	1
$x^2 + 1$	$x^2 + 1$	1	x^2	x	$x^2 + x + 1$	$x + 1$	$x^2 + x$

$\langle x^3 + x + 1 \rangle$ implies $x^3 + \langle x^3 + x + 1 \rangle = x + 1 + \langle x^3 + x + 1 \rangle$. Thus, we may multiply both sides by x to obtain

$$x^4 + \langle x^3 + x + 1 \rangle = x^2 + x + \langle x^3 + x + 1 \rangle.$$

Similarly,

$$(x^2 + x + \langle x^3 + x + 1 \rangle) \cdot (x + \langle x^3 + x + 1 \rangle)$$
$$= x^3 + x^2 + \langle x^3 + x + 1 \rangle$$
$$= x^2 + x + 1 + \langle x^3 + x + 1 \rangle.$$

A partial multiplication table for this field is given in Table 17.1. To simplify the notation, we indicate a coset by its representative only. (Complete the table yourself. Keep in mind that x^3 can be replaced by $x + 1$ and x^4 by $x^2 + x$.) ◆

EXAMPLE 11 Since $x^2 + 1$ has no zero in Z_3, it is irreducible over Z_3. Thus, $Z_3[x]/\langle x^2 + 1 \rangle$ is a field. Analogous to Example 12 in Chapter 14, $Z_3[x]/\langle x^2 + 1 \rangle = \{ax + b + \langle x^2 + 1 \rangle \mid a, b \in Z_3\}$. Thus, this field has nine elements. A multiplication table for this field can be obtained from Table 13.1 by replacing i by x. (Why does this work?) ◆

UNIQUE FACTORIZATION IN Z[x]

As a further application of the ideas presented in this chapter, we next prove that $Z[x]$ has an important factorization property. In Chapter 18, we will study this property in greater depth. The first proof of Theorem 17.6 was given by Gauss. In reading this theorem and its proof, keep in mind that the units in $Z[x]$ are precisely $f(x) = 1$ and $f(x) = -1$ (see Exercise 25 in Chapter 12), the irreducible polynomials of degree 0 over Z are precisely those of the form $f(x) = p$ and $f(x) = -p$ where p is a prime, and every nonconstant polynomial from $Z[x]$ that is irreducible over Z is primitive (see Exercise 3).

Theorem 17.6 **Unique Factorization in Z[x]**

> *Every polynomial in Z[x] that is not the zero polynomial or a unit in Z[x] can be written in the form $b_1 b_2 \cdots b_s\, p_1(x) p_2(x) \cdots p_m(x)$, where the b_i's are irreducible polynomials of degree 0, and the $p_i(x)$'s are irreducible polynomials of positive degree. Furthermore, if*
>
> $$b_1 b_2 \cdots b_s\, p_1(x) p_2(x) \cdots p_m(x) =$$
> $$c_1 c_2 \cdots c_t\, q_1(x) q_2(x) \cdots q_n(x),$$
>
> *where the b's and c's are irreducible polynomials of degree 0, and the p(x)'s and q(x)'s are irreducible polynomials of positive degree, then $s = t$, $m = n$, and, after renumbering the c's and q(x)'s, we have $b_i = \pm c_i$ for $i = 1, \ldots, s$; and $p_i(x) = \pm q_i(x)$ for $i = 1, \ldots, m$.*

PROOF. Let $f(x)$ be a nonzero, nonunit polynomial from $Z[x]$. If deg $f(x) = 0$, then $f(x)$ is constant and the result follows from the Fundamental Theorem of Arithmetic. If deg $f(x) > 0$, let b denote the content of $f(x)$, and let $b_1 b_2 \cdots b_s$ be the factorization of b as a product of primes. Then, $f(x) = b_1 b_2 \cdots b_s\, f_1(x)$, where $f_1(x)$ belongs to $Z[x]$, is primitive and deg $f_1(x) = \deg f(x)$. Thus, to prove the existence portion of the theorem, it suffices to show that a primitive polynomial $f(x)$ of positive degree can be written as a product of irreducible polynomials of positive degree. We proceed by induction on deg $f(x)$. If deg $f(x) = 1$, then $f(x)$ is already irreducible and we are done. Now suppose that every primitive polynomial of degree less than deg $f(x)$ can be written as a product of irreducibles of positive degree. If $f(x)$ is irreducible, there is nothing to prove. Otherwise, $f(x) = g(x)h(x)$, where both $g(x)$ and $h(x)$ are primitive and have degree less than that of $f(x)$. Thus, by induction, both $g(x)$ and $h(x)$ can be written as a product of irreducibles of positive degree. Clearly, then, $f(x)$ is also such a product.

To prove the uniqueness portion of the theorem, suppose that $f(x) = b_1 b_2 \cdots b_s\, p_1(x) p_2(x) \cdots p_m(x) = c_1 c_2 \cdots c_t\, q_1(x) q_2(x) \cdots q_n(x)$, where the b's and c's are irreducible polynomials of degree 0, and the $p(x)$'s and $q(x)$'s are irreducible polynomials of positive degree. Let $b = b_1 b_2 \cdots b_s$ and $c = c_1 c_2 \cdots c_t$. Since the $p(x)$'s and $q(x)$'s are primitive, it follows from Gauss's Lemma that $p_1(x) p_2(x) \cdots p_m(x)$ and $q_1(x) q_2(x) \cdots q_n(x)$ are primitive. Hence, both b and c must equal plus or minus the content of $f(x)$ and, therefore, are equal in absolute value. It then follows from the Fundamental Theorem of Arithmetic that $s = t$ and, after renumbering, $b_i = \pm c_i$ for $i = 1, 2, \ldots, s$. Thus, by canceling the constant terms in the two factorizations for $f(x)$, we have $p_1(x)p_2(x) \cdots p_m(x) = \pm q_1(x) q_2(x) \cdots q_n(x)$. Now, viewing the $p(x)$'s and $q(x)$'s as elements of $Q[x]$ and noting that $p_1(x)$ divides $q_1(x) \cdots q_n(x)$, it follows from Corollary 2 of Theorem 17.5 and induction (see Exercise 27) that $p_1(x) \mid q_i(x)$ for some i. By renumbering, we may assume $i = 1$. Then, since $q_1(x)$ is irreducible, we have $q_1(x) = (r/s)p_1(x)$, where $r, s \in Z$. However, because both $q_1(x)$ and $p_1(x)$ are primitive, we must have $r/s = \pm 1$. So, $q_1(x) = \pm p_1(x)$. Also, after canceling, we have $p_2(x) \cdots p_m(x) = \pm q_2(x) \cdots q_n(x)$. Now, we may repeat the argument above with $p_2(x)$ in place of $p_1(x)$. If $m < n$, after m such steps we would have 1 on the left and a nonconstant polynomial on the right. Clearly, this is impossible. On the other hand, if $m > n$, after n steps we would have ± 1 on the right and a nonconstant polynomial on the left—another impossibility. So, $m = n$ and $p_i(x) = \pm q_i(x)$ after suitable renumbering of the $q(x)$'s. ∎

WEIRD DICE: AN APPLICATION OF UNIQUE FACTORIZATION

EXAMPLE 12 Consider an ordinary pair of dice whose faces are labeled 1 through 6. The probability of rolling a sum of 2 is 1/36, the probability of rolling a sum of 3 is 2/36, and so on. In a 1978 issue of *Scientific American* [1], Martin Gardner remarked that if one were to label the six faces of one cube with the integers 1, 2, 2, 3, 3, 4 and the six faces of another cube with the integers 1, 3, 4, 5, 6, 8, then the probability of obtaining any particular sum with these dice (called *Sicherman dice*) would be the same as the probability of rolling that sum with ordinary dice (that is, 1/36 for a 2, 2/36 for a 3, and so on). See Figure 17.1. In this example, we show how the Sicherman labels can be derived, and that they are the only possible such labels besides 1 through 6. To do so, we utilize the fact that $Z[x]$ has the unique factorization property.

To begin, let us ask ourselves how we may obtain a sum of 6, say, with an ordinary pair of dice. Well, there are five possibilities for the two faces: (5, 1), (4, 2), (3, 3), (2, 4), and (1, 5). Next we consider the product of the two polynomials created by using the ordinary dice labels as exponents:

$$(x^6 + x^5 + x^4 + x^3 + x^2 + x)(x^6 + x^5 + x^4 + x^3 + x^2 + x).$$

•	2	3	4	5	6	7
•ː	3	4	5	6	7	8
•ˑ•	4	5	6	7	8	9
•• ••	5	6	7	8	9	10
•• •ː•	6	7	8	9	10	11
•• •• ••	7	8	9	10	11	12

•	2	3	3	4	4	5
•ː	4	5	5	6	6	7
•ˑ•	5	6	6	7	7	8
•• ••	6	7	7	8	8	9
•• •ː•	7	8	8	9	9	10
•• •• ••	9	10	10	11	11	12

Figure 17.1

Observe that we pick up the term x^6 in this product in precisely the following ways: $x^5 \cdot x^1, x^4 \cdot x^2, x^3 \cdot x^3, x^2 \cdot x^4, x^1 \cdot x^5$. Notice the correspondence between pairs of labels whose sums are 6 and pairs of terms whose products are x^6. This correspondence is one-to-one, and it is valid for all sums and all dice—including the Sicherman dice and any other dice that yield the desired probabilities. So, let $a_1, a_2, a_3, a_4, a_5, a_6$ and $b_1, b_2, b_3, b_4, b_5, b_6$ be any two lists of positive integer labels for a pair of cubes with the property that the probability of rolling any particular sum with these dice (let us call them *weird dice*) is the same as the probability of rolling that sum with ordinary dice labeled 1 through 6. Using our observation about products of polynomials, this means that

$$(x^6 + x^5 + x^4 + x^3 + x^2 + x)(x^6 + x^5 + x^4 + x^3 + x^2 + x)$$

$$= (x^{a_1} + x^{a_2} + x^{a_3} + x^{a_4} + x^{a_5} + x^{a_6}) \cdot$$

$$(x^{b_1} + x^{b_2} + x^{b_3} + x^{b_4} + x^{b_5} + x^{b_6}). \quad (1)$$

Now all we have to do is solve this equation for the a's and b's. Here is where unique factorization in $Z[x]$ comes in. The polynomial $x^6 + x^5 + x^4 + x^3 + x^2 + x$ factors uniquely into irreducibles as

$$x(x + 1)(x^2 + x + 1)(x^2 - x + 1)$$

so that the left-hand side of Equation (1) has the irreducible factorization

$$x^2(x + 1)^2(x^2 + x + 1)^2(x^2 - x + 1)^2.$$

So, by Theorem 17.6, this means that these factors are the only possible irreducible factors of $P(x) = x^{a_1} + x^{a_2} + x^{a_3} + x^{a_4} + x^{a_5} + x^{a_6}$. Thus, $P(x)$ has the form

$$x^q(x + 1)^r(x^2 + x + 1)^t(x^2 - x + 1)^u,$$

where $0 \le q, r, t, u \le 2$.

To restrict further the possibilities for these four parameters, we evaluate $P(1)$ in two ways. $P(1) = 1^{a_1} + 1^{a_2} + \cdots + 1^{a_6} = 6$ and $P(1) = 1^q 2^r 3^t 1^u$. Clearly, this means that $r = 1$ and $t = 1$. What about q? Evaluating $P(0)$ in two ways shows that $q \ne 0$. On the other hand, if $q = 2$, the smallest possible sum one could roll with the corresponding labels for dice would be 3. Since this violates our assumption, we have now reduced our list of possibilities for $q, r, t,$ and u to $q = 1, r = 1, t = 1,$ and $u = 0, 1, 2$. Let's consider each of these possibilities in turn.

When $u = 0$, $P(x) = x^4 + x^3 + x^3 + x^2 + x^2 + x$, so the die labels are 4, 3, 3, 2, 2, 1—a Sicherman die.

When $u = 1$, $P(x) = x^6 + x^5 + x^4 + x^3 + x^2 + x$, so the die labels are 6, 5, 4, 3, 2, 1—an ordinary die.

When $u = 2$, $P(x) = x^8 + x^6 + x^5 + x^4 + x^3 + x$, so the die labels are 8, 6, 5, 4, 3, 1—the other Sicherman die.

This proves that the Sicherman dice do give the same probabilities as ordinary dice *and* that they are the *only* other pair of dice that have this property.
◆

EXERCISES ■■

No matter how good you are at something, there's always about a million people better than you.

HOMER SIMPSON

1. Verify the assertion made in Example 2.

2. Suppose that D is an integral domain and F is a field containing D. If $f(x) \in D[x]$ and $f(x)$ is irreducible over F but reducible over D, what can you say about the factorization of $f(x)$ over D?

3. Show that a nonconstant polynomial from $Z[x]$ that is irreducible over Z is primitive. (This exercise is referred to in this chapter.)

4. Suppose that $f(x) = x^n + a_{n-1}x^{n-1} + \cdots + a_0 \in Z[x]$. If r is rational and $x - r$ divides $f(x)$, show that r is an integer.

5. Let F be a field and let a be a nonzero element of F.
   a. If $af(x)$ is irreducible over F, prove that $f(x)$ is irreducible over F.
   b. If $f(ax)$ is irreducible over F, prove that $f(x)$ is irreducible over F.
   c. If $f(x + a)$ is irreducible over F, prove that $f(x)$ is irreducible over F.
   d. Use part c to prove that $8x^3 - 6x + 1$ is irreducible over Q.

   (This exercise is referred to in this chapter.)

6. Suppose that $f(x) \in Z_p[x]$ and is irreducible over Z_p, where p is a prime. If $\deg f(x) = n$, prove that $Z_p[x]/\langle f(x) \rangle$ is a field with p^n elements.

7. Construct a field of order 25.

8. Construct a field of order 27.

9. Show that $x^3 + x^2 + x + 1$ is reducible over Q. Does this fact contradict the corollary to Theorem 17.4?

10. Determine which of the polynomials below is (are) irreducible over Q.
    a. $x^5 + 9x^4 + 12x^2 + 6$
    b. $x^4 + x + 1$
    c. $x^4 + 3x^2 + 3$
    d. $x^5 + 5x^2 + 1$
    e. $(5/2)x^5 + (9/2)x^4 + 15x^3 + (3/7)x^2 + 6x + 3/14$

11. Show that $x^4 + 1$ is irreducible over Q but reducible over $\mathbf{R}$. (This exercise is referred to in this chapter.)

12. Show that $x^2 + x + 4$ is irreducible over Z_{11}.

13. Let $f(x) = x^3 + 6 \in Z_7[x]$. Write $f(x)$ as a product of irreducible polynomials over Z_7.

14. Let $f(x) = x^3 + x^2 + x + 1 \in Z_2[x]$. Write $f(x)$ as a product of irreducible polynomials over Z_2.

15. Let p be a prime.
    a. Show that the number of reducible polynomials over Z_p of the form $x^2 + ax + b$ is $p(p + 1)/2$.
    b. Determine the number of reducible quadratic polynomials over Z_p.

16. Let p be a prime.
    a. Determine the number of irreducible polynomials over Z_p of the form $x^2 + ax + b$.
    b. Determine the number of irreducible quadratic polynomials over Z_p.

17. Show that for every prime p there exists a field of order p^2.

18. Prove that, for every positive integer n, there are infinitely many polynomials of degree n in $Z[x]$ that are irreducible over Q.

19. Show that the field given in Example 11 in this chapter is isomorphic to the field given in Example 9 in Chapter 13.

20. Let $f(x) \in Z_p[x]$. Prove that if $f(x)$ has no factor of the form $x^2 + ax + b$, then it has no quadratic factor over Z_p.

21. Find all monic irreducible polynomials of degree 2 over Z_3.

22. Given that π is not the zero of a polynomial (nonzero) with rational coefficients, prove that π^2 cannot be written in the form $a\pi + b$, where a and b are rational.

23. Find all the zeros and their multiplicities of $x^5 + 4x^4 + 4x^3 - x^2 - 4x + 1$ over Z_5.

24. Find all zeros of $f(x) = 3x^2 + x + 4$ over Z_7 by substitution. Find all zeros of $f(x)$ by using the Quadratic Formula $(-b \pm \sqrt{b^2 - 4ac})(2a)^{-1}$.

(all calculations are done in Z_7). Do your answers agree? Should they? Find all zeros of $g(x) = 2x^2 + x + 3$ over Z_5 by substitution. Try the Quadratic Formula on $g(x)$. Do your answers agree? State necessary and sufficient conditions for the Quadratic Formula to yield the zeros of a quadratic from $Z_p[x]$, where p is a prime greater than 2.

25. (Rational Root Theorem) Let

$$f(x) = a_n x^n + a_{n-1} x^{n-1} + \cdots + a_0 \in Z[x]$$

   and $a_n \neq 0$. Prove that if r and s are relatively prime integers and $f(r/s) = 0$, then $r \mid a_0$ and $s \mid a_n$.

26. Let F be a field and $f(x) \in F[x]$. Show that, as far as deciding upon the irreducibility of $f(x)$ over F is concerned, we may assume that $f(x)$ is monic. (This assumption is useful when one uses a computer to check for irreducibility.)

27. Let F be a field and let $p(x), a_1(x), a_2(x), \ldots, a_k(x) \in F[x]$, where $p(x)$ is irreducible over F. If $p(x) \mid a_1(x)a_2(x) \cdots a_k(x)$, show that $p(x)$ divides some $a_i(x)$. (This exercise is referred to in the proof of Theorem 17.6.)

28. Explain how the Mod p Irreducibility Test (Theorem 17.3) can be used to test members of $Q[x]$ for irreducibility.

29. Show that $x^4 + 1$ is reducible over Z_p for every prime p. (This exercise is referred to in this chapter.)

30. If p is a prime, prove that $x^{p-1} - x^{p-2} + x^{p-3} - \cdots - x + 1$ is irreducible over Q.

31. Let F be a field and let $p(x)$ be irreducible over F. If E is a field that contains F and there is an element a in E such that $p(a) = 0$, show that the mapping $\phi: F[x] \rightarrow E$ given by $f(x) \rightarrow f(a)$ is a ring homomorphism with kernel $\langle p(x) \rangle$. (This exercise is referred to in Chapter 20.)

32. Prove that the ideal $\langle x^2 + 1 \rangle$ is prime in $Z[x]$ but not maximal in $Z[x]$.

33. Let F be a field and let $p(x)$ be irreducible over F. Show that $\{a + \langle p(x) \rangle \mid a \in F\}$ is a subfield of $F[x]/\langle p(x) \rangle$ isomorphic to F. (This exercise is referred to in Chapter 20.)

34. Carry out the analysis given in Example 12 for a pair of tetrahedrons instead of a pair of cubes. (Define ordinary tetrahedral dice as the ones labeled 1 through 4.)

35. Suppose in Example 12 that we begin with n ($n > 2$) ordinary dice each labeled 1 through 6, instead of just two. Show that the only possible labels that produce the same probabilities as n ordinary dice are the labels 1 through 6 and the Sicherman labels.

36. Show that one two-sided die labeled with 1 and 4 and another 18-sided die labeled with 1, 2, 2, 3, 3, 3, 4, 4, 4, 5, 5, 5, 6, 6, 6, 7, 7, 8 yield the same

probabilities as an ordinary pair of cubes labeled 1 through 6. Carry out an analysis similar to that given in Example 12 to derive these labels.

37. In the game of Monopoly, would the probabilities of landing on various properties be different if the game were played with Sicherman dice instead of ordinary dice? Why?

COMPUTER EXERCISES ▪▪

The experiment serves two purposes, often independent one from the other: it allows the observation of new facts, hitherto either unsuspected, or not yet well defined; and it determines whether a working hypothesis fits the world of observable facts.

RENÉ J. DUBOS

Software for the computer exercises in this chapter is available at the website:

http://www.d.umn.edu/~jgallian

1. This software implements the Mod p Irreducibility Test. Use it to test the polynomials in the examples given in this chapter and the polynomials given in Exercise 10 for irreducibility.

2. Use software such as Mathematica, Maple, or GAP to express $x^n - 1$ as a product of irreducible polynomials with integer coefficients for $n = 4$, 8, 12, and 20. On the basis of these data, make a conjecture about the coefficients of the irreducible factors of $x^n - 1$. Test your conjecture for $n = 105$. Does your conjecture hold up?

3. Use software such as Mathematica, Maple, or GAP to express $x^{p^n} - x$ as a product of irreducibles over Z_p for several choices of the prime p and n. On the basis of these data, make a conjecture relating the degrees of the irreducible factors of $x^{p^n} - x$ and n.

Reference

1. Martin Gardner, "Mathematical Games," *Scientific American* 238/2 (1978): 19–32.

Suggested Readings

Duane Broline, "Renumbering the Faces of Dice," *Mathematics Magazine* 52 (1979): 312–315.

In this article, the author extends the analysis we carried out in Example 12 to dice in the shape of Platonic solids.

J. A. Gallian and D. J. Rusin, "Cyclotomic Polynomials and Nonstandard Dice," *Discrete Mathematics* 27 (1979): 245–259.

> Here Example 12 is generalized to the case of n dice each with m labels for all n and m greater than 1.

Randall Swift and Brian Fowler, "Relabeling Dice," *Mathematics Magazine* 72 (1999): 204–208.

> The authors use the method presented in this chapter to derive positive integer labels for a pair of dice that are not six-sided but give the same probabilities for the sum of the faces as a pair of cubes labeled 1 through 6.

Divisibility in Integral Domains

*Give me a fruitful error anytime, full of seeds, bursting
with its own corrections. You can keep your sterile truth
for yourself.*

VILFREDO PARETO

IRREDUCIBLES, PRIMES

In the preceding two chapters, we focused on factoring polynomials over the integers or a field. Several of those results—unique factorization in $Z[x]$ and the division algorithm for $F[x]$, for instance—are natural counterparts to theorems about the integers. In this chapter and the next, we examine factoring in a more abstract setting.

DEFINITION   *Associates, Irreducibles, Primes*

Elements a and b of an integral domain D are called *associates* if $a = ub$, where u is a unit of D. A nonzero element a of an integral domain D is called an *irreducible* if a is not a unit and, whenever $b, c \in D$ with $a = bc$, then b or c is a unit. A nonzero element a of an integral domain D is called a *prime* if a is not a unit and $a \mid bc$ implies $a \mid b$ or $a \mid c$.

Roughly speaking, an irreducible is an element that can be factored only in a trivial way. Notice that an element a is a prime if and only if $\langle a \rangle$ is a prime ideal.

Relating the definitions above to the integers may seem a bit confusing, since in Chapter 0 we defined a positive integer to be a prime if it satisfies our definition of an irreducible, and we proved that a prime integer satisfies the definition of a prime in an integral domain

(Euclid's Lemma). The source of the confusion is that in the case of the integers, the concepts of irreducibles and primes are equivalent, but in general, as we will soon see, they are not.

The distinction between primes and irreducibles is best illustrated by integral domains of the form $Z[\sqrt{d}] = \{a + b\sqrt{d} \mid a, b \in Z\}$, where d is not 1 and is not divisible by the square of a prime. (These rings are of fundamental importance in number theory.) To analyze these rings, we need a convenient method of determining their units, irreducibles, and primes. To do this, we define a function N, called the *norm*, from $Z[\sqrt{d}]$ into the nonnegative integers by $N(a + b\sqrt{d}) = |a^2 - db^2|$. We leave it to the reader (Exercise 1) to verify the following four properties: $N(x) = 0$ if and only if $x = 0$; $N(xy) = N(x)N(y)$ for all x and y; x is a unit if and only if $N(x) = 1$; and, if $N(x)$ is prime, then x is irreducible in $Z[\sqrt{d}]$.

EXAMPLE 1   We exhibit an irreducible in $Z[\sqrt{-3}]$ that is not prime. Here, $N(a + b\sqrt{-3}) = a^2 + 3b^2$. Consider $1 + \sqrt{-3}$. Suppose that we can factor this as xy, where neither x nor y is a unit. Then $N(xy) = N(x)N(y) = N(1 + \sqrt{-3}) = 4$, and it follows that $N(x) = 2$. But there are no integers a and b that satisfy $a^2 + 3b^2 = 2$. Thus, x or y is a unit and $1 + \sqrt{-3}$ is an irreducible. To verify that it is not prime, we observe that $(1 + \sqrt{-3})(1 - \sqrt{-3}) = 4 = 2 \cdot 2$, so that $1 + \sqrt{-3}$ divides $2 \cdot 2$. On the other hand, for integers a and b to exist so that $2 = (1 + \sqrt{-3})(a + b\sqrt{-3}) = (a - 3b) + (a + b)\sqrt{-3}$, we must have $a - 3b = 2$ and $a + b = 0$, which is impossible.   ◆

Showing that an element of a ring of the form $Z[\sqrt{d}]$ is irreducible is more difficult when $d > 1$. The next example illustrates one method of doing this. The example also shows that the converse of the fourth property above for the norm is not true. That is, it shows that x may be irreducible even if $N(x)$ is not prime.

EXAMPLE 2   The element 7 is irreducible in the ring $Z[\sqrt{5}]$. To verify this assertion, suppose that $7 = xy$, where neither x nor y is a unit. Then $49 = N(7) = N(x) N(y)$, and since x is not a unit, we cannot have $N(x) = 1$. This leaves only the case $N(x) = 7$. Let $x = a + b\sqrt{5}$. Then there are integers a and b satisfying $|a^2 - 5b^2| = 7$. This means that $a^2 - 5b^2 = \pm 7$. Viewing this equation modulo 7 and trying all possible cases for a and b reveals that the only solutions are $a = 0 = b$. But this means that both a and b are divisible by 7, and this implies that $|a^2 - 5b^2| = 7$ is divisible by 49, which is false.   ◆

Example 1 raises the question of whether or not there is an integral domain containing a prime that is not an irreducible. The answer: no.

Theorem 18.1   **Prime Implies Irreducible**

> *In an integral domain, every prime is an irreducible.*

PROOF.   Suppose that a is a prime in an integral domain and $a = bc$. We must show that b or c is a unit. By the definition of prime, we know that $a \mid b$ or $a \mid c$. Say $at = b$. Then $1b = b = at = (bc)t = b(ct)$ and, by cancellation, $1 = ct$. Thus, c is a unit.   ∎

Recall that a principal ideal domain is an integral domain in which every ideal has the form $\langle a \rangle$. The next theorem reveals a circumstance in which primes and irreducibles are equivalent.

Theorem 18.2   **PID Implies Irreducible Equals Prime**

> *In a principal ideal domain, an element is an irreducible if and only if it is a prime.*

PROOF.   Theorem 18.1 shows that primes are irreducibles. To prove the converse, let a be an irreducible element of a principal ideal domain D and suppose that $a \mid bc$. We must show that $a \mid b$ or $a \mid c$. Consider the ideal $I = \{ax + by \mid x, y \in D\}$ and let $\langle d \rangle = I$. Since $a \in I$, we can write $a = dr$, and because a is irreducible, d is a unit or r is a unit. If d is a unit, then $I = D$ and we may write $1 = ax + by$. Then $c = acx + bcy$, and since a divides both terms on the right, a also divides c.

On the other hand, if r is a unit, then $\langle a \rangle = \langle d \rangle = I$, and, because $b \in I$, there is an element t in D such that $at = b$. Thus, a divides b.   ∎

It is an easy consequence of the respective division algorithms for Z and $F[x]$, where F is a field, that Z and $F[x]$ are principal ideal domains (see Exercise 39 in Chapter 14 and Theorem 16.3). Our next example shows, however, that one of the most familiar rings is not a principal ideal domain.

EXAMPLE 3   We show that $Z[x]$ is not a principal ideal domain. Consider the ideal $I = \{f(x) \in Z[x] \mid f(0) \text{ is even}\}$. We claim that I is not of the form $\langle h(x) \rangle$. If this were so, there would be $f(x)$ and $g(x)$ in $Z[x]$ such that $2 = h(x)f(x)$ and $x = h(x)g(x)$, since both 2 and x belong to I. By the degree rule (Exercise 17 in Chapter 16), $0 = \deg 2 = \deg h(x) + \deg f(x)$, so that $h(x)$ is a constant polynomial. To determine which constant, we observe that $2 = h(1)f(1)$. Thus, $h(1) = \pm 1$ or ± 2. Since 1 is not in I, we must have $h(x) = \pm 2$. But then $x = \pm 2g(x)$, which is nonsense. ◆

We have previously proved that the integral domains Z and $Z[x]$ have important factorization properties: Every integer greater than 1 can be uniquely factored as a product of irreducibles (that is, primes), and every nonzero, nonunit polynomial can be uniquely factored as a product of irreducible polynomials. It is natural to ask whether all integral domains have this property. The question of unique factorization in integral domains first arose with the efforts to solve a famous problem in number theory that goes by the name Fermat's Last Theorem.

HISTORICAL DISCUSSION OF FERMAT'S LAST THEOREM

There are infinitely many nonzero integers x, y, z that satisfy the equation $x^2 + y^2 = z^2$. But what about the equation $x^3 + y^3 = z^3$ or, more generally, $x^n + y^n = z^n$, where n is an integer greater than 2 and x, y, z are nonzero integers? Well, no one has ever found a single solution of this equation, and for more than three centuries many have tried to prove that there is none. The tremendous effort put forth by the likes of Euler, Legendre, Abel, Gauss, Dirichlet, Cauchy, Kummer, Kronecker, and Hilbert to prove that there are no solutions to this equation has greatly influenced the development of ring theory.

About a thousand years ago, Arab mathematicians gave an incorrect proof that there were no solutions when $n = 3$. The problem lay dormant until 1637, when the French mathematician Pierre de Fermat (1601–1665) wrote in the margin of a book, ". . . it is impossible to separate a cube into two cubes, a fourth power into two fourth powers, or, generally, any power above the second into two powers of the same degree: I have discovered a truly marvelous demonstration [of this general theorem] which this margin is too narrow to contain."

Because Fermat gave no proof, many mathematicians tried to prove the result. The case where $n = 3$ was done by Euler in 1770, although his proof was incomplete. The case where $n = 4$ is elementary and was done by Fermat himself. The case where $n = 5$ was done in 1825 by Dirichlet, who had just turned 20, and by Legendre, who was past 70. Since the validity of the case for a particular integer implies the validity for all multiples of that integer, the next case of interest was $n = 7$. This case resisted the efforts of the best mathematicians until it was done by Gabriel Lamé in 1839. In 1847, Lamé stirred excitement by announcing that he had completely solved the problem. His approach was to factor the expression $x^p + y^p$, where p is an odd prime, into

$$(x + y)(x + \alpha y) \cdots (x + \alpha^{p-1} y),$$

where α is the complex number $\cos(2\pi/p) + i \sin(2\pi/p)$. Thus, his factorization took place in the ring $Z[\alpha] = \{a_0 + a_1\alpha + \cdots + a_{p-1}\alpha^{p-1} \mid a_i \in Z\}$. But Lamé made the mistake of assuming that, in such a ring, factorization into the product of irreducibles is unique. In fact, three years earlier, Ernst Eduard Kummer had proved that this is not always the case. Undaunted by the failure of unique factorization, Kummer began developing a theory to "save" factorization by creating a new type of number. Within a few weeks of Lamé's announcement, Kummer had shown that Fermat's Last Theorem is true for all primes of a special type. This proved that the theorem was true for all exponents less than 100, prime or not, except for 37, 59, 67, and 74. Kummer's work has led to the theory of ideals as we know it today.

Over the centuries, many proposed proofs have not held up under scrutiny. The famous number theorist Edmund Landau received so many of these that he had a form printed with "On page ____, lines ____ to ____, you will find there is a mistake." Martin Gardner, "Mathematical Games" columnist of *Scientific American,* had postcards printed to decline requests from readers asking him to examine their proofs.

Recent discoveries tying Fermat's Last Theorem closely to modern mathematical theories gave hope that these theories might eventually lead to a proof. In March 1988, newspapers and scientific publications worldwide carried news of a proof by Yoichi Miyaoka (see Figure 18.1). Within weeks, however, Miyaoka's proof was shown to be invalid. In June 1993, excitement spread through the mathematics community with the announcement that Andrew Wiles of Princeton University had proved Fermat's Last Theorem (see Figure 18.2). The Princeton mathematics department chairperson was quoted as saying, "When we heard it, people started walking on air." But once again a proof did not hold up under scrutiny. This story does have a happy ending. The mathematical community has agreed on the validity of the revised proof given by Wiles and Richard Taylor in September of 1994.

Figure 18.1

In view of the fact that so many eminent mathematicians were unable to prove Fermat's Last Theorem, despite the availability of the vastly powerful theories, it seems highly improbable that Fermat had a correct proof. Most likely, he made the error that his successors made of assuming that the properties of integers, such as unique factorization, carry over to integral domains in general.

Figure 18.2

UNIQUE FACTORIZATION DOMAINS

We now have the necessary terminology to formalize the idea of unique factorization.

> **DEFINITION** *Unique Factorization Domain (UFD)*
>
> An integral domain D is a *unique factorization domain* if
>
> 1. every nonzero element of D that is not a unit can be written as a product of irreducibles of D, and
> 2. the factorization into irreducibles is unique up to associates and the order in which the factors appear.

Another way to formulate part 2 of this definition is the following: If $p_1{}^{n_1} p_2{}^{n_2} \cdots p_r{}^{n_r}$ and $q_1{}^{m_1} q_2{}^{m_2} \cdots q_s{}^{m_s}$ are two factorizations of some element as a product of irreducibles, where no two of the p_i's are associates and no two of the q_j's are associates, then $r = s$, and each $p_i{}^{n_i}$ is an associate of one and only one $q_j{}^{m_j}$.

Of course, the Fundamental Theorem of Arithmetic tells us that the ring of integers is a unique factorization domain, and Theorem 17.6

says that $Z[x]$ is a unique factorization domain. In fact, as we shall soon see, most of the integral domains we have encountered are unique factorization domains.

Before proving our next theorem, we need the ascending chain condition for ideals.

Lemma **Ascending Chain Condition for a PID**

> *In a principal ideal domain, any strictly increasing chain of ideals $I_1 \subset I_2 \subset \cdot \ \cdot \ \cdot$ must be finite in length.*

PROOF. Let $I_1 \subset I_2 \subset \cdots$ be a chain of strictly increasing ideals in an integral domain D, and let I be the union of all the ideals in this chain. We leave it as an exercise (Exercise 3) to verify that I is an ideal of D.

Then, since D is a principal ideal domain, there is an element a in D such that $I = \langle a \rangle$. Because $a \in I$ and $I = \cup I_k$, a belongs to some member of the chain, say $a \in I_n$. Clearly, then, for any member I_i of the chain, we have $I_i \subseteq I = \langle a \rangle \subseteq I_n$, so that I_n must be the last member of the chain. ■ ■

Theorem 18.3 **PID Implies UFD**

> *Every principal ideal domain is a unique factorization domain.*

PROOF. Let D be a principal ideal domain and let a_0 be any nonzero nonunit in D. We will show that a_0 is a product of irreducibles (the product might consist of only one factor). We begin by showing that a_0 has at least one irreducible factor. If a_0 is irreducible, we are done. Thus, we may assume that $a_0 = b_1 a_1$, where neither b_1 nor a_1 is a unit and a_1 is nonzero. If a_1 is not irreducible, then we can write $a_1 = b_2 a_2$, where neither b_2 nor a_2 is a unit and a_2 is nonzero. Continuing in this fashion, we obtain a sequence $b_1, b_2, \ldots$ of elements that are not units in D and a sequence $a_0, a_1, a_2, \ldots$ of nonzero elements of D with $a_n = b_{n+1} a_{n+1}$ for each n. Hence, $\langle a_0 \rangle \subset \langle a_1 \rangle \subset \cdots$ is a strictly increasing chain of ideals (see Exercise 5), which, by the preceding lemma, must be finite, say, $\langle a_0 \rangle \subset \langle a_1 \rangle \subset \cdots \subset \langle a_r \rangle$. In particular, a_r is an irreducible factor of a_0. This argument shows that every nonzero nonunit in D has at least one irreducible factor.

Now write $a_0 = p_1c_1$, where p_1 is irreducible and c_1 is not a unit. If c_1 is not irreducible, then we can write $c_1 = p_2c_2$, where p_2 is irreducible and c_2 is not a unit. Continuing in this fashion, we obtain, as before, a strictly increasing sequence $\langle a_0 \rangle \subset \langle c_1 \rangle \subset \langle c_2 \rangle \subset \cdots$, which must end in a finite number of steps. Let us say that the sequence ends with $\langle c_s \rangle$. Then c_s is irreducible and $a_0 = p_1p_2 \cdots p_sc_s$, where each p_i is also irreducible. This completes the proof that every nonzero nonunit of a principal ideal domain is a product of irreducibles.

It remains to be shown that the factorization is unique up to associates and the order in which the factors appear. To do this, suppose that some element a of D can be written

$$a = p_1p_2 \cdots p_r = q_1q_2 \cdots q_s,$$

where the p's and q's are irreducible and repetition is permitted. We induct on r. If $r = 1$, then a is irreducible and, clearly, $s = 1$ and $p_1 = q_1$. So we may assume that any element that can be expressed as a product of fewer than r irreducible factors can be so expressed in only one way (up to order and associates). Since p_1 divides $q_1q_2 \cdots q_s$, it must divide some q_i (see Exercise 29), say, $p_1 \mid q_1$. Then, $q_1 = up_1$, where u is a unit of D. Since

$$up_1p_2 \cdots p_r = uq_1q_2 \cdots q_s = q_1(uq_2) \cdots q_s$$

and

$$up_1 = q_1,$$

we have, by cancellation,

$$p_2 \cdots p_r = (uq_2) \cdots q_s.$$

The induction hypothesis now tells us that these two factorizations are identical up to associates and the order in which the factors appear. Hence, the same is true about the two factorizations of a. ■■

In the existence portion of the proof of Theorem 18.3, the only way we used the fact that the integral domain D is a principal ideal domain was to say that D has the property that there is no infinite, strictly increasing chain of ideals in D. An integral domain with this property is called a *Noetherian domain*, in honor of Emmy Noether, who inaugurated the use of chain conditions in algebra. Noetherian domains are of the utmost importance in algebraic geometry. One reason for this is that, for many important rings R, the polynomial ring $R[x]$ is a Noetherian domain but not a principal ideal domain. One such example is $Z[x]$. In particular, $Z[x]$ shows that a UFD need not be a PID (see Example 3).

As an immediate corollary of Theorem 18.3, we have the following fact.

Corollary   *F[x]* **Is a UFD**

> *Let F be a field. Then F[x] is a unique factorization domain.*

PROOF.   By Theorem 16.3, $F[x]$ is a principal ideal domain. So, $F[x]$ is a unique factorization domain, as well.   ■■

As an application of the preceding corollary, we give an elegant proof, due to Richard Singer, of Eisenstein's Criterion (Theorem 17.4).

EXAMPLE 4   Let
$$f(x) = a_n x^n + a_{n-1} x^{n-1} + \cdots + a_0 \in Z[x],$$
and suppose that p is prime such that
$$p \nmid a_n, p \mid a_{n-1}, \ldots, p \mid a_0 \quad \text{and} \quad p^2 \nmid a_0.$$

We will prove that $f(x)$ is irreducible over Q. If $f(x)$ is reducible over Q, we know by Theorem 17.2 that there exist elements $g(x)$ and $h(x)$ in $Z[x]$ such that $f(x) = g(x)h(x)$ and $1 \le \deg g(x), \deg h(x) < n$. Let $\overline{f}(x)$, $\overline{g}(x)$, and $\overline{h}(x)$ be the polynomials in $Z_p[x]$ obtained from $f(x)$, $g(x)$, and $h(x)$ by reducing all coefficients modulo p. Then, since p divides all the coefficients of $f(x)$ except a_n, we have $\overline{a}_n x^n = \overline{f}(x) = \overline{g}(x)\overline{h}(x)$. Since Z_p is a field, $Z_p[x]$ is a unique factorization domain. Thus, $x \mid \overline{g}(x)$ and $x \mid \overline{h}(x)$. So, $\overline{g}(0) = \overline{h}(0) = 0$ and, therefore, $p \mid g(0)$ and $p \mid h(0)$. But, then, $p^2 \mid g(0)h(0) = f(0) = a_0$, which is a contradiction.   ◆

EUCLIDEAN DOMAINS

Another important kind of integral domain is a Euclidean domain.

DEFINITION   *Euclidean Domain*

> An integral domain D is called a *Euclidean domain* if there is a function d (called the *measure*) from the nonzero elements of D to the nonnegative integers such that
>
> **1.** $d(a) \le d(ab)$ for all nonzero a, b in D; and
> **2.** if $a, b \in D, b \ne 0$, then there exist elements q and r in D such that $a = bq + r$, where $r = 0$ or $d(r) < d(b)$.

EXAMPLE 5 The ring Z is a Euclidean domain with $d(a) = |a|$ (the absolute value of a). ◆

EXAMPLE 6 Let F be a field. Then $F[x]$ is a Euclidean domain with $d(f(x)) = \deg f(x)$ (see Theorem 16.2). ◆

Examples 5 and 6 illustrate just one of many similarities between the rings Z and $F[x]$. Additional similarities are summarized in Table 18.1.

EXAMPLE 7 The ring of Gaussian integers

$$Z[i] = \{a + bi \mid a, b \in Z\}$$

is a Euclidean domain with $d(a + bi) = a^2 + b^2$. Unlike the previous two examples, in this example the function d does not obviously satisfy the necessary conditions. That $d(x) \le d(xy)$ for $x, y \in Z[i]$ follows

TABLE 18.1 Similarities Between Z and $F[x]$

Z		$F[x]$
Form of elements:	↔	Form of elements:
$a_n 10^n + a_{n-1} 10^{n-1} + \cdots + a_1 10 + a_0$		$a_n x^n + a_{n-1} x^{n-1} + \cdots + a_1 x + a_0$
Euclidean domain:	↔	Euclidean domain:
$d(a) = |a|$		$d(f(x)) = \deg f(x)$
Units:		Units:
a is a unit if and only if $|a| = 1$		$f(x)$ is a unit if and only if $\deg f(x) = 0$
Division algorithm:	↔	Division algorithm:
For $a, b \in Z, b \ne 0$, there exist $q, r \in Z$ such that $a = bq + r, 0 \le r < |b|$		For $f(x), g(x) \in F[x], g(x) \ne 0$, there exist $q(x), r(x) \in F[x]$ such that $f(x) = g(x)q(x) + r(x), 0 \le \deg r(x) <$ $\deg g(x)$ or $r(x) = 0$
PID:	↔	PID:
Every nonzero ideal $I = \langle a \rangle$, where $a \ne 0$ and $|a|$ is minimum		Every nonzero ideal $I = \langle f(x) \rangle$, where $\deg f(x)$ is minimum
Prime:	↔	Irreducible:
No nontrivial factors		No nontrivial factors
UFD:	↔	UFD:
Every element is a "unique" product of primes		Every element is a "unique" product of irreducibles

directly from the fact that $d(xy) = d(x)d(y)$ (Exercise 7). If $x, y \in Z[i]$ and $y \neq 0$, then $xy^{-1} \in Q[i]$, the field of quotients of $Z[i]$ (Exercise 51 in Chapter 15). Say $xy^{-1} = s + ti$, where $s, t \in Q$. Now let m be the integer nearest s, and let n be the integer nearest t. (These integers may not be uniquely determined, but that does not matter.) Thus, $|m - s| \leq 1/2$ and $|n - t| \leq 1/2$. Then

$$xy^{-1} = s + ti = (m - m + s) + (n - n + t)i$$

$$= (m + ni) + [(s - m) + (t - n)i].$$

So,

$$x = (m + ni)y + [(s - m) + (t - n)i]y.$$

We claim that the division condition of the definition of a Euclidean domain is satisfied with $q = m + ni$ and

$$r = [(s - m) + (t - n)i]y.$$

Clearly, q belongs to $Z[i]$, and since $r = x - qy$, so does r. Finally,

$$d(r) = d([(s - m) + (t - n)i])d(y)$$

$$= [(s - m)^2 + (t - n)^2]d(y)$$

$$\leq \left(\frac{1}{4} + \frac{1}{4}\right)d(y) < d(y). \quad \blacklozenge$$

Theorem 18.4   ED (Euclidean Domain) Implies PID

> *Every Euclidean domain is a principal ideal domain.*

PROOF.   Let D be a Euclidean domain and I a nonzero ideal of D. Among all the nonzero elements of I, let a be such that $d(a)$ is a minimum. Then $I = \langle a \rangle$. For, if $b \in I$, there are elements q and r such that $b = aq + r$, where $r = 0$ or $d(r) < d(a)$. But $r = b - aq \in I$, so $d(r)$ cannot be less than $d(a)$. Thus, $r = 0$ and $b \in \langle a \rangle$. Finally, the zero ideal is $\langle 0 \rangle$.   ■■

Although it is not easy to verify, we remark that there are principal ideal domains that are not Euclidean domains. The first such example was given by T. Motzkin in 1949. A more accessible account of Motzkin's result can be found in [2].

As an immediate consequence of Theorems 18.3 and 18.4, we have the following important result.

Corollary   **ED Implies UFD**

> *Every Euclidean domain is a unique factorization domain.*

We may summarize our theorems and remarks as follows:

$$ED \Rightarrow PID \Rightarrow UFD$$

$$UFD \nRightarrow PID \nRightarrow ED$$

(You can remember these implications by listing the types alphabetically.)

In Chapter 17, we proved that $Z[x]$ is a unique factorization domain. Since Z is a unique factorization domain, the next theorem is a broad generalization of this fact. The proof is similar to that of the special case, and we therefore omit it.

Theorem 18.5   **D a UFD Implies D[x] a UFD**

> *If D is a unique factorization domain, then D[x] is a unique factorization domain.*

We conclude this chapter with an example of an integral domain that is not a unique factorization domain.

EXAMPLE 8   The ring $Z[\sqrt{-5}] = \{a + b\sqrt{-5} \mid a, b \in Z\}$ is an integral domain but not a unique factorization domain. It is straightforward that $Z[\sqrt{-5}]$ is an integral domain (see Exercise 9 in Chapter 13). To verify that unique factorization does not hold, we mimic the method used in Example 1 with $N(a + b\sqrt{-5}) = a^2 + 5b^2$. Since $N(xy) = N(x)N(y)$ and $N(x) = 1$ if and only if x is a unit (see Exercise 1), it follows that the only units of $Z[\sqrt{-5}]$ are ± 1.

Now consider the following factorizations:

$$46 = 2 \cdot 23,$$
$$46 = (1 + 3\sqrt{-5})(1 - 3\sqrt{-5}).$$

We claim that each of these four factors is irreducible over $Z[\sqrt{-5}]$. Suppose that, say, $2 = xy$, where $x, y \in Z[\sqrt{-5}]$ and neither is a unit. Then $4 = N(2) = N(x)N(y)$ and, therefore, $N(x) = N(y) = 2$, which is impossible. Likewise, if $23 = xy$ were a nontrivial factorization, then $N(x) = 23$. Thus, there would be integers a and b such that $a^2 + 5b^2 = 23$. Clearly, no such integers exist. The same argument applies to $1 \pm 3\sqrt{-5}$. ◆

In light of Examples 7 and 8, one can't help but wonder for which $d < 0$ is $Z[\sqrt{d}]$ a unique factorization domain. The answer is only when $d = -1$ or -2 (see [1, p. 297]). The case where $d = -1$ was first proved, naturally enough, by Gauss.

EXERCISES

I tell them that if they will occupy themselves with the study of mathematics they will find in it the best remedy against lust of the flesh.

THOMAS MANN, The Magic Mountain

1. For the ring $Z[\sqrt{d}] = \{a + b\sqrt{d} \mid a, b \in Z\}$, where $d \neq 1$ and d is not divisible by the square of a prime, prove that the norm $N(a + b\sqrt{d}) = |a^2 - db^2|$ satisfies the four assertions made preceding Example 1. (This exercise is referred to in this chapter.)

2. In an integral domain, show that a and b are associates if and only if $\langle a \rangle = \langle b \rangle$.

3. Show that the union of a chain $I_1 \subset I_2 \subset \cdots$ of ideals of a ring R is an ideal of R. (This exercise is referred to in this chapter.)

4. In an integral domain, show that the product of an irreducible and a unit is an irreducible.

5. Suppose that a and b belong to an integral domain, $b \neq 0$, and a is not a unit. Show that $\langle ab \rangle$ is a proper subset of $\langle b \rangle$. (This exercise is referred to in this chapter.)

6. Let D be an integral domain. Define $a \sim b$ if a and b are associates. Show that this defines an equivalence relation on D.

7. In the notation of Example 7, show that $d(xy) = d(x)d(y)$.

8. Let D be a Euclidean domain with measure d. Prove that u is a unit in D if and only if $d(u) = d(1)$.

9. Let D be a Euclidean domain with measure d. Show that if a and b are associates in D, then $d(a) = d(b)$.

10. Let D be a principal ideal domain and let $p \in D$. Prove that $\langle p \rangle$ is a maximal ideal in D if and only if p is irreducible.

11. Trace through the argument given in Example 7 to find q and r in $Z[i]$ such that $3 - 4i = (2 + 5i)q + r$ and $d(r) < d(2 + 5i)$.

12. Let D be a principal ideal domain. Show that every proper ideal of D is contained in a maximal ideal of D.

13. In $Z[\sqrt{-5}]$, show that 21 does not factor uniquely as a product of irreducibles.

14. Show that $1 - i$ is an irreducible in $Z[i]$.

15. Show that $Z[\sqrt{-6}]$ is not a unique factorization domain. (*Hint:* Factor 10 in two ways.) Why does this show that $Z[\sqrt{-6}]$ is not a principal ideal domain?

16. Give an example of a unique factorization domain with a subdomain that does not have a unique factorization.

17. In $Z[i]$, show that 3 is irreducible but 2 and 5 are not.

18. Prove that 7 is irreducible in $Z[\sqrt{6}]$, even though $N(7)$ is not prime.

19. Prove that if p is a prime in Z that can be written in the form $a^2 + b^2$, then $a + bi$ is irreducible in $Z[i]$. Find three primes that have this property and the corresponding irreducibles.

20. Prove that $Z[\sqrt{-3}]$ is not a principal ideal domain.

21. In $Z[\sqrt{-5}]$, prove that $1 + 3\sqrt{-5}$ is irreducible but not prime.

22. In $Z[\sqrt{5}]$, prove that both 2 and $1 + \sqrt{5}$ are irreducible but not prime.

23. Prove that $Z[\sqrt{5}]$ is not a unique factorization domain.

24. Let a and b be idempotents in a commutative ring. Show that each of the following is also an idempotent: ab, $a - ab$, $a + b - ab$, $a + b - 2ab$.

25. Let d be an integer less than -1 that is not divisible by the square of a prime. Prove that the only units of $Z[\sqrt{d}]$ are $+1$ and -1.

26. If a and b belong to $Z[\sqrt{d}]$, where d is not divisible by the square of a prime and ab is a unit, prove that a and b are units.

27. Prove or disprove that if D is a principal ideal domain, then $D[x]$ is a principal ideal domain.

28. Determine the units in $Z[i]$.

29. Let p be a prime in an integral domain. If $p \mid a_1 a_2 \cdots a_n$, prove that p divides some a_i. (This exercise is referred to in this chapter.)

30. Show that $3x^2 + 4x + 3 \in Z_5[x]$ factors as $(3x + 2)(x + 4)$ and $(4x + 1)(2x + 3)$. Explain why this does not contradict the corollary of Theorem 18.3.

31. Let D be a principal ideal domain and p an irreducible element of D. Prove that $D/\langle p \rangle$ is a field.

32. Show that an integral domain with the property that every strictly decreasing chain of ideals $I_1 \supset I_2 \supset \cdots$ must be finite in length is a field.

33. An ideal A of a commutative ring R with unity is said to be *finitely generated* if there exist elements $a_1, a_2, \ldots, a_n$ of A such that $A = \langle a_1, a_2, \ldots, a_n \rangle$. An integral domain R is said to satisfy the *ascending chain condition* if every strictly increasing chain of ideals $I_1 \subset I_2 \subset \cdots$ must be finite in length. Show that an integral domain R satisfies the ascending chain condition if and only if every ideal of R is finitely generated.

34. Prove or disprove that a subdomain of a Euclidean domain is a Euclidean domain.

35. Show that for any nontrivial ideal I of $Z[i]$, $Z[i]/I$ is finite.

36. Find the inverse of $1 + \sqrt{2}$ in $Z[\sqrt{2}]$. What is the multiplicative order of $1 + \sqrt{2}$?

37. In $Z[\sqrt{-7}]$, show that $N(6 + 2\sqrt{-7}) = N(1 + 3\sqrt{-7})$ but $6 + 2\sqrt{-7}$ and $1 + 3\sqrt{-7}$ are not associates.

38. Let $R = Z \oplus Z \oplus \cdots$ (the collection of all sequences of integers under componentwise addition and multiplication). Show that R has ideals I_1, I_2, I_3, . . . with the property that $I_1 \subset I_2 \subset I_3 \subset \cdots$. (Thus R does not have the ascending chain condition.)

39. Prove that in a unique factorization domain an element is irreducible if and only if it is prime.

COMPUTER EXERCISE ▪▪

In the fields of observation chance favors only the prepared mind.

LOUIS PASTEUR

Software for the computer exercise in this chapter is available at the website:

http://www.d.umn.edu/~jgallian

1. In the ring $Z[i]$ (where $i^2 = -1$), this software determines when a positive integer n is a prime in $Z[i]$. Run the program for several cases and formulate a conjecture based on your data.

References

1. H. M. Stark, *An Introduction to Number Theory,* Chicago, Ill.: Markham, 1970.

2. J. C. Wilson, "A Principal Ideal Ring That Is Not a Euclidean Ring," *Mathematics Magazine* 46 (1973): 74–78.

Suggested Readings

Oscar Campoli, "A Principal Ideal Domain That Is Not a Euclidean Domain," *The American Mathematical Monthly* 95 (1988): 868–871.

The author shows that $\{a + b\theta \mid a, b \in Z, \theta = (1 + \sqrt{-19})/2\}$ is a PID that is not an ED.

Gina Kolata, "At Last, Shout of 'Eureka!' in Age-Old Math Mystery," *The New York Times,* June 24, 1993.
This front-page article reports on Andrew Wiles's announced proof of Fermat's Last Theorem.

C. Krauthhammer, "The Joy of Math, or Fermat's Revenge," *Time,* April 18, 1988; 92.
The demise of Miyaoka's proof of Fermat's Last Theorem is charmingly lamented.

Sahib Singh, "Non-Euclidean Domains: An Example," *Mathematics Magazine* 49 (1976): 243.
This article gives a short proof that $Z[\sqrt{-n}] = \{a + b\sqrt{-n} \mid a, b \in Z\}$ is an integral domain that is not Euclidean when $n > 2$ and $-n = 2$ or 3 modulo 4.

Simon Singh and Kenneth Ribet, "Fermat's Last Stand," *Scientific American* 277 (1997): 68–73.
This article gives an accessible description of Andrew Wiles's proof of Fermat's Last Theorem.

Sophie Germain

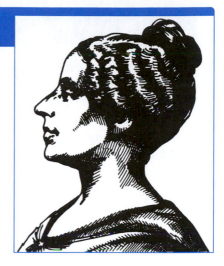

One of the very few women to overcome the prejudice and discrimination that tended to exclude women from the pursuit of higher mathematics in her time was Sophie Germain.

SOPHIE GERMAIN was born in Paris on April 1, 1776. She educated herself by reading the works of Newton and Euler in Latin and the lecture notes of Lagrange. In 1804, Germain wrote to Gauss about her work in number theory but used the pseudonym Monsieur LeBlanc because she feared that Gauss would not take seriously the efforts of a woman. Gauss gave Germain's results high praise and a few years later, upon learning her true identity, wrote to her:

> But how to describe to you my admiration and astonishment at seeing my esteemed correspondent Mr. LeBlanc metamorphose himself into this illustrious personage who gives such a brilliant example of what I would find it difficult to believe. A taste for the abstract sciences in general and above all the mysteries of numbers is excessively rare: it is not a subject which strikes everyone; the enchanting charms of this sublime science reveal themselves only to those who have the courage to go deeply into it. But when a person of the sex which, according to our customs and prejudices, must encounter infinitely more difficulties than men to familiarize herself with these thorny researches, succeeds nevertheless in surmounting these obstacles and penetrating the most obscure parts of them, then without doubt she must have the noblest courage, quite extraordinary talents, and a superior genius.

Germain is best known for her work on Fermat's Last Theorem. She died on June 27, 1831, in Paris.

For more information about Germain, visit:

http://www-groups.dcs.st-and.ac.uk/~history

Andrew Wiles

For spectacular contributions to number theory and related fields, for major advances on fundamental conjectures, and for settling Fermat's Last Theorem.

CITATION FOR THE WOLF PRIZE

IN 1993, ANDREW WILES of Princeton electrified the mathematics community by announcing that he had proved Fermat's Last Theorem after seven years of effort. His proof, which ran 200 pages, relied heavily on ring theory and group theory. Because of Wiles's solid reputation and because his approach was based on deep results that had already shed much light on the problem, many experts in the field believed that Wiles had succeeded where so many others had failed. Wiles's achievement was reported in newspapers and magazines around the world. *The New York Times* ran a front-page story on it, and one TV network announced it on the evening news. Wiles even made *People* magazine's list of the 25 most intriguing people of 1993! In San Francisco a group of mathematicians rented a 1200-seat movie theater and sold tickets for $5.00 each for public lectures on the proof. Scalpers received as much as $25.00 a ticket for the sold-out event.

The bubble soon burst when experts had an opportunity to scrutinize Wiles's manuscript. By December, Wiles released a statement saying he was working to resolve a gap in the proof. In September of 1994, a paper by Wiles and Richard Taylor, a former student of Wiles, circumvented the gap in the original proof. Since then, many experts have checked the proof and have found no errors. One mathematician was quoted as saying, "The exuberance is back." In 1997, Wiles's proof was the subject of a PBS Nova program.

Wiles was born in 1953 in Cambridge, England. He obtained his bachelor's degree at Oxford and his doctoral degree at Cambridge University in 1980. He was a professor at Harvard before accepting his present position at Princeton in 1982, and he has received many prestigious awards.

To find more information about Wiles, visit:

http://www-groups.dcs.st-and.ac.uk/~history/

SUPPLEMENTARY EXERCISES FOR CHAPTERS 15–18 — ▪▪

> *The intelligence is proved not by ease of learning, but by understanding what we learn.*
>
> JOSEPH WHITNEY

True/false questions for Chapters 15–18 are available on the Web at

http://www.d.umn.edu/~jgallian/TF

1. Suppose that F is a field and there is a ring homomorphism from Z onto F. Show that F is isomorphic to Z_p for some prime p.

2. Let $Q[\sqrt{2}] = \{r + s\sqrt{2} \mid r, s \in Q\}$. Determine all ring automorphisms of $Q[\sqrt{2}]$.

3. (Second Isomorphism Theorem for Rings) Let A be a subring of R and let B be an ideal of R. Show that $A \cap B$ is an ideal of A and that $A/(A \cap B)$ is isomorphic to $(A + B)/B$. (Recall that $A + B = \{a + b \mid a \in A, b \in B\}$.)

4. (Third Isomorphism Theorem for Rings) Let A and B be ideals of a ring R with $B \subseteq A$. Show that A/B is an ideal of R/B and $(R/B)/(A/B)$ is isomorphic to R/A.

5. Let $f(x)$ and $g(x)$ be irreducible polynomials over a field F. If $f(x)$ and $g(x)$ are not associates, prove that $F[x]/\langle f(x)g(x)\rangle$ is isomorphic to $F[x]/\langle f(x)\rangle \oplus F[x]/\langle g(x)\rangle$.

6. (Chinese Remainder Theorem for Rings) If R is a commutative ring and I and J are two proper ideals with $I + J = R$, prove that $R/(I \cap J)$ is isomorphic to $R/I \oplus R/J$. Explain why Exercise 5 is a special case of this theorem.

7. Prove that the set of all polynomials all of whose coefficients are even is a prime ideal in $Z[x]$.

8. Let $R = Z[\sqrt{-5}]$ and let $I = \{a + b\sqrt{-5} \mid a, b \in Z, a - b \text{ is even}\}$. Show that I is a maximal ideal of R.

9. Let R be a ring with unity and let a be a unit in R. Show that the mapping from R into itself given by $x \to axa^{-1}$ is a ring automorphism.

10. Let $a + b\sqrt{-5}$ belong to $Z[\sqrt{-5}]$ with $b \neq 0$. Show that 2 does not belong to $\langle a + b\sqrt{-5}\rangle$.

11. Show that $Z[i]/\langle 2 + i\rangle$ is a field. How many elements does it have?

12. Is the homomorphic image of a principal ideal domain a principal ideal domain?

13. In $Z[\sqrt{2}] = \{a + b\sqrt{2} \mid a, b \in Z\}$, show that every element of the form $(3 + 2\sqrt{2})^n$ is a unit, where n is a positive integer.

14. Let p be a prime. Show that there is exactly one ring homomorphism from Z_m to Z_{p^k} if p^k does not divide m, and exactly two ring homomorphisms from Z_m to Z_{p^k} if p^k does divide m.

15. Recall that a is an idempotent if $a^2 = a$. Show that if $1 + k$ is an idempotent in Z_n, then $n - k$ is an idempotent in Z_n.

16. Show that Z_n (where $n > 1$) always has an even number of idempotents. (The number is 2^d, where d is the number of distinct prime divisors of n.)

17. Show that the equation $x^2 + y^2 = 2003$ has no solutions in the integers.

18. Prove that if both k and $k + 1$ are idempotents in Z_n and $k \neq 0$, then $n = 2k$.

19. Prove that $x^4 + 15x^3 + 7$ is irreducible over Q.

20. For any integers m and n, prove that the polynomial $x^3 + (5m + 1)x + 5n + 1$ is irreducible over Z.

21. Prove that $\langle \sqrt{2} \rangle$ is a maximal ideal in $Z[\sqrt{2}]$. How many elements are in the ring $Z[\sqrt{2}]/\langle \sqrt{2} \rangle$?

22. Prove that $Z[\sqrt{-2}]$ and $Z[\sqrt{2}]$ are unique factorization domains. (*Hint:* Mimic Example 7 in Chapter 18.)

23. Is $\langle 3 \rangle$ a maximal ideal in $Z[i]$?

24. Express both 13 and $5 + i$ as products of irreducibles from $Z[i]$.

25. Let $R = \{a/b \mid a, b \in Z, 3 \nmid b\}$. Prove that R is an integral domain. Find its field of quotients.

26. Give an example of a ring that contains a subring isomorphic to Z and a subring isomorphic to Z_3.

27. Show that $Z[i]/\langle 3 \rangle$ is not ring-isomorphic to $Z_3 \oplus Z_3$.

28. For any $n > 1$, prove that $R = \left\{ \begin{bmatrix} a & 0 \\ 0 & b \end{bmatrix} \middle| a, b \in Z_n \right\}$ is ring-isomorphic to $Z_n \oplus Z_n$.

29. Suppose that R is a commutative ring and I is an ideal of R. Prove that $R[x]/I[x]$ is isomorphic to $(R/I)[x]$.

30. Find an ideal I of $Z_8[x]$ such that the factor ring $Z_8[x]/I$ is a field.

31. Find an ideal I of $Z_8[x]$ such that the factor ring $Z_8[x]/I$ is an integral domain but not a field.

32. For any $f(x) \in Z_p[x]$, show that $f(x^p) = (f(x))^p$.

33. Find an ideal I of $Z[x]$ such that $Z[x]/I$ is ring-isomorphic to Z_3.

PART 4

Fields

 For online student resources,
visit this textbook's website at
math.college.hmco.com/students

19 | Vector Spaces

*Still round the corner there may wait
A new road or a secret gate.*

J. R. R. TOLKIEN, *Lord of the Rings*

DEFINITION AND EXAMPLES

Abstract algebra has three basic components: groups, rings, and fields. Thus far we have covered groups and rings in some detail, and we have touched on the notion of a field. To explore fields more deeply, we need some rudiments of vector space theory that are covered in a linear algebra course. In this chapter, we provide a concise review of this material.

DEFINITION *Vector Space*

A set V is said to be a *vector space* over a field F if V is an Abelian group under addition (denoted by $+$) and, if for each $a \in F$ and $v \in V$, there is an element av in V such that the following conditions hold for all a, b in F and all u, v in V.

1. $a(v + u) = av + au$
2. $(a + b)v = av + bv$
3. $a(bv) = (ab)v$
4. $1v = v$

The members of a vector space are called *vectors*. The members of the field are called *scalars*. The operation that combines a scalar a and a vector v to form the vector av is called *scalar multiplication*. In general, we will denote vectors by letters from the end of the alphabet, such as u, v, w, and scalars by letters from the beginning of the alphabet, such as a, b, c.

343

EXAMPLE 1 The set $\mathbf{R}^n = \{(a_1, a_2, \ldots, a_n) \mid a_i \in \mathbf{R}\}$ is a vector space over $\mathbf{R}$. Here the operations are the obvious ones.

$$(a_1, a_2, \ldots, a_n) + (b_1, b_2, \ldots, b_n) = (a_1 + b_1, a_2 + b_2, \ldots, a_n + b_n)$$

and

$$b(a_1, a_2, \ldots, a_n) = (ba_1, ba_2, \ldots, ba_n). \qquad \blacklozenge$$

EXAMPLE 2 The set $M_2(Q)$ of 2×2 matrices with entries from Q is a vector space over Q. The operations are

$$\begin{bmatrix} a_1 & a_2 \\ a_3 & a_4 \end{bmatrix} + \begin{bmatrix} b_1 & b_2 \\ b_3 & b_4 \end{bmatrix} = \begin{bmatrix} a_1 + b_1 & a_2 + b_2 \\ a_3 + b_3 & a_4 + b_4 \end{bmatrix}$$

and

$$b\begin{bmatrix} a_1 & a_2 \\ a_3 & a_4 \end{bmatrix} = \begin{bmatrix} ba_1 & ba_2 \\ ba_3 & ba_4 \end{bmatrix}. \qquad \blacklozenge$$

EXAMPLE 3 The set $Z_p[x]$ of polynomials with coefficients from Z_p is a vector space over Z_p, where p is a prime. $\qquad \blacklozenge$

EXAMPLE 4 The set of complex numbers $\mathbf{C} = \{a + bi \mid a, b \in \mathbf{R}\}$ is a vector space over $\mathbf{R}$. The operations are the usual addition and multiplication of complex numbers. $\qquad \blacklozenge$

The next example is a generalization of Example 4. Although it appears rather trivial, it is of the utmost importance in the theory of fields.

EXAMPLE 5 Let E be a field and let F be a subfield of E. Then E is a vector space over F. The operations are the operations of E. $\qquad \blacklozenge$

SUBSPACES

Of course, there is a natural analog of subgroup and subring.

DEFINITION *Subspace*

Let V be a vector space over a field F and let U be a subset of V. We say that U is a *subspace* of V if U is also a vector space over F under the operations of V.

EXAMPLE 6   The set $\{a_2x^2 + a_1x + a_0 \mid a_0, a_1, a_2 \in \mathbf{R}\}$ is a subspace of the vector space of all polynomials with real coefficients over $\mathbf{R}$.   ◆

EXAMPLE 7   Let V be a vector space over F and let $v_1, v_2, \ldots, v_n$ be (not necessarily distinct) elements of V. Then the subset

$$\langle v_1, v_2, \ldots, v_n \rangle = \{a_1v_1 + a_2v_2 + \cdots + a_nv_n \mid a_1, a_2, \ldots, a_n \in F\}$$

is called the *subspace of V spanned by* $v_1, v_2, \ldots, v_n$. Any summand of the form $a_1v_1 + a_2v_2 + \cdots + a_nv_n$ is called a *linear combination of* $v_1, v_2, \ldots, v_n$. If $\langle v_1, v_2, \ldots, v_n \rangle = V$, we say that $\{v_1, v_2, \ldots, v_n\}$ *spans V.*   ◆

LINEAR INDEPENDENCE

The next definition is the heart of the theory.

DEFINITION   ***Linearly Dependent, Linearly Independent***

> A set S of vectors is said to be *linearly dependent* over the field F if there are vectors $v_1, v_2, \ldots, v_n$ from S and elements $a_1, a_2, \ldots, a_n$ from F, not all zero, such that $a_1v_1 + a_2v_2 + \cdots + a_nv_n = 0$. A set of vectors that is not linearly dependent over F is called *linearly independent over F.*

EXAMPLE 8   In $\mathbf{R}^3$ the vectors $(1, 0, 0)$, $(1, 0, 1)$, and $(1, 1, 1)$ are linearly independent over $\mathbf{R}$. To verify this, assume that there are real numbers a, b, and c such that $a(1, 0, 0) + b(1, 0, 1) + c(1, 1, 1) = (0, 0, 0)$. Then $(a + b + c, c, b + c) = (0, 0, 0)$. From this we see that $a = b = c = 0$.   ◆

Certain kinds of linearly independent sets play a crucial role in the theory of vector spaces.

DEFINITION   ***Basis***

> Let V be a vector space over F. A subset B of V is called a *basis* for V if B is linearly independent over F and every element of V is a linear combination of elements of B.

The motivation for this definition is twofold. First, if B is a basis for a vector space V, then every member of V is a unique linear combination of the elements of B (see Exercise 19). Second, with every vector space spanned by finitely many vectors, we can use the notion of basis to associate a unique integer that tells us much about the vector space. (In fact, this integer and the field completely determine the vector space up to isomorphism—see Exercise 31.)

EXAMPLE 9 The set $V = \left\{ \begin{bmatrix} a & a+b \\ a+b & b \end{bmatrix} \middle| a, b \in \mathbf{R} \right\}$

is a vector space over $\mathbf{R}$ (see Exercise 17). We claim that the set $B = \left\{ \begin{bmatrix} 1 & 1 \\ 1 & 0 \end{bmatrix}, \begin{bmatrix} 0 & 1 \\ 1 & 1 \end{bmatrix} \right\}$ is a basis for V over $\mathbf{R}$. To prove that the set B is linearly independent, suppose that there are real numbers a and b such that

$$a \begin{bmatrix} 1 & 1 \\ 1 & 0 \end{bmatrix} + b \begin{bmatrix} 0 & 1 \\ 1 & 1 \end{bmatrix} = \begin{bmatrix} 0 & 0 \\ 0 & 0 \end{bmatrix}.$$

This gives $\begin{bmatrix} a & a+b \\ a+b & b \end{bmatrix} = \begin{bmatrix} 0 & 0 \\ 0 & 0 \end{bmatrix}$, so that $a = b = 0$. On the other hand, since every member of V has the form

$$\begin{bmatrix} a & a+b \\ a+b & b \end{bmatrix} = a \begin{bmatrix} 1 & 1 \\ 1 & 0 \end{bmatrix} + b \begin{bmatrix} 0 & 1 \\ 1 & 1 \end{bmatrix},$$

we see that B spans V.   ◆

We now come to the main result of this chapter.

Theorem 19.1   Invariance of Basis Size

> If $\{u_1, u_2, \ldots, u_m\}$ and $\{w_1, w_2, \ldots, w_n\}$ are both bases of a vector space V, then $m = n$.

PROOF.   Suppose that $m \neq n$. To be specific, let us say that $m < n$. Consider the set $\{w_1, u_1, u_2, \ldots, u_m\}$. Since the u's span V, we know that w_1 is a linear combination of the u's, say, $w_1 = a_1 u_1 + a_2 u_2 + \cdots + a_m u_m$. Clearly, not all the a's are 0. For convenience, say $a_1 \neq 0$. Then $\{w_1, u_2, \ldots, u_m\}$ spans V (see Exercise 21). Next, consider the set $\{w_1, w_2, u_2, \ldots, u_m\}$. This time, w_2 is a linear combination of $w_1, u_2, \ldots, u_m$, say, $w_2 = b_1 w_1 + b_2 u_2 + \cdots + b_m u_m$. Then at least one of $b_2, \ldots, b_m$ is nonzero, for otherwise the w's are not linearly independent. Let us say $b_2 \neq 0$. Then $w_1, w_2, u_3, \ldots, u_m$ span V. Continuing in this fashion, we see that $\{w_1, w_2, \ldots, w_m\}$ spans V. But then w_{m+1} is a linear combination of $w_1, w_2, \ldots, w_m$ and, therefore, the set $\{w_1, \ldots, w_n\}$ is not linearly independent. This contradiction finishes the proof.   ■■

Theorem 19.1 shows that any two finite bases for a vector space have the same size. Of course, not all vector spaces have finite bases. However, there is no vector space that has a finite basis and an infinite basis (see Exercise 25).

DEFINITION   *Dimension*

A vector space that has a basis consisting of n elements is said to have *dimension n*. For completeness, the trivial vector space $\{0\}$ is said to be spanned by the empty set and to have dimension 0.

Although it requires a bit of set theory that is beyond the scope of this text, it can be shown that every vector space has a basis. A vector space that has a finite basis is called *finite dimensional*; otherwise, it is called *infinite dimensional*.

EXERCISES   ■■

Somebody who thinks logically is a nice contrast to the real world.

The Law of Thumb

1. Verify that each of the sets in Examples 1–4 satisfies the axioms for a vector space. Find a basis for each of the vector spaces in Examples 1–4.

2. (Subspace Test) Prove that a nonempty subset U of a vector space V over a field F is a subspace of V if, for every u and u' in U and every a in F, $u + u' \in U$ and $au \in U$.

3. Verify that the set in Example 6 is a subspace. Find a basis for this subspace. Is $\{x^2 + x + 1, x + 5, 3\}$ a basis?

4. Verify that the set $\langle v_1, v_2, \ldots, v_n \rangle$ defined in Example 7 is a subspace.

5. Determine whether or not the set $\{(2, -1, 0), (1, 2, 5), (7, -1, 5)\}$ is linearly independent over $\mathbf{R}$.

6. Determine whether or not the set

$$\left\{ \begin{bmatrix} 2 & 1 \\ 1 & 0 \end{bmatrix}, \begin{bmatrix} 0 & 1 \\ 1 & 2 \end{bmatrix}, \begin{bmatrix} 1 & 1 \\ 1 & 1 \end{bmatrix} \right\}$$

is linearly independent over Z_5.

7. If $\{u, v, w\}$ is a linearly independent subset of a vector space, show that $\{u, u + v, u + v + w\}$ is also linearly independent.

8. If $\{v_1, v_2, \ldots, v_n\}$ is a linearly dependent set of vectors, prove that one of these vectors is a linear combination of the others.

9. (Every spanning collection contains a basis.) If $\{v_1, v_2, \ldots, v_n\}$ spans a vector space V, prove that some subset of the v's is a basis for V.

10. (Every independent set is contained in a basis.) Let V be a finite-dimensional vector space and let $\{v_1, v_2, \ldots, v_n\}$ be a linearly independent subset of V. Show that there are vectors $w_1, w_2, \ldots, w_m$ such that $\{v_1, v_2, \ldots, v_n, w_1, \ldots, w_m\}$ is a basis for V.

11. If V is a vector space over F of dimension 5 and U and W are subspaces of V of dimension 3, prove that $U \cap W \neq \{0\}$. Generalize.

12. Show that the solution set to a system of equations of the form

$$a_{11}x_1 + \cdots + a_{1n}x_n = 0$$
$$a_{21}x_1 + \cdots + a_{2n}x_n = 0$$
$$\vdots \qquad\qquad \vdots \qquad \vdots$$
$$a_{m1}x_1 + \cdots + a_{mn}x_n = 0,$$

where the a's are real, is a subspace of $\mathbf{R}^n$.

13. Let V be the set of all polynomials over Q of degree 2 together with the zero polynomial. Is V a vector space over Q?

14. Let $V = \mathbf{R}^3$ and $W = \{(a, b, c) \in V \mid a^2 + b^2 = c^2\}$. Is W a subspace of V? If so, what is its dimension?

15. Let $V = \mathbf{R}^3$ and $W = \{(a, b, c) \in V \mid a + b = c\}$. Is W a subspace of V? If so, what is its dimension?

16. Let $V = \left\{ \begin{bmatrix} a & b \\ b & c \end{bmatrix} \Big| \, a, b, c \in Q \right\}$. Prove that V is a vector space over Q and find a basis for V over Q.

17. Verify that the set V in Example 9 is a vector space over $\mathbf{R}$.

18. Let $P = \{(a, b, c) \mid a, b, c \in \mathbf{R}, a = 2b + 3c\}$. Prove that P is a subspace of $\mathbf{R}^3$. Find a basis for P. Give a geometric description of P.

19. Let B be a finite subset of a vector space V. Show that B is a basis for V if and only if every member of V is a unique linear combination of the elements of B. (This exercise is referred to in this chapter and in Chapter 20.)

20. If U is a proper subspace of a finite-dimensional vector space V, show that the dimension of U is less than the dimension of V.

21. Referring to the proof of Theorem 19.1, prove that $\{w_1, u_2, \ldots, u_m\}$ spans V.

22. If V is a vector space of dimension n over the field Z_p, how many elements are in V?

23. Let $S = \{(a, b, c, d) \mid a, b, c, d \in \mathbf{R}, a = c, d = a + b\}$. Find a basis for S.

24. Let U and W be subspaces of a vector space V. Show that $U \cap W$ is a subspace of V and that $U + W = \{u + w \mid u \in U, w \in W\}$ is a subspace of V.

25. If a vector space has one basis that contains infinitely many elements, prove that every basis contains infinitely many elements. (This exercise is referred to in this chapter.)

26. Let $u = (2, 3, 1)$, $v = (1, 3, 0)$, and $w = (2, -3, 3)$. Since $\frac{1}{2}u - \frac{2}{3}v - \frac{1}{6}w = (0, 0, 0)$, can we conclude that the set $\{u, v, w\}$ is linearly dependent over Z_7?

27. Define the vector space analog of group homomorphism and ring homomorphism. Such a mapping is called a *linear transformation*. Define the vector space analog of group isomorphism and ring isomorphism.

28. Let T be a linear transformation from V to W. Prove that the image of V under T is a subspace of W.

29. Let T be a linear transformation of a vector space V. Prove that $\{v \in V \mid T(v) = 0\}$, the *kernel* of T, is a subspace of V.

30. Let T be a linear transformation of V onto W. If $\{v_1, v_2, \ldots, v_n\}$ spans V, show that $\{T(v_1), T(v_2), \ldots, T(v_n)\}$ spans W.

31. If V is a vector space over F of dimension n, prove that V is isomorphic as a vector space to $F^n = \{(a_1, a_2, \ldots, a_n) \mid a_i \in F\}$. (This exercise is referred to in this chapter.)

32. Let V be a vector space over an infinite field. Prove that V is not the union of finitely many proper subspaces of V.

Emil Artin

For Artin, to be a mathematician meant to participate in a great common effort, to continue work begun thousands of years ago, to shed new light on old discoveries, to seek new ways to prepare the developments of the future. Whatever standards we use, he was a great mathematician.

RICHARD BRAUER, *Bulletin of the American Mathematical Society*

EMIL ARTIN was one of the leading mathematicians of the twentieth century and a major contributor to linear algebra and abstract algebra. Artin was born on March 3, 1898, in Vienna, Austria, and grew up in what was recently known as Czechoslovakia. He received a Ph.D. in 1921 from the University of Leipzig. From 1923 until he emigrated to America in 1937, he was a professor at the University of Hamburg. After one year at Notre Dame, Artin went to Indiana University. In 1946, he moved to Princeton, where he stayed until 1958. The last four years of his career were spent where it began, at Hamburg.

Artin's mathematics is both deep and broad. He made contributions to number theory, group theory, ring theory, field theory, Galois theory, geometric algebra, algebraic topology, and the theory of braids—a field he invented. Artin received the American Mathematical Society's Cole Prize in number theory, and he solved one of the 23 famous problems posed by the eminent mathematician David Hilbert in 1900.

Artin was an outstanding teacher of mathematics at all levels, from freshman calculus to seminars for colleagues. Many of his Ph.D. students as well as his son Michael have become leading mathematicians. Through his research, teaching, and books, Artin exerted great influence among his contemporaries. He died of a heart attack, at the age of 64, in 1962.

For more information about Artin, visit:

http://www-groups.dcs.st-and.ac.uk/~history/

Olga Taussky-Todd

"Olga Taussky-Todd was a distinguished and prolific mathematician who wrote about 300 papers."

EDITH LUCHINS AND MARY ANN MCLOUGHLIN,
Notices of the American Mathematical Society, 1996

OLGA TAUSSKY-TODD was born on August 30, 1906, in Olmütz in the Austro-Hungarian Empire. Taussky-Todd received her doctoral degree in 1930 from the University of Vienna. In the early 1930s she was hired as an assistant at the University of Göttingen to edit books on the work of David Hilbert. She also edited lecture notes of Emil Artin and assisted Richard Courant. She spent 1934 and 1935 at Bryn Mawr and the next two years at Girton College in Cambridge, England. In 1937, she taught at the University of London. Olga Taussky married the mathematician John Todd in 1938. In 1947, she moved to the United States and took a job at the National Bureau of Standards' National Applied Mathematics Laboratory. In 1957, she became the first woman to teach at the California Institute of Technology as well as the first woman to receive tenure and a full professorship in mathematics, physics, or astronomy there. Thirteen Caltech Ph.D. students wrote their Ph.D. theses under her direction.

In addition to her influential contributions to linear algebra, Taussky-Todd did important work in number theory.

Taussky-Todd received many honors and awards. She was elected a Fellow of the American Association for the Advancement of Science and vice president of the American Mathematical Society. In 1990, Caltech established an instructorship named in her honor. Taussky-Todd died on October 7, 1995, at the age of 89.

For more information about Taussky-Todd, visit:

http://www-groups.dcs.st-and.ac.uk/~history

http://www-scottlan.edu/lriddle/women/women.html

20

Extension Fields

In many respects this [Kronecker's Theorem] is the fundamental theorem of algebra.

RICHARD A. DEAN, *Elements of Abstract Algebra*

THE FUNDAMENTAL THEOREM OF FIELD THEORY

In our work on rings, we came across a number of fields, both finite and infinite. Indeed, we saw that $Z_3[x]/\langle x^2 + 1 \rangle$ is a field of order 9, whereas $\mathbf{R}[x]/\langle x^2 + 1 \rangle$ is a field isomorphic to the complex numbers. In the next three chapters, we take up, in a systematic way, the subject of fields.

DEFINITION   *Extension Field*

A field E is an *extension field* of a field F if $F \subseteq E$ and the operations of F are those of E restricted to F.

Cauchy's observation in 1847 that $\mathbf{R}[x]/\langle x^2 + 1 \rangle$ is a field that contains a zero of $x^2 + 1$ prepared the way for the following sweeping generalization of that fact.

Theorem 20.1   Fundamental Theorem of Field Theory (Kronecker's Theorem, 1887)

Let F be a field and let $f(x)$ be a nonconstant polynomial in $F[x]$. Then there is an extension field E of F in which $f(x)$ has a zero.

PROOF. Since $F[x]$ is a unique factorization domain, $f(x)$ has an irreducible factor, say, $p(x)$. Clearly, it suffices to construct an extension field E of F in which $p(x)$ has a zero. Our candidate for E is $F[x]/\langle p(x) \rangle$. We already know that this is a field from Corollary 1 of Theorem 17.5. Also, since the mapping of $\phi: F \to E$ given by $\phi(a) = a + \langle p(x) \rangle$ is one-to-one and preserves both operations, E has a subfield isomorphic to F. We may think of E as containing F if we simply identify the coset $a + \langle p(x) \rangle$ with its unique coset representative a that belongs to F [that is, think of $a + \langle p(x) \rangle$ as just a and vice versa; see Exercise 33 in Chapter 17].

Finally, to show that $p(x)$ has a zero in E, write

$$p(x) = a_n x^n + a_{n-1} x^{n-1} + \cdots + a_0.$$

Then, in E, $x + \langle p(x) \rangle$ is a zero of $p(x)$, because

$$
\begin{aligned}
p(x + \langle p(x) \rangle) &= a_n (x + \langle p(x) \rangle)^n + a_{n-1}(x + \langle p(x) \rangle)^{n-1} + \cdots + a_0 \\
&= a_n(x^n + \langle p(x) \rangle) + a_{n-1}(x^{n-1} + \langle p(x) \rangle) + \cdots + a_0 \\
&= a_n x^n + a_{n-1} x^{n-1} + \cdots + a_0 + \langle p(x) \rangle \\
&= p(x) + \langle p(x) \rangle = 0 + \langle p(x) \rangle.
\end{aligned}
$$
■■

EXAMPLE 1 Let $f(x) = x^2 + 1 \in Q[x]$. Then, in $E = Q[x]/\langle x^2 + 1 \rangle$, we have

$$
\begin{aligned}
f(x + \langle x^2 + 1 \rangle) &= (x + \langle x^2 + 1 \rangle)^2 + 1 \\
&= x^2 + \langle x^2 + 1 \rangle + 1 \\
&= x^2 + 1 + \langle x^2 + 1 \rangle \\
&= 0 + \langle x^2 + 1 \rangle.
\end{aligned}
$$

Of course, the polynomial $x^2 + 1$ has the complex number $\sqrt{-1}$ as a zero, but the point we wish to emphasize here is that we have constructed a field that contains the rational numbers and a zero for the polynomial $x^2 + 1$ by using only the rational numbers. No knowledge of complex numbers is necessary. Our method utilizes only the field we are given. ◆

EXAMPLE 2 Let $f(x) = x^5 + 2x^2 + 2x + 2 \in Z_3[x]$. Then, the irreducible factorization of $f(x)$ over Z_3 is $(x^2 + 1)(x^3 + 2x + 2)$. So, to find an extension E of Z_3 in which $f(x)$ has a zero, we may take $E = Z_3[x]/\langle x^2 + 1 \rangle$, a field with nine elements, or $E = Z_3[x]/\langle x^3 + 2x + 2 \rangle$, a field with 27 elements. ◆

Since every integral domain is contained in its field of quotients (Theorem 15.6), we see that every nonconstant polynomial with coefficients from an integral domain always has a zero in some field containing the ring of coefficients. The next example shows that this is not true for commutative rings in general.

EXAMPLE 3   Let $f(x) = 2x + 1 \in Z_4[x]$. Then $f(x)$ has no zero in any ring containing Z_4 as a subring, because if β were a zero in such a ring, then $0 = 2\beta + 1$, and therefore $0 = 2(2\beta + 1) = 2(2\beta) + 2 = (2 \cdot 2)\beta + 2 = 0 \cdot \beta + 2 = 2$. But $0 \neq 2$ in Z_4.   ◆

SPLITTING FIELDS

To motivate the next definition and theorem, let's return to Example 1 for a moment. For notational convenience, in $Q[x]/\langle x^2 + 1 \rangle$, let $\alpha = x + \langle x^2 + 1 \rangle$. Then, since α and $-\alpha$ are both zeros of $x^2 + 1$, it should be the case that $x^2 + 1 = (x - \alpha)(x + \alpha)$. Let's check this out. First note that

$$(x - \alpha)(x + \alpha) = x^2 - \alpha^2 = x^2 - (x^2 + \langle x^2 + 1 \rangle).$$

At the same time,

$$x^2 + \langle x^2 + 1 \rangle = -1 + \langle x^2 + 1 \rangle$$

and we have agreed to identify -1 and $-1 + \langle x^2 + 1 \rangle$, so

$$(x - \alpha)(x + \alpha) = x^2 - (-1) = x^2 + 1.$$

This shows that $x^2 + 1$ can be written as a product of linear factors in some extension of Q. That was easy and you might argue coincidental. The polynomial given in Example 2 presents a greater challenge. Is there an extension of Z_3 in which that polynomial factors as a product of linear factors? Yes, there is. But first a definition.

DEFINITION   *Splitting Field*

Let E be an extension field of F and let $f(x) \in F[x]$. We say that $f(x)$ *splits* in E if $f(x)$ can be factored as a product of linear factors in $E[x]$. We call E a *splitting field for $f(x)$ over F* if $f(x)$ splits in E but in no proper subfield of E.

Note that a splitting field of a polynomial over a field depends not only on the polynomial but on the field as well. Indeed, a splitting field of $f(x)$ over F is just a smallest extension field of F in which $f(x)$ splits. The next example illustrates how a splitting field of a polynomial $f(x)$ over field F depends on F.

EXAMPLE 4   Consider the polynomial $f(x) = x^2 + 1 \in Q[x]$. Since $x^2 + 1 = (x + \sqrt{-1})(x - \sqrt{-1})$, we see that $f(x)$ splits in **C**, but a splitting field over Q is $Q(i) = \{r + si \mid r, s \in Q\}$. A splitting field for $x^2 + 1$ over **R** is **C**. Likewise, $x^2 - 2 \in Q[x]$ splits in **R**, but a splitting field over Q is $Q(\sqrt{2}) = \{r + s\sqrt{2} \mid r, s \in Q\}$.   ◆

There is a useful analogy between the definition of a splitting field and the definition of an irreducible polynomial. Just as it makes no sense to say "$f(x)$ is irreducible," it makes no sense to say "E is a splitting field for $f(x)$." In each case, the underlying field must be specified; that is, one must say "$f(x)$ is irreducible over F" and "E is a splitting field for $f(x)$ over F."

The following notation is convenient. Let F be a field and let a_1, $a_2, \ldots, a_n$ be elements of some extension E of F. We use $F(a_1, a_2, \ldots, a_n)$ to denote the smallest subfield of E that contains F and the set $\{a_1, a_2, \ldots, a_n\}$. It is an easy exercise to show that $F(a_1, a_2, \ldots, a_n)$ is the intersection of all subfields of E that contain F and the set $\{a_1, a_2, \ldots, a_n\}$.

Notice that if $f(x) \in F[x]$ and $f(x)$ factors as

$$b(x - a_1)(x - a_2) \cdots (x - a_n)$$

over some extension E of F, then $F(a_1, \ldots, a_n)$ is a splitting field for $f(x)$ over F in E.

This notation appears to be inconsistent with the notation that we used in earlier chapters. For example, we denoted the set $\{a + b\sqrt{2} \mid a, b \in Z\}$ by $Z[\sqrt{2}]$ and the set $\{a + b\sqrt{2} \mid a, b \in Q\}$ by $Q(\sqrt{2})$. The difference is that $Z[\sqrt{2}]$ is merely a ring, whereas $Q(\sqrt{2})$ is a field. In general, parentheses are used when one wishes to indicate that the set is a field, although no harm would be done by using, say, $Q[\sqrt{2}]$ to denote $\{a + b\sqrt{2} \mid a, b \in Q\}$ if we were concerned with its ring properties only. After all, notation is just a convenient way to convey information. Using parentheses rather than brackets simply conveys a bit more information about the set.

Theorem 20.2   Existence of Splitting Fields

> *Let F be a field and let $f(x)$ be a nonconstant element of $F[x]$. Then there exists a splitting field E for $f(x)$ over F.*

PROOF.   We proceed by induction on deg $f(x)$. If deg $f(x) = 1$, then $f(x)$ is linear. Now suppose that the statement is true for all fields and all polynomials of degree less than that of $f(x)$. By Theorem 20.1, there is an extension E of F in which $f(x)$ has a zero, say, a_1. Then we may write $f(x) = (x - a_1)g(x)$, where $g(x) \in E[x]$. Since deg $g(x) <$ deg $f(x)$, by induction, there is a field K that contains E and all the zeros of $g(x)$, say, $a_2, \ldots, a_n$. Clearly, then, a splitting field for $f(x)$ over F is $F(a_1, a_2, \ldots, a_n)$.   ■ ■

EXAMPLE 5   Consider

$$f(x) = x^4 - x^2 - 2 = (x^2 - 2)(x^2 + 1)$$

over Q. Obviously, the zeros of $f(x)$ are $\pm\sqrt{2}$ and $\pm i$. So a splitting field for $f(x)$ over Q is

$$Q(\sqrt{2}, i) = Q(\sqrt{2})(i) = \{\alpha + \beta i \mid \alpha, \beta \in Q(\sqrt{2})\}$$

$$= \{(a + b\sqrt{2}) + (c + d\sqrt{2})i \mid a, b, c, d \in Q\}. \quad \blacklozenge$$

EXAMPLE 6   Consider $f(x) = x^2 + x + 2$ over Z_3. Then $Z_3(i) = \{a + bi \mid a, b \in Z_3\}$ (see Example 9 in Chapter 13) is a splitting field for $f(x)$ over Z_3 because

$$f(x) = [x - (1 + i)][x - (1 - i)].$$

At the same time, we know by the proof of Kronecker's Theorem that the element $x + \langle x^2 + x + 2 \rangle$ of

$$F = Z_3[x]/\langle x^2 + x + 2 \rangle$$

is a zero of $f(x)$. Since $f(x)$ has degree 2, it follows from the Factor Theorem (Corollary 2 of Theorem 16.2) that the other zero of $f(x)$ must also be in F. Thus, $f(x)$ splits in F, and because F has only nine elements, it is obvious that F is also a splitting field of $f(x)$ over Z_3. But how do we factor $f(x)$ in F? Factoring $f(x)$ in F is confusing because we are using the symbol x in two distinct ways: It is used as a placeholder to write the polynomial $f(x)$, and it is used to create the coset representatives of the elements of F. This confusion can be avoided by simply identifying the coset $1 + \langle x^2 + x + 2 \rangle$ with the element 1 in Z_3 and denoting the coset $x + \langle x^2 + x + 2 \rangle$ by β. With this identification, the field $Z_3[x]/\langle x^2 + x + 2 \rangle$ can be represented as $\{0, 1, 2, \beta, 2\beta, \beta + 1, 2\beta + 1, \beta + 2, 2\beta + 2\}$. These elements are added and multiplied just as polynomials are, except that we use the observation that $x^2 + x + 2 + \langle x^2 + x + 2 \rangle = 0$ implies that $\beta^2 + \beta + 2 = 0$, so that $\beta^2 = -\beta - 2 = 2\beta + 1$. For example, $(2\beta + 1)(\beta + 2) =$

$2\beta^2 + 5\beta + 2 = 2(2\beta + 1) + 5\beta + 2 = 9\beta + 4 = 1$. To obtain the factorization of $f(x)$ in F, we simply long divide, as follows:

$$
\begin{array}{r}
x + (\beta + 1) \\
x - \beta \overline{\smash{)}\ x^2 + x + 2} \\
\underline{x^2 - \beta x} \\
(\beta + 1)x + 2 \\
\underline{(\beta + 1)x - (\beta + 1)\beta} \\
(\beta + 1)\beta + 2 = \beta^2 + \beta + 2 = 0.
\end{array}
$$

So, $x^2 + x + 2 = (x - \beta)(x + \beta + 1)$. Thus, we have found two splitting fields for $x^2 + x + 2$ over Z_3, one of the form $F(a)$ and one of the form $F[x]/\langle p(x) \rangle$ [where $F = Z_3$ and $p(x) = x^2 + x + 2$]. ◆

The next theorem shows how the fields $F(a)$ and $F[x]/\langle p(x) \rangle$ are related in the case where $p(x)$ is irreducible over F and a is a zero of $p(x)$ in some extension of F.

Theorem 20.3 $F(a) \approx F[x]/\langle p(x) \rangle$

> *Let F be a field and let $p(x) \in F[x]$ be irreducible over F. If a is a zero of $p(x)$ in some extension E of F, then $F(a)$ is isomorphic to $F[x]/\langle p(x) \rangle$. Furthermore, if $\deg p(x) = n$, then every member of $F(a)$ can be uniquely expressed in the form*
>
> $$c_{n-1}a^{n-1} + c_{n-2}a^{n-2} + \cdots + c_1 a + c_0,$$
>
> *where $c_0, c_1, \ldots, c_{n-1} \in F$.*

PROOF. Consider the function ϕ from $F[x]$ to $F(a)$ given by $\phi(f(x)) = f(a)$. Clearly, ϕ is a ring homomorphism. We claim that Ker $\phi = \langle p(x) \rangle$. (This is Exercise 31 in Chapter 17.) Since $p(a) = 0$, we have $\langle p(x) \rangle \subseteq$ Ker ϕ. On the other hand, we know by Theorem 17.5 that $\langle p(x) \rangle$ is a maximal ideal in $F[x]$. So, because Ker $\phi \neq F[x]$ [it does not contain the constant polynomial $f(x) = 1$], we have Ker $\phi = \langle p(x) \rangle$. At this point it follows from the First Isomorphism Theorem for Rings and Corollary 1 of Theorem 17.5 that $\phi(F[x])$ is a subfield of $F(a)$. Noting that $\phi(F[x])$ contains both F and a and recalling that $F(a)$ is the smallest such field, we have $F[x]/\langle p(x) \rangle \approx \phi(F[x]) = F(a)$.

The final assertion of the theorem follows from the fact that every element of $F[x]/\langle p(x) \rangle$ can be expressed uniquely in the form

$$c_{n-1}x^{n-1} + \cdots + c_0 + \langle p(x) \rangle,$$

where $c_0, \ldots, c_{n-1} \in F$ (see Exercise 21 in Chapter 16) and the natural isomorphism from $F[x]/\langle p(x)\rangle$ to $F(a)$ carries $c_k x^k + \langle p(x)\rangle$ to $c_k a^k$.   ∎

As an immediate corollary of Theorem 20.3, we have the following attractive result.

Corollary $F(a) \approx F(b)$

> *Let F be a field and let $p(x) \in F[x]$ be irreducible over F. If a is a zero of $p(x)$ in some extension E of F and b is a zero of $p(x)$ in some extension E' of F, then the fields $F(a)$ and $F(b)$ are isomorphic.*

PROOF.   From Theorem 20.3, we have

$$F(a) \approx F[x]/\langle p(x)\rangle \approx F(b).$$ ∎

Recall that a basis for an n-dimensional vector space over a field F is a set of n vectors $v_1, v_2, \ldots, v_n$ with the property that every member of the vector space can be expressed uniquely in the form $a_1 v_1 + a_2 v_2 + \cdots + a_n v_n$, where the a's belong to F (Exercise 19 in Chapter 19). So, in the language of vector spaces, the latter portion of Theorem 20.3 says that if a is a zero of an irreducible polynomial over F of degree n, then the set $\{1, a, \ldots, a^{n-1}\}$ is a basis for $F(a)$ over F.

Theorem 20.3 often provides a convenient way of describing the elements of a field.

EXAMPLE 7   Consider the irreducible polynomial $f(x) = x^6 - 2$ over Q. Since $\sqrt[6]{2}$ is a zero of $f(x)$, we know from Theorem 20.3 that the set $\{1, 2^{1/6}, 2^{2/6}, 2^{3/6}, 2^{4/6}, 2^{5/6}\}$ is a basis for $Q(\sqrt[6]{2})$ over Q. Thus,

$$Q(\sqrt[6]{2}) = \{a_0 + a_1 2^{1/6} + a_2 2^{2/6} + a_3 2^{3/6} + a_4 2^{4/6} + a_5 2^{5/6} \mid a_i \in Q\}.$$

This field is isomorphic to $Q[x]/\langle x^6 - 2\rangle$.   ◆

In 1882, Ferdinand Lindemann (1852–1939) proved that π is not the zero of any polynomial in $Q[x]$. Because of this important result, Theorem 20.3 does not apply to $Q(\pi)$ (see Exercise 11).

In Example 6, we produced two splitting fields for the polynomial $x^2 + x + 2$ over Z_3. Likewise, it is an easy exercise to show that both $Q[x]/\langle x^2 + 1\rangle$ and $Q(i) = \{r + si \mid r, s \in Q\}$ are splitting fields of the polynomial $x^2 + 1$ over Q. But are these different-looking splitting fields algebraically different? Not really. We conclude our discussion of splitting fields by proving that splitting fields are unique up to isomorphism. To make it easier to apply induction, we will prove a more general result.

We begin by observing first that any field isomorphism ϕ from F to F' has a natural extension from $F[x]$ to $F'[x]$ given by $c_n x^n + c_{n-1} x^{n-1} + \cdots + c_1 x + c_0 \to \phi(c_n)x^n + \phi(c_{n-1})x^{n-1} + \cdots + \phi(c_1)x + \phi(c_0)$. Since this mapping agrees with ϕ on F, it is convenient and natural to use ϕ to denote this mapping as well.

Lemma

> *Let F be a field, let $p(x) \in F[x]$ be irreducible over F, and let a be a zero of $p(x)$ in some extension of F. If ϕ is a field isomorphism from F to F' and b is a zero of $\phi(p(x))$ in some extension of F', then there is an isomorphism from $F(a)$ to $F'(b)$ that agrees with ϕ on F and carries a to b.*

PROOF.   First observe that since $p(x)$ is irreducible over F, $\phi(p(x))$ is irreducible over F'. It is straightforward to check that the mapping from $F[x]/\langle p(x)\rangle$ to $F'[x]/\langle \phi(p(x))\rangle$ given by

$$f(x) + \langle p(x)\rangle \to \phi(f(x)) + \langle \phi(p(x))\rangle$$

is a field isomorphism. By a slight abuse of notation, we denote this mapping by ϕ also. (If you object, put a bar over the ϕ.) From the proof of Theorem 20.3, we know that there is an isomorphism α from $F(a)$ to $F[x]/\langle p(x)\rangle$ that is the identity on F and carries a to $x + \langle p(x)\rangle$. Similarly, there is an isomorphism β from $F'[x]/\langle \phi(p(x))\rangle$ to $F'(b)$ that is the identity on F' and carries $x + \langle \phi(p(x))\rangle$ to b. Thus, $\beta\phi\alpha$ is the desired mapping. See Figure 20.1. ■ ■

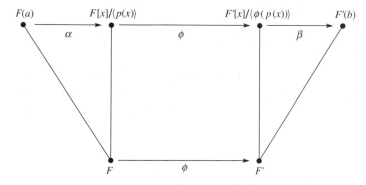

Figure 20.1

Theorem 20.4   Extending ϕ: $F \rightarrow F'$

> *Let ϕ be an isomorphism from a field F to a field F' and let $f(x) \in F[x]$. If E is a splitting field for $f(x)$ over F and E' is a splitting field for $\phi(f(x))$ over F', then there is an isomorphism from E to E' that agrees with ϕ on F.*

PROOF.   We induct on deg $f(x)$. If deg $f(x) = 1$, then $E = F$ and $E' = F'$, so that ϕ itself is the desired mapping. If deg $f(x) > 1$, let $p(x)$ be an irreducible factor of $f(x)$, let a be a zero of $p(x)$ in E, and let b be a zero of $\phi(p(x))$ in E'. By the preceding lemma, there is an isomorphism α from $F(a)$ to $F'(b)$ that agrees with ϕ on F and carries a to b. Now write $f(x) = (x - a)g(x)$, where $g(x) \in F(a)[x]$. Then E is a splitting field for $g(x)$ over $F(a)$ and E' is a splitting field for $\alpha(g(x))$ over $F'(b)$. Since deg $g(x) <$ deg $f(x)$, there is an isomorphism from E to E' that agrees with α on $F(a)$ and therefore with ϕ on F.   ■■

Corollary   Splitting Fields Are Unique

> *Let F be a field and let $f(x) \in F[x]$. Then any two splitting fields of $f(x)$ over F are isomorphic.*

PROOF.   Suppose that E and E' are splitting fields of $f(x)$ over F. The result follows immediately from Theorem 20.4 by letting ϕ be the identity from F to F.   ■■

In light of the corollary above, we may refer to "the" splitting field of a polynomial over F without ambiguity.

Even though $x^6 - 2$ has a zero in $Q(\sqrt[6]{2})$, it does not split in $Q(\sqrt[6]{2})$. The splitting field is easy to obtain, however.

EXAMPLE 8   The Splitting Field of $x^n - a$ over Q

Let a be a positive rational number and let ω be a primitive nth root of unity (see Example 2 in Chapter 16). Then each of

$$a^{1/n}, \omega a^{1/n}, \omega^2 a^{1/n}, \ldots, \omega^{n-1} a^{1/n}$$

is a zero of $x^n - a$ in $Q(\sqrt[n]{a}, \omega)$.   ◆

ZEROS OF AN IRREDUCIBLE POLYNOMIAL

Now that we know that every nonconstant polynomial over a field splits in some extension, we ask whether irreducible polynomials must split in some special way. Yes, they do. To discover how, we borrow something whose origins are in calculus.

DEFINITION

Let $f(x) = a_n x^n + a_{n-1} x^{n-1} + \cdots + a_1 x + a_0$ belong to $F[x]$. The *derivative* of $f(x)$, denoted by $f'(x)$, is the polynomial $n a_n x^{n-1} + (n-1) a_{n-1} x^{n-2} + \cdots + a_1$ in $F[x]$.

Notice that our definition does not involve the notion of a limit. The standard rules for handling sums and products of functions in calculus carry over to arbitrary fields as well.

Lemma   **Properties of the Derivative**

Let $f(x)$ and $g(x) \in F[x]$ and let $a \in F$. Then

1. $(f(x) + g(x))' = f'(x) + g'(x)$
2. $(af(x))' = af'(x)$
3. $(f(x)g(x))' = f(x)g'(x) + g(x)f'(x).$

PROOF. Parts 1 and 2 follow from straightforward applications of the definition. Using part 1 and induction on $\deg f(x)$, part 3 reduces to the special case in which $f(x) = a_n x^n$. This also follows directly from the definition. ∎

Before addressing the question of the nature of the zeros of an irreducible polynomial, we establish a general result concerning zeros of multiplicity greater than 1. Such zeros are called *multiple* zeros.

Theorem 20.5   **Criterion for Multiple Zeros**

A polynomial $f(x)$ over a field F has a multiple zero in some extension E if and only if $f(x)$ and $f'(x)$ have a common factor of positive degree in $F[x]$.

PROOF. If a is a multiple zero of $f(x)$ in some extension E, then there is a $g(x)$ in $E[x]$ such that $f(x) = (x - a)^2 g(x)$. Since $f'(x) =$

$(x - a)^2g'(x) + 2(x - a)g(x)$, we see that $f'(a) = 0$. Thus $x - a$ is a factor of both $f(x)$ and $f'(x)$ in the extension E of F. Now if $f(x)$ and $f'(x)$ have no common divisor of positive degree in $F[x]$, there are polynomials $h(x)$ and $k(x)$ in $F[x]$ such that $f(x)h(x) + f'(x)k(x) = 1$ (see Exercise 39 in Chapter 16). Viewing $f(x)h(x) + f'(x)k(x)$ as an element of $E[x]$, we see also that $x - a$ is a factor of 1. Since this is nonsense, $f(x)$ and $f'(x)$ must have a common divisor of positive degree in $F[x]$.

Conversely, suppose that $f(x)$ and $f'(x)$ have a common factor of positive degree. Let a be a zero of the common factor. Then a is a zero of $f(x)$ and $f'(x)$. Since a is a zero of $f(x)$, there is a polynomial $q(x)$ such that $f(x) = (x - a)q(x)$. Then $f'(x) = (x - a)q'(x) + q(x)$ and $0 = f'(a) = q(a)$. Thus, $x - a$ is a factor of $q(x)$ and a is a multiple zero of $f(x)$. ∎ ∎

Theorem 20.6   Zeros of an Irreducible

> *Let $f(x)$ be an irreducible polynomial over a field F. If F has characteristic 0, then $f(x)$ has no multiple zeros. If F has characteristic $p \neq 0$, then $f(x)$ has a multiple zero only if it is of the form $f(x) = g(x^p)$ for some $g(x)$ in $F[x]$.*

PROOF.   If $f(x)$ has a multiple zero, then, by Theorem 20.5, $f(x)$ and $f'(x)$ have a common divisor of positive degree in $F[x]$. Since the only divisor of positive degree of $f(x)$ in $F[x]$ is $f(x)$ itself (up to associates), we see that $f(x)$ divides $f'(x)$. Because a polynomial over a field cannot divide a polynomial of smaller degree, we must have $f'(x) = 0$.

Now what does it mean to say that $f'(x) = 0$? If we write $f(x) = a_nx^n + a_{n-1}x^{n-1} + \cdots + a_1x + a_0$, then $f'(x) = na_nx^{n-1} + (n - 1)a_{n-1}x^{n-2} + \cdots + a_1$. Thus, $f'(x) = 0$ only when $ka_k = 0$ for $k = 1, \ldots, n$.

So, when char $F = 0$, we have $f(x) = a_0$, which is not an irreducible polynomial. This contradicts the hypothesis that $f(x)$ is irreducible over F. Thus, $f(x)$ has no multiple zeros.

When char $F = p \neq 0$, we have $a_k = 0$ when p does not divide k. Thus, the only powers of x that appear in the sum $a_nx^n + \cdots + a_1x + a_0$ are those of the form $x^{pj} = (x^p)^j$. It follows that $f(x) = g(x^p)$ for some $g(x) \in F[x]$. [For example, if $f(x) = x^{4p} + 3x^{2p} + x^p + 1$, then $g(x) = x^4 + 3x^2 + x + 1$.] ∎∎

Theorem 20.6 shows that an irreducible polynomial over a field of characteristic 0 cannot have multiple zeros. The desire to extend this result to a larger class of fields motivates the following definition.

DEFINITION

A field F is called *perfect* if F has characteristic 0 or if F has characteristic p and $F^p = \{a^p \mid a \in F\} = F$.

The most important family of perfect fields of characteristic p is the finite fields.

Theorem 20.7 Finite Fields Are Perfect

Every finite field is perfect.

PROOF. Let F be a finite field of characteristic p. Consider the mapping ϕ from F to F defined by $\phi(x) = x^p$ for all $x \in F$. We claim that ϕ is a field automorphism. Obviously, $\phi(ab) = (ab)^p = a^p b^p = \phi(a)\phi(b)$. Moreover, $\phi(a + b) = (a + b)^p = a^p + \binom{p}{1} a^{p-1}b + \binom{p}{2} a^{p-2}b^2 + \cdots + \binom{p}{p-1} ab^{p-1} + b^p = a^p + b^p$, since each $\binom{p}{i}$ is divisible by p. Finally, since $x^p \neq 0$ when $x \neq 0$, Ker $\phi = \{0\}$. Thus, ϕ is one-to-one and, since F is finite, ϕ is onto. This proves that $F^p = F$. ∎∎

Theorem 20.8 Criterion for No Multiple Zeros

If $f(x)$ is an irreducible polynomial over a perfect field F, then $f(x)$ has no multiple zeros.

PROOF. The case where F has characteristic 0 has been done. So let us assume that $f(x) \in F[x]$ is irreducible over a perfect field F of characteristic p and that $f(x)$ has multiple zeros. From Theorem 20.6 we know that $f(x) = g(x^p)$ for some $g(x) \in F[x]$, say, $g(x) = a_n x^n + a_{n-1}x^{n-1} + \cdots + a_1 x + a_0$. Since $F^p = F$, each a_i in F can be written in the form b_i^p for some b_i in F. So, using Exercise 41**a** in Chapter 13, we have

$$f(x) = g(x^p) = b_n{}^p x^{pn} + b_{n-1}{}^p x^{p(n-1)} + \cdots + b_1{}^p x^p + b_0{}^p$$
$$= (b_n x^n + b_{n-1}x^{n-1} + \cdots + b_1 x + b_0)^p = (h(x))^p$$

where $h(x) \in F[x]$. But then $f(x)$ is not irreducible. ∎

The next theorem shows that when an irreducible polynomial does have multiple zeros, there is something striking about the multiplicities.

Theorem 20.9   **Zeros of an Irreducible over a Splitting Field**

> *Let $f(x)$ be an irreducible polynomial over a field F and let E be a splitting field of $f(x)$ over F. Then all the zeros of $f(x)$ in E have the same multiplicity.*

PROOF.   Let a and b be distinct zeros of $f(x)$ in E. If a has multiplicity m, then in $E[x]$ we may write $f(x) = (x - a)^m g(x)$. It follows from the lemma preceding Theorem 20.4 and from Theorem 20.4 that there is a field isomorphism ϕ from E to itself that carries a to b and acts as the identity on F. Thus,

$$f(x) = \phi(f(x)) = (x - b)^m \phi(g(x))$$

and we see that the multiplicity of b is greater than or equal to the multiplicity of a. By interchanging the roles of a and b, we observe that the multiplicity of a is greater than or equal to the multiplicity of b. So, we have proved that a and b have the same multiplicity. ∎

As an immediate corollary of Theorem 20.9 we have the following appealing result.

Corollary   **Factorization of an Irreducible over a Splitting Field**

> *Let $f(x)$ be an irreducible polynomial over a field F and let E be a splitting field of $f(x)$. Then $f(x)$ has the form*
> $$a(x - a_1)^n(x - a_2)^n \cdots (x - a_t)^n$$
> *where $a_1, a_2, \ldots, a_t$ are distinct elements of E and $a \in F$.*

We conclude this chapter by giving an example of an irreducible polynomial over a field that does have a multiple zero. In particular, notice that the field we use is not perfect.

EXAMPLE 9   Let $F = Z_2(t)$ be the field of quotients of the ring $Z_2[t]$ of polynomials in the indeterminate t with coefficients from Z_2. (We must introduce a letter other than x, since the members of F are going to be our coefficients for the elements in $F[x]$.) Consider $f(x) = x^2 - t \in F[x]$. To see that $f(x)$ is irreducible over F, it suffices to show that it has no zeros in F. Well, suppose that $h(t)/k(t)$ is a zero of $f(x)$. Then $(h(t)/k(t))^2 = t$, and therefore $(h(t))^2 = t(k(t))^2$. Since $h(t), k(t) \in Z_2[t]$, we then have $h(t^2) = tk(t^2)$ (see Exercise 41 in Chapter 13). But deg $h(t^2)$ is even, whereas deg $tk(t^2)$ is odd. So, $f(x)$ is irreducible over F.

Finally, since t is a constant in $F[x]$ and the characteristic of F is 2, we have $f'(x) = 0$, so that $f'(x)$ and $f(x)$ have $f(x)$ as a common factor. So, by Theorem 20.5, $f(x)$ has a multiple zero in some extension of F. (Indeed, it has a single zero of multiplicity 2 in $K = F[x]/\langle x^2 - t\rangle$.)   ◆

EXERCISES ▬▬▬▬▬▬▬▬▬▬▬▬▬▬▬▬▬▬▬▬▬▬▬▬▬▬▬▬▬▬▬▬ ▪▪

I have yet to see any problem, however complicated, which, when you looked at it in the right way, did not become still more complicated.

PAUL ANDERSON, New Scientist

1. Describe the elements of $Q(\sqrt[3]{5})$.
2. Show that $Q(\sqrt{2}, \sqrt{3}) = Q(\sqrt{2} + \sqrt{3})$.
3. Find the splitting field of $x^3 - 1$ over Q. Express your answer in the form $Q(a)$.
4. Find the splitting field of $x^4 + 1$ over Q.
5. Find the splitting field of
$$x^4 + x^2 + 1 = (x^2 + x + 1)(x^2 - x + 1)$$
   over Q.
6. Let $a, b \in \mathbf{R}$ with $b \neq 0$. Show that $\mathbf{R}(a + bi) = \mathbf{C}$.
7. Find a polynomial $p(x)$ in $Q[x]$ such that $Q(\sqrt{1+\sqrt{5}})$ is ring-isomorphic to $Q[x]/\langle p(x)\rangle$.
8. Let $F = Z_2$ and let $f(x) = x^3 + x + 1 \in F[x]$. Suppose that a is a zero of $f(x)$ in some extension of F. How many elements does $F(a)$ have? Express each member of $F(a)$ in terms of a. Write out a complete multiplication table for $F(a)$.

9. Let $F(a)$ be the field described in Exercise 8. Express each of a^5, a^{-2}, and a^{100} in the form $c_2 a^2 + c_1 a + c_0$.

10. Let $F(a)$ be the field described in Exercise 8. Show that a^2 and $a^2 + a$ are zeros of $x^3 + x + 1$.

11. Describe the elements in $Q(\pi)$.

12. Let $F = Q(\pi^3)$. Find a basis for $F(\pi)$ over F.

13. Write $x^7 - x$ as a product of linear factors over Z_3. Do the same for $x^{10} - x$.

14. Find all ring automorphisms of $Q(\sqrt[3]{5})$.

15. Let F be a field of characteristic p and let $f(x) = x^p - a \in F[x]$. Show that $f(x)$ is irreducible over F or $f(x)$ splits in F.

16. Suppose that β is a zero of $f(x) = x^4 + x + 1$ in some field extension E of Z_2. Write $f(x)$ as a product of linear factors in $E[x]$.

17. Find a, b, c in Q such that
$$(1 + \sqrt[3]{4})/(2 - \sqrt[3]{2}) = a + b\sqrt[3]{2} + c\sqrt[3]{4}.$$

Note that such a, b, c exist, since
$$(1 + \sqrt[3]{4})/(2 - \sqrt[3]{2}) \in Q(\sqrt[3]{2}) = \{a + b\sqrt[3]{2} + c\sqrt[3]{4} \mid a, b, c \in Q\}.$$

18. Express $(3 + 4\sqrt{2})^{-1}$ in the form $a + b\sqrt{2}$, where $a, b \in Q$.

19. Show that $Q(4 - i) = Q(1 + i)$, where $i = \sqrt{-1}$.

20. Let F be a field, and let a and b belong to F with $a \neq 0$. If c belongs to some extension of F, prove that $F(c) = F(ac + b)$. (F "absorbs" its own elements.)

21. Let $f(x) \in F[x]$ and let $a \in F$. Show that $f(x)$ and $f(x + a)$ have the same splitting field over F.

22. Recall that two polynomials $f(x)$ and $g(x)$ from $F[x]$ are said to be relatively prime if there is no polynomial of positive degree in $F[x]$ that divides both $f(x)$ and $g(x)$. Show that if $f(x)$ and $g(x)$ are relatively prime in $F[x]$, they are relatively prime in $K[x]$, where K is any extension of F.

23. Determine all of the subfields of $Q(\sqrt{2})$.

24. Let E be an extension of F and let a and b belong to E. Prove that $F(a, b) = F(a)(b) = F(b)(a)$.

25. Write $x^3 + 2x + 1$ as a product of linear polynomials over some field extension of Z_3.

26. Express $x^8 - x$ as a product of irreducibles over Z_2.

27. Prove or disprove that $Q(\sqrt{3})$ and $Q(\sqrt{-3})$ are ring-isomorphic.

28. For any prime p, find a field of characteristic p that is not perfect.

29. If β is a zero of $x^2 + x + 2$ over Z_5, find the other zero.

30. Show that $x^4 + x + 1$ over Z_2 does not have any multiple zeros in any field extension of Z_2.

31. Show that $x^{21} + 2x^8 + 1$ does not have multiple zeros in any extension of Z_3.

32. Show that $x^{21} + 2x^9 + 1$ has multiple zeros in some extension of Z_3.

33. Let F be a field of characteristic $p \neq 0$. Show that the polynomial $f(x) = x^{p^n} - x$ over F has distinct zeros.

34. Find the splitting field for $f(x) = (x^2 + x + 2)(x^2 + 2x + 2)$ over $Z_3[x]$. Write $f(x)$ as a product of linear factors.

Leopold Kronecker

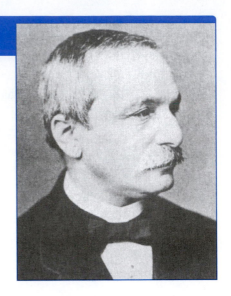

He [Kronecker] wove together the three strands of his greatest interests—the theory of numbers, the theory of equations and elliptic functions—into one beautiful pattern.

E. T. BELL

LEOPOLD KRONECKER was born on December 7, 1823, in Leignitz, Prussia. As a schoolboy, he received special instruction from the great algebraist Kummer. Kronecker entered the University of Berlin in 1841 and completed his Ph.D. dissertation in 1845 on the units in a certain ring.

Kronecker devoted the years 1845–1853 to business affairs, relegating mathematics to a hobby. Thereafter, being well-off financially, he spent most of his time doing research in algebra and number theory. Kronecker was one of the early advocates of the abstract approach to algebra. He innovatively applied rings and fields in his investigations of algebraic numbers, established the Fundamental Theorem of Finite Abelian Groups, and was the first mathematician to master Galois's theory of fields.

Kronecker advocated constructive methods for all proofs and definitions. He believed that all mathematics should be based on relationships among integers. He went so far as to say to Lindemann, who proved that π is transcendental, that irrational numbers do not exist. His most famous remark on the matter was, "God made the integers, all the rest is the work of man." Henri Poincaré once remarked that Kronecker was able to produce fine work in number theory and algebra only by temporarily forgetting his own philosophy.

Kronecker died on December 29, 1891, at the age of 68.

For more information about Kronecker, visit:

http://www-groups.dcs.st-and.ac.uk/~history/

21

Algebraic Extensions

Banach once told me, "Good mathematicians see analogies between theorems or theories, the very best ones see analogies between analogies."

S. M. ULAM, *Adventures of a Mathematician*

CHARACTERIZATION OF EXTENSIONS

In Chapter 20, we saw that every element in the field $Q(\sqrt{2})$ has the particularly simple form $a + b\sqrt{2}$, where a and b are rational. On the other hand, the elements of $Q(\pi)$ have the more complicated form

$$(a_n\pi^n + a_{n-1}\pi^{n-1} + \cdots + a_0)/(b_m\pi^m + b_{m-1}\pi^{m-1} + \cdots + b_0),$$

where the a's and b's are rational. The fields of the first type have a great deal of structure. This structure is the subject of this chapter.

DEFINITION *Types of Extensions*

Let E be an extension field of a field F and let $a \in E$. We call *a al-gebraic over F* if a is the zero of some nonzero polynomial in $F[x]$. If a is not algebraic over F, it is called *transcendental over F.* An extension E of F is called an *algebraic* extension of F if every element of E is algebraic over F. If E is not an algebraic extension of F, it is called a *transcendental* extension of F. An extension of F of the form $F(a)$ is called a *simple* extension of F.

Leonhard Euler used the term *transcendental* for numbers that are not algebraic because "they transcended the power of algebraic methods." Although Euler made this distinction in 1744, it wasn't until 1844 that the existence of transcendental numbers over Q was proved by Joseph Liouville. Charles Hermite proved that e is transcendental over Q in 1873, and Lindemann showed that π is transcendental over Q in 1882. To this day, it is not known whether $\pi + e$ is transcendental over Q. With a precise definition of "almost all," it can be shown that almost all real numbers are transcendental over Q.

Theorem 21.1 shows why we make the distinction between elements that are algebraic over a field and elements that are transcendental over a field. Recall that $F(x)$ is the field of quotients of $F[x]$; that is,

$$F(x) = \{f(x)/g(x) \mid f(x), g(x) \in F[x], g(x) \neq 0\}.$$

Theorem 21.1   Characterization of Extensions

Let E be an extension field of the field F and let $a \in E$. If a is transcendental over F, then $F(a) \approx F(x)$. If a is algebraic over F, then $F(a) \approx F[x]/\langle p(x)\rangle$, where $p(x)$ is a polynomial in $F[x]$ of minimum degree such that $p(a) = 0$. Moreover, $p(x)$ is irreducible over F.

PROOF.   Consider the homomorphism $\phi:F[x] \rightarrow F(a)$ given by $f(x) \rightarrow f(a)$. If a is transcendental over F, then Ker $\phi = \{0\}$, and so we may extend ϕ to an isomorphism $\overline{\phi}:F(x) \rightarrow F(a)$ by defining $\overline{\phi}(f(x)/g(x)) = f(a)/g(a)$.

If a is algebraic over F, then Ker $\phi \neq \{0\}$; and, by Theorem 16.4, there is a polynomial $p(x)$ in $F[x]$ such that Ker $\phi = \langle p(x)\rangle$ and $p(x)$ has minimum degree among all nonzero elements of Ker ϕ. Thus, $p(a) = 0$ and, since $p(x)$ is a polynomial of minimum degree with this property, it is irreducible over F.   ■ ■

The proof of Theorem 21.1 can readily be adapted to yield the next two results also. The details are left to the reader (see Exercise 1).

Theorem 21.2   Uniqueness Property

If a is algebraic over a field F, then there is a unique monic irreducible polynomial $p(x)$ in $F[x]$ such that $p(a) = 0$.

The polynomial with the property specified in Theorem 21.2 is called the *minimal polynomial for a over F.*

***Theorem 21.3*   Divisibility Property**

> *Let a be algebraic over F, and let p(x) be the minimal*
> *polynomial for a over F. If f(x) ∈ F[x] and f(a) = 0, then*
> *p(x) divides f(x) in F[x].*

If E is an extension field of F, we may view E as a vector space over F (that is, the elements of E are the vectors and the elements of F are the scalars). We are then able to use such notions as dimension and basis in our discussion.

FINITE EXTENSIONS

DEFINITION   *Degree of an Extension*

Let E be an extension field of a field F. We say that E *has degree n over F* and write $[E:F] = n$ if E has dimension n as a vector space over F. If $[E:F]$ is finite, E is called a *finite extension* of F; otherwise, we say that E is an *infinite extension* of F.

Figure 21.1 illustrates a convenient method of depicting the degree of a field extension over a field.

EXAMPLE 1   The field of complex numbers has degree 2 over the reals, since $\{1, i\}$ is a basis. The field of complex numbers is an infinite extension of the rationals.   ◆

EXAMPLE 2   If a is algebraic over F and its minimal polynomial over F has degree n, then, by Theorem 20.3, we know that $\{1, a, \ldots, a^{n-1}\}$ is a basis for $F(a)$ over F; and, therefore, $[F(a):F] = n$. In this case, we say that a has *degree n over F*.   ◆

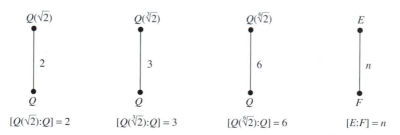

Figure 21.1

Theorem 21.4   Finite Implies Algebraic

> *If E is a finite extension of F, then E is an algebraic extension of F.*

PROOF.   Suppose that $[E:F] = n$ and $a \in E$. Then the set $\{1, a, \ldots, a^n\}$ is linearly dependent over F; that is, there are elements $c_0, c_1, \ldots, c_n$ in F, not all zero, such that

$$c_n a^n + c_{n-1} a^{n-1} + \cdots + c_1 a + c_0 = 0.$$

Clearly, then, a is a zero of the nonzero polynomial

$$f(x) = c_n x^n + c_{n-1} x^{n-1} + \cdots + c_1 x + c_0. \qquad \blacksquare\blacksquare$$

The converse of Theorem 21.4 is not true, for otherwise, the degrees of the elements of every algebraic extension of E over F would be bounded. But $Q(\sqrt{2}, \sqrt[3]{2}, \sqrt[4]{2}, \ldots)$ is an algebraic extension of Q that contains elements of every degree over Q (see Exercise 3).

The next theorem is the field theory counterpart of Lagrange's Theorem for finite groups. Like all counting theorems, it has far-reaching consequences.

Theorem 21.5 $[K:F] = [K:E][E:F]$

> *Let K be a finite extension field of the field E and let E be a finite extension field of the field F. Then K is a finite extension field of F and $[K:F] = [K:E][E:F]$.*

PROOF.   Let $X = \{x_1, x_2, \ldots, x_n\}$ be a basis for K over E, and let $Y = \{y_1, y_2, \ldots, y_m\}$ be a basis for E over F. It suffices to prove that

$$YX = \{y_j x_i \mid 1 \leq j \leq m, 1 \leq i \leq n\}$$

is a basis for K over F. To do this, let $a \in K$. Then there are elements $b_1, b_2, \ldots, b_n \in E$ such that

$$a = b_1 x_1 + b_2 x_2 + \cdots + b_n x_n.$$

And, for each $i = 1, \ldots, n$, there are elements $c_{i1}, c_{i2}, \ldots, c_{im} \in F$ such that

$$b_i = c_{i1} y_1 + c_{i2} y_2 + \cdots + c_{im} y_m.$$

Thus,

$$a = \sum_{i=1}^{n} b_i x_i = \sum_{i=1}^{n} \left(\sum_{j=1}^{m} c_{ij} y_j \right) x_i = \sum_{i,j} c_{ij}(y_j x_i).$$

This proves that YX spans K over F.

Now suppose there are elements c_{ij} in F such that

$$0 = \sum_{i,j} c_{ij}(y_j x_i) = \sum_{i} \sum_{j} (c_{ij} y_j) x_i.$$

Then, since each $c_{ij} y_j \in E$ and X is a basis for K over E, we have

$$\sum_{j} c_{ij} y_j = 0$$

for each i. But each $c_{ij} \in F$ and Y is a basis for E over F, so each $c_{ij} = 0$. This proves that the set YX is linearly independent over F. ∎

Using the fact that for any field extension L of a field J, $[L{:}J] = n$ if and only if L is isomorphic to J^n as vector spaces (see Exercise 29), we may give a concise conceptual proof of Theorem 21.5, as follows. Let $[K{:}E] = n$ and $[E{:}F] = m$. Then $K \approx E^n$ and $E \approx F^m$, so that $K \approx E^n \approx (F^m)^n \approx F^{mn}$. Thus, $[K{:}F] = mn$.

The content of Theorem 21.5 can be pictured as in Figure 21.2. Examples 3, 4, and 5 show how Theorem 21.5 is often utilized.

EXAMPLE 3   Since $\{1, \sqrt{3}\}$ is a basis for $Q(\sqrt{3}, \sqrt{5})$ over $Q(\sqrt{5})$ and $\{1, \sqrt{5}\}$ is a basis for $Q(\sqrt{5})$ over Q, the proof of Theorem 21.5 shows that $\{1, \sqrt{3}, \sqrt{5}, \sqrt{15}\}$ is a basis for $Q(\sqrt{3}, \sqrt{5})$ over Q. (See Figure 21.3.) ◆

EXAMPLE 4   Consider $Q(\sqrt[3]{2}, \sqrt[4]{3})$. Then $[Q(\sqrt[3]{2}, \sqrt[4]{3}) : Q] = 12$. For, clearly, $[Q(\sqrt[3]{2}, \sqrt[4]{3}) : Q] = [Q(\sqrt[3]{2}, \sqrt[4]{3}) : Q(\sqrt[3]{2})] [Q(\sqrt[3]{2}) : Q]$ and $[Q(\sqrt[3]{2}, \sqrt[4]{3}) : Q] = [Q(\sqrt[3]{2}, \sqrt[4]{3}) : Q(\sqrt[4]{3})] [Q(\sqrt[4]{3}) : Q]$ show that both $3 = [Q(\sqrt[3]{2}) : Q]$ and $4 = [Q(\sqrt[4]{3}) : Q]$ divide $[Q(\sqrt[3]{2}, \sqrt[4]{3}) : Q]$. Thus, $[Q(\sqrt[3]{2}, \sqrt[4]{3}) : Q] \geq 12$. On the other hand, $[Q(\sqrt[3]{2}, \sqrt[4]{3}) : Q(\sqrt[3]{2})]$ is at most 4, since $\sqrt[4]{3}$ is a zero of $x^4 - 3 \in Q(\sqrt[3]{2})[x]$. Therefore, $[Q(\sqrt[3]{2}, \sqrt[4]{3}) : Q] = [Q(\sqrt[3]{2}, \sqrt[4]{3}) : Q(\sqrt[3]{2})][Q(\sqrt[3]{2}) : Q] \leq 4 \cdot 3 = 12$. (See Figure 21.4.) ◆

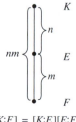

$$[K{:}F] = [K{:}E][E{:}F]$$

Figure 21.2

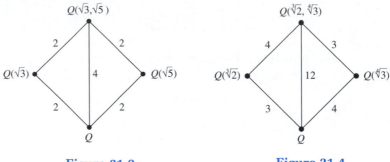

Figure 21.3   **Figure 21.4**

Theorem 21.5 can sometimes be used to show that a field does not contain a particular element.

EXAMPLE 5   Recall from Example 7 in Chapter 17 that $h(x) = 15x^4 - 10x^2 + 9x + 21$ is irreducible over Q. Let β be a zero of $h(x)$ in some extension of Q. Then, even though we don't know what β is, we can still prove that $\sqrt[3]{2}$ is not an element of $Q(\beta)$. For, if so, then $Q \subset Q(\sqrt[3]{2}) \subseteq Q(\beta)$ and $4 = [Q(\beta) : Q] = [Q(\beta) : Q(\sqrt[3]{2})][Q(\sqrt[3]{2}) : Q]$ implies that 3 divides 4. Notice that this argument cannot be used to show that $\sqrt{2}$ is not contained in $Q(\beta)$.   ◆

EXAMPLE 6   Consider $Q(\sqrt{3}, \sqrt{5})$. We claim that $Q(\sqrt{3}, \sqrt{5}) = Q(\sqrt{3} + \sqrt{5})$. The inclusion $Q(\sqrt{3} + \sqrt{5}) \subseteq Q(\sqrt{3}, \sqrt{5})$ is clear. Now note that since $(\sqrt{3} + \sqrt{5})^{-1} = \dfrac{1}{\sqrt{3} + \sqrt{5}} \cdot \dfrac{\sqrt{3} - \sqrt{5}}{\sqrt{3} - \sqrt{5}} = -\dfrac{1}{2}(\sqrt{3} - \sqrt{5})$, we know that $\sqrt{3} - \sqrt{5}$ belongs to $Q\,(\sqrt{3} + \sqrt{5})$. It follows that $[(\sqrt{3} + \sqrt{5}) + (\sqrt{3} - \sqrt{5})]/2 = \sqrt{3}$ and $[(\sqrt{3} + \sqrt{5}) - (\sqrt{3} - \sqrt{5})]/2 = \sqrt{5}$ both belong to $Q(\sqrt{3} + \sqrt{5})$, and therefore $Q(\sqrt{3}, \sqrt{5}) \subseteq Q(\sqrt{3} + \sqrt{5})$.   ◆

EXAMPLE 7   It follows from Example 6 and Theorem 20.3 that the minimal polynomial for $\sqrt{3} + \sqrt{5}$ over Q has degree 4. How can we find this polynomial? We begin with $x = \sqrt{3} + \sqrt{5}$. Then $x^2 = 3 + 2\sqrt{15} + 5$. From this we obtain $x^2 - 8 = 2\sqrt{15}$ and, by squaring both sides, $x^4 - 16x + 64 = 60$. Thus, $\sqrt{3} + \sqrt{5}$ is a zero of $x^4 - 16x + 4$. We know that this is the minimal polynomial of $\sqrt{3} + \sqrt{5}$ over Q since it is monic and has degree 4.   ◆

Example 6 shows that an extension obtained by adjoining two elements to a field can sometimes be obtained by adjoining a single element to the field. Our next theorem shows that, under certain conditions, this can always be done.

Theorem 21.6 **Primitive Element Theorem (Steinitz, 1910)**

> *If F is a field of characteristic 0, and a and b are algebraic over F, then there is an element c in F(a, b) such that F(a,b) = F(c).*

PROOF. Let $p(x)$ and $q(x)$ be the minimal polynomials over F for a and b, respectively. In some extension K of F, let $a_1, a_2, \ldots, a_m$ and $b_1, b_2, \ldots, b_n$ be the distinct zeros of $p(x)$ and $q(x)$, respectively, where $a = a_1$ and $b = b_1$. Among the infinitely many elements of F, choose an element d not equal to $(a_i - a)/(b - b_j)$ for all $i \geq 1$ and all $j > 1$. In particular, $a_i \neq a + d(b - b_j)$ for $j > 1$.

We shall show that $c = a + db$ has the property that $F(a, b) = F(c)$. Certainly, $F(c) \subseteq F(a, b)$. To verify that $F(a, b) \subseteq F(c)$, it suffices to prove that $b \in F(c)$, for then b, c, and d belong to $F(c)$ and $a = c - bd$. Consider the polynomials $q(x)$ and $r(x) = p(c - dx)$ [that is, $r(x)$ is obtained by substituting $c - dx$ for x in $p(x)$] over $F(c)$. Since both $q(b) = 0$ and $r(b) = p(c - db) = p(a) = 0$, both $q(x)$ and $r(x)$ are divisible by the minimal polynomial $s(x)$ for b over $F(c)$ (see Theorem 21.3). Because $s(x) \in F(c)[x]$, we may complete the proof by proving that $s(x) = x - b$. Since $s(x)$ is a common divisor of $q(x)$ and $r(x)$, the only possible zeros of $s(x)$ in K are the zeros of $q(x)$ that are also zeros of $r(x)$. But $r(b_j) = p(c - db_j) = p(a + db - db_j) = p(a + d(b - b_j))$ and d was chosen such that $a + d(b - b_j) \neq a_i$ for $j > 1$. It follows that b is the only zero of $s(x)$ in $K[x]$ and, therefore, $s(x) = (x - b)^u$. Since $s(x)$ is irreducible and F has characteristic 0, Theorem 20.6 guarantees that $u = 1$. ∎

In the terminology introduced earlier, it follows from Theorem 21.6 and induction that any finite extension of a field of characteristic 0 is a simple extension. An element a with the property that $E = F(a)$ is called a *primitive element* of E.

PROPERTIES OF ALGEBRAIC EXTENSIONS

Theorem 21.7 **Algebraic over Algebraic Is Algebraic**

> *If K is an algebraic extension of E and E is an algebraic extension of F, then K is an algebraic extension of F.*

PROOF. Let $a \in K$. It suffices to show that a belongs to some finite extension of F. Since a is algebraic over E, we know that a is the zero of some irreducible polynomial in $E[x]$, say, $p(x) = b_n x^n + \cdots + b_0$. Now we construct a tower of field extensions of F, as follows:

$$F_0 = F(b_0),$$

$$F_1 = F_0(b_1), \ldots, F_n = F_{n-1}(b_n).$$

In particular,

$$F_n = F(b_0, b_1, \ldots, b_n),$$

so that $p(x) \in F_n[x]$. Thus, $[F_n(a){:}F_n] = n$; and, because each b_i is algebraic over F, we know that each $[F_{i+1}{:}F_i]$ is finite. So,

$$[F_n(a){:}F] = [F_n(a){:}F_n][F_n{:}F_{n-1}] \cdots [F_1{:}F_0][F_0{:}F]$$

is finite. (See Figure 21.5.)   ■ ■

Corollary   **Subfield of Algebraic Elements**

> *Let E be an extension field of the field F. Then the set of all elements of E that are algebraic over F is a subfield of E.*

PROOF.   Suppose that $a, b \in E$ are algebraic over F and $b \neq 0$. To show that $a + b$, $a - b$, ab, and a/b are algebraic over F, it suffices to show that $[F(a, b){:}F]$ is finite, since each of these four elements belongs to $F(a, b)$. But note that

$$[F(a, b){:}F] = [F(a, b){:}F(b)][F(b){:}F].$$

Also, since a is algebraic over F, it is certainly algebraic over $F(b)$. Thus, both $[F(a, b){:}F(b)]$ and $[F(b){:}F]$ are finite.   ■ ■

For any extension E of a field F, the subfield of E of the elements that are algebraic over F is called the *algebraic closure of F in E*.

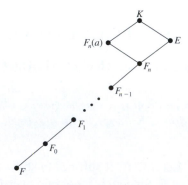

Figure 21.5

One might wonder if there is such a thing as a maximal algebraic extension of a field *F*—that is, whether there is an algebraic extension *E* of *F* that has no proper algebraic extensions. For such an *E* to exist, it is necessary that every polynomial in *E*[*x*] splits in *E*. Otherwise, it follows from Kronecker's Theorem that *E* would have a proper algebraic extension. This condition is also sufficient. If every member of *E*[*x*] splits in *E*, and *K* is an algebraic extension of *E*, then every member of *K* is a zero of some element of *E*[*x*]. But the zeros of elements of *E*[*x*] are in *E*. A field that has no proper algebraic extension is called *algebraically closed*. In 1910, Ernst Steinitz proved that every field *F* has a unique (up to isomorphism) algebraic extension that is algebraically closed. This field is called the *algebraic closure of F*. A proof of this result requires a sophisticated set theory background.

In 1799, Gauss, at the age of 22, proved that **C** is algebraically closed. This fact was considered so important at the time that it was called "The Fundamental Theorem of Algebra." Over a 50-year period, Gauss found three additional proofs of the Fundamental Theorem. Today more than 100 proofs exist. In view of the ascendancy of abstract algebra in this century, a more appropriate phrase for Gauss's result would be "The Fundamental Theorem of Classical Algebra."

EXERCISES ∎∎

It matters not what goal you seek
Its secret here reposes:
You've got to dig from week to week
To get Results or Roses.

EDGAR GUEST

1. Prove Theorem 21.2 and Theorem 21.3.

2. Let *E* be the algebraic closure of *F*. Show that every polynomial in *F*[*x*] splits in *E*.

3. Prove that $Q(\sqrt{2}, \sqrt[3]{2}, \sqrt[4]{2}, \ldots)$ is an algebraic extension of *Q* but not a finite extension of *Q*. (This exercise is referred to in this chapter.)

4. Let *E* be an algebraic extension of *F*. If every polynomial in *F*[*x*] splits in *E*, show that *E* is algebraically closed.

5. Suppose that *F* is a field and every irreducible polynomial in *F*[*x*] is linear. Show that *F* is algebraically closed.

6. Suppose that $f(x)$ and $g(x)$ are irreducible over *F* and deg $f(x)$ and deg $g(x)$ are relatively prime. If *a* is a zero of $f(x)$ in some extension of *F*, show that $g(x)$ is irreducible over $F(a)$.

7. Let *a* and *b* belong to *Q* with $b \neq 0$. Show that $Q(\sqrt{a}) = Q(\sqrt{b})$ if and only if there exists some $c \in Q$ such that $a = bc^2$.

8. Find the degree and a basis for $Q(\sqrt{3} + \sqrt{5})$ over $Q(\sqrt{15})$. Find the degree and a basis for $Q(\sqrt{2}, \sqrt[3]{2}, \sqrt[4]{2})$ over Q.

9. Suppose that E is an extension of F of prime degree. Show that, for every a in E, $F(a) = F$ or $F(a) = E$.

10. Let a be a complex number that is algebraic over Q and let $p(x)$ denote the minimal polynomial for a over Q. Show that $\sqrt{a}$ is algebraic over Q.

11. Suppose that E is an extension of F and $a, b \in E$. If a is algebraic over F of degree m, and b is algebraic over F of degree n, where m and n are relatively prime, show that $[F(a, b):F] = mn$.

12. Find an example of a field F and elements a and b from some extension field such that $F(a, b) \neq F(a)$, $F(a, b) \neq F(b)$, and $[F(a, b):F] < [F(a):F][F(b):F]$.

13. Let K be a field extension of F and let $a \in K$. Show that $[F(a):F(a^3)] \leq 3$. Find examples to illustrate that $[F(a):F(a^3)]$ can be 1, 2, or 3.

14. Find the minimal polynomial for $\sqrt{-3} + \sqrt{2}$ over Q.

15. Let K be an extension of F. Suppose that E_1 and E_2 are contained in K and are extensions of F. If $[E_1:F]$ and $[E_2:F]$ are both prime, show that $E_1 = E_2$ or $E_1 \cap E_2 = F$.

16. Find the minimal polynomial for $\sqrt[3]{2} + \sqrt[3]{4}$ over Q.

17. Let E be a finite extension of $\mathbf{R}$. Use the fact that $\mathbf{C}$ is algebraically closed to prove that $E = \mathbf{C}$ or $E = \mathbf{R}$.

18. Suppose that $[E:Q] = 2$. Show that there is an integer d such that $E = Q(\sqrt{d})$ and d is not divisible by the square of any prime.

19. Suppose that $p(x) \in F[x]$ and E is a finite extension of F. If $p(x)$ is irreducible over F and deg $p(x)$ and $[E:F]$ are relatively prime, show that $p(x)$ is irreducible over E.

20. Let E be a field extension of F. Show that $[E:F]$ is finite if and only if $E = F(a_1, a_2, \ldots, a_n)$, where $a_1, a_2, \ldots, a_n$ are algebraic over F.

21. If α and β are real numbers and α and β are transcendental over Q, show that either $\alpha\beta$ or $\alpha + \beta$ is also transcendental over Q.

22. Let $f(x)$ be a nonzero element of $F[x]$. If a belongs to some extension of F and $f(a)$ is algebraic over F, prove that a is algebraic over F.

23. Let $f(x) = ax^2 + bx + c \in Q[x]$. Find a primitive element for the splitting field for $f(x)$ over Q.

24. Find the splitting field for $x^4 - x^2 - 2$ over Z_3.

25. Let $f(x) \in F[x]$. If deg $f(x) = 2$ and a is a zero of $f(x)$ in some extension of F, prove that $F(a)$ is the splitting field for $f(x)$ over F.

26. Let a be a complex zero of $x^2 + x + 1$ over Q. Prove that $Q(\sqrt{a}) = Q(a)$.

27. If F is a field and the multiplicative group of nonzero elements of F is cyclic, prove that F is finite.

28. Let a be a complex number that is algebraic over Q and let r be a rational number. Show that a^r is algebraic over Q.

29. Prove that, if K is a field extension of F, then $[K:F] = n$ if and only if K is isomorphic to F^n as vector spaces. (See Exercise 27 in Chapter 19 for the appropriate definition. This exercise is referred to in this chapter.)

30. Let a be a positive real number and let n be an integer greater than 1. Prove or disprove that $[Q(a^{1/n}):Q] = n$.

31. Let a and b belong to some extension of F and let b be algebraic over F. Prove that $[F(a, b):F(a)] \leq [F(a, b):F]$.

32. Let $f(x)$ and $g(x)$ be irreducible polynomials over a field F and let a and b belong to some extension E of F. If a is a zero of $f(x)$ and b is a zero of $g(x)$, show that $f(x)$ is irreducible over $F(b)$ if and only if $g(x)$ is irreducible over $F(a)$.

33. Let β be a zero of $f(x) = x^5 + 2x + 4$ (see Example 8 in Chapter 17). Show that none of $\sqrt{2}, \sqrt[3]{2}, \sqrt[4]{2}$ belongs to $Q(\beta)$.

34. Prove that $Q(\sqrt{2}, \sqrt[3]{2}) = Q(\sqrt[6]{2})$.

35. Let a and b be rational numbers. Show that $Q(\sqrt{a}, \sqrt{b}) = Q(\sqrt{a} + \sqrt{b})$.

Suggested Readings

R. L. Roth, "On Extensions of Q by Square Roots," *American Mathematical Monthly* 78 (1971): 392–393.

In this paper, it is proved that if $p_1, p_2, \ldots, p_n$ are distinct primes, then $[Q(\sqrt{p_1}, \sqrt{p_2}, \ldots, \sqrt{p_n}):Q] = 2^n$.

Paul B. Yale, "Automorphisms of the Complex Numbers," *Mathematics Magazine* 39 (1966): 135–141.

This award-winning expository paper is devoted to various results on automorphisms of the complex numbers.

Irving Kaplansky

He got to the top of the heap by being a first-rate doer and expositor of algebra.

PAUL R. HALMOS, *I Have a Photographic Memory*

IRVING KAPLANSKY was born on March 22, 1917, in Toronto, Canada, a few years after his parents emigrated from Poland. Although his parents thought he would pursue a career in music, Kaplansky knew early on that mathematics was what he wanted to do. As an undergraduate at the University of Toronto, Kaplansky was a member of the winning team in the first William Lowell Putnam competition, a mathematical contest for United States and Canadian college students. Kaplansky received a B.A. degree from Toronto in 1938 and an M.A. in 1939. In 1939, he entered Harvard University for his doctorate as the first recipient of a Putnam Fellowship. After receiving his Ph.D. from Harvard in 1941, Kaplansky stayed on as Benjamin Peirce instructor until 1944. After one year at Columbia University, he went to the University of Chicago, where he remained until his retirement in 1984. He then became the director of the Mathematical Sciences Research Institute at the University of California, Berkeley.

Kaplansky's interests are broad, including areas such as ring theory, group theory, field theory, Galois theory, ergodic theory, algebras, metric spaces, number theory, statistics, and probability. He is also regarded as a first-rate mathematical stylist.

Among the many honors received by Kaplansky are election to both the National Academy of Sciences and the American Academy of Arts and Sciences, election to the presidency of the American Mathematical Society, and the 1989 Steele Prize for cumulative influence from the American Mathematical Society. The Steele Prize citation says, in part, ". . . he has made striking changes in mathematics and has inspired generations of younger mathematicians."

For more information about Kaplansky, visit:

http://www-groups.dcs.st-and.ac.uk/~history/

22 Finite Fields

This theory [of finite fields] is of considerable interest in its own right and it provides a particularly beautiful example of how the general theory of the preceding chapters fits together to provide a rather detailed description of all finite fields.

RICHARD A. DEAN, *Elements of Abstract Algebra*

CLASSIFICATION OF FINITE FIELDS

In this, our final chapter on field theory, we take up one of the most beautiful and important areas of abstract algebra—finite fields. Finite fields were first introduced by Galois in 1830 in his proof of the unsolvability of the general quintic equation. When Cayley invented matrices a few decades later, it was natural to investigate groups of matrices over finite fields. To this day, matrix groups over finite fields are among the most important classes of groups. In the past 50 years, there have been important applications of finite fields in computer science, coding theory, information theory, and cryptography. But, besides the many uses of finite fields in pure and applied mathematics, there is yet another good reason for studying them. They are just plain fun!

The most striking fact about finite fields is the restricted nature of their order and structure. We have already seen that every finite field has prime-power order (Exercise 43 in Chapter 13). A converse of sorts is also true.

Theorem 22.1 Classification of Finite Fields

For each prime p and each positive integer n, there is, up to isomorphism, a unique finite field of order p^n.

PROOF.   Consider the splitting field E of $f(x) = x^{p^n} - x$ over Z_p. We will show that $|E| = p^n$. Since $f(x)$ splits in E, we know that $f(x)$ has exactly p^n zeros in E, counting multiplicity. Moreover, by Theorem 20.5, every zero of $f(x)$ has multiplicity 1. Thus, $f(x)$ has p^n distinct zeros in E. On the other hand, the set of zeros of $f(x)$ in E is closed under addition, subtraction, multiplication, and division by nonzero elements (see Exercise 31), so that the set of zeros of $f(x)$ is itself a field extension of Z_p in which $f(x)$ splits. Thus, the set of zeros of $f(x)$ is E and, therefore, $|E| = p^n$.

To show that there is a unique field for each prime-power, suppose that K is any field of order p^n. Then K has a subfield isomorphic to Z_p (generated by 1), and, because the nonzero elements of K form a multiplicative group of order $p^n - 1$, every element of K is a zero of $f(x) = x^{p^n} - x$ (see Exercise 21). So, K must be a splitting field for $f(x)$ over Z_p. By the corollary to Theorem 20.4, there is only one such field up to isomorphism.   ■■

The existence portion of Theorem 22.1 appeared in the works of Galois and Gauss in the first third of the nineteenth century. Rigorous proofs were given by Dedekind in 1857 and by Jordan in 1870 in his classic book on group theory. The uniqueness portion of the theorem was proved by E. H. Moore in an 1893 paper concerning finite groups. The mathematics historian E. T. Bell once said that this paper by Moore marked the beginning of abstract algebra in America.

Because there is only one field for each prime-power p^n, we may unambiguously denote it by GF(p^n), in honor of Galois, and call it the *Galois field of order p^n.*

STRUCTURE OF FINITE FIELDS

The next theorem tells us the additive and multiplicative group structure of a field of order p^n.

Theorem 22.2   **Structure of Finite Fields**

> As a group under addition, GF(p^n) is isomorphic to
>
> $$\underbrace{Z_p \oplus Z_p \oplus \cdots \oplus Z_p.}_{n \text{ factors}}$$
>
> As a group under multiplication, the set of nonzero elements of GF(p^n) is isomorphic to Z_{p^n-1} (and is, therefore, cyclic).

PROOF. Since GF(p^n) has characteristic p (Theorem 13.3), we know that $px = 0$ for all x in GF(p^n). Thus, every nonzero element of GF(p^n) has additive order p. Clearly, then, under addition, GF(p^n) is isomorphic to a direct product of n copies of Z_p.

To see that the multiplicative group GF(p^n)* of nonzero elements of GF(p^n) is cyclic, we first note, by Exercise 11 in Chapter 11, that it is isomorphic to a direct product of the form $Z_{n_1} \oplus Z_{n_2} \oplus \cdots \oplus Z_{n_k}$, where each n_{i+1} divides n_i. So, for any element $a = (a_1, a_2, \ldots, a_k)$ in this product, we have

$$a^{n_1} = (n_1 a_1, n_1 a_2, \ldots, n_1 a_k) = (0, 0, \ldots, 0).$$

(Remember, the operation is componentwise addition.) Thus, the polynomial $x^{n_1} - 1$ has $p^n - 1$ zeros in GF(p^n). Since the number of zeros of a polynomial over a field cannot exceed the degree of the polynomial (Corollary 3 of Theorem 16.2), we know that $p^n - 1 \leq n_1$. On the other hand, since GF(p^n)* has a subgroup isomorphic to Z_{n_1}, we also have $n_1 \leq p^n - 1$. It follows, then, that GF(p^n)* is isomorphic to Z_{p^n-1}. ∎

Since $Z_p \oplus Z_p \oplus \cdots \oplus Z_p$ is a vector space over Z_p with $\{(1, 0, \ldots, 0), (0, 1, 0, \ldots, 0), \ldots, (0, 0, \ldots, 1)\}$ as a basis, we have the following useful and aesthetically appealing formula.

Corollary 1

$$[\text{GF}(p^n){:}\text{GF}(p)] = n$$

Corollary 2   GF(p^n) Contains an Element of Degree n

> *Let a be a generator of the group of nonzero elements of GF(p^n) under multiplication. Then a is algebraic over GF(p) of degree n.*

PROOF.   Observe that $[\text{GF}(p)(a){:}\text{GF}(p)] = [\text{GF}(p^n){:}\text{GF}(p)] = n$. ∎

EXAMPLE 1   Let's examine the field GF(16) in detail. Since $x^4 + x + 1$ is irreducible over Z_2, we know that

$$\text{GF}(16) \approx \{ax^3 + bx^2 + cx + d + \langle x^4 + x + 1\rangle \mid a, b, c, d \in Z_2\}.$$

Thus, we may think of GF(16) as the set

$$F = \{ax^3 + bx^2 + cx + d \mid a, b, c, d \in Z_2\},$$

where addition is done as in $Z_2[x]$, but multiplication is done modulo $x^4 + x + 1$. For example,

$$(x^3 + x^2 + x + 1)(x^3 + x) = x^3 + x^2,$$

since the remainder upon dividing

$$(x^3 + x^2 + x + 1)(x^3 + x) = x^6 + x^5 + x^2 + x$$

by $x^4 + x + 1$ in $Z_2[x]$ is $x^3 + x^2$. An easier way to perform the same calculation is to observe that in this context $x^4 + x + 1$ is 0, so

$$x^4 = -x - 1 = x + 1,$$
$$x^5 = x^2 + x,$$

and

$$x^6 = x^3 + x^2.$$

Thus,

$$x^6 + x^5 + x^2 + x = (x^3 + x^2) + (x^2 + x) + x^2 + x = x^3 + x^2.$$

Another way to simplify the multiplication process is to make use of the fact that the nonzero elements of GF(16) form a cyclic group of order 15. To take advantage of this, we must first find a generator of this group. Since any element F^* must have a multiplicative order that divides 15, all we need to do is find an element α in F^* such that $\alpha^3 \neq 1$ and $\alpha^5 \neq 1$. Obviously, x has these properties. So, we may think of GF(16) as the set $\{0, 1, x, x^2, \ldots, x^{14}\}$, where $x^{15} = 1$. This makes multiplication in F trivial, but, unfortunately, it makes addition more difficult. For example, $x^{10} \cdot x^7 = x^{17} = x^2$, but what is $x^{10} + x^7$? So, we face a dilemma. If we write the elements of F^* in the additive form $ax^3 + bx^2 + cx + d$, then addition is easy and multiplication is hard. On the other hand, if we write the elements of F^* in the multiplicative form x^i, then multiplication is easy and addition is hard. Can we have the best of both? Yes, we can. All we need to do is use the relation $x^4 = x + 1$ to make a two-way conversion table, as in Table 22.1.

So, we see from Table 22.1 that

$$x^{10} + x^7 = (x^2 + x + 1) + (x^3 + x + 1)$$
$$= x^3 + x^2 = x^6$$

and

$$(x^3 + x^2 + 1)(x^3 + x^2 + x + 1) = x^{13} \cdot x^{12}$$
$$= x^{25} = x^{10} = x^2 + x + 1. \qquad \blacklozenge$$

Don't be misled by the preceding example into believing that the element x is always a generator for the cyclic multiplicative group of nonzero elements. It is not. (See Exercise 13.) Although any two

TABLE 22.1 Conversion Table for Addition and Multiplication in GF(16)

Multiplicative Form to Additive Form		Additive Form to Multiplicative Form	
1	1	1	1
x	x	x	x
x^2	x^2	$x + 1$	x^4
x^3	x^3	x^2	x^2
x^4	$x + 1$	$x^2 + x$	x^5
x^5	$x^2 + x$	$x^2 + 1$	x^8
x^6	$x^3 + x^2$	$x^2 + x + 1$	x^{10}
x^7	$x^3 + x + 1$	x^3	x^3
x^8	$x^2 + 1$	$x^3 + x^2$	x^6
x^9	$x^3 + x$	$x^3 + x$	x^9
x^{10}	$x^2 + x + 1$	$x^3 + 1$	x^{14}
x^{11}	$x^3 + x^2 + x$	$x^3 + x^2 + x$	x^{11}
x^{12}	$x^3 + x^2 + x + 1$	$x^3 + x^2 + 1$	x^{13}
x^{13}	$x^3 + x^2 + 1$	$x^3 + x + 1$	x^7
x^{14}	$x^3 + 1$	$x^3 + x^2 + x + 1$	x^{12}

irreducible polynomials of the same degree over $Z_p[x]$ yield isomorphic fields, some are better than others for computational purposes.

EXAMPLE 2 Consider $f(x) = x^3 + x^2 + 1$ over Z_2. We will show how to write $f(x)$ as the product of linear factors. Let $F = Z_2[x]/\langle f(x) \rangle$ and let a be a zero of $f(x)$ in F. Then $|F| = 8$ and $|F^*| = 7$. So, by Corollary 2 to Theorem 7.1, we know that $|a| = 7$. Thus, by Theorem 20.3,

$$F = \{0, 1, a, a^2, a^3, a^4, a^5, a^6\}$$
$$= \{0, 1, a, a + 1, a^2, a^2 + a + 1, a^2 + 1, a^2 + a\}.$$

We know that a is one zero of $f(x)$, and we can test the other elements of F to see if they are zeros. We can simplify the calculations by using the fact that $a^3 + a^2 + 1 = 0$ to make a conversion table for the two forms of writing the elements of F. Because char $F = 2$, we know that $a^3 = a^2 + 1$. Then,

$$a^4 = a^3 + a = (a^2 + 1) + a = a^2 + a + 1,$$
$$a^5 = a^3 + a^2 + a = (a^2 + 1) + a^2 + a = a + 1,$$
$$a^6 = a^2 + a,$$
$$a^7 = 1.$$

Now let's see if a^2 is a zero of $f(x)$.

$$f(a^2) = (a^2)^3 + (a^2)^2 + 1 = a^6 + a^4 + 1$$
$$= (a^2 + a) + (a^2 + a + 1) + 1 = 0.$$

So, yes, it is. Next we try a^3.

$$f(a^3) = (a^3)^3 + (a^3)^2 + 1 = a^9 + a^6 + 1$$
$$= a^2 + (a^2 + a) + 1 = a + 1 \neq 0.$$

Now a^4.

$$f(a^4) = (a^4)^3 + (a^4)^2 + 1 = a^{12} + a^8 + 1$$
$$= a^5 + a + 1 = (a + 1) + a + 1 = 0.$$

So, a^4 is our remaining zero. Thus, $f(x) = (x - a)(x - a^2)(x - a^4) = (x + a)(x + a^2)(x + a^4)$, since char $F = 2$. ◆

SUBFIELDS OF A FINITE FIELD

Theorem 22.1 gives us a complete description of all finite fields. The following theorem gives us a complete description of all the subfields of a finite field. Notice the close analogy between this theorem and Theorem 4.3, which describes the subgroups of a finite cyclic group.

Theorem 22.3   Subfields of a Finite Field

> *For each divisor m of n, $\mathrm{GF}(p^n)$ has a unique subfield of order p^m. Moreover, these are the only subfields of $\mathrm{GF}(p^n)$.*

PROOF.   To show the existence portion of the theorem, suppose that m divides n. Then, since

$$p^n - 1 = (p^m - 1)(p^{n-m} + p^{n-2m} + \cdots + p^m + 1),$$

we see that $p^m - 1$ divides $p^n - 1$. For simplicity, write $p^n - 1 = (p^m - 1)t$. Let $K = \{x \in \mathrm{GF}(p^n) | x^{p^m} = x\}$. We leave it as an easy exercise for the reader to show that K is a subfield of $\mathrm{GF}(p^n)$. (Exercise 19). Since the polynomial $x^{p^m} - x$ has at most p^m zeros in $\mathrm{GF}(p^n)$, we have $|K| \leq p^m$. Let $\langle a \rangle = \mathrm{GF}(p^n)^*$. Then $|a^t| = p^m - 1$ and since $(a^t)^{p^m - 1} = 1$, it follows that $a^t \in K$. So, K is a subfield of $\mathrm{GF}(p^n)$ of order p^m.

The uniqueness portion of the theorem follows from the observation that if $\mathrm{GF}(p^n)$ had two distinct subfields of order p^m, then the polynomial $x^{p^m} - x$ would have more than p^m zeros in $\mathrm{GF}(p^n)$. This contradicts Corollary 3 of Theorem 16.2.

Finally, suppose that F is a subfield of GF(p^n). Then F is isomorphic to GF(p^m) for some m and, by Theorem 21.5,

$$n = [\text{GF}(p^n):\text{GF}(p)]$$
$$= [\text{GF}(p^n):\text{GF}(p^m)][\text{GF}(p^m):\text{GF}(p)]$$
$$= [\text{GF}(p^n):\text{GF}(p^m)]m.$$

Thus, m divides n.   ▨ ■

Theorems 22.2 and 22.3, together with Theorem 4.3, make the task of finding the subfields of a finite field a simple exercise in arithmetic.

EXAMPLE 3   Let F be the field of order 16 given in Example 1. Then there are exactly three subfields of F, and their orders are 2, 4, and 16. Obviously, the subfield of order 2 is $\{0, 1\}$ and the subfield of order 16 is F itself. To find the subfield of order 4, we merely observe that the three nonzero elements of this subfield must be the cyclic subgroup of $F^* = \langle x \rangle$ of order 3. So the subfield of order 4 is

$$\{0, 1, x^5, x^{10}\} = \{0, 1, x^2 + x, x^2 + x + 1\}. \qquad \blacklozenge$$

EXAMPLE 4   If F is a field of order $3^6 = 729$ and α is a generator of F^*, then the subfields of F are

1. $\text{GF}(3) = \{0\} \cup \langle \alpha^{364} \rangle = \{0, 1, 2\}$
2. $\text{GF}(9) = \{0\} \cup \langle \alpha^{91} \rangle$
3. $\text{GF}(27) = \{0\} \cup \langle \alpha^{28} \rangle$
4. $\text{GF}(729) = \{0\} \cup \langle \alpha \rangle. \qquad \blacklozenge$

EXAMPLE 5   The subfield lattice of GF(2^{24}) is

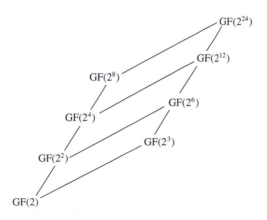

$\blacklozenge$

EXERCISES ▪▪

No pressure, no diamonds.

MARY CASE

1. Find [GF(729):GF(9)] and [GF(64):GF(8)].
2. If m divides n, show that $[GF(p^n):GF(p^m)] = n/m$.
3. Let K be a finite extension field of a finite field F. Show that there is an element a in K such that $K = F(a)$.
4. How many elements of the cyclic group GF(81)* are generators?
5. Let $f(x)$ be a cubic irreducible over Z_2. Prove that the splitting field of $f(x)$ over Z_2 has order 8.
6. Prove that the rings $Z_3[x]/\langle x^2 + x + 2 \rangle$ and $Z_3[x]/\langle x^2 + 2x + 2 \rangle$ are isomorphic.
7. Show that the *Frobenius mapping* ϕ:GF(p^n) $\to$ GF(p^n), given by $a \to a^p$, is a ring automorphism of order n (that is, ϕ^n is the identity mapping). (This exercise is referred to in Chapter 32.)
8. Determine the possible finite fields whose largest proper subfield is GF(2^5).
9. Prove that the degree of any irreducible factor of $x^8 - x$ over Z_2 is 1 or 3.
10. Find the smallest field that has exactly 6 subfields.
11. Show that x is a generator of the cyclic group $(Z_3[x]/\langle x^3 + 2x + 1 \rangle)$*.
12. Without actually calculating $|x|$, explain why x is a generator of the cyclic group $(Z_2[x]/\langle x^5 + x^3 + 1 \rangle)$*.
13. Show that x is not a generator of the cyclic group $(Z_3[x]/\langle x^3 + 2x + 2 \rangle)$*. Find one such generator.
14. If $f(x)$ is a cubic irreducible polynomial over Z_3, prove that either x or $2x$ is a generator for the cyclic group $Z_3[x]/\langle f(x) \rangle$*.
15. Prove the uniqueness portion of Theorem 22.3 using a group-theoretic argument.
16. Suppose that α and β belong to GF(81)*, with $|\alpha| = 5$ and $|\beta| = 16$. Show that $\alpha\beta$ is a generator of GF(81)*.
17. Construct a field of order 9 and carry out the analysis as in Example 1, including the conversion table.
18. Show that any finite subgroup of the multiplicative group of a field is cyclic.
19. Show that the set K in the proof of Theorem 22.3 is a subfield.
20. If $g(x)$ is irreducible over GF(p) and $g(x)$ divides $x^{p^n} - x$, prove that deg $g(x)$ divides n.
21. Use a purely group-theoretic argument to show that if F is a field of order p^n, then every element of F* is a zero of $x^{p^n} - x$. (This exercise is referred to in the proof of Theorem 22.1.)
22. Draw the subfield lattices of GF(3^{18}) and of GF(2^{30}).

23. How does the subfield lattice of GF(2^{30}) compare with the subfield lattice of GF(3^{30})?

24. If $p(x)$ is a polynomial in $Z_p[x]$ with no multiple zeros, show that $p(x)$ divides $x^{p^n} - x$ for some n.

25. Suppose that p is a prime and $p \neq 2$. Let a be a nonsquare in GF(p)—that is, a does not have the form b^2 for any b in GF(p). Show that a is a nonsquare in GF(p^n) if n is odd and that a is a square in GF(p^n) if n is even.

26. Let $f(x)$ be a cubic irreducible over Z_p, where p is a prime. Prove that the splitting field of $f(x)$ over Z_p has order p^3 or p^6.

27. Show that every element of GF(p^n) can be written in the form a^p for some unique a in GF(p^n).

28. Suppose that F is a field of order 1024 and $F^* = \langle \alpha \rangle$. List the elements of each subfield of F.

29. Suppose that F is a field of order 125 and $F^* = \langle \alpha \rangle$. Show that $\alpha^{62} = -1$.

30. Show that no finite field is algebraically closed.

31. Let E be the splitting field of $f(x) = x^{p^n} - x$ over Z_p. Show that the set of zeros of $f(x)$ in E is closed under addition, subtraction, multiplication, and division (by nonzero elements). (This exercise is referred to in the proof of Theorem 22.1.)

32. Suppose that L and K are subfields of GF(p^n). If L has p^s elements and K has p^t elements, how many elements does $L \cap K$ have?

33. Give an example to show that the mapping $a \rightarrow a^p$ need not be an automorphism for arbitrary fields of prime characteristic p.

COMPUTER EXERCISE ▬▬▬▬▬▬▬▬▬▬▬▬▬▬▬▬▬▬ ■■

Hardware: the parts of a computer that can be kicked.

JEFF PESIS

Software for the computer exercise in this chapter is available at the website:

http://www.d.umn.edu/~jgallian

1. This software tests cubic polynomials over Z_p for irreducibility. When a polynomial $f(x)$ is irreducible, the software finds a generator for the cyclic group of nonzero elements of the field $Z_p[x]/\langle f(x) \rangle$ and creates a conversion table for addition and multiplication similar to Table 22.1.

Suggested Reading

Judy L. Smith and J. A. Gallian, "Factoring Finite Factor Rings," *Mathematics Magazine* 58 (1985): 93–95.

This paper gives an algorithm for finding the group of units of the ring $F[x]/\langle g(x)^m \rangle$.

L. E. Dickson

One of the books [written by L. E. Dickson] is his major, three-volume History of the Theory of Numbers *which would be a life's work by itself for a more ordinary man.*

      A. A. ALBERT, *Bulletin of the American Mathematical Society*

LEONARD EUGENE DICKSON was born in Independence, Iowa, on January 22, 1874. In 1896, he received a Ph.D., the first Ph.D. to be awarded in mathematics at the University of Chicago. After spending a few years at the University of California and the University of Texas, he was appointed to the faculty at Chicago and remained there until his retirement in 1939.

Dickson was one of the most prolific mathematicians of the twentieth century, writing 267 research papers and 18 books. His principal interests were matrix groups, finite fields, algebra, and number theory.

Dickson had a disdainful attitude toward applicable mathematics; he would often say, "Thank God that number theory is unsullied by any applications." He also had a sense of humor. Dickson would often mention his honeymoon: "It was a great success," he said, "except that I only got two research papers written."

Dickson received many honors in his career. He was the first to be awarded the prize from the American Association for the Advancement of Science for the most notable contribution to the advancement of science, and the first to receive the Cole Prize in algebra from the American Mathematical Society. The University of Chicago has research instructorships named after him. Dickson died on January 17, 1954.

For more information about Dickson, visit:

http://www-groups.dcs.st-and.ac.uk/~history/

23 Geometric Constructions

At the age of eleven, I began Euclid. . . . This was one of the great events of my life, as dazzling as first love.
BERTRAND RUSSELL

HISTORICAL DISCUSSION OF GEOMETRIC CONSTRUCTIONS

The ancient Greeks were fond of geometric constructions. They were especially interested in constructions that could be achieved using only a straightedge without markings and a compass. They knew, for example, that any angle can be bisected, and they knew how to construct an equilateral triangle, a square, a regular pentagon, and a regular hexagon. But they did not know how to trisect every angle or how to construct a regular seven-sided polygon (heptagon). Another problem that they attempted was the duplication of the cube—that is, given any cube, they tried to construct a new cube having twice the volume of the given one using only an unmarked straightedge and a compass. Legend has it that the ancient Athenians were told by the oracle at Delos that a plague would end if they constructed a new altar to Apollo in the shape of a cube with double the volume of the old altar, which was also a cube. Besides "doubling the cube," the Greeks also attempted to "square the circle"—to construct a square with area equal to that of a given circle. They knew how to solve all these problems using other means, such as a compass and a straightedge with two marks, or an unmarked straightedge and a spiral, but they could not achieve any of the constructions with a compass and an unmarked straightedge alone. These problems vexed mathematicians for over 2000 years. The resolution of these perplexities was made possible when they were transferred from questions of geometry to questions of algebra in the nineteenth century.

The first of the famous problems of antiquity to be solved was that of the construction of regular polygons. It had been known since Euclid that regular polygons with a number of sides of the form 2^k, $2^k \cdot 3$, $2^k \cdot 5$, and $2^k \cdot 3 \cdot 5$ could be constructed, and it was believed that no others were possible. In 1796, while still a teenager, Gauss proved that the 17-sided regular polygon is constructible. In 1801, Gauss asserted that a regular polygon of n sides is constructible if and only if n has the form $2^k p_1 p_2 \cdots p_i$, where the p's are distinct primes of the form $2^{2^s} + 1$ and $k \geq 0$.

Thus, regular polygons with 3, 4, 5, 6, 8, 10, 12, 15, 16, 17, and 20 sides are possible to construct, whereas those with 7, 9, 11, 13, 14, 18, and 19 sides are not. How these constructions can be effected is another matter. One person spent 10 years trying to determine a way to construct the 65,537-sided polygon.

Gauss's result on the constructibility of regular n-gons eliminated another of the famous unsolved problems because the ability to trisect a 60° angle enables one to construct a regular 9-gon. Thus, there is no method for trisecting a 60° angle with an unmarked straightedge and a compass. In 1837, Wantzel proved that it was not possible to double the cube. The problem of the squaring of a circle resisted all attempts until 1882, when Ferdinand Lindemann proved that π is transcendental since, as we will show, all constructible numbers are algebraic.

CONSTRUCTIBLE NUMBERS

With the field theory we now have, it is an easy matter to solve the following problem: Given an unmarked straightedge, a compass, and a unit length, what other lengths can be constructed? To begin, we call a real number α *constructible* if, by means of an unmarked straightedge, a compass, and a line segment of length 1, we can construct a line segment of length $|\alpha|$ in a finite number of steps. It follows from plane geometry that if α and β ($\beta \neq 0$) are constructible numbers, then so are $\alpha + \beta$, $\alpha - \beta$, $\alpha \cdot \beta$, and α/β. (See the exercises for hints.) Thus, the set of constructible numbers contains Q and is a subfield of the real numbers. What we desire is an algebraic characterization of this field. To derive such a characterization, let F be any subfield of the reals. Call the subset $\{(x, y) \in R^2 \mid x, y \in F\}$ of the real plane the *plane of F*, call any line joining two points in the plane of F a *line in F*, and call any circle whose center is in the plane of F and whose radius is in F a *circle in F*. Then a line in F has an equation of the form

$$ax + by + c = 0, \qquad \text{where } a, b, c \in F,$$

and a circle in F has an equation of the form

$$x^2 + y^2 + ax + by + c = 0, \qquad \text{where } a, b, c \in F.$$

In particular, note that to find the point of intersection of a pair of lines in F or the points of intersection of a line in F and a circle in F, one need only solve a linear or quadratic equation in F. We now come to the crucial question. Starting with points in the plane of some field F, which points in the real plane can be obtained with an unmarked straightedge and a compass? Well, there are only three ways to construct points, starting with points in the plane of F.

1. Intersect two lines in F.
2. Intersect a circle in F and a line in F.
3. Intersect two circles in F.

In case 1, we do not obtain any new points, because two lines in F intersect in a point in the plane of F. In case 2, the point of intersection is the solution to either a linear equation in F or a quadratic equation in F. So, the point lies in the plane of F or in the plane of $F(\sqrt{\alpha})$, where $\alpha \in F$ and α is positive. In case 3, no new points are obtained, because, if the two circles are given by $x^2 + y^2 + ax + by + c = 0$ and $x^2 + y^2 + a'x + b'y + c' = 0$, then we have $(a - a')x + (b - b')y + (c - c') = 0$, which is a line in F. So, the points of intersection are in F.

It follows, then, that the only points in the real plane that can be constructed from the plane of a field F are those whose coordinates lie in fields of the form $F(\sqrt{\alpha})$, where $\alpha \in F$ and α is positive. Of course, we can start over with $F_1 = F(\sqrt{\alpha})$ and construct points whose coordinates lie in fields of the form $F_2 = F_1(\sqrt{\beta})$, where $\beta \in F_1$ and β is positive. Continuing in this fashion, we see that a real number c is constructible if and only if there is a series of fields $Q = F_1 \subseteq F_2 \subseteq \cdots \subseteq F_n \subseteq \mathbf{R}$ such that $F_{i+1} = F_i(\sqrt{\alpha_i})$, where $\alpha_i \in F_i$ and $c \in F_n$. Since $[F_{i+1}:F_i] = 1$ or 2, we see by Theorem 21.5 that if c is constructible, then $[Q(c):Q] = 2^k$ for some nonnegative integer k.

We now dispatch the problems that plagued the Greeks. Consider doubling the cube of volume 1. The enlarged cube would have an edge of length $\sqrt[3]{2}$. But $[Q(\sqrt[3]{2}):Q] = 3$, so such a cube cannot be constructed.

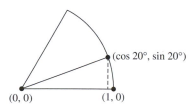

$(\cos 20°, \sin 20°)$

$(0, 0)$ $(1, 0)$

Figure 23.1

Next consider the possibility of trisecting a 60° angle. If it were possible to trisect an angle of 60°, then cos 20° would be constructible. (See Figure 23.1.) In particular, $[Q(\cos 20°):Q] = 2^k$ for some k. Now, using the trigonometric identity $\cos 3\theta = 4 \cos^3 \theta - 3 \cos \theta$, with $\theta = 20°$, we see that $1/2 = 4 \cos^3 20° - 3 \cos 20°$, so that cos 20° is a zero of $8x^3 - 6x - 1$. But, since $8x^3 - 6x - 1$ is irreducible over Q (see Exercise 13), we must also have $[Q(\cos 20°):Q] = 3$. This contradiction shows that trisecting a 60° angle is impossible.

The remaining problems are relegated to the reader as Exercises 14, 15, and 17.

ANGLE-TRISECTORS AND CIRCLE-SQUARERS

Down through the centuries, hundreds of people have claimed to have achieved one or more of the impossible constructions. In 1775, the Paris Academy, so overwhelmed with these claims, passed a resolution to no longer examine these claims or claims of machines purported to exhibit perpetual motion. Although it has been more than 100 years since the last of the constructions was shown to be impossible, there continues to be a steady parade of people who claim to have done one or more of them. Most of these people have heard that this is impossible but have refused to believe it. One person insisted that he could trisect any angle with a straightedge alone [2, p. 158]. Another found his trisection in 1973 after 12,000 hours of work [2, p. 80]. One got his from God [2, p. 73]. In 1971, a person with a Ph.D. in mathematics asserted that he had a valid trisection method [2, p. 127]. Many people have claimed the hat trick: trisecting the angle, doubling the cube, and squaring the circle. Two men who did this in 1961 succeeded in having their accomplishment noted in the *Congressional Record* [2, p. 110]. Occasionally, newspapers and magazines have run stories about "doing the impossible," often giving the impression that the construction may be valid. Many angle-trisectors and circle-squarers have had their work published at their own expense and distributed to colleges and universities. One had his printed in four languages! There are two delightful books written by mathematicians about their encounters with these people. The books are full of wit, charm, and humor ([1] and [2]).

EXERCISES ▪▪

Only prove to me that it is impossible, and I will set about it this very evening.

SPOKEN BY A MEMBER OF THE AUDIENCE AFTER DE MORGAN GAVE A
LECTURE ON THE IMPOSSIBILITY OF SQUARING THE CIRCLE.

1. If a and b are constructible numbers, give a geometric proof that $a + b$ and $a - b$ are constructible.

2. If *a* and *b* are constructible, give a geometric proof that *ab* is constructible. (*Hint:* Consider the following figure. Notice that all segments in the figure can be made with an unmarked straightedge and a compass.)

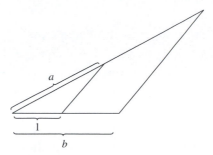

3. Prove that if *c* is a constructible number, then so is $\sqrt{|c|}$. (*Hint:* Consider the following semicircle with diameter $1 + |c|$.) (This exercise is referred to in Chapter 33.)

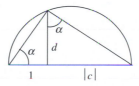

4. If *a* and *b* (*b* ≠ 0) are constructible numbers, give a geometric proof that *a*/*b* is constructible. (*Hint:* Consider the following figure.)

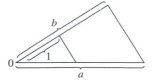

5. Prove that sin θ is constructible if and only if cos θ is constructible.

6. Prove that an angle θ is constructible if and only if sin θ is constructible.

7. Prove that cos 2θ is constructible if and only if cos θ is constructible.

8. Prove that 30° is a constructible angle.

9. Prove that a 45° angle can be trisected with an unmarked straightedge and a compass.

10. Prove that a 40° angle is not constructible.

11. Show that the point of intersection of two lines in the plane of a field *F* lies in the plane of *F*.

12. Show that the points of intersection of a circle in the plane of a field F and a line in the plane of F are points in the plane of F or in the plane of $F(\sqrt{\alpha})$, where $\alpha \in F$ and α is positive. Give an example of a circle and a line in the plane of Q whose points of intersection are not in the plane of Q.

13. Prove that $8x^3 - 6x - 1$ is irreducible over Q.

14. Use the fact that $8 \cos^3(2\pi/7) + 4 \cos^2(2\pi/7) - 4 \cos(2\pi/7) - 1 = 0$ to prove that a regular seven-sided polygon is not constructible with an unmarked straightedge and a compass.

15. Show that a regular 9-gon cannot be constructed with an unmarked straightedge and a compass.

16. Show that if a regular n-gon is constructible, then so is a regular $2n$-gon.

17. (Squaring the Circle) Show that it is impossible to construct, with an unmarked straightedge and a compass, a square whose area equals that of a circle of radius 1. You may use the fact that π is transcendental over Q.

18. Use the fact that $4 \cos^2(2\pi/5) + 2 \cos(2\pi/5) - 1 = 0$ to prove that a regular pentagon is constructible.

19. Can the cube be "tripled"?

20. Can the cube be "quadrupled"?

21. Can the circle be "cubed"?

22. If a, b, and c are constructible, show that the real roots of $ax^2 + bx + c$ are constructible.

References

1. Augustus De Morgan, *A Budget of Paradoxes,* 2nd ed., Salem, N.H.: Ayer, 1915.

2. Underwood Dudley, *A Budget of Trisections,* New York: Springer-Verlag, 1987.

SUPPLEMENTARY EXERCISES FOR CHAPTERS 19–23 ▬ ▪▪

Difficulties strengthen the mind, as labor does the body.

SENECA

True/false questions for Chapters 19–23 are available on the Web at

http://www.d.umn.edu/~jgallian/TF

1. Show that $x^{50} - 1$ has no multiple zeros in any extension of Z_3.

2. Suppose that $p(x)$ is a quadratic polynomial with rational coefficients and is irreducible over Q. Show that $p(x)$ has two zeros in $Q[x]/\langle p(x) \rangle$.

3. Let F be a finite field of order q and let a be a nonzero element in F. If n divides $q - 1$, prove that the equation $x^n = a$ has either no solutions in F or n distinct solutions in F.

4. Without using the Primitive Element Theorem, prove that if $[K{:}F]$ is prime, then K has a primitive element.

5. Let a be a complex zero of $x^2 + x + 1$. Express $(5a^2 + 2)/a$ in the form $c + ba$, where c and b are rational.

6. Describe the elements of the extension $Q(\sqrt[4]{2})$ over the field $Q(\sqrt{2})$.

7. If $[F(a){:}F] = 5$, find $[F(a^3){:}F]$.

8. If $p(x) \in F[x]$ and deg $p(x) = n$, show that the splitting field for $p(x)$ over F has degree at most $n!$.

9. Let a be a nonzero algebraic element over F of degree n. Show that a^{-1} is also algebraic over F of degree n.

10. Prove that $\pi^2 - 1$ is algebraic over $Q(\pi^3)$.

11. If ab is algebraic over F and $b \neq 0$, prove that a is algebraic over $F(b)$.

12. Let E be an algebraic extension of a field F. If R is a ring and $E \supseteq R \supseteq F$, show that R must be a field.

13. If a is transcendental over F, show that every element of $F(a)$ that is not in F is transcendental over F.

14. What is the order of the splitting field of $x^5 + x^4 + 1 = (x^2 + x + 1)(x^3 + x + 1)$ over Z_2?

15. Show that a finite extension of a finite field is a simple extension.

16. Let R be an integral domain that contains a field F as a subring. If R is finite-dimensional when viewed as a vector space over F, prove that R is a field.

17. Show that it is impossible to find a basis for the vector space of $n \times n$ $(n > 1)$ matrices such that each pair of elements in the basis commutes under multiplication.

18. Let $P_n = \{a_n x^n + a_{n-1} x^{n-1} + \cdots + a_1 x + a_0 \mid$ each a_i is a real number$\}$. Is it possible to have a basis for P_n such that every element of the basis has x as a factor?

19. Find a basis for the vector space $\{f \in P_3 \mid f(0) = 0\}$. (See Exercise 18 for notation.)

20. Given that f is a polynomial of degree n in P_n, show that $\{f, f', f'', \ldots, f^{(n)}\}$ is a basis for P_n. ($f^{(k)}$ denotes the kth derivative of f.)

21. Suppose that K is a field extension of a field F of characteristic $p \neq 0$. Let $L = \{a \in K \mid a^{p^n} \in F$ for some nonnegative integer $n\}$. Prove that L is a subfield of K that contains F.

22. In which fields does $x^n - x$ have a multiple zero?

23. If $a \neq 0$ belongs to a field F and $x^n - a$ splits in some extension E of F, prove that E contains all the nth roots of unity.

PART 5

Special
Topics

For online student resources,
visit this textbook's website at
math.college.hmco.com/students

Sylow Theorems

CONJUGACY CLASSES

In this chapter, we derive several important arithmetic relationships between a group and certain of its subgroups. Recall from Chapter 7 that Lagrange's Theorem was proved by showing that cosets of a subgroup partition the group. Another fruitful method of partitioning the elements of a group is by way of conjugacy classes.

DEFINITION Conjugacy Class of a

Let a and b be elements of a group G. We say that a and b are *conjugate* in G (and call b a *conjugate* of a) if $xax^{-1} = b$ for some x in G. The *conjugacy class of a* is the set $\mathrm{cl}(a) = \{xax^{-1} \mid x \in G\}$.

We leave it to the reader (Exercise 1) to prove that conjugacy is an equivalence relation on G, and that the conjugacy class of a is the equivalence class of a under conjugacy. Thus, we may partition any group into disjoint conjugacy classes. Let's look at one example. In D_4 we have

$$\mathrm{cl}(H) = \{R_0 H R_0^{-1}, R_{90} H R_{90}^{-1}, R_{180} H R_{180}^{-1}, R_{270} H R_{270}^{-1},$$
$$HHH^{-1}, VHV^{-1}, DHD^{-1}, D'HD'^{-1}\} = \{H, V\}.$$

Similarly, one may verify that

$$\text{cl}(R_0) = \{R_0\},$$
$$\text{cl}(R_{90}) = \{R_{90}, R_{270}\} = \text{cl}(R_{270}),$$
$$\text{cl}(R_{180}) = \{R_{180}\},$$
$$\text{cl}(V) = \{V, H\} = \text{cl}(H),$$
$$\text{cl}(D) = \{D, D'\} = \text{cl}(D').$$

Theorem 24.1 gives an arithmetic relationship between the size of the conjugacy class of a and the size of $C(a)$, the centralizer of a.

Theorem 24.1   The Number of Conjugates of a

> *Let G be a finite group and let a be an element of G. Then,*
> $|\text{cl}(a)| = |G:C(a)|$.

PROOF.   Consider the function T that sends the coset $xC(a)$ to the conjugate xax^{-1} of a. A routine calculation shows that T is well defined, is one-to-one, and maps the set of left cosets onto the conjugacy class of a. Thus, the number of conjugates of a is the index of the centralizer of a.   ■ ■

Corollary 1 $|\text{cl}(a)|$ Divides $|G|$

> *In a finite group, $|\text{cl}(a)|$ divides $|G|$.*

THE CLASS EQUATION

Since the conjugacy classes partition a group, the following important counting principle is a corollary to Theorem 24.1.

Corollary 2   The Class Equation

> *For any finite group G,*
>
> $$|G| = \Sigma \, |G:C(a)|,$$
>
> *where the sum runs over one element a from each conjugacy class of G.*

In finite group theory, counting principles such as this corollary are powerful tools.[†] Theorem 24.2 is the single most important fact about finite groups of prime-power order (a group of order p^n, where p is a prime, is called a *p-group*).

Theorem 24.2   *p*-Groups Have Nontrivial Centers

Let G be a nontrivial finite group whose order is a power of a prime p. Then Z(G) has more than one element.

PROOF.   First observe that $\text{cl}(a) = \{a\}$ if and only if $a \in Z(G)$ (see Exercise 4). Thus, by culling out these elements, we may write the class equation in the form

$$|G| = |Z(G)| + \Sigma|G:C(a)|,$$

where the sum runs over representatives of all conjugacy classes with more than one element (this set may be empty). But $|G:C(a)| = |G|/|C(a)|$, so each term in $\Sigma|G:C(a)|$ has the form p^k with $k \geq 1$. Hence,

$$|G| - \Sigma|G:C(a)| = |Z(G)|,$$

where each term on the left is divisible by p. It follows, then, that p also divides $|Z(G)|$, and hence $|Z(G)| \neq 1$.   ■ ■

Corollary   Groups of Order p^2 Are Abelian

If $|G| = p^2$, where p is prime, then G is Abelian.

PROOF.   By Theorem 24.2 and Lagrange's Theorem, $|Z(G)| = p$ or p^2. If $|Z(G)| = p^2$, then $G = Z(G)$ and G is Abelian. If $|Z(G)| = p$, then $|G/Z(G)| = p$, so that $G/Z(G)$ is cyclic. But, then, by Theorem 9.3, G is Abelian.   ■ ■

THE PROBABILITY THAT TWO ELEMENTS COMMUTE

Before proceeding to the main goal of this chapter, we pause for an interesting application of Theorem 24.1 and the class equation. (Our

---

[†]"Never underestimate a theorem that counts something." John Fraleigh, *A First Course in Abstract Algebra*.

discussion is based on [1] and [2].) Suppose we select two elements at random (with replacement) from a finite group. What is the probability that these two elements commute? Well, suppose that G is a finite group of order n. Then the probability $\Pr(G)$ that two elements selected at random from G commute is $|K|/n^2$, where $K = \{(x, y) \in G \oplus G \mid xy = yx\}$. Now notice that for each $x \in G$ we have $(x, y) \in K$ if and only if $y \in C(x)$. Thus,

$$|K| = \sum_{x \in G} |C(x)|.$$

Also, it follows from Theorem 24.1 that if x and y are in the same conjugacy class, then $|C(x)| = |C(y)|$ (see Exercise 50). If, for example, $\mathrm{cl}(a) = \{a_1, a_2, \ldots, a_t\}$, then

$$|C(a_1)| + |C(a_2)| + \cdots + |C(a_t)| = t|C(a)|$$

$$= |G{:}C(a)|\,|C(a)| = |G| = n.$$

So, by choosing one representative from each conjugacy class, say, x_1, $x_2, \ldots, x_m$, we have

$$|K| = \sum_{x \in G} |C(x)| = \sum_{i=1}^{m} |G{:}C(x_i)||C(x_i)| = m \cdot n.$$

Thus, the answer to our question is $mn/n^2 = m/n$, where m is the number of conjugacy classes in G and n is the number of elements of G.

Obviously, when G is non-Abelian, $\Pr(G)$ is less than 1. But how much less than 1? Clearly, the more conjugacy classes there are, the larger $\Pr(G)$ is. Consequently, $\Pr(G)$ is large when the sizes of the conjugacy classes are small. Noting that $|\mathrm{cl}(a)| = 1$ if and only if $a \in Z(G)$, we obtain the maximum number of conjugacy classes when $|Z(G)|$ is as large as possible and all other conjugacy classes have exactly two elements in each. Since G is non-Abelian, it follows from Theorem 9.3 that $|G/Z(G)| \geq 4$ and, therefore, $|Z(G)| \leq |G|/4$. Thus, in the extreme case, we would have $|Z(G)| = |G|/4$, and the remaining $(3/4)|G|$ elements would be distributed in conjugacy classes with two elements each. So, in a non-Abelian group, the number of conjugacy classes is no more than $|G|/4 + (1/2)(3/4)|G|$, and $\Pr(G)$ is less than or equal to 5/8. The dihedral group D_4 is an example of a group that has probability equal to 5/8.

THE SYLOW THEOREMS

Now to the Sylow theorems. Recall that the converse of Lagrange's Theorem is false; that is, if G is a group of order m and n divides m, G

need *not* have a subgroup of order *n*. Our next theorem is a partial converse of Lagrange's Theorem. It, as well as Theorem 24.2, was first proved by the Norwegian mathematician Ludwig Sylow (1832–1918). Sylow's Theorem and Lagrange's Theorem are the two most important results in finite group theory. The first gives a sufficient condition for the existence of subgroups, and the second gives a necessary condition.

Theorem 24.3   Existence of Subgroups of Prime-Power Order (Sylow's First Theorem, 1872)

> *Let G be a finite group and let p be a prime. If p^k divides $|G|$, then G has at least one subgroup of order p^k.*

PROOF.   We proceed by induction on $|G|$. If $|G| = 1$, Theorem 24.3 is trivially true. Now assume that the statement is true for all groups of order less than $|G|$. If G has a proper subgroup H such that p^k divides $|H|$, then, by our inductive assumption, H has a subgroup of order p^k and we are done. Thus, we may henceforth assume that p^k does not divide the order of any proper subgroup of G. Next, consider the class equation for G in the form

$$|G| = |Z(G)| + \Sigma|G{:}C(a)|,$$

where we sum over a representative of each conjugacy class cl(*a*), where $a \notin Z(G)$. Since p^k divides $|G| = |G{:}C(a)|\,|C(a)|$ and p^k does not divide $|C(a)|$, we know that p must divide $|G{:}C(a)|$ for all $a \notin Z(G)$. It then follows from the class equation that p divides $|Z(G)|$. The Fundamental Theorem of Finite Abelian Groups (Theorem 11.1), or Theorem 9.5, then guarantees that $Z(G)$ contains an element of order p, say x. Since x is in the center of G, $\langle x \rangle$ is a normal subgroup of G, and we may form the factor group $G/\langle x \rangle$. Now observe that p^{k-1} divides $|G/\langle x \rangle|$. Thus, by the induction hypothesis, $G/\langle x \rangle$ has a subgroup of order p^{k-1} and, by Exercise 49 in Chapter 10, this subgroup has the form $H/\langle x \rangle$, where H is a subgroup of G. Finally, note that $|H/\langle x \rangle| = p^{k-1}$ and $|\langle x \rangle| = p$ imply that $|H| = p^k$, and this completes the proof.   ■ ■

Let's be sure we understand exactly what Sylow's First Theorem means. Say we have a group G of order $2^3 \cdot 3^2 \cdot 5^4 \cdot 7$. Then Sylow's First Theorem says that G must have at least one subgroup of each of the following orders: 2, 4, 8, 3, 9, 5, 25, 125, 625, and 7. On the other hand, Sylow's First Theorem tells us nothing about the possible existence of subgroups of order 6, 10, 15, 30, or any other divisor of $|G|$ that has two or more distinct prime factors. Because certain subgroups guaranteed by Sylow's First Theorem play a central role in the theory of finite groups, they are given a special name.

DEFINITION   *Sylow p-Subgroup*

> Let G be a finite group and let p be a prime divisor of $|G|$. If p^k divides $|G|$ and p^{k+1} does not divide $|G|$, then any subgroup of G of order p^k is called a *Sylow p-subgroup of G.*

So, returning to our group G of order $2^3 \cdot 3^2 \cdot 5^4 \cdot 7$, we call any subgroup of order 8 a Sylow 2-subgroup of G, any subgroup of order 625 a Sylow 5-subgroup of G, and so on. Notice that a Sylow p-subgroup of G is a subgroup whose order is the largest power of p consistent with Lagrange's Theorem.

Since any subgroup of order p is cyclic, we have the following generalization of Theorem 9.5, first proved by Cauchy in 1845. His proof ran nine pages!

Corollary   **Cauchy's Theorem**

> *Let G be a finite group and let p be a prime that divides the order of G. Then G has an element of order p.*

Sylow's First Theorem is so fundamental to finite group theory that many different proofs of it have been published over the years [our proof is essentially the one given by Georg Frobenius (1849–1917) in 1895]. Likewise, there are scores of generalizations of Sylow's Theorem.

Observe that the corollary to the Fundamental Theorem of Finite Abelian Groups and Sylow's First Theorem show that the converse of Lagrange's Theorem is true for all finite Abelian groups and all finite groups of prime-power order.

There are two more Sylow theorems that are extremely valuable tools in finite group theory. But first we introduce a new term.

DEFINITION   *Conjugate Subgroups*

> Let H and K be subgroups of a group G. We say that H and K are *conjugate* in G if there is an element g in G such that $H = gKg^{-1}$.

Recall from Chapter 7 that if G is a finite group of permutations on a set S and $i \in S$, then $\text{orb}_G(i) = \{\phi(i) \mid \phi \in G\}$ and $|\text{orb}_G(i)|$ divides $|G|$.

Theorem 24.4   Sylow's Second Theorem

> *If H is a subgroup of a finite group G and |H| is a power of a prime p, then H is contained in some Sylow p-subgroup of G.*

PROOF.   Let K be a Sylow p-subgroup of G and let $C = \{K_1, K_2, \ldots, K_n\}$ with $K = K_1$ be the set of all conjugates of K in G. Since conjugation is an automorphism, each element of C is a Sylow p-subgroup of G. Let S_C denote the group of all permutations of C. For each $g \in G$, define $\phi_g : C \rightarrow C$ by $\phi_g(K_i) = gK_ig^{-1}$. It is easy to show that each $\phi_g \in S_C$.

Now define a mapping $T:G \rightarrow S_C$ by $T(g) = \phi_g$. Since $\phi_{gh}(K_i) = (gh)K_i(gh)^{-1} = g(hK_ih^{-1})g^{-1} = g\phi_h(K_i)g^{-1} = \phi_g(\phi_h(K_i)) = (\phi_g\phi_h)(K_i)$, we have $\phi_{gh} = \phi_g\phi_h$, and therefore T is a homomorphism from G to S_C.

Next consider $T(H)$, the image of H under T. Since $|H|$ is a power of p, so is $|T(H)|$ (see property 6 of Theorem 10.2). Thus, by the Orbit-Stabilizer Theorem (Theorem 7.3), for each i, $|\text{orb}_{T(H)}(K_i)|$ divides $|T(H)|$, so that $|\text{orb}_{T(H)}(K_i)|$ is a power of p. Now we ask: Under what condition does $|\text{orb}_{T(H)}(K_i)| = 1$? Well, $|\text{orb}_{T(H)}(K_i)| = 1$ means that $\phi_g(K_i) = gK_ig^{-1} = K_i$ for all $g \in H$; that is, $|\text{orb}_{T(H)}(K_i)| = 1$ if and only if $H \leq N(K_i)$. But the only elements of $N(K_i)$ that have orders that are powers of p are those of K_i (see Exercise 9). Thus, $|\text{orb}_{T(H)}(K_i)| = 1$ if and only if $H \leq K_i$.

So, to complete the proof, all we need to do is show that for some i, $|\text{orb}_{T(H)}(K_i)| = 1$. Analogous to Theorem 24.1, we have $|C| = |G:N(K)|$ (see Exercise 19). And since $|G:K| = |G:N(K)||N(K):K|$ is not divisible by p, neither is $|C|$. Because the orbits partition C, $|C|$ is the sum of powers of p. If no orbit has size 1, then p divides each summand and, therefore, p divides $|C|$, which is a contradiction. Thus, there is an orbit of size 1, and the proof is complete.   ∎∎

Theorem 24.5   Sylow's Third Theorem

> *The number of Sylow p-subgroups of G is equal to 1 modulo p and divides |G|. Furthermore, any two Sylow p-subgroups of G are conjugate.*

PROOF.   Let K be any Sylow p-subgroup of G and let $C = \{K_1, K_2, \ldots, K_n\}$ with $K = K_1$ be the set of all conjugates of K in G. We first prove that $n \bmod p = 1$.

Let S_C and T be as in the proof of Theorem 24.4. This time we consider $T(K)$, the image of K under T. As before, we have $|\mathrm{orb}_{T(K)}(K_i)|$ is a power of p for each i and $|\mathrm{orb}_{T(K)}(K_i)| = 1$ if and only if $K \leq K_i$. Thus, $|\mathrm{orb}_{T(K)}(K_1)| = 1$ and $|\mathrm{orb}_{T(K)}(K_i)|$ is a power of p greater than 1 for all $i \neq 1$. Since the orbits partition C, it follows that, modulo p, $n = |C| = 1$.

Next we show that every Sylow p-subgroup of G belongs to C. To do this, suppose that H is a Sylow p-subgroup of G that is not in C. Let S_C and T be as in the proof of Theorem 24.4, and this time consider $T(H)$. As in the previous paragraph, $|C|$ is the sum of the orbits' sizes under the action of $T(H)$. However, no orbit has size 1, since H is not in C. Thus, $|C|$ is a sum of terms each divisible by p, so that, modulo p, $n = |C| = 0$. This contradiction proves that H belongs to C, and that n is the number of Sylow p-subgroups of G.

Finally, that n divides $|G|$ follows directly from the fact that $n = |G:N(K)|$ (see Exercise 19). ■ ■

It is convenient to let n_p denote the number of Sylow p-subgroups of a group. Observe that the first portion of Sylow's Third Theorem is a counting principle.[†] As an important consequence of Sylow's Third Theorem, we have the following corollary.

***Corollary*   A Unique Sylow p-Subgroup Is Normal**

> *A Sylow p-subgroup of a finite group G is a normal subgroup of G if and only if it is the only Sylow p-subgroup of G.*

We illustrate Sylow's Third Theorem with two examples.

EXAMPLE 1   Consider the Sylow 2-subgroups of S_3. They are $\{(1), (12)\}$, $\{(1), (23)\}$, and $\{(1), (13)\}$. According to Sylow's Third Theorem, we should be able to obtain the latter two of these from the first by conjugation. Indeed,

$$(13)\{(1), (12)\}(13)^{-1} = \{(1), (23)\},$$
$$(23)\{(1), (12)\}(23)^{-1} = \{(1), (13)\}.$$ ◆

EXAMPLE 2   Consider the Sylow 3-subgroups of A_4. They are $\{\alpha_1, \alpha_5, \alpha_9\}$, $\{\alpha_1, \alpha_6, \alpha_{11}\}$, $\{\alpha_1, \alpha_7, \alpha_{12}\}$, and $\{\alpha_1, \alpha_8, \alpha_{10}\}$. (See the table on page 105.) Then,

---

[†]"Whenever you can, count." Sir Francis Galton (1822–1911), *The World of Mathematics*.

$$\alpha_2\{\alpha_1, \alpha_5, \alpha_9\}\alpha_2^{-1} = \{\alpha_1, \alpha_7, \alpha_{12}\},$$
$$\alpha_3\{\alpha_1, \alpha_5, \alpha_9\}\alpha_3^{-1} = \{\alpha_1, \alpha_8, \alpha_{10}\},$$
$$\alpha_4\{\alpha_1, \alpha_5, \alpha_9\}\alpha_4^{-1} = \{\alpha_1, \alpha_6, \alpha_{11}\}.$$

Thus, the number of Sylow 3-subgroups is 1 modulo 3, and the four Sylow 3-subgroups are conjugate. ◆

Figure 24.1 shows the subgroup lattices for S_3 and A_4. We have connected the Sylow p-groups with dashed circles to indicate that they belong to one orbit under conjugation. Notice that the three subgroups of order 2 in A_4 are contained in a Sylow 2-group, as required by Sylow's Second Theorem. As it happens, these three subgroups also belong to one orbit under conjugation, but this is not a consequence of Sylow's Third Theorem.

In contrast to the two preceding examples, observe that the dihedral group of order 12 has seven subgroups of order 2, but that conjugating $\{R_0, R_{180}\}$ does not yield any of the other six. (Why?)

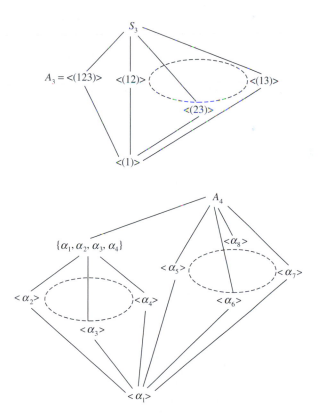

Figure 24.1   Lattices of subgroups for S_3 and A_4.

APPLICATIONS OF SYLOW THEOREMS

A few numerical examples will make the Sylow theorems come to life. Say G is a group of order 40. What do the Sylow theorems tell us about G? A great deal! Since 1 is the only divisor of 40 that is congruent to 1 modulo 5, we know that G has exactly one subgroup of order 5, and therefore it is normal. Similarly, G has either one or five subgroups of order 8. If there is only one subgroup of order 8, it is normal. If there are five subgroups of order 8, none is normal and all five can be obtained by starting with any particular one, say H, and computing xHx^{-1} for various x's. Finally, if we let K denote the normal subgroup of order 5 and let H denote any subgroup of order 8, then $G = HK$. (See Exercise 7, Supplementary Exercises for Chapters 5–8.) If H happens to be normal, we can say even more: $G = H \times K$.

What about a group G of order 30? It must have either one or six subgroups of order 5 and one or 10 subgroups of order 3. However, G cannot have both six subgroups of order 5 *and* 10 subgroups of order 3 (for then G would have more than 30 elements). Thus, G has one subgroup of order 3 and one of order 5, and at least one of these is normal in G. It follows, then, that the product of a subgroup of order 3 and one of order 5 is a group of order 15 that is both cyclic (Exercise 23) and normal (Exercise 7 in Chapter 9) in G. [This, in turn, implies that *both* the subgroup of order 3 and the subgroup of order 5 are normal in G (Exercise 53 in Chapter 9).] So, if we let y be a generator of the cyclic subgroup of order 15 and let x be an element of order 2 (the existence of which is guaranteed by Cauchy's Theorem), we see that

$$G = \{x^i y^j \mid 0 \le i \le 1, 0 \le j \le 14\}.$$

Note that in these two examples we were able to deduce all of this information from knowing only the order of the group—so many conclusions from one assumption! This is the beauty of finite group theory.

In Chapter 7 we saw that the only group (up to isomorphism) of prime order p is Z_p. As a further illustration of the power of the Sylow theorems, we next give a sufficient condition that guarantees that a group of order pq, where p and q are primes, must be Z_{pq}.

Thereom 24.6   Cyclic Groups of Order pq

> *If G is a group of order pq, where p and q are primes, $p < q$, and p does not divide $q - 1$, then G is cyclic. In particular, G is isomorphic to Z_{pq}.*

PROOF.   Let H be a Sylow p-subgroup of G and let K be a Sylow q-subgroup of G. Sylow's Third Theorem states that the number of

Sylow p-subgroups of G is of the form $1 + kp$ and divides pq. So $1 + kp = 1, p, q$, or pq. From this and the fact that $p \nmid q - 1$, it follows that $k = 0$, and therefore H is the only Sylow p-subgroup of G.

Similarly, there is only one Sylow q-subgroup of G. Thus, by the corollary to Theorem 24.5, H and K are normal subgroups of G. Let $H = \langle x \rangle$ and $K = \langle y \rangle$. To show that G is cyclic, it suffices to show that x and y commute, for then $|xy| = |x||y| = pq$. But observe that, since H and K are normal, we have

$$xyx^{-1}y^{-1} = (xyx^{-1})y^{-1} \in Ky^{-1} = K$$

and

$$xyx^{-1}y^{-1} = x(yx^{-1}y^{-1}) \in xH = H.$$

Thus, $xyx^{-1}y^{-1} \in K \cap H = \{e\}$, and hence $xy = yx$. ∎

Theorem 24.6 demonstrates the power of the Sylow theorems in classifying the finite groups whose orders have small numbers of prime factors. Similar results exist for groups of orders p^2q, p^2q^2, p^3, and p^4, where p and q are prime.

For your amusement, Figure 24.2 lists the number of nonisomorphic groups with orders at most 100. Note in particular the large number of groups of order 64. Also observe that, generally speaking, it is not the size of the group that gives rise to a large number of groups of that size but the number of prime factors involved. In all, there are 1047 nonisomorphic groups with 100 or fewer elements. Contrast this with the fact reported in 1989 that there are 2328 groups of order 128 and 56,092 groups of order 256 [3]. Estimates put the number of groups of order 512 at more than one million.

Order	1	2	3	4	5	6	7	8	9	10	11	12	13	14	15	16	17	18	19	20
Number	1	1	1	2	1	2	1	5	2	2	1	5	1	2	1	14	1	5	1	5

Order	21	22	23	24	25	26	27	28	29	30	31	32	33	34	35	36	37	38	39	40
Number	2	2	1	15	2	2	5	4	1	4	1	51	1	2	1	14	1	2	2	14

Order	41	42	43	44	45	46	47	48	49	50	51	52	53	54	55	56	57	58	59	60
Number	1	6	1	4	2	2	1	52	2	5	1	5	1	15	2	13	2	2	1	13

Order	61	62	63	64	65	66	67	68	69	70	71	72	73	74	75	76	77	78	79	80
Number	1	2	4	267	1	4	1	5	1	4	1	50	1	2	3	4	1	6	1	52

Order	81	82	83	84	85	86	87	88	89	90	91	92	93	94	95	96	97	98	99	100
Number	15	2	1	15	1	2	1	12	1	10	1	4	2	2	1	230	1	5	2	16

Figure 24.2 The number of groups of a given order up to 100.

As a final application of the Sylow theorems, you might enjoy seeing a determination of the groups of order 99, 66, and 255. In fact, our arguments serve as a good review of much of our work in group theory.

EXAMPLE 3   Determination of the Groups of Order 99
Suppose that G is a group of order 99. Let H be a Sylow 3-subgroup of G and let K be a Sylow 11-subgroup of G. Since 1 is the only positive divisor of 99 that is equal to 1 modulo 11, we know from Sylow's Third Theorem and its corollary that K is normal in G. Similarly, H is normal in G. It follows, by the argument used in the proof of Theorem 24.6, that elements from H and K commute, and therefore $G = H \times K$. Since both H and K are Abelian, G is also Abelian. Thus, G is isomorphic to Z_{99} or $Z_3 \oplus Z_{33}$.   ◆

EXAMPLE 4   Determination of the Groups of Order 66
Suppose that G is a group of order 66. Let H be a Sylow 3-subgroup of G and let K be a Sylow 11-subgroup of G. Since 1 is the only positive divisor of 66 that is equal to 1 modulo 11, we know that K is normal in G. Thus, HK is a subgroup of G of order 33 (Exercise 51 in Chapter 9 and Exercise 7, Supplementary Exercises for Chapters 5–8). Since any group of order 33 is cyclic (Theorem 24.6), we may write $HK = \langle x \rangle$. Next, let $y \in G$ and $|y| = 2$. Since $\langle x \rangle$ has index 2 in G, we know it is normal. So $yxy^{-1} = x^i$ for some i from 1 to 32. Then, $yx = x^i y$ and, since every member of G is of the form $x^s y^t$, the structure of G is completely determined by the value of i. We claim that there are only four possibilities for i. To prove this, observe that $|x^i| = |x|$ (Exercise 5, Supplementary Exercises for Chapters 1–4). Thus, i and 33 are relatively prime. But also, since y has order 2,

$$x = y^{-1}(yxy^{-1})y = y^{-1}x^i y = yx^i y^{-1} = (yxy^{-1})^i = (x^i)^i = x^{i^2}.$$

So $x^{i^2-1} = e$ and therefore 33 divides $i^2 - 1$. From this it follows that 11 divides $i \pm 1$, and therefore $i = 0 \pm 1$, $i = 11 \pm 1$, $i = 22 \pm 1$, or $i = 33 \pm 1$. Putting this together with the other information we have about i, we see that $i = 1, 10, 23,$ or 32. This proves that there are at most four groups of order 66.

To prove that there are exactly four such groups, we simply observe that Z_{66}, D_{33}, $D_{11} \oplus Z_3$, and $D_3 \oplus Z_{11}$ each has order 66 and that no two are isomorphic. For example, $D_{11} \oplus Z_3$ has 11 elements of order 2, whereas $D_3 \oplus Z_{11}$ has only three elements of order 2. (See Exercises 25–28 of the Supplementary Exercises for Chapters 5–8.)   ◆

EXAMPLE 5   The Only Group of Order 255 is Z_{255}
Let G be a group of order $255 = 3 \cdot 5 \cdot 17$, and let H be a Sylow 17-subgroup of G. By Sylow's Third Theorem, H is the only Sylow

17-subgroup of G, so $N(H) = G$. By Example 15 in Chapter 10, $|N(H)/C(H)|$ divides $|\text{Aut}(H)| = |\text{Aut}(Z_{17})|$. By Theorem 6.5, $|\text{Aut}(Z_{17})|$ $= |U(17)| = 16$. Since $|N(H)/C(H)|$ must divide 255 and 16, we have $|N(H)/C(H)| = 1$. Thus, $C(H) = G$. This means that every element of G commutes with every element of H, and, therefore, $H \subseteq Z(G)$. Thus, 17 divides $|Z(G)|$, which in turn divides 255. So $|Z(G)| = 17, 51, 85$, or 255 and $|G/Z(G)| = 15, 5, 3$, or 1. But the only groups of order 15, 5, 3, or 1 are the cyclic ones, so we know that $G/Z(G)$ is cyclic. Now the "G/Z theorem" (Theorem 9.3) shows that G is Abelian, and the Fundamental Theorem of Finite Abelian Groups tells us that G is cyclic. ◆

EXERCISES

I have always grown from my problems and challenges, from the things that don't work out. That's when I've really learned.

CAROL BURNETT

1. Show that conjugacy is an equivalence relation on a group.
2. Calculate all conjugacy classes for the quaternions (see Exercise 4, Supplementary Exercises for Chapters 1–4).
3. Show that the function T defined in the proof of Theorem 24.1 is well defined, one-to-one, and maps the set of left cosets onto the conjugacy class of a.
4. Show that $\text{cl}(a) = \{a\}$ if and only if $a \in Z(G)$.
5. If $|G| = 36$ and G is non-Abelian, prove that G has more than one Sylow 2-subgroup or more than one Sylow 3-subgroup.
6. Exhibit a Sylow 2-subgroup of S_4. Describe an isomorphism from this group to D_4.
7. Suppose that G is a group of order 48. Show that the intersection of any two distinct Sylow 2-subgroups of G has order 8.
8. Find all the Sylow 3-subgroups of A_4.
9. Let K be a Sylow p-subgroup of a finite group G. Prove that if $x \in N(K)$ and the order of x is a power of p, then $x \in K$. (This exercise is referred to in this chapter.)
10. Let H be a Sylow p-subgroup of G. Prove that H is the only Sylow p-subgroup of G contained in $N(H)$.
11. Suppose that G is a group of order 168. If G has more than one Sylow 7-subgroup, exactly how many does it have?
12. Show that every group of order 56 has a proper nontrivial normal subgroup.
13. What is the smallest composite (that is, nonprime and greater than 1) integer n such that there is a unique group of order n?

14. Let G be a noncyclic group of order 21. How many Sylow 3-subgroups does G have?

15. Prove that a noncyclic group of order 21 must have 14 elements of order 3.

16. How many Sylow 5-subgroups of S_5 are there? Exhibit two.

17. How many Sylow 3-subgroups of S_5 are there? Exhibit five.

18. Prove that a group of order 175 is Abelian.

19. Let H be a subgroup of a group G. Prove that the number of conjugates of H in G is $|G:N(H)|$. *Hint:* Mimic the proof of Theorem 24.1. (This exercise is referred to in this chapter.)

20. Generalize the argument given in Example 3 to obtain a theorem about groups of order p^2q, where p and q are distinct primes.

21. What is the smallest possible odd integer that can be the order of a non-Abelian group?

22. Prove that a group of order 375 has a subgroup of order 15.

23. Without using Theorem 24.6, prove that a group of order 15 is cyclic. (This exercise is referred to in the discussion about groups of order 30.)

24. Prove that a group of order 105 contains a subgroup of order 35.

25. Prove that a group of order 595 has a normal Sylow 17-subgroup.

26. Let G be a group of order 60. Show that G has exactly four elements of order 5 or exactly 24 elements of order 5. Which of these cases holds for A_5?

27. Show that the center of a group of order 60 cannot have order 4.

28. Suppose that G is a group of order 60 and G has a normal subgroup N of order 2. Show that

    a. G has normal subgroups of orders 6, 10, and 30.

    b. G has subgroups of orders 12 and 20.

    c. G has a cyclic subgroup of order 30.

29. Let G be a group of order 60. If the Sylow 3-subgroup is normal, show that the Sylow 5-subgroup is normal.

30. Show that if G is a group of order 168 that has a normal subgroup of order 4, then G has a normal subgroup of order 28.

31. Suppose that p is prime and $|G| = p^n$. Show that G has normal subgroups of order p^k for all k between 1 and n (inclusive).

32. Suppose that G is a group of order p^n, where p is prime, and G has exactly one subgroup for each divisor of p^n. Show that G is cyclic.

33. Suppose that p is prime and $|G| = p^n$. If H is a proper subgroup of G, prove that $N(H) > H$. (This exercise is referred to in Chapter 25.)

34. Suppose that G is a finite group and that all its Sylow subgroups are normal. Show that G is a direct product of its Sylow subgroups.

35. Let G be a finite group and let H be a normal Sylow p-subgroup of G. Show that $\alpha(H) = H$ for all automorphisms α of G.

36. If H is a normal subgroup of a finite group G and $|H| = p^k$ for some prime p, show that H is contained in every Sylow p-subgroup of G.

37. Let H and K denote a Sylow 3-subgroup and a Sylow 5-subgroup of a group, respectively. Suppose that $|H| = 3$ and $|K| = 5$. If 3 divides $|N(K)|$, show that 5 divides $|N(H)|$.

38. Let G be a group of order p^2q^2, where p and q are distinct primes, $q \nmid p^2 - 1$, and $p \nmid q^2 - 1$. Prove that G is Abelian. List three pairs of primes that satisfy these conditions.

39. Let H be a normal subgroup of a group G. Show that H is the union of the conjugacy classes of the elements of H. Is this true when H is not normal in G?

40. Let p be prime. If the order of every element of a finite group G is a power of p, prove that $|G|$ is a power of p. (Such a group is called a *p-group*.)

41. For each prime p, prove that all Sylow p-subgroups of a finite group are isomorphic.

42. Suppose that K is a normal subgroup of a finite group G and S is a Sylow p-subgroup of G. Prove that $K \cap S$ is a Sylow p-subgroup of K.

43. If G is a group of odd order and $x \in G$, show that x^{-1} is not in $cl(x)$.

44. Determine the groups of order 45.

45. Show that there are at most three nonisomorphic groups of order 21.

46. Prove that if H is a normal subgroup of index p^2 where p is prime, then $G' \subseteq H$ (see Exercise 3 in the Supplementary Exercises for Chapters 5–8 for a description of G').

47. Show that Z_2 is the only group that has exactly two conjugacy classes.

48. What is the probability that a randomly selected element from D_4 commutes with V?

49. Let G be a finite group and let $a \in G$. Express the probability that a randomly selected element from G commutes with a in terms of orders of subgroups of G.

50. Prove that if x and y are in the same conjugacy class of a group, then $|C(x)| = |C(y)|$. (This exercise is referred to in the discussion on the probability that two elements from a group commute.)

51. Find $Pr(D_4)$, $Pr(S_3)$, and $Pr(A_4)$.

52. Prove that $Pr(G \oplus H) = Pr(G) \cdot Pr(H)$.

53. Let R be a finite noncommutative ring. Show that the probability that two randomly chosen elements from R commute is at most $\frac{5}{8}$. [*Hint:* Mimic the group case and use the fact that the additive group $R/C(R)$ is not cyclic.]

References

1. W. H. Gustafson, "What Is the Probability That Two Group Elements Commute?" *The American Mathematical Monthly* 80 (1973): 1031–1034.
2. Desmond MacHale, "How Commutative Can a Non-Commutative Group Be?" *The Mathematical Gazette* 58 (1974): 199–202.
3. E. A. O'Brien, "The Groups of Order Dividing 256," *Bulletin of the Australian Mathematical Society* 39 (1989): 159–160.
4. H. Paley and P. Weichsel, *A First Course in Abstract Algebra*, New York: Holt, Rinehart & Winston, 1966.

Suggested Readings

J. A. Gallian and D. Moulton, "When Is Z_n the Only Group of Order n?" *Elemente der Mathematik* 48 (1993): 118–120.

It is shown that Z_n is the only group of order n if and only if n and $\phi(n)$ are relatively prime.

S. W. Golomb, "A Symmetry Criterion for Conjugacy in Finite Groups," *Mathematics Magazine* 69 (1996): 373–375.

This easy-to-read article gives a simple method for determining the conjugacy classes of a group from its multiplication table.

W. H. Gustafson, "What Is the Probability That Two Group Elements Commute?" *The American Mathematical Monthly* 80 (1973): 1031–1034.

This paper is concerned with the problem posed in the title. It is shown that for all finite non-Abelian groups and certain infinite non-Abelian groups, the probability that two elements from a group commute is at most 5/8. The paper concludes with several exercises.

Dieter Jungnickel, "On the Uniqueness of the Cyclic Group of Order n," *The American Mathematical Monthly* 99 (1992): 545–546.

In this article, it is shown that Z_n is the only group of order n if and only if n and $\phi(n)$ are relatively prime.

Desmond MacHale, "Commutativity in Finite Rings," *The American Mathematical Monthly* 83 (1976): 30–32.

In this easy-to-read paper, it is shown that the probability that two elements from a finite noncommutative ring commute is at most 5/8. Several properties of $\Pr(G)$ when G is a finite group are stated. For example, if $H \le G$, then $\Pr(G) \le \Pr(H)$. Also, there is no group G such that $7/16 < \Pr(G) < 1/2$.

Suggested Video

Conjugacy and Fixed Points, Films for the Humanities & Sciences, FFH 6356, 25 minutes, Box 2053, Princeton, NJ 08543-2053, 1-800-257-5126.

This video uses models to show the geometric significance of conjugation in the symmetry group of the regular octahedron and the symmetry group of a square.

Ludwig Sylow

Sylow's Theorem is 100 years old. In the course of a century this remarkable theorem has been the basis for the construction of numerous theories.

L. A. SHEMETKOV

LUDWIG SYLOW (pronounced "SEE-loe") was born on December 12, 1832, in Christiania (now Oslo), Norway. While a student at Christiania University, Sylow won a gold medal for competitive problem solving. In 1855, he became a high school teacher; despite the long hours required by his teaching duties, Sylow found time to study the papers of Abel. During the school year 1862–1863, Sylow received a temporary appointment at Christiania University and gave lectures on Galois theory and permutation groups. Among his students that year was the great mathematician Sophus Lie (pronounced "Lee"), after whom Lie algebras and Lie groups are named. From 1873 to 1881, Sylow, with some help from Lie, prepared a new edition of Abel's works. In 1902, Sylow and Elling Holst published Abel's correspondence.

Sylow's great discovery, Sylow's Theorem, came in 1872. Upon learning of Sylow's result, C. Jordan called it "one of the essential points in the theory of permutations." The result took on greater importance when the theory of abstract groups flowered in the late 19th century and early 20th century.

In 1869, Sylow was offered a professorship at Christiania University, but turned it down. Upon Sylow's retirement from high school teaching at age 65, Lie mounted a successful campaign to establish a chair for Sylow at Christiania University. Sylow held this position until his death on September 7, 1918.

To find more information about Sylow, visit:

http://www-groups.dcs.st-and.ac.uk/~history

25 Finite Simple Groups

It is a widely held opinion that the problem of classify-
ing finite simple groups is close to a complete solution.
This will certainly be one of the great achievements of
mathematics of this century.

NATHAN JACOBSON

HISTORICAL BACKGROUND

We now come to the El Dorado of finite group theory—the simple
groups. Simple group theory is a vast and difficult subject; we call it
the El Dorado of group theory because of the enormous effort put
forth by hundreds of mathematicians during recent years to discover
and classify all finite simple groups. Let's begin our discussion with
the definition of a simple group and some historical background.

DEFINITION *Simple Group*

A group is *simple* if its only normal subgroups are the identity
subgroup and the group itself.

The notion of a simple group was introduced by Galois about 175
years ago. The simplicity of A_5, the group of even permutations on
five symbols, played a crucial role in his proof that there is not a solution by radicals of the general fifth-degree polynomial (that is, there is
no "quintic formula"). But what makes simple groups important in the
theory of groups? They are important because they play a role in
group theory somewhat analogous to that of primes in number theory
or the elements in chemistry; that is, they serve as the building blocks
for all groups. These building blocks may be determined in the following way. Given a finite group G, choose a normal subgroup G_1 of
$G = G_0$ of largest order. Then the factor group G_0/G_1 is simple, and
we next choose a normal subgroup G_2 of G_1 of largest order. Then

419

G_1/G_2 is also simple, and we continue in this fashion until we arrive at $G_n = \{e\}$. The simple groups $G_0/G_1, G_1/G_2, \ldots, G_{n-1}/G_n$ are called the *composition factors* of G. More than 100 years ago, Jordan and Hölder proved that these factors are independent of the choices of the normal subgroups made in the process described. In a certain sense, a group can be reconstructed from its composition factors, and many of the properties of a group are determined by the nature of its composition factors. This and the fact that many questions about finite groups can be reduced (by induction) to questions about simple groups make clear the importance of determining all finite simple groups.

Just which groups are the simple ones? The Abelian simple groups are precisely Z_n, where $n = 1$ or n is prime. This follows directly from the corollary in Chapter 11. In contrast, it is extremely difficult to describe the non-Abelian simple groups. The best we can do here is to give a few examples and mention a few words about their discovery. It was Galois in 1831 who first observed that A_n is simple for all $n \geq 5$. The next discoveries were made by Jordan in 1870, when he found four infinite families of simple matrix groups over the field Z_p, where p is prime. One such family is the factor group $SL(n, Z_p)/Z(SL(n, Z_p))$, except when $n = 2$ and $p = 2$ or 3. Between the years 1892 and 1905, the American mathematician Leonard Dickson (see Chapter 22 for a biography) generalized Jordan's results to arbitrary finite fields and discovered several new infinite families of simple groups. About the same time, it was shown by G. A. Miller and F. N. Cole that a family of five groups first described by E. Mathieu in 1861 were in fact simple groups. Since these five groups were constructed by ad hoc methods that did not yield infinitely many possibilities, like A_n or the matrix groups over finite fields, they were called "sporadic."

The next important discoveries came in the 1950s. In that decade, many new infinite families of simple groups were found, and the initial steps down the long and winding road that led to the complete classification of all finite simple groups were taken. The first step was Richard Brauer's observation that the centralizer of an element of order 2 was an important tool for studying simple groups. A few years later, John Thompson, in his Ph.D. dissertation, introduced the crucial idea of studying the normalizers of various subgroups of prime-power order.

In the early 1960s came the momentous Feit-Thompson Theorem, which says that a non-Abelian simple group must have even order. This property was first conjectured around 1900 by one of the pioneers of modern group-theoretic methods, the Englishman William

Burnside (see Chapter 29 for a biography). The proof of the Feit-Thompson Theorem filled an entire issue of a journal [2], 255 pages in all (see Figure 25.1). This result provided the impetus to classify the finite simple groups—that is, a program to discover all finite simple groups and *prove* that there are no more to be found. Throughout the 1960s, the methods introduced in the Feit-Thompson proof were generalized and improved with great success by several mathematicians. Moreover, between 1966 and 1975, 19 new sporadic simple groups were discovered. Despite many spectacular achievements, research in simple group theory in the 1960s was haphazard, and the decade ended with many people believing that the classification would never be completed. (The pessimists feared that the sporadic simple groups would foil all attempts. The anonymously written "song" in Figure 25.1 captures the spirit of the times.) Others, more optimistic, were predicting that it would be accomplished in the 1990s.

The 1970s began with Thompson receiving the Fields Medal for his fundamental contributions to simple group theory. This honor is the highest recognition a mathematician can receive (more information about the Fields Medal is given near the end of this chapter). Within a few years, three major events took place that ultimately led to the classification. First, Thompson published what is regarded as the single most important paper in simple group theory—the N-group paper. Here, Thompson introduced many fundamental techniques and supplied a model for the classification of a broad family of simple groups. Second, Daniel Gorenstein produced an elaborate outline for the classification, which he delivered in a series of lectures at the University of Chicago in 1972. Here a program for the overall proof was laid out. The army of researchers now had a battle plan and a commander-in-chief. But this army still needed more and better weapons. Thus came the third critical development: the involvement of Michael Aschbacher. In a dazzling series of papers, Aschbacher combined his own insight with the methods of Thompson, which had been generalized throughout the 1960s, and a geometric approach pioneered by Bernd Fischer to achieve one brilliant result after another in rapid succession. In fact, so much progress was made by Aschbacher and others that, by 1976, it was clear to nearly everyone involved that enough techniques had been developed to complete the classification. Only details remained.

The 1980s were ushered in with Aschbacher following in the footsteps of Feit and Thompson by winning the American Mathematical Society's Cole Prize in algebra (see the last section of this chapter).

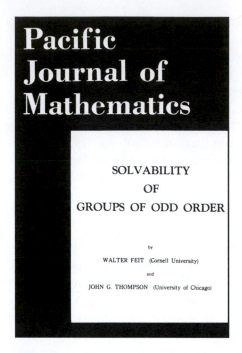

Oh, what are the orders of all simple groups?
I speak of the honest ones, not of the loops.
It seems that old Burnside their orders has
    guessed
Except for the cyclic ones, even the rest.

CHORUS: Finding all groups that are simple
            is no simple task.

Groups made up with permutes will produce
    some more:
For A_n is simple, if n exceeds 4.
Then, there was Sir Matthew who came into
    view
Exhibiting groups of an order quite new.

Still others have come on to study this thing.
Of Artin and Chevalley now we shall sing.
With matrices finite they made quite a list
The question is: Could there be others
    they've missed?

Suzuki and Ree then maintained it's the case
That these methods had not reached the end of
    the chase.
They wrote down some matrices, just four by
    four.
That made up a simple group. Why not make
    more?

And then came the opus of Thompson and
    Feit
Which shed on the problem remarkable light.
A group, when the order won't factor by two
Is cyclic or solvable. That's what is true.

Suzuki and Ree had caused eyebrows to raise,
But the theoreticians they just couldn't faze.
Their groups were not new: if you added a
    twist,
You could get them from old ones with a flick
    of the wrist.

Figure 25.1

Still, some hardy souls felt a thorn in their
    side.
For the five groups of Mathieu all reason
    defied;
Not A_n, not twisted, and not Chevalley,
They called them sporadic and filed them
    away.

Are Mathieu groups creatures of heaven or
    hell?
Zvonimir Janko determined to tell.
He found out that nobody wanted to know:
The masters had missed 1 7 5 5 6 0.

The floodgates were opened! New groups
    were the rage!
(And twelve or more sprouted, to greet the
    new age.)
By Janko and Conway and Fischer and Held
McLaughlin, Suzuki, and Higman, and Sims.

No doubt you noted the last lines don't
    rhyme.
Well, that is, quite simply, a sign of the time.
There's chaos, not order, among simple
    groups;
And maybe we'd better go back to the loops.

A week later, Robert L. Griess made the spectacular announcement that he had constructed the "Monster."[†] The Monster is the largest of the sporadic simple groups. In fact, it has vastly more elements than there are atoms on the earth! Its order is

$$808,017,424,794,512,875,886,459,904,961,710,757,005,754,$$
$$368,000,000,000$$

(hence, the name). This is approximately 8×10^{53}. The Monster is a group of rotations in 196,883 dimensions. Thus, each element can be expressed as a $196,883 \times 196,883$ matrix.

At the annual meeting of the American Mathematical Society in 1981, Gorenstein announced that the "Twenty-five Years' War" to classify all the finite simple groups was over. Group theorists at long last had a list of all finite simple groups and a proof that the list was complete. The proof was spread out over hundreds of papers—both published and unpublished—and ran more than 10,000 pages in length. Because of the proof's extreme length and complexity, and the fact that some key parts of it had not been published, there was some concern in the mathematics community that the classification was not a certainty. By the end of the decade, group theorists had concluded that there was indeed a gap in the unpublished work that would be difficult to rectify. In the mid 1990s, Aschbacher and Stephen Smith began work on this problem. In 2004, at the annual meeting of the American Mathematical Society, Aschbacher announced that he and Smith had completed the classification. Their paper is about 1200 pages in length. Aschbacher concluded his remarks by saying that he would not bet his house that the proof is now error free. Several people who played a central role in the classification are working on a "second generation" proof that will be much shorter and more comprehensible.

[†]The name was coined by John H. Conway.

NONSIMPLICITY TESTS

In view of the fact that simple groups are the building blocks for all groups, it is surprising how scarce the non-Abelian simple groups are. For example, A_5 is the only one whose order is less than 168; there are only five non-Abelian simple groups of order less than 1000 and only 56 of order less than 1,000,000. In this section, we give a few theorems that are useful in proving that a particular integer is not the order of a non-Abelian simple group. Our first such result is an easy arithmetic test that comes from combining Sylow's Third Theorem and the fact that groups of prime-power order have nontrivial centers.

Theorem 25.1   **Sylow Test for Nonsimplicity**

> *Let n be a positive integer that is not prime, and let p be a prime divisor of n. If 1 is the only divisor of n that is equal to 1 modulo p, then there does not exist a simple group of order n.*

PROOF.   If n is a prime-power, then a group of order n has a nontrivial center and, therefore, is not simple. If n is not a prime-power, then every Sylow subgroup is proper, and, by Sylow's Third Theorem, we know that the number of Sylow p-subgroups of a group of order n is equal to 1 modulo p and divides n. Since 1 is the only such number, the Sylow p-subgroup is unique, and therefore, by the corollary to Sylow's Third Theorem, it is normal.   ■ ■

How good is this test? Well, a computer program that applies this criterion to all the nonprime integers between 1 and 200 would leave only the following integers as possible orders of finite simple groups: 12, 24, 30, 36, 48, 56, 60, 72, 80, 90, 96, 105, 108, 112, 120, 132, 144, 150, 160, 168, 180, and 192. (In fact, computer experiments have revealed that for large intervals, say, 500 or more, this test eliminates more than 90% of the nonprime integers as possible orders of simple groups. See [3] for more on this.)

Our next test rules out 30, 90, and 150.

Theorem 25.2   **2 · Odd Test**

> *An integer of the form 2 · n, where n is an odd number greater than 1, is not the order of a simple group.*

PROOF. Let G be a group of order $2n$, where n is odd and greater than 1. Recall from the proof of Cayley's Theorem (Theorem 6.1) that the mapping $g \rightarrow T_g$ is an isomorphism from G to a permutation group on the elements of G [where $T_g(x) = gx$ for all x in G]. Since $|G| = 2n$, Cauchy's Theorem guarantees that there is an element g in G of order 2. Then, when the permutation T_g is written in disjoint cycle form, each cycle must have length 1 or 2; otherwise, $|g| \neq 2$. But T_g can contain no 1-cycles, because the 1-cycle (x) would mean $x = T_g(x) = gx$, so $g = e$. Thus, in cycle form, T_g consists of exactly n transpositions, where n is odd. Therefore, T_g is an odd permutation. This means that the set of even permutations in the image of G is a normal subgroup of index 2. (See Exercise 19 in Chapter 5 and Exercise 7 in Chapter 9.) Hence, G is not simple. ∎

The next theorem is a broad generalization of Cayley's Theorem. We will make heavy use of its two corollaries.

Theorem 25.3 Generalized Cayley Theorem

> Let G be a group and let H be a subgroup of G. Let S be the group of all permutations of the left cosets of H in G. Then there is a homomorphism from G into S whose kernel lies in H and contains every normal subgroup of G that is contained in H.

PROOF. For each $g \in G$, define a permutation T_g of the left cosets of H by $T_g(xH) = gxH$. As in the proof of Cayley's Theorem, it is easy to verify that the mapping of $\alpha:g \rightarrow T_g$ is a homomorphism from G into S.

Now, if $g \in \text{Ker } \alpha$, then T_g is the identity map, so $H = T_g(H) = gH$, and, therefore, g belongs to H. Thus, Ker $\alpha \subseteq H$. On the other hand, if K is normal in G and $K \subseteq H$, then for any $k \in K$ and any x in G, there is an element k' in K such that $kx = xk'$. Thus,

$$T_k(xH) = kxH = xk'H = xH$$

and, therefore, T_k is the identity permutation. This means that $k \in \text{Ker } \alpha$. We have proved, then, that every normal subgroup of G contained in H is also contained in Ker α. ∎

As a consequence of Theorem 25.3, we obtain the following very powerful arithmetic test for nonsimplicity.

Corollary 1   **Index Theorem**

> *If G is a finite group and H is a proper subgroup of G such that |G| does not divide |G:H|!, then H contains a nontrivial normal subgroup of G. In particular, G is not simple.*

PROOF.   Let α be the homomorphism given in Theorem 25.3. Then Ker α is a normal subgroup of G contained in H, and $G/\text{Ker } \alpha$ is isomorphic to a subgroup of S. Thus, $|G/\text{Ker } \alpha| = |G|/|\text{Ker } \alpha|$ divides $|S| = |G:H|!$. Since $|G|$ does not divide $|G:H|!$, the order of Ker α must be greater than 1.   ■ ■

Corollary 2   **Embedding Theorem**

> *If a finite non-Abelian simple group G has a subgroup of index n, then G is isomorphic to a subgroup of A_n.*

PROOF.   Let H be the subgroup of index n and let S_n be the group of all permutations of the n left cosets of H in G. By the Generalized Cayley Theorem, there is a nontrivial homomorphism from G into S_n. Since G is simple and the kernel of a homomorphism is a normal subgroup of G, we see that the mapping from G into S_n is one-to-one, so that G is isomorphic to some subgroup of S_n. Recall from Exercise 19 in Chapter 5 that any subgroup of S_n consists of even permutations only or half even and half odd. If G were isomorphic to a subgroup of the latter type, the even permutations would be a normal subgroup of index 2 (see Exercise 7 in Chapter 9), which would contradict the fact that G is simple. Thus, G is isomorphic to a subgroup of A_n.   ■ ■

Using the Index Theorem with the largest Sylow subgroup for H reduces our list of possible orders of non-Abelian simple groups still further. For example, let G be any group of order $80 = 16 \cdot 5$. We may choose H to be a subgroup of order 16. Since 80 is not a divisor of 5!, there is no simple group of order 80. The same argument applies to 12, 24, 36, 48, 96, 108, 160, and 192, leaving only 56, 60, 72, 105, 112, 120, 132, 144, 168, and 180 as possible orders of non-Abelian simple groups up to 200. Let's consider these orders. Quite often we may use a counting argument to eliminate an integer. Consider 56. By Sylow's Theorem, we know that a simple group of order $56 = 8 \cdot 7$

would contain eight Sylow 7-subgroups and seven Sylow 2-subgroups. Now, any two Sylow p-subgroups that have order p must intersect in only the identity. So the union of the eight Sylow 7-subgroups yields 48 elements of order 7, and the union of any two Sylow 2-subgroups gives at least $8 + 8 - 4 = 12$ new elements. But there are only 56 elements in all. This contradiction shows that there is not a simple group of order 56. An analogous argument also eliminates the integers 105 and 132.

So, our list of possible orders of non-Abelian simple groups up to 200 is down to 60, 72, 112, 120, 144, 168, and 180. Of these, 60 and 168 do correspond to simple groups. The others can be eliminated with a bit of razzle-dazzle.

The easiest case to handle is $112 = 2^4 \cdot 7$. Suppose there were a simple group G of order 112. A Sylow 2-subgroup of G must have index 7. So, by the Embedding Theorem, G is isomorphic to a subgroup of A_7. But 112 does not divide $|A_7|$, which is a contradiction.

Another easy case is 72. Recall from Exercise 19 in Chapter 24 that if we denote the number of Sylow p-subgroups of a group G by n_p, then $n_p = |G : N(H)|$, where H is any Sylow p-subgroup of G, and $n_p \bmod p = 1$. It follows, then, that in a simple group of order 72, we have $n_3 = 4$, which is impossible, since 72 does not divide 4!.

Next consider the possibility of a simple group G of order $144 = 9 \cdot 16$. By the Sylow theorems, we know that $n_3 = 4$ or 16 and $n_2 \geq 3$. The Index Theorem rules out the case where $n_3 = 4$, so we know that there are 16 Sylow 3-subgroups. Now, if every pair of Sylow 3-subgroups had only the identity in common, a straightforward counting argument would produce more than 144 elements. So, let H and H' be a pair of Sylow 3-subgroups whose intersection has order 3. Then $H \cap H'$ is a subgroup of both H and H' and, by the corollary to Theorem 24.2 (or by Exercise 33 in Chapter 24), we see that $N(H \cap H')$ must contain both H and H' and, therefore, the set HH'. (HH' need not be a subgroup.) Thus,

$$|N(H \cap H')| \geq |HH'| = \frac{|H|\,|H'|}{|H \cap H'|} = \frac{9 \cdot 9}{3} = 27.$$

Now, we have three arithmetic conditions on $k = |N(H \cap H')|$. We know that 9 divides k; k divides 144; and $k \geq 27$. Clearly, then, $k \geq 36$, and so $|G : N(H \cap H')| \leq 4$. The Index Theorem now gives us the desired contradiction.

Finally, suppose that G is a non-Abelian simple group of order $180 = 2^2 \cdot 3^2 \cdot 5$. Then $n_5 = 6$ or 36 and $n_3 = 10$. First, assume that $n_5 = 36$. Then G has $36 \cdot 4 = 144$ elements of order 5. Now, if each pair of the Sylow 3-subgroups intersects in only the identity, then there are 80 more elements in the group, which is a contradiction. So, we may assume that there are two Sylow 3-subgroups L_3 and L_3' whose intersection has order 3. Then, as was the case for order 144, we have

$$|N(L_3 \cap L_3')| \geq |L_3 L_3'| = \frac{9 \cdot 9}{3} = 27.$$

Thus,

$$|N(L_3 \cap L_3')| = 9 \cdot k,$$

where $k \geq 3$ and k divides 20. Clearly, then,

$$|N(L_3 \cap L_3')| \geq 36$$

and therefore

$$|G : N(L_3 \cap L_3')| \leq 5.$$

The Index Theorem now gives us another contradiction. Hence, we may assume that $n_5 = 6$. In this case, we let H be the normalizer of a Sylow 5-subgroup of G. By Sylow's Third Theorem, we have $6 = |G:H|$, so that $|H| = 30$. In Chapter 24, we proved that every group of order 30 has an element of order 15. On the other hand, since $n_5 = 6$, G has a subgroup of index 6 and the Embedding Theorem tells us that G is isomorphic to a subgroup of A_6. But A_6 has no element of order 15. (See Exercise 7 in Chapter 5.)

Unfortunately, the argument for 120 is fairly long and complicated. However, no new techniques are required to do it. We leave this as an exercise (Exercise 17). Some hints are given in the answer section.

THE SIMPLICITY OF A_5

Once 120 has been disposed of, we will have shown that the only integers between 1 and 200 that can be the orders of simple groups are 60 and 168. For completeness, we will now prove that A_5, which has order 60, is a simple group. A similar argument can be used to show that the factor group $SL(2, Z_7)/Z(SL(2, Z_7))$ is a simple group of order 168. [This group is denoted by $PSL(2, Z_7)$.]

If A_5 had a nontrivial proper normal subgroup H, then $|H| = 2, 3, 4, 5, 6, 10, 12, 15, 20,$ or 30. By Exercise 43 in Chapter 5, A_5 has 24 elements of order 5, 20 elements of order 3, and no elements of order 15. Now, if $|H| = 3, 6, 12,$ or 15, then $|A_5/H|$ is relatively prime to 3, and by Exercise 55 in Chapter 9, H would have to contain all 20 elements of order 3. If $|H| = 5, 10,$ or 20, then $|A_5/H|$ is relatively prime to 5, and, therefore, H would have to contain the 24 elements of order 5. If $|H| = 30$, then $|A_5/H|$ is relatively prime to both 3 and 5, and so H would have to contain all the elements of orders 3 and 5. Finally, if $|H| = 2$ or 4, then $|A_5/H| = 30$ or 15. But we know from our results in Chapter 24 that any group of order 30 or 15 has an element of order 15. However, since A_5 contains no such element, neither does A_5/H. This proves that A_5 is simple.

The simplicity of A_5 was known to Galois in 1830, although the first formal proof was done by Jordan in 1870. A few years later, Felix Klein showed that the group of rotations of a regular icosahedron is simple and, therefore, isomorphic to A_5 (see Exercise 27). Since then it has frequently been called the *icosahedral group*. Klein was the first to prove that there is a simple group of order 168.

The problem of determining which integers in a certain interval are possible orders for finite simple groups goes back to 1892, when Hölder went up to 200. His arguments for the integers 144 and 180 alone used up 10 pages. By 1975, this investigation had been pushed to well beyond 1,000,000. See [4] for a detailed account of this endeavor. Of course, now that all finite simple groups have been classified, this problem is merely a historical curiosity.

THE FIELDS MEDAL

The highest award for mathematical achievement is the Fields Medal. Two to four such awards are bestowed at the opening session of the International Congress of Mathematicians, held once every four years. Although the Fields Medal is considered by most mathematicians to be the equivalent of the Nobel Prize, there are great differences between these awards. Besides the huge disparity in publicity and monetary value associated with the two honors, the Fields Medal is restricted to those under 40 years of age.[†] This tradition stems from John Charles

---

[†]"Take the sum of human achievement in action, in science, in art, in literature—subtract the work of the men above forty, and while we should miss great treasures, even priceless treasures, we would practically be where we are today. . . . The effective, moving, vitalizing work of the world is done between the ages of twenty-five and forty." Sir William Osler (1849–1919), *Life of Sir William Osler,* vol. I, chap. 24 (The Fixed Period).

Fields's stipulation, in his will establishing the medal, that the awards should be "an encouragement for further achievement." This restriction precluded Andrew Wiles from winning the Fields Medal for his proof of Fermat's Last Theorem.

More details about the Fields Medal can be found in [1].

THE COLE PRIZE

Approximately every five years, beginning in 1928, the American Mathematical Society awards one or two Cole Prizes for research in algebra and one or two Cole Prizes for research in algebraic number theory. The prize was founded in honor of Frank Nelson Cole on the occasion of his retirement as secretary of the American Mathematical Society. In view of the fact that Cole was one of the first people interested in simple groups, it is interesting to note that no fewer than six recipients of the prize—Dickson, Chevalley, Brauer, Feit, Thompson, and Aschbacher—have made fundamental contributions to simple group theory at some time in their careers.

EXERCISES ▬▬

If you don't learn from your mistakes, there's no sense making them.

HERBERT V. PROCHNOW

1. Prove that there is no simple group of order $210 = 2 \cdot 3 \cdot 5 \cdot 7$.
2. Prove that there is no simple group of order $280 = 2^3 \cdot 5 \cdot 7$.
3. Prove that there is no simple group of order $216 = 2^3 \cdot 3^3$.
4. Prove that there is no simple group of order $300 = 2^2 \cdot 3 \cdot 5^2$.
5. Prove that there is no simple group of order $525 = 3 \cdot 5^2 \cdot 7$.
6. Prove that there is no simple group of order $540 = 2^2 \cdot 3^3 \cdot 5$.
7. Prove that there is no simple group of order $528 = 2^4 \cdot 3 \cdot 11$.
8. Prove that there is no simple group of order $315 = 3^2 \cdot 5 \cdot 7$.
9. Prove that there is no simple group of order $396 = 2^2 \cdot 3^2 \cdot 11$.
10. Prove that there is no simple group of order n, where $201 \le n \le 235$.
11. Without using the Generalized Cayley Theorem or its corollaries, prove that there is no simple group of order 112.
12. Without using the "2 · odd" test, prove that there is no simple group of order 210.
13. You may have noticed that all the "hard integers" are even. Choose three odd integers between 200 and 1000. Show that none of these is the order of a simple group unless it is prime.

14. Show that there is no simple group of order pqr, where p, q, and r are primes (p, q, and r need not be distinct).

15. Show that A_5 does not contain a subgroup of order 30, 20, or 15.

16. Show that S_5 does not contain a subgroup of order 40 or 30.

17. Prove that there is no simple group of order $120 = 2^3 \cdot 3 \cdot 5$. (This exercise is referred to in this chapter.)

18. Prove that if G is a finite group and H is a proper normal subgroup of largest order, then G/H is simple.

19. Suppose that H is a subgroup of a finite group G and that $|H|$ and $(|G:H| - 1)!$ are relatively prime. Prove that H is normal in G. What does this tell you about a subgroup of index 2 in a finite group?

20. Suppose that p is the smallest prime that divides $|G|$. Show that any subgroup of index p in G is normal in G.

21. Prove that the only nontrivial proper normal subgroup of S_5 is A_5. (This exercise is referred to in Chapter 32.)

22. Prove that a simple group of order 60 has a subgroup of order 6 and a subgroup of order 10.

23. Show that $PSL(2, Z_7) = SL(2, Z_7)/Z(SL(2, Z_7))$, which has order 168, is a simple group. (This exercise is referred to in this chapter.)

24. Show that the permutations (12) and (12345) generate S_5.

25. Suppose that a subgroup H of S_5 contains a 5-cycle and a 2-cycle. Show that $H = S_5$. (This exercise is referred to in Chapter 32.)

26. Suppose that G is a finite simple group and contains subgroups H and K such that $|G:H|$ and $|G:K|$ are prime. Show that $|H| = |K|$.

27. Show that (up to isomorphism) A_5 is the only simple group of order 60. (This exercise is referred to in this chapter.)

28. Prove that a simple group cannot have a subgroup of index 4.

COMPUTER EXERCISES ■■

One machine can do the work of fifty ordinary men. No machine can do the work of one extraordinary man.

ELBERT HUBBARD, Roycraft Dictionary and Book of Epigrams

Software for the computer exercises in this chapter is available at the website:
http://www.d.umn.edu/~jgallian

1. This software uses a counter M to keep track of how many integers Theorem 25.1 eliminates in any given interval. Run the program for the following intervals: 1–100; 501–600; 5001–5100; 10,001–10,100. How does M seem to behave as the sizes of the integers grow?

2. This software uses a counter M to keep track of how many integers the Index Theorem eliminates in any given interval of integers. Run the program for the same intervals as in Exercise 1. How does M seem to behave as the sizes of the integers grow?

References

1. H. Edwards, "A Short History of the Fields Medal," *The Mathematical Intelligencer* 1 (1978): 127–129.
2. W. Feit and J. G. Thompson, "Solvability of Groups of Odd Order," *Pacific Journal of Mathematics* 13 (1963): 775–1029.
3. J. A. Gallian, "Computers in Group Theory," *Mathematics Magazine* 49 (1976): 69–73.
4. J. A. Gallian, "The Search for Finite Simple Groups," *Mathematics Magazine* 49 (1976): 163–179.

Suggested Readings

G. Cornell, N. Pele, and M. Wage, "Simple Groups of Orders Less Than 1000," *Journal of Undergraduate Research* 5 (1973): 77–86.

In this charming article, three undergraduate students use slightly more theory than was given in this chapter to show that the only integers less than 1000 that could be orders of simple groups arc 60, 168, 320, 504, 660, and 720. All but the last one are orders of simple groups. The proof that there is no simple group of order 720 is omitted because it is significantly beyond most undergraduates.

K. David, "Using Commutators to Prove A_5 Is Simple," *The American Mathematical Monthly* 94 (1987): 775–776.

This note gives an elementary proof that A_5 is simple using commutators.

J. A. Gallian, "The Search for Finite Simple Groups," *Mathematics Magazine* 49 (1976): 163–179.

A historical account is given of the search for finite simple groups.

Martin Gardner, "The Capture of the Monster: A Mathematical Group with a Ridiculous Number of Elements," *Scientific American* 242 (6) (1980): 20–32.

This article gives an elementary introduction to groups and a discussion of simple groups, including the "Monster."

Daniel Gorenstein, "The Enormous Theorem," *Scientific American* 253 (6) (1985): 104–115.

You won't find an article on a complex subject better written for the layperson than this one. Gorenstein, the driving force behind the classification, uses concrete examples, analogies, and nontechnical terms to make the difficult subject matter of simple groups accessible.

Sandra M. Lepsi, "PSL(2, Z_7) Is Simple, by Counting," *Pi Mu Epsilon Journal,* Fall (1993): 576–578.

The author shows that the group $SL(2, Z_7)/Z(SL(2, Z_7))$ of order 168 is simple using a counting argument.

Michael Aschbacher

> *Fresh out of graduate school, he [Aschbacher] had just entered the field, and from that moment he became the driving force behind my program. In rapid succession he proved one astonishing theorem after another. Although there were many other major contributors to this final assault, Aschbacher alone was responsible for shrinking my projected 30-year timetable to a mere 10 years.*

DANIEL GORENSTEIN,
Scientific American

MICHAEL ASCHBACHER was born on April 8, 1944, in Little Rock, Arkansas. Shortly after his birth, his family moved to Illinois, where his father was a professor of accounting and his mother was a high school English teacher. When he was nine years old, his family moved to East Lansing, Michigan; six years later, they moved to Los Angeles.

After high school, Aschbacher enrolled at the California Institute of Technology. In addition to his schoolwork, he passed the first four actuary exams and was employed for a few years as an actuary, full-time in the summers and part-time during the academic year. Two of the Cal Tech mathematicians who influenced him were Marshall Hall and Donald Knuth. In his senior year, Aschbacher took abstract algebra but showed little interest in the course. Accordingly, he received a grade of C.

In 1966, Aschbacher went to the University of Wisconsin for a Ph.D. degree. He completed his dissertation in 1969, and, after spending one year as an assistant professor at the University of Illinois, he returned to Caltech and quickly moved up to the rank of professor.

Aschbacher's dissertation work in the area of combinatorial geometries had led him to consider certain group-theoretic questions. Gradually, he turned his attention more and more to purely group-theoretic problems, particularly those bearing on the classification of finite simple groups. The 1980 Cole Prize Selection Committee said of one of his papers, *"[It] lifted the subject to a new plateau and brought the classification within reach."* Aschbacher has been elected to the National Academy of Sciences, the American Academy of Sciences, and the vice presidency of the American Mathematical Society.

Daniel Gorenstein

Gorenstein was one of the most influential mathematicians of the last few decades.

MICHAEL ASCHBACHER,
Notices of the American Mathematical Society 39 (1992): 1190

DANIEL GORENSTEIN was born in Boston on January 1, 1923. Upon graduating from Harvard in 1943 during World War II, Gorenstein was offered an instructorship at Harvard to teach mathematics to army personnel. After the war ended, he began graduate work at Harvard. He received his Ph.D. degree in 1951, working in algebraic geometry under Oscar Zariski. It was in his dissertation that he introduced the class of rings that is now named after him. In 1951, Gorenstein took a position at Clark University in Worcester, Massachusetts, where he stayed until moving to Northeastern University in 1964. From 1969 until his death on August 26, 1992 he was at Rutgers University.

In 1957, Gorenstein switched from algebraic geometry to finite groups, learning the basic material from I. N. Herstein while collaborating with him over the next few years. A milestone in Gorenstein's development as a group theorist came during 1960–1961, when he was invited to participate in a "Group Theory Year" at the University of Chicago. It was there that Gorenstein, assimilating the revolutionary techniques then being developed by John Thompson, began his fundamental work that contributed to the classification of finite simple groups.

Through his pioneering research papers, his dynamic lectures, his numerous personal contacts, and his influential book on finite groups, Gorenstein became the leader in the 25-year effort, by hundreds of mathematicians, that led to the classification of the finite simple groups.

Among the honors received by Gorenstein are the Steele Prize from the American Mathematical Society and election to membership in the National Academy of Sciences and the American Academy of Arts and Sciences.

To find more information about Gorenstein, visit:

http://www-groups.dcs.st-and.ac.uk/~history/

John Thompson

There seemed to be no limit to his power.

DANIEL GORENSTEIN

JOHN G. THOMPSON was born on October 13, 1932, in Ottawa, Kansas. In 1951, he entered Yale University as a divinity student, but switched to mathematics in his sophomore year. In 1955, he began graduate school at the University of Chicago and obtained his Ph.D. degree four years later. After one year on the faculty at Harvard, Thompson returned to Chicago. He remained there until 1968, when he moved to Cambridge University in England. In 1993, Thompson accepted an appointment at the University of Florida.

Thompson's brilliance was evident early. In his dissertation, he verified a 50-year-old conjecture about finite groups possessing a certain kind of automorphism. (An article about his achievement appeared in *The New York Times*.) The novel methods Thompson used in his dissertation foreshadowed the revolutionary ideas he would later introduce in the Feit-Thompson paper and the classification of minimal simple groups (i.e., simple groups that contain no proper non-Abelian simple subgroups). The assimilation and extension of Thompson's methods by others throughout the 1960s and 1970s ultimately led to the classification of finite simple groups.

In the late 1970s, Thompson made significant contributions to coding theory, the theory of finite projective planes, and the theory of modular functions. His recent work on Galois groups is considered the most important in the field in the last half of the 20th century.

Among Thompson's many honors are the Cole Prize in algebra and the Fields Medal. He was elected to the National Academy of Sciences in 1967, the Royal Society of London in 1979, and the American Academy of Arts and Sciences in 1998. In 2000, President Clinton presented Thompson the National Medal of Science.

To find more information about Thompson, visit:

http://www-groups.dcs.st-and.ac.uk/~history/

26 Generators and Relations

One cannot escape the feeling that these mathematical formulae have an independent existence and an intelligence of their own, that they are wiser than we are, wiser even than their discoverers, that we get more out of them than we originally put into them.

HENRICH HERTZ

MOTIVATION

In this chapter, we present a convenient way to define a group with certain prescribed properties. Simply put, we begin with a set of elements that we want to generate the group, and a set of equations (called *relations*) that specify the conditions that these generators are to satisfy. Among all such possible groups, we will select one that is as large as possible. This will uniquely determine the group up to isomorphism.

To provide motivation for the theory involved, we begin with a concrete example. Consider D_4, the group of symmetries of a square. Recall that $R = R_{90}$ and H, a reflection across a horizontal axis, generate the group. Observe that R and H are related in the following ways:

$$R^4 = H^2 = (RH)^2 = R_0 \qquad \text{(the identity).} \qquad (1)$$

Other relations between R and H, such as $HR = R^3H$ and $RHR = H$, also exist, but they can be derived from those given in Equation (1). For example, $(RH)^2 = R_0$ yields $HR = R^{-1}H^{-1}$, and $R^4 = H^2 = R_0$ yields $R^{-1} = R^3$ and $H^{-1} = H$. So, $HR = R^3H$. In fact, every relation between R and H can be derived from those given in Equation (1).

Thus, D_4 is a group that is generated by a pair of elements a and b subject to the relations $a^4 = b^2 = (ab)^2 = e$ and such that all other relations between a and b can be derived from these relations. This last stipulation is necessary because the subgroup $\{R_0, R_{180}, H, V\}$ of D_4 is generated by the two elements $a = R_{180}$ and $b = H$ that satisfy the relations $a^4 = b^2 = (ab)^2 = e$. However, the "extra" relation $a^2 = e$ satisfied by this subgroup cannot be derived from the original ones (since $R_{90}^2 \neq R_0$). It is natural to ask whether this description of D_4 applies to some other group as well. The answer is no. Any other group generated by two elements α and β satisfying only the relations $\alpha^4 = \beta^2 = (\alpha\beta)^2 = e$, and those that can be derived from these relations, is isomorphic to D_4.

Similarly, one can show that the group $Z_4 \oplus Z_2$ is generated by two elements a and b such that $a^4 = b^2 = e$ and $ab = ba$, and any other relation between a and b can be derived from these relations. The purpose of this chapter is to show that this procedure can be reversed; that is, we can begin with any set of generators and relations among the generators and construct a group that is uniquely described by these generators and relations, subject to the stipulation that all other relations among the generators can be derived from the original ones.

DEFINITIONS AND NOTATION

We begin with some definitions and notation. For any set $S = \{a, b, c, \ldots\}$ of distinct symbols, we create a new set $S^{-1} = \{a^{-1}, b^{-1}, c^{-1}, \ldots\}$ by replacing each x in S by x^{-1}. Define the set $W(S)$ to be the collection of all formal finite strings of the form $x_1 x_2 \cdots x_k$, where each $x_i \in S \cup S^{-1}$. The elements of $W(S)$ are called *words from S*. We also permit the string with no elements to be in $W(S)$. This word is called the *empty word* and is denoted by e.

We may define a binary operation on the set $W(S)$ by juxtaposition; that is, if $x_1 x_2 \cdots x_k$ and $y_1 y_2 \cdots y_t$ belong to $W(S)$, then so does $x_1 x_2 \cdots x_k y_1 y_2 \cdots y_t$. Observe that this operation is associative and the empty word is the identity. Also, notice that a word such as aa^{-1} is not the identity, because we are treating the elements of $W(S)$ as formal symbols with no implied meaning.

At this stage we have everything we need to make a group out of $W(S)$ except inverses. Here a difficulty arises, since it seems reasonable that the inverse of the word ab, say, should be $b^{-1}a^{-1}$. But $abb^{-1}a^{-1}$ is not the empty word! You may recall that we faced a similar obstacle long ago when we carried out the construction of the field of quotients of an integral domain. There we had formal symbols of the form a/b and we wanted the inverse of a/b to be b/a. But their product, $ab/(ba)$, was a formal symbol that was not the same as the formal symbol $1/1$, the identity. So, we proceed here as we did there—by way of equivalence classes.

DEFINITION   *Equivalence Classes of Words*

For any pair of elements u and v of $W(S)$, we say that u is related to v if v can be obtained from u by a finite sequence of insertions or deletions of words of the form xx^{-1} or $x^{-1}x$, where $x \in S$.

We leave it as an exercise to show that this relation is an equivalence relation on $W(S)$. (See Exercise 1.)

EXAMPLE 1   Let $S = \{a, b, c\}$. Then $acc^{-1}b$ is equivalent to ab; $aab^{-1}bbaccc^{-1}$ is equivalent to $aabac$; the word $a^{-1}aabb^{-1}a^{-1}$ is equivalent to the empty word; and the word $ca^{-1}b$ is equivalent to $cc^{-1}caa^{-1}a^{-1}bbca^{-1}ac^{-1}b^{-1}$. Note, however, that $cac^{-1}b$ is not equivalent to ab. ◆

FREE GROUP

Theorem 26.1   Equivalence Classes Form a Group

> Let S be a set of distinct symbols. For any word u in $W(S)$, let $\overline{u}$ denote the set of all words in $W(S)$ equivalent to u (that is, $\overline{u}$ is the equivalence class containing u). Then the set of all equivalence classes of elements of $W(S)$ is a group under the operation $\overline{u} \cdot \overline{v} = \overline{uv}$.

PROOF.   This proof is left to the reader.   ■ ■

The group defined in Theorem 26.1 is called a *free group on S*. Theorem 26.2 shows why free groups are important.

Theorem 26.2   The Universal Mapping Property

> Every group is a homomorphic image of a free group.

PROOF.   Let G be a group and let S be a set of generators for G. (Such a set exists, because we may take S to be G itself.) Now let F be the free group on S. Unfortunately, since our notation for any word in $W(S)$ also denotes an element of G, we have created a notational problem for ourselves. So, to distinguish between these two cases, we will denote the word $x_1x_2 \cdots x_n$ in $W(S)$ by $(x_1x_2 \cdots x_n)_F$ and the product $x_1x_2 \cdots x_n$ in G by $(x_1x_2 \cdots x_n)_G$. As before, $\overline{x_1x_2 \cdots x_n}$ denotes the equivalence class in F containing the word $(x_1x_2 \cdots x_n)_F$ in $W(S)$. Notice that $(\overline{x_1x_2 \cdots x_n})$ and $(x_1x_2 \cdots x_n)_G$ may be entirely different elements, since the operations on F and G are different.

Now consider the mapping from F into G given by

$$\phi((\overline{x_1x_2 \cdots x_n})) = (x_1x_2 \cdots x_n)_G.$$

[All we are doing is taking a product in F and viewing it as a product in G. For example, if G is the cyclic group of order 4 generated by a, then

$$\phi((\overline{aaaaa})) = (aaaaa)_G = a.]$$

Clearly, ϕ is well defined, for inserting or deleting expressions of the form xx^{-1} or $x^{-1}x$ in elements of $W(S)$ corresponds to inserting or deleting the identity in G. To check that ϕ is operation-preserving, observe that

$$\phi((\overline{x_1x_2 \cdots x_n})\,(\overline{y_1y_2 \cdots y_m})) = \phi((\overline{x_1x_2 \cdots x_ny_1y_2 \cdots y_m}))$$
$$= (x_1x_2 \cdots x_ny_1y_2 \cdots y_m)_G$$
$$= (x_1x_2 \cdots x_n)_G(y_1y_2 \cdots y_m)_G.$$

Finally, ϕ is onto G because S generates G.   ■ ■

The following corollary is an immediate consequence of Theorem 26.2 and the First Isomorphism Theorem for Groups.

Corollary   **Universal Factor Group Property**

> *Every group is isomorphic to a factor group of a free group.*

GENERATORS AND RELATIONS

We have now laid the foundation for defining a group by way of generators and relations. Before giving the definition, we will illustrate the basic idea with an example.

EXAMPLE 2   Let F be the free group on the set $\{a, b\}$ and let N be the smallest normal subgroup of F containing the set $\{a^4, b^2, (ab)^2\}$. We will show that F/N is isomorphic to D_4. We begin by observing that the mapping ϕ from F onto D_4, which takes a to R_{90} and b to H (horizontal reflection), defines a homomorphism whose kernel contains N. Thus, $F/\text{Ker } \phi$ is isomorphic to D_4. On the other hand, we claim that the set

$$K = \{N, aN, a^2N, a^3N, bN, abN, a^2bN, a^3bN\}$$

of left cosets of N is F/N itself. To see this, notice that every member of F/N can be generated by starting with N and successively multiplying on the left by various combinations of a's and b's. So, it suffices to show that K is closed under multiplication on the left by a and b. It is trivial that K is closed under left multiplication by a. For b, we will do only one of the eight cases. The others can be done in a similar fashion. Consider $b(aN)$. Since $b^2, abab, a^4 \in N$ and $Nb = bN$, we have $baN = baNb^2 = babNb = a^{-1}(abab)Nb = a^{-1}Nb = a^{-1}a^4Nb = a^3Nb = a^3bN$. Upon completion of the other cases (Exercise 3), we know that F/N has at most eight elements. At the same time, we know that $F/\text{Ker } \phi$ has exactly eight elements. Since $F/\text{Ker } \phi$ is a factor group of F/N [indeed, $F/\text{Ker } \phi \approx (F/N)/(\text{Ker } \phi/N)$], it follows that F/N also has eight elements and $F/N = F/\text{Ker } \phi \approx D_4$. ◆

DEFINITION   *Generators and Relations*

Let G be a group generated by some set $A = \{a_1, a_2, \ldots, a_n\}$ and let F be the free group on A. Let $W = \{w_1, w_2, \ldots, w_t\}$ be a subset of F and let N be the smallest normal subgroup of F containing W. We say that G is *given by the generators* $a_1, a_2, \ldots, a_n$ *and the relations* $w_1 = w_2 = \cdots = w_t = e$ if there is an isomorphism from F/N onto G that carries a_iN to a_i.

The notation for this situation is

$$G = \langle a_1, a_2, \ldots, a_n \mid w_1 = w_2 = \cdots = w_t = e \rangle.$$

As a matter of convenience, we have restricted the number of generators and relations in our definition to be finite. This restriction is not necessary, however. Also, it is often more convenient to write a relation in implicit form. For example, the relation $a^{-1}b^{-3}ab = e$ is often written as $ab = b^3a$. In practice, one does not bother writing down the normal subgroup N that contains the relations. Instead, one just manipulates the generators and treats anything in N as the identity, as our notation suggests. Rather than saying that G is given by

$$\langle a_1, a_2, \ldots, a_n \mid w_1 = w_2 = \cdots = w_t = e \rangle,$$

many authors prefer to say that G has the *presentation*

$$\langle a_1, a_2, \ldots, a_n \mid w_1 = w_2 = \cdots = w_t = e\rangle.$$

Notice that a free group is "free" of relations; that is, the equivalence class containing the empty word is the only relation. We mention in passing the fact that a subgroup of a free group is also a free group. Free groups are of fundamental importance in a branch of algebra known as combinatorial group theory.

EXAMPLE 3 The discussion in Example 2 can now be summed up by writing

$$D_4 = \langle a, b \mid a^4 = b^2 = (ab)^2 = e\rangle. \qquad \blacklozenge$$

EXAMPLE 4 The group of integers is the free group on one letter; that is, $Z \approx \langle a\rangle$. (This is the only nontrivial Abelian group that is free.) ◆

The next theorem formalizes the argument used in Example 2 to prove that the group defined there has eight elements.

Theorem 26.3 (Dyck, 1882)

Let

$$G = \langle a_1, a_2, \ldots, a_n \mid w_1 = w_2 = \cdots = w_t = e\rangle$$

and let

$$\overline{G} = \langle a_1, a_2, \ldots, a_n \mid w_1 = w_2 = \cdots = w_t$$
$$= w_{t+1} = \cdots = w_{t+k} = e\rangle.$$

Then $\overline{G}$ is a homomorphic image of G.

PROOF. See Exercise 5. ∎

Corollary Largest Group Satisfying Defining Relations

If K is a group satisfying the defining relations of a finite group G and $|K| \geq |G|$, then K is isomorphic to G.

PROOF. See Exercise 5. ∎

EXAMPLE 5 Quaternions
Consider the group $G = \langle a, b \mid a^2 = b^2 = (ab)^2\rangle$. What does G look like? Formally, of course, G is isomorphic to F/N, where F is free on

$\{a, b\}$ and N is the smallest normal subgroup of F containing $b^{-2}a^2$ and $(ab)^{-2}a^2$. But as we have already said, we need not use N. We just think of the elements of G as words in a and b, where $a^2 = b^2 = (ab)^2$. Now, let $H = \langle b \rangle$ and $S = \{H, aH\}$. Then, just as in Example 2, it follows that S is closed under multiplication by a and b from the left. So, as in Example 2, we have $G = H \cup aH$. Thus, we can determine the elements of G once we know exactly how many elements there are in H. (Here again, the three relations come in.) To do this, first observe that $b^2 = (ab)^2 = abab$ implies $b = aba$. Then $a^2 = b^2 = (aba)(aba) = aba^2ba = ab^4a$ and therefore $b^4 = e$. Hence, H has at most four elements, and therefore G has at most eight—namely, e, b, b^2, b^3, a, ab, ab^2, and ab^3. It is conceivable, however, that not all of these eight elements are distinct. For example, $Z_2 \oplus Z_2$ satisfies the defining relations and has only four elements. Perhaps it is the largest group satisfying the relations. How can we show that the eight elements listed above are distinct? Well, consider the group $\overline{G}$ generated by the matrices

$$A = \begin{bmatrix} 0 & 1 \\ -1 & 0 \end{bmatrix} \quad \text{and} \quad B = \begin{bmatrix} 0 & i \\ i & 0 \end{bmatrix},$$

where $i = \sqrt{-1}$. Direct calculations show that in $\overline{G}$ the elements e, B, B^2, B^3, A, AB, AB^2, and AB^3 are distinct and that $\overline{G}$ satisfies the relations $A^2 = B^2 = (AB)^2$. So, it follows from the corollary to Dyck's Theorem that G has order 8. ◆

The next example illustrates why, in Examples 2 and 5, it is necessary to show that the eight elements listed for the group are distinct.

EXAMPLE 6   Let

$$G = \langle a, b \mid a^3 = b^9 = e, a^{-1}ba = b^{-1} \rangle.$$

Once again, we let $H = \langle b \rangle$ and observe that $G = H \cup aH \cup a^2H$. Thus,

$$G = \{a^i b^j \mid 0 \le i \le 2, 0 \le j \le 8\},$$

and therefore G has at most 27 elements. But this time we will not be able to find some concrete group of order 27 satisfying the same relations that G does, for notice that $b^{-1} = a^{-1}ba$ implies

$$b = (a^{-1}ba)^{-1} = a^{-1}b^{-1}a.$$

Hence,

$$b = ebe = a^{-3}ba^3 = a^{-2}(a^{-1}ba)a^2 = a^{-2}b^{-1}a^2$$

$$= a^{-1}(a^{-1}b^{-1}a)a = a^{-1}ba = b^{-1}.$$

So, the original three relations imply the additional relation $b^2 = e$. But $b^2 = e = b^9$ further implies $b = e$. It follows, then, that G has at most three distinct elements—namely, e, a, and a^2. But Z_3 satisfies the defining relations with $a = 1$ and $b = 0$. So, $|G| = 3$. ◆

We hope Example 6 convinces you of the fact that, once a list of the elements of the group given by a set of generators and relations has been obtained, one must further verify that this list has no duplications. Typically, this is accomplished by exhibiting a specific group that satisfies the given set of generators and relations and that has the same size as the list. Obviously, experience plays a role here.

Here is a fun example adapted from [1].

EXAMPLE 7 Let G be the group with the 26 letters of the alphabet as generators. For relations we take strings $A = B$, where A and B are words in some fixed reference, say [2], and have the same pronunciation but different meanings (such words are called homophones). For example, $buy = by = bye$, $hour = our$, $lead = led$, $whole = hole$. From these strings and cancellation, we obtain $u = e = h = a = w = \emptyset$ ($\emptyset$ is the identity string). With these examples in mind, we ask, What is the group given by these generators and relations? Surprisingly, the answer is the infinite cyclic group generated by v. To verify this, one must show that every letter except v is equivalent to $\emptyset$ and that there are no two homophones that contain a different number of v's. The former can easily be done with common words. For example, from $inn = in$, $plumb = plum$, and $knot = not$, we see that $n = b = k = \emptyset$. From $too = to$ we have $o = \emptyset$. Thus, $w = \emptyset$ as well. That there are no two homophones in [2] that have a different number of v's can be verified by simply checking all cases. In contrast, the reference *Handbook of Homophones* by W. C. Townsend, available at the website http://www.peak.org/~jeremy/dictionary/chapters, lists felt/veldt as homophones. Of course, including these makes the group trivial. ◆

CLASSIFICATION OF GROUPS OF ORDER UP TO 15

The next theorem illustrates the utility of the ideas presented in this chapter.

Theorem 26.4   Classification of Groups of Order 8 (Cayley, 1859)

> *Up to isomorphism, there are only five groups of order 8: Z_8, $Z_4 \oplus Z_2$, $Z_2 \oplus Z_2 \oplus Z_2$, D_4, and the quaternions.*

PROOF. The Fundamental Theorem of Finite Abelian Groups takes care of the Abelian cases. Now, let G be a non-Abelian group of order 8. Also, let $G_1 = \langle a, b \mid a^4 = b^2 = (ab)^2 = e \rangle$ and let $G_2 = \langle a, b \mid a^2 = b^2 = (ab)^2 \rangle$. We know from the preceding examples that G_1 is isomorphic to D_4 and G_2 is isomorphic to the quaternions. Thus, it suffices to show that G must satisfy the defining relations for G_1 or G_2. It follows from Exercise 33 in Chapter 2 and Lagrange's Theorem that G has an element of order 4; call it a. Then, if b is any element of G not in $\langle a \rangle$, we know that

$$G = \langle a \rangle \cup \langle a \rangle b = \{e, a, a^2, a^3, b, ab, a^2b, a^3b\}.$$

Consider the element b^2 of G. Which of the eight elements of G can it be? Not b, ab, a^2b, or a^3b, by cancellation. Not a, for b^2 commutes with b and a does not. Not a^3, for the same reason. Thus, $b^2 = e$ or $b^2 = a^2$. Suppose $b^2 = e$. Since $\langle a \rangle$ is a normal subgroup of G, we know that $bab^{-1} \in \langle a \rangle$. From this and the fact that $|bab^{-1}| = |a|$, we then conclude that $bab^{-1} = a$ or $bab^{-1} = a^{-1}$. The first relation would mean that G is Abelian, so we know that $bab^{-1} = a^{-1}$. But then, since $b^2 = e$, we have $(ab)^2 = e$, and therefore G satisfies the defining relations for G_1.

Finally, if $b^2 = a^2$ holds instead of $b^2 = e$, we can show by an argument like that already given that $(ab)^2 = a^2$, and therefore G satisfies the defining relations for G_2. ■ ■

TABLE 26.1 Classification of Groups of Order Up to 15

Order	Abelian Groups	Non-Abelian Groups
1	Z_1	
2	Z_2	
3	Z_3	
4	$Z_4, Z_2 \oplus Z_2$	
5	Z_5	
6	Z_6	D_3
7	Z_7	
8	$Z_8, Z_4 \oplus Z_2, Z_2 \oplus Z_2 \oplus Z_2$	D_4, Q_4
9	$Z_9, Z_3 \oplus Z_3$	
10	Z_{10}	D_5
11	Z_{11}	
12	$Z_{12}, Z_6 \oplus Z_2$	D_6, A_4, Q_6
13	Z_{13}	
14	Z_{14}	D_7
15	Z_{15}	

The classification of the groups of order 8, together with our results on groups of order p^2, $2p$, and pq from Chapter 24, allow us to classify the groups of order up to 15, with the exception of those of order 12. We already know four groups of order 12—namely, Z_{12}, $Z_6 \oplus Z_2$, D_6, and A_4. An argument along the lines of Theorem 26.4 can be given to show that there is only one more group of order 12. This group, called the *dicyclic group of order 12* and denoted by Q_6, has presentation $\langle a, b \mid a^6 = e, a^3 = b^2, b^{-1}ab = a^{-1} \rangle$. Table 26.1 lists the groups of order at most 15. We use Q_4 to denote the quaternions (see Example 5 in this chapter).

CHARACTERIZATION OF DIHEDRAL GROUPS

As another nice application of generators and relations, we will now give a characterization of the dihedral groups that has been known for more than 100 years. For $n \geq 3$, we have used D_n to denote the group of symmetries of a regular n-gon. Imitating Example 2, one can show that $D_n \approx \langle a, b \mid a^n = b^2 = (ab)^2 = e \rangle$ (see Exercise 9). By analogy, these generators and relations serve to define D_1 and D_2 also. (These are also called dihedral groups.) Finally, we define the infinite dihedral group D_∞ as $\langle a, b \mid a^2 = b^2 = e \rangle$. The elements of D_∞ can be listed as e, a, b, ab, ba, $(ab)a$, $(ba)b$, $(ab)^2$, $(ba)^2$, $(ab)^2a$, $(ba)^2b$, $(ab)^3$, $(ba)^3$, ...

Theorem 26.5   **Characterization of Dihedral Groups**

> *Any group generated by a pair of elements of order 2 is dihedral.*

PROOF.   Let G be a group generated by a pair of distinct elements of order 2, say, a and b. We consider the order of ab. If $|ab| = \infty$, then G is infinite and satisfies the relations of D_∞. We will show that G is isomorphic to D_∞. By Dyck's Theorem, G is isomorphic to some factor group of D_∞, say, D_∞/H. Now, suppose $h \in H$ and $h \neq e$. Since every element of D_∞ has one of the forms $(ab)^i$, $(ba)^i$, $(ab)^ia$, or $(ba)^ib$, by symmetry, we may assume that $h = (ab)^i$ or $h = (ab)^ia$. If $h = (ab)^i$, we will show that D_∞/H satisfies the relations for D_i given in Exercise 9. Since $(ab)^i$ is in H, we have

$$H = (ab)^iH = (abH)^i,$$

so that $(abH)^{-1} = (abH)^{i-1}$. But

$$(ab)^{-1}H = b^{-1}a^{-1}H = baH,$$

and it follows that

$$aHabHaH = baH = (abH)^{-1}.$$

Thus,

$$D_\infty/H = \langle aH, bH \rangle = \langle aH, abH \rangle$$

(see Exercise 7), and D_∞/H satisfies the defining relations for D_i (use Exercise 9 with $x = aH$ and $y = abH$). In particular, G is finite—an impossibility.

If $h = (ab)^i a$, then

$$H = (ab)^i aH = (ab)^i HaH,$$

and therefore

$$(abH)^i = (ab)^i H = (aH)^{-1} = a^{-1}H = aH.$$

It follows that

$$\langle aH, bH \rangle = \langle aH, abH \rangle \subseteq \langle abH \rangle.$$

However,

$$(abH)^{2i} = (aH)^2 = a^2H = H,$$

so that D_∞/H is again finite. This contradiction forces $H = \{e\}$ and G to be isomorphic to D_∞.

Finally, suppose that $|ab| = n$. Since $G = \langle a, b \rangle = \langle a, ab \rangle$, we can show that G is isomorphic to D_n by proving that $b(ab)b = (ab)^{-1}$, which is the same as $ba = (ab)^{-1}$ (see Exercise 9). But $(ab)^{-1} = b^{-1}a^{-1} = ba$, since a and b have order 2. ∎

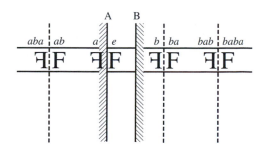

Figure 26.1 The group D_∞—reflections in parallel mirrors.

REALIZING THE DIHEDRAL GROUPS
WITH MIRRORS

A geometric realization of D_∞ can be obtained by placing two mirrors facing each other in a parallel position, as shown in Figure 26.1. If we let a and b denote reflections in mirrors A and B, respectively, then ab, viewed as the composition of a and b, represents a translation through twice the distance between the two mirrors to the left, and ba is the translation through the same distance to the right.

The finite dihedral groups can also be realized with a pair of mirrors. For example, if we place a pair of mirrors facing each other at a $45°$ angle, we obtain the group D_4. Notice that in Figure 26.2, the effect of reflecting an object in mirror A, then mirror B, is a rotation of twice the angle between the two mirrors (that is, $90°$).

In Figure 26.3, we see a portion of the pattern produced by reflections in a pair of mirrors set at a $1°$ angle. The corresponding group is D_{180}. In general, reflections in a pair of mirrors set at the angle $180°/n$ correspond to the group D_n. As n becomes larger and larger, the mirrors approach a parallel position. In the limiting case, we have the group D_∞.

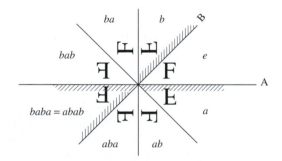

Figure 26.2   The group D_4—reflections in mirrors at a $45°$ angle.

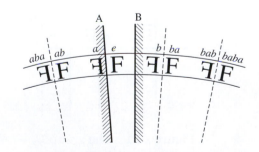

Figure 26.3   The group D_{180}—reflections in mirrors at a $1°$ angle.

We conclude this chapter by commenting on the advantages and disadvantages of using generators and relations to define groups. The principal advantage is that in many situations—particularly in knot theory, algebraic topology, and geometry—groups defined by way of generators and relations arise in a natural way. Within group theory itself, it is often convenient to construct examples and counterexamples with generators and relations. Among the disadvantages of defining a group by generators and relations is the fact that it is often difficult to decide whether or not the group is finite, or even whether or not a particular element is the identity. Furthermore, the same group can be defined with entirely different sets of generators and relations, and, given two groups defined by generators and relations, it is often extremely difficult to decide whether or not these two groups are isomorphic. Nowadays, these questions are frequently tackled with the aid of a computer.

EXERCISES

It don't come easy.

TITLE OF A SONG BY RINGO STARR, MAY 1971

1. Let S be a set of distinct symbols. Show that the relation defined on $W(S)$ in this chapter is an equivalence relation.

2. Let n be an even integer. Prove that $D_n/Z(D_n)$ is isomorphic to $D_{n/2}$.

3. Verify that the set K in Example 2 is closed under multiplication on the left by b.

4. Show that $\langle a, b \mid a^5 = b^2 = e, ba = a^2b \rangle$ is isomorphic to Z_2.

5. Prove Theorem 26.3 and its corollary.

6. Let G be the group $\{\pm 1, \pm i, \pm j, \pm k\}$ with multiplication defined as in Exercise 48 in Chapter 9. Show that G is isomorphic to $\langle a, b \mid a^2 = b^2 = (ab)^2 \rangle$. (Hence, the name "quaternions.")

7. In any group, show that $\langle a, b \rangle = \langle a, ab \rangle$. (This exercise is referred to in the proof of Theorem 26.5.)

8. Let $\alpha = (12)(34)$ and $\beta = (24)$. Show that the group generated by α and β is isomorphic to D_4.

9. Prove that $G = \langle x, y \mid x^2 = y^n = e, xyx = y^{-1} \rangle$ is isomorphic to D_n. (This exercise is referred to in the proof of Theorem 26.5.)

10. What is the minimum number of generators needed for $Z_2 \oplus Z_2 \oplus Z_2$? Find a set of generators and relations for this group.

11. Suppose that $x^2 = y^2 = e$ and $yz = zxy$. Show that $xy = yx$.

12. Let $G = \langle a, b \mid a^2 = b^4 = e, ab = b^3a \rangle$.
    a. Express $a^3b^2abab^3$ in the form b^ia^j.
    b. Express b^3abab^3a in the form b^ia^j.

13. Let $G = \langle a, b \mid a^2 = b^2 = (ab)^2 \rangle$.

    a. Express b^2abab^3 in the form $b^i a^j$.

    b. Express b^3abab^3a in the form $b^i a^j$.

14. Let G be the group defined by the following table. Show that G is isomorphic to D_n.

	1	2	3	4	5	6	$\cdots$	2n
1	1	2	3	4	5	6	$\cdots$	2n
2	2	1	2n	2n − 1	2n − 2	2n − 3	$\cdots$	3
3	3	4	5	6	7	8	$\cdots$	2
4	4	3	2	1	2n	2n − 1	$\cdots$	5
5	5	6	7	8	9	10	$\cdots$	4
6	6	5	4	3	2	1	$\cdots$	7
$\vdots$	$\vdots$	$\vdots$	$\vdots$	$\vdots$	$\vdots$	$\vdots$	$\vdots$	$\vdots$
2n	2n	2n − 1	2n − 2	2n − 3	2n − 4	2n − 5	$\cdots$	1

15. Let $G = \langle x, y \mid x^8 = y^2 = e, yxyx^3 = e \rangle$. Show that $|G| \leq 16$. Assuming that $|G| = 16$, find the center of G and the order of xy.

16. Classify all groups of order up to 11.

17. Let G be defined by some set of generators and relations. Show that every factor group of G satisfies the relations defining G.

18. Let $G = \langle s, t \mid sts = tst \rangle$. Show that the permutations (23) and (13) satisfy the defining relations of G. Explain why this proves that G is non-Abelian.

19. In $D_{12} = \langle x, y \mid x^2 = y^{12} = e, xyx = y^{-1} \rangle$, prove that the subgroup $H = \langle x, y^3 \rangle$ (which is isomorphic to D_4) is not a normal subgroup.

20. Let $G = \langle x, y \mid x^{2n} = e, x^n = y^2, y^{-1}xy = x^{-1} \rangle$. Show that $Z(G) = \{e, x^n\}$. Assuming that $|G| = 4n$, show that $G/Z(G)$ is isomorphic to D_n. (The group G is called the *dicyclic* group of order $4n$.)

21. Let $G = \langle a, b \mid a^6 = b^3 = e, b^{-1}ab = a^3 \rangle$. How many elements does G have? To what familiar group is G isomorphic?

22. Let $G = \langle x, y \mid x^4 = y^4 = e, xyxy^{-1} = e \rangle$. Show that $|G| \leq 16$. Assuming that $|G| = 16$, find the center of G and show that $G/\langle y^2 \rangle$ is isomorphic to D_4.

23. Determine the orders of the elements of D_∞.

24. Let $G = \left\{ \begin{bmatrix} 1 & a & b \\ 0 & 1 & c \\ 0 & 0 & 1 \end{bmatrix} \,\middle|\, a, b, c \in Z_2 \right\}$. Prove that G is isomorphic to D_4.

25. Let $G = \langle a, b, c, d \mid ab = c, bc = d, cd = a, da = b \rangle$. Determine $|G|$.

26. Referring to Example 7 in this chapter, show as many letters as you can that are equivalent to $\emptyset$.

27. Suppose that a group of order 8 has exactly five elements of order 2. Identify the group.

References

1. J.-F. Mestre, R. Schoof, L. Washington, D. Zagier, "Quotient homophones des groupes libres [Homophonic quotients of free groups]," *Experimental Mathematics* 2 (1993): 153–155.
2. H. C. Whitford, *A Dictionary of American Homophones and Homographs,* New York: Teachers College Press, 1962.

Suggested Readings

Alexander H. Fran, Jr. and David Singmaster, *Handbook of Cubik Math,* Hillside, N. J.: Enslow, 1982.

This book is replete with the group-theoretic aspects of the Magic Cube. It uses permutation group theory and generators and relations to discuss the solutions to the cube and related results. The book has numerous challenging exercises stated in group-theoretic terms.

Lee Neuwirth, "The Theory of Knots," *Scientific American* 240 (1979): 110–124.

This article shows how a group can be associated with a knotted string. Mathematically, a knot is just a one-dimensional curve situated in three-dimensional space. The theory of knots—a branch of topology—seeks to classify and analyze the different ways of embedding such a curve. Around the beginning of the 20th century, Henri Poincaré observed that important geometric characteristics of knots could be described in terms of group generators and relations—the so-called knot group. Among other knots, Neuwirth describes the construction of the knot group for the cloverleaf knot pictured. One set of generators and relations for this group is $\langle x, y, z \mid xy = yz, zx = yz \rangle$.

The cloverleaf knot.

David Peifer, "An Introduction to Combinatorial Group Theory and the Word Problem," *Mathematics Magazine* 70 (1997): 3–10.

This article discusses some fundamental ideas and problems regarding groups given by presentations.

Marshall Hall, Jr.

Professor Hall was a mathematician in the broadest sense of the word but with a predilection for group theory, geometry and combinatorics.

HANS ZASSENHAUS, *Notices of the American Mathematical Society*

MARSHALL HALL, JR., was born on September 17, 1910, in St. Louis, Missouri. He demonstrated interest in mathematics at the age of 11 when he constructed a seven-place table of logarithms for the positive integers up to 1000. He completed a B.A. degree in 1932 at Yale. After spending a year at Cambridge University, where he worked with Philip Hall, Harold Davenport, and G. H. Hardy, he returned to Yale for his Ph.D. degree. After a year at the Institute for Advanced Study, he returned to Yale as an instructor until the outbreak of World War II, when he joined Naval Intelligence. During the war, Hall had significant success in deciphering both the Japanese codes and the German Enigma messages. These successes helped to turn the tide of the war. In 1945, Hall returned to Yale for a year before going to Ohio State University. Hall spent the years from 1959 to 1981 at Caltech and the years from 1981 until his death on July 4, 1990, at Emory University.

Hall's highly regarded books on group theory and combinatorial theory are classics. His mathematical legacy includes more than 120 research papers on group theory, coding theory, and design theory. His 1943 paper on projective planes ranks among the most cited papers in mathematics. Several fundamental concepts as well as a sporadic simple group are identified with Hall's name. One of Hall's most celebrated results is his solution to the "Burnside Problem" for exponent 6—that is, a finitely generated group in which the order of every element divides 6, must be finite. Hall influenced both John Thompson and Michael Aschbacher, two of finite group theory's greatest contributors. It was Hall who suggested Thompson's Ph.D. dissertation problem. Hall's Ph.D. students at Caltech included Donald Knuth and Robert McEliece.

To find more information about Hall, visit:

http://www–groups.dcs.st–and.ac.uk/~history/

27

Symmetry Groups

It [group theory] provides a sensitive instrument for investigating symmetry, one of the most pervasive and elemental phenomena of the real world.

M. I. KARGAPOLOV AND JU. I. MARZLJAKOV,
Fundamentals of the Theory of Groups

ISOMETRIES

In the early chapters of this book, we briefly discussed symmetry groups. In this chapter and the next, we examine this fundamentally important concept in some detail. It is convenient to begin such a discussion with the definition of an isometry (from the Greek *isometros*, meaning "equal measure") in $\mathbf{R}^n$.

DEFINITION *Isometry*

An *isometry* of n-dimensional space $\mathbf{R}^n$ is a function from $\mathbf{R}^n$ onto $\mathbf{R}^n$ that preserves distance.

In other words, a function T from $\mathbf{R}^n$ onto $\mathbf{R}^n$ is an isometry if, for every pair of points p and q in $\mathbf{R}^n$, the distance from $T(p)$ to $T(q)$ is the same as the distance from p to q. With this definition, we may now make precise the definition of the symmetry group of an n-dimensional figure.

DEFINITION *Symmetry Group of a Figure in $\mathbf{R}^n$*

Let F be a set of points in $\mathbf{R}^n$. The *symmetry group of F in $\mathbf{R}^n$* is the set of all isometries of $\mathbf{R}^n$ that carry F onto itself. The group operation is function composition.

It is important to realize that the symmetry group of an object depends not only on the object, but also on the space in which we view it. For example, the symmetry group of a line segment in $\mathbf{R}^1$ has order 2, the symmetry group of a line segment considered as a set of points in $\mathbf{R}^2$ has order 4, and the symmetry group of a line segment viewed as a set of points in $\mathbf{R}^3$ has infinite order (see Exercise 9).

Although we have formulated our definitions for all finite dimensions, our chief interest will be the two-dimensional case. It has been known since 1831 that every isometry of $\mathbf{R}^2$ is one of four types: rotation, reflection, translation, and glide-reflection (see [1, p. 46]). Rotation about a point in a plane needs no explanation. A *reflection across a line L* is that transformation that leaves every point of L fixed and takes every point Q, not on L, to the point Q' so that L is the perpendicular bisector of the line segment from Q to Q' (see Figure 27.1). The line L is called the *axis of reflection*. In an xy-coordinate plane, the transformation $(x, y) \rightarrow (x, -y)$ is a reflection across the x-axis, whereas $(x, y) \rightarrow (y, x)$ is a reflection across the line $y = x$. Some authors call an axis of reflective symmetry L a *mirror* because L acts like a two-sided mirror; that is, the image of a point Q in a mirror placed on the line L is, in fact, the image of Q under the reflection across the line L. Reflections are called *opposite* isometries because they reverse orientation. For example, the reflected image of a clockwise spiral is a counterclockwise spiral. Similarly, the reflected image of a right hand is a left hand. (See Figure 27.1.)

A *translation* is simply a function that carries all points the same distance in the same direction. For example, if p and q are points in a plane and T is a translation, then the two vectors joining p to $T(p)$ and q to $T(q)$ have the same length and direction. A *glide-reflection* is the product of a translation and a reflection across the line containing the translation vector. In Figure 27.2, the vector gives the direction and length of the translation, and is contained in the axis of reflection. A glide-reflection is also an opposite isometry. Successive footprints in wet sand are related by a glide-reflection.

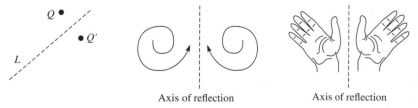

Axis of reflection   Axis of reflection

Figure 27.1   Reflected images.

Figure 27.2   Glide-reflection.

Figure 27.3   *Aurelia insulinda,* an organism whose plane symmetry group is Z_4.

CLASSIFICATION OF FINITE PLANE SYMMETRY GROUPS

Our first goal in this chapter is to classify all finite plane symmetry groups. As we have seen in earlier chapters, the dihedral group D_n is the plane symmetry group of a regular n-gon. (For convenience, call D_2 the plane symmetry group of a nonsquare rectangle and D_1 the plane symmetry group of the letter "V." In particular, $D_2 \approx Z_2 \oplus Z_2$ and $D_1 \approx Z_2$.) The cyclic groups Z_n are easily seen to be plane symmetry groups also. Figure 27.3 is an illustration of an organism whose plane symmetry group consists of four rotations and is isomorphic to Z_4. The surprising fact is that the cyclic groups and dihedral groups are the only finite plane symmetry groups. The famous mathematician Hermann Weyl attributes the following theorem to Leonardo da Vinci (1452–1519).

Theorem 27.1   **Finite Symmetry Groups in the Plane**

> *The only finite plane symmetry groups are Z_n and D_n.*

PROOF.   Let G be a finite plane symmetry group of some figure. We first observe that G cannot contain a translation or a glide-reflection, because in either case G would be infinite. We next use the fact from geometry [2; 366] that a finite group of rotations must have a common center, say P. Now observing that the composition of two reflections preserves orientation, we know that such a composition is a translation or rotation. When the two reflections have parallel axes of reflection, there is no fixed point (see Exercise 10 in the Supplementary Exercises for Chapters 1–4) so the composition is a translation. Thus, every two reflections in G have reflection axes that intersect in some point. Suppose that σ and ρ are two distinct reflections in G. Then because $\sigma\rho$ preserves orientation, we know that $\sigma\rho$ is a rotation. This means that their axes of reflection must intersect at point P. So, we have shown that all the elements of G have the common fixed point P.

For convenience, let us denote a rotation about P of σ degrees by R_σ. Now, among all rotations in G, let β be the smallest positive angle of rotation. (Such an angle exists, since G is finite and R_{360} belongs to G.) We claim that every rotation in G is some power of R_β. To see this, suppose that R_σ is in G. We may assume $0° < \sigma \le 360°$. Then, $\beta \le \sigma$ and there is some integer t such that $t\beta \le \sigma < (t + 1)\beta$. But, then $R_{\sigma - t\beta} = R_\sigma(R_\beta)^{-t}$ is in G and $0 \le \sigma - t\beta < \beta$. Since β represents the smallest positive angle of rotation among the elements of G, we must have $\sigma - t\beta = 0$, and therefore, $R_\sigma = (R_\beta)^t$. This verifies the claim.

For convenience, let us say that $|R_\beta| = n$. Now, if G has no reflections, we have proved that $G = \langle R_\beta \rangle \approx Z_n$. If G has at least one reflection, say α, then

$$\alpha,\ \alpha R_\beta,\ \alpha(R_\beta)^2, \ldots, \alpha(R_\beta)^{n-1}$$

are also reflections. Furthermore, this is the entire set of reflections of G. For if γ is any reflection in G, then $\alpha\gamma$ is a rotation, and so $\alpha\gamma = (R_\beta)^k$ for some k. Thus, $\gamma = \alpha^{-1}(R_\beta)^k = \alpha(R_\beta)^k$. So

$$G = \{R_0, R_\beta, (R_\beta)^2, \ldots, (R_\beta)^{n-1}, \alpha, \alpha R_\beta, \alpha(R_\beta)^2, \ldots, \alpha(R_\beta)^{n-1}\},$$

and G is generated by the pair of reflections α and αR_β. Hence, by our characterization of the dihedral groups (Theorem 26.5), G is the dihedral group D_n.   ■ ■

CLASSIFICATION OF FINITE GROUPS OF ROTATIONS IN $\mathbf{R}^3$

One might think that the set of all possible finite symmetry groups in three dimensions would be much more diverse than is the case for two dimensions. Surprisingly, this is not the case. For example, moving to three dimensions introduces only three new groups of rotations. This observation was first made by the physicist and mineralogist Auguste Bravais in 1849, in his study of possible structures of crystals.

Theorem 27.2   **Finite Groups of Rotations in $\mathbf{R}^3$**

> *Up to isomorphism, the finite groups of rotations in $\mathbf{R}^3$ are Z_n, D_n, A_4, S_4, and A_5.*

Theorem 27.2, together with the Orbit-Stabilizer Theorem (Theorem 7.3), makes easy work of determining the group of rotations of an object in $\mathbf{R}^3$.

EXAMPLE 1   We determine the group G of rotations of the solid in Figure 27.4, which is composed of six congruent squares and eight congruent equilateral triangles. We begin by singling out any one of the squares. Obviously, there are four rotations that map this square to itself, and the designated square can be rotated to the location of any of the other five. So, by the Orbit-Stabilizer Theorem (Theorem 7.3), the rotation group has order $4 \cdot 6 = 24$. By Theorem 27.2, G is one of Z_{24}, D_{12}, and S_4. But each of the first two groups has exactly two elements of order 4, whereas G has more than two. So, G is isomorphic to S_4.   ◆

The group of rotations of a tetrahedron (the *tetrahedral group*) is isomorphic to A_4; the group of rotations of a cube or an octahedron (the *octahedral group*) is isomorphic to S_4; the group of rotations of a dodecahedron or an icosahedron (the *icosahedral group*) is isomorphic to A_5. (Coxeter [1, pp. 271–273] specifies which portions of the polyhedra are being permuted in each case.) These five solids are illustrated in Figure 27.5.

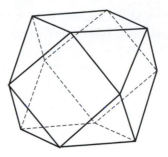

Figure 27.4

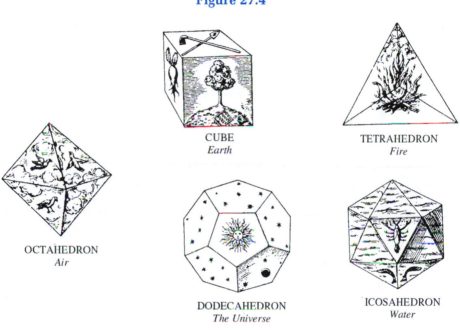

CUBE
Earth

TETRAHEDRON
Fire

OCTAHEDRON
Air

DODECAHEDRON
The Universe

ICOSAHEDRON
Water

Figure 27.5   The five regular solids as depicted by Johannes Kepler in *Harmonices Mundi, Book II* (1619).

EXERCISES ▪▪

Perhaps the most valuable result of all education is the ability to make your-self do the thing you have to do, when it ought to be done, whether you like it or not.

THOMAS HENRY HUXLEY, Technical Education

1. Show that an isometry of $\mathbf{R}^n$ is one-to-one.
2. Show that the translations in $\mathbf{R}^n$ form a group.
3. Exhibit a plane figure whose plane symmetry group is Z_5.

4. Show that the group of rotations in $\mathbf{R}^3$ of a 3-prism (that is, a prism with equilateral ends, as in the following figure) is isomorphic to D_3.

5. What is the order of the (entire) symmetry group in $\mathbf{R}^3$ of a 3-prism?

6. What is the order of the symmetry group in $\mathbf{R}^3$ of a 4-prism (a box with square ends that is not a cube)?

7. What is the order of the symmetry group in $\mathbf{R}^3$ of an n-prism?

8. Show that the symmetry group in $\mathbf{R}^3$ of a box of dimensions $2'' \times 3'' \times 4''$ is isomorphic to $Z_2 \oplus Z_2 \oplus Z_2$.

9. Describe the symmetry group of a line segment viewed as
   a. a subset of $\mathbf{R}^1$,
   b. a subset of $\mathbf{R}^2$,
   c. a subset of $\mathbf{R}^3$.

   (This exercise is referred to in this chapter.)

10. (From the "Ask Marilyn" column in *Parade Magazine*, December 11, 1994.) The letters of the alphabet can be sorted into the following categories:
    1. FGJLNPQRSZ
    2. BCDEK
    3. AMTUVWY
    4. HIOX

    What defines the categories?

11. Exactly how many elements of order 4 does the group in Example 1 have?

12. Why is inversion not listed as one of the four kinds of isometries in $\mathbf{R}^2$?

13. Explain why inversion through a point in $\mathbf{R}^3$ cannot be realized by a rotation in $\mathbf{R}^3$.

14. Reflection in a line L in $\mathbf{R}^3$ is the isometry that takes each point Q to the point Q' with the property that L is a perpendicular bisector of the line segment joining Q and Q'. Describe a rotation that has this same effect.

15. In $\mathbf{R}^2$, a rotation fixes a point; in $\mathbf{R}^3$, a rotation fixes a line. In $\mathbf{R}^4$, what does a rotation fix? Generalize these observations to $\mathbf{R}^n$.

16. Show that an isometry of a plane preserves angles.

17. Show that an isometry of a plane is completely determined by the image of three noncollinear points.

18. Suppose that an isometry of a plane leaves three noncollinear points fixed. Which isometry is it?

19. Suppose that an isometry of a plane fixes exactly one point. What type of isometry must it be?

20. Suppose that A and B are rotations of $180°$ about the points a and b, respectively. What is A followed by B? How is the composite motion related to the points a and b?

21. If the composition of a rotation and a translation carries the point p to the point p' and the point q to the point q', describe the center of the rotation formed by the composition in terms of $q, q', p,$ and p'.

References

1. H. S. M. Coxeter, *Introduction to Geometry*, 2nd ed., New York: Wiley, 1969.

2. Marvin Jay Greenberg, *Euclidean and non-Euclidean Geometries: Development and History*, 3rd ed., New York: W. H. Freeman, 1993.

Suggested Readings[†]

Lorraine Foster, "On the Symmetry Group of the Dodecahedron," *Mathematics Magazine* 63 (1990): 106–107.

> It is shown that the group of rotations of a dodecahedron and the group of rotations of an icosahedron are both A_5.

J. Rosen, *Symmetry Discovered,* Cambridge: Cambridge University Press, 1975.

> This excellent book was written for first- or second-year college students. It includes sections on group theory, spatial symmetry, temporal symmetry, and color symmetry, and chapters on symmetry in nature and the uses of symmetry in science.

Andrew Watson, "The Mathematics of Symmetry," *New Scientist,* October (1990): 45–50.

> This article discusses how chemists use group theory to understand molecular structure and how physicists use it to study the fundamental forces and particles.

_____

[†]See also the Suggested Reading for Chapter 1.

Frieze Groups and Crystallographic Groups

Symmetry, considered as a law of regular composition of structural objects, is similar to harmony. More precisely, symmetry is one of its components, while the other component is dissymmetry. In our opinion the whole esthetics of scientific and artistic creativity lies in the ability to feel this where others fail to perceive it.

A. V. SHUBNIKOV AND V. A. KOPTSIK,
Symmetry in Science and Art

THE FRIEZE GROUPS

In this chapter, we discuss an interesting collection of infinite symmetry groups that arise from periodic designs in a plane. There are two types of such groups. The *discrete frieze groups* are the plane symmetry groups of patterns whose subgroups of translations are isomorphic to Z. These kinds of designs are the ones used for decorative strips and for patterns on jewelry, as illustrated in Figure 28.1. In mathematics, familiar examples include the graphs of $y = \sin x$, $y = \tan x$, $y = |\sin x|$, and $|y| = \sin x$. After we analyze the discrete frieze groups, we examine the discrete symmetry groups of plane patterns whose subgroups of translations are isomorphic to $Z \oplus Z$.

In previous chapters, it was our custom to view two isomorphic groups as the same group, since we could not distinguish between them algebraically. In the case of the frieze groups, we will soon see that, although some of them are isomorphic as groups (that is, algebraically the same), geometrically they are quite different. To emphasize this difference, we will treat them separately. In each of the following cases, the given pattern extends infinitely far in both directions. A proof that there are exactly seven types of frieze patterns is given in the appendix of [6].

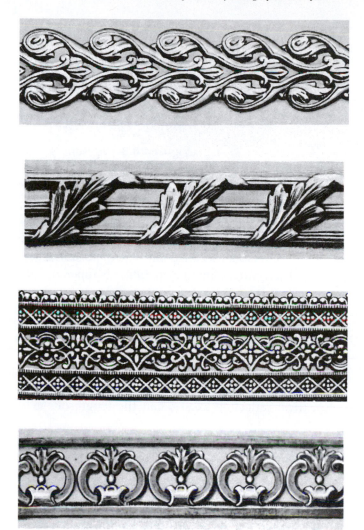

Figure 28.1 Frieze patterns.

The symmetry group of pattern I (Figure 28.2) consists of translations only. Letting x denote a translation to the right of one unit (that is, the distance between two consecutive R's), we may write the symmetry group of pattern I as

$$F_1 = \{x^n \mid n \in Z\}.$$

The group for pattern II (Figure 28.3), like that of pattern I, is infinitely cyclic. Letting x denote a glide-reflection, we may write the symmetry group of pattern II as

$$F_2 = \{x^n \mid n \in Z\}.$$

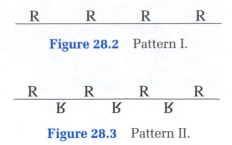

R   R   R   R

Figure 28.2 Pattern I.

R   R   R   R
   Я   Я   Я

Figure 28.3 Pattern II.

Notice that the translation subgroup of pattern II is just $\langle x^2 \rangle$.

The symmetry group for pattern III (Figure 28.4) is generated by a translation x and a reflection y across the dashed vertical line. (There are infinitely many axes of reflective symmetry, including those midway between consecutive pairs of opposite-facing R's. Any one will do.) The entire group (the operation is function composition) is

$$F_3 = \{x^n y^m \mid n \in Z, m = 0 \text{ or } 1\}.$$

Note that the two elements xy and y have order 2, they generate F_3, and their product $(xy)y = x$ has infinite order. Thus, by Theorem 26.5, F_3 is the infinite dihedral group. A geometric fact about pattern III worth mentioning is that the distance between consecutive pairs of vertical reflection axes is half the length of the smallest translation vector.

In pattern IV (Figure 28.5), the symmetry group F_4 is generated by a translation x and a rotation y of 180° about a point p midway between consecutive R's (such a rotation is often called a *half-turn*). This group, like F_3, is also infinite dihedral. (Another rotation point lies between a top and bottom R. As in pattern III, the distance between consecutive points of rotational symmetry is half the length of the smallest translation vector.)

$$F_4 = \{x^n y^m \mid n \in Z, m = 0 \text{ or } 1\}$$

ЯR   ЯR   Я R   ЯR   ЯR

Figure 28.4 Pattern III.

R   R   R   R
Я   Я   *p* Я   Я

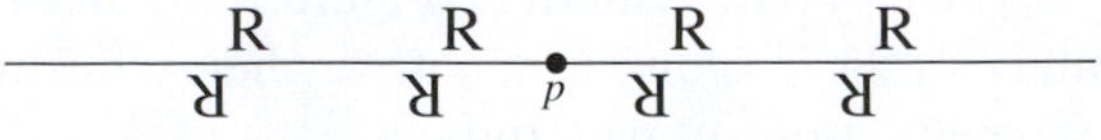

Figure 28.5 Pattern IV.

The symmetry group F_5 for pattern V (Figure 28.6) is yet another infinite dihedral group generated by a glide-reflection x and a rotation y of 180° about the point p. Notice that pattern V has vertical reflection symmetry xy. The rotation points are midway between the vertical reflection axes.

$$F_5 = \{x^n y^m \mid n \in Z, m = 0 \text{ or } 1\}$$

The symmetry group F_6 for pattern VI (Figure 28.7) is generated by a translation x and a horizontal reflection y. The elements are

$$F_6 = \{x^n y^m \mid n \in Z, m = 0 \text{ or } 1\}$$

Note that, since x and y commute, F_6 is not infinite dihedral. In fact, F_6 is isomorphic to $Z \oplus Z_2$. Pattern VI is invariant under a glide-reflection also, but in this case the glide-reflection is called *trivial*, since it is the product of x and y. (Conversely, a glide-reflection is *nontrivial* if its translation component and reflection component are not elements of the symmetry group.)

The symmetry group F_7 of pattern VII (Figure 28.8) is generated by a translation x, a horizontal reflection y, and a vertical reflection z. It is isomorphic to the direct product of the infinite dihedral group and Z_2. The product of y and z is a 180° rotation.

$$F_7 = \{x^n y^m z^k \mid n \in Z, m = 0 \text{ or } 1, k = 0 \text{ or } 1\}$$

The preceding discussion is summarized in Figure 28.9. Figure 28.10 provides an identification algorithm for the frieze patterns.

In describing the seven frieze groups, we have not explicitly said how multiplication is done algebraically. However, each group element corresponds to some isometry, so multiplication is the same as function composition. Thus, we can always use the geometry to determine the product of any particular string of elements.

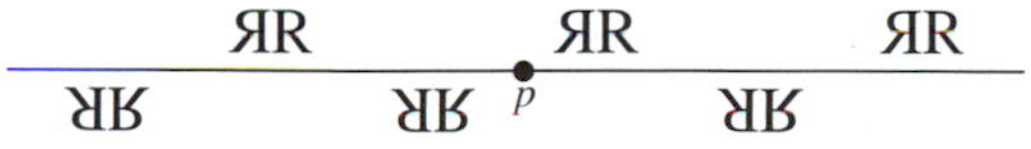

Figure 28.6 Pattern V.

Figure 28.7 Pattern VI.

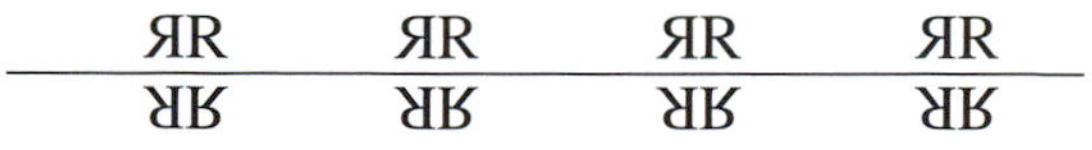

Figure 28.8 Pattern VII.

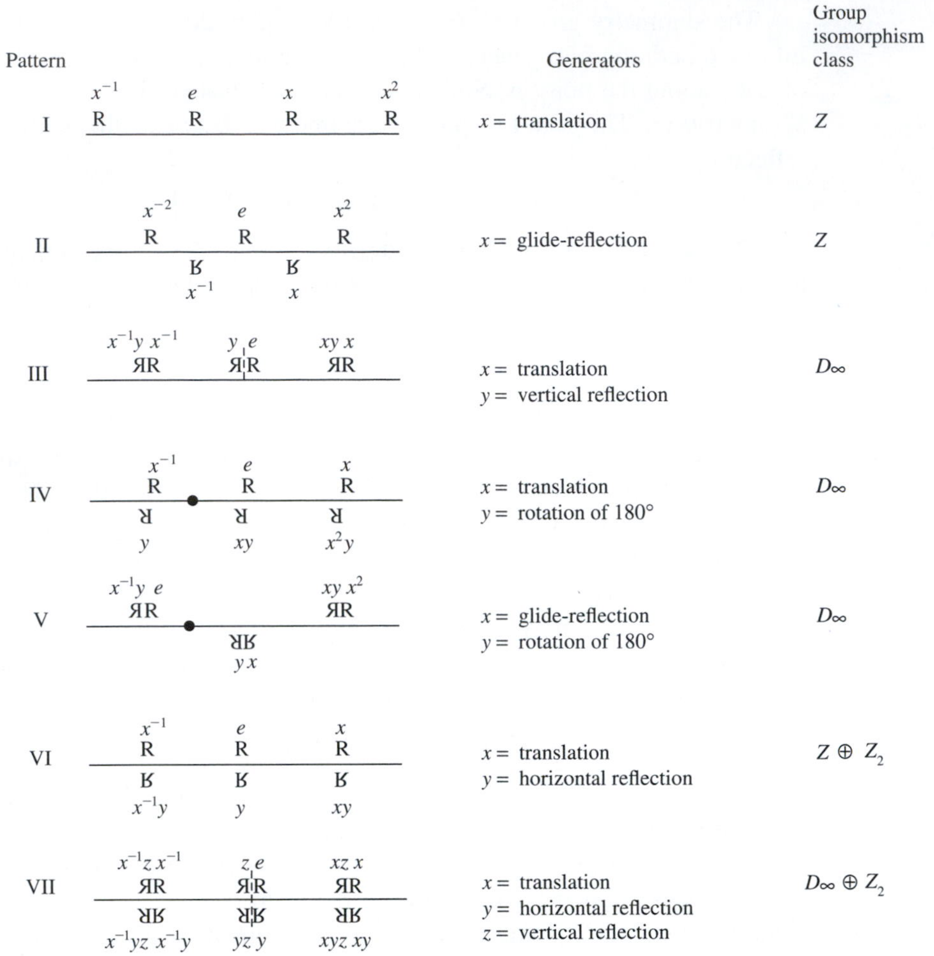

Pattern		Generators	Group isomorphism class

I

$$x^{-1} \quad\quad e \quad\quad x \quad\quad x^2$$
$$\text{R} \quad\quad \text{R} \quad\quad \text{R} \quad\quad \text{R}$$

x = translation

$\mathbf{Z}$

II

$$x^{-2} \quad\quad e \quad\quad x^2$$
$$\text{R} \quad\quad \text{R} \quad\quad \text{R}$$
$$\qquad \text{Я} \qquad\qquad \text{Я}$$
$$\qquad x^{-1} \qquad\qquad x$$

x = glide-reflection

$\mathbf{Z}$

III

$$x^{-1}y\ x^{-1} \quad\quad y\ \ e \quad\quad xy\ x$$
$$\text{ЯR} \quad\quad\quad \text{Я|R} \quad\quad\quad \text{ЯR}$$

x = translation
y = vertical reflection

D_∞

IV

$$x^{-1} \quad\quad e \quad\quad x$$
$$\text{R} \quad\quad \text{R} \quad\quad \text{R}$$
$$\text{Я} \quad\quad \text{Я} \quad\quad \text{Я}$$
$$y \quad\quad xy \quad\quad x^2y$$

x = translation
y = rotation of 180°

D_∞

V

$$x^{-1}y\ e \quad\quad\quad xy\ x^2$$
$$\text{ЯR} \quad\quad\quad\quad \text{ЯR}$$
$$\qquad\quad \text{ЯЯ}$$
$$\qquad\quad y\ x$$

x = glide-reflection
y = rotation of 180°

D_∞

VI

$$x^{-1} \quad\quad e \quad\quad x$$
$$\text{R} \quad\quad \text{R} \quad\quad \text{R}$$
$$\text{Я} \quad\quad \text{Я} \quad\quad \text{Я}$$
$$x^{-1}y \quad\quad y \quad\quad xy$$

x = translation
y = horizontal reflection

$\mathbf{Z} \oplus \mathbf{Z}_2$

VII

$$x^{-1}z\ x^{-1} \quad\quad z\ \ e \quad\quad xz\ x$$
$$\text{ЯR} \quad\quad\quad \text{Я|R} \quad\quad\quad \text{ЯR}$$
$$\text{ЯЯ} \quad\quad\quad \text{ЯЯ} \quad\quad\quad \text{ЯЯ}$$
$$x^{-1}yz\ x^{-1}y \quad\quad yz\ y \quad\quad xyz\ xy$$

x = translation
y = horizontal reflection
z = vertical reflection

$D_\infty \oplus \mathbf{Z}_2$

Figure 28.9   The seven frieze patterns and their groups of symmetries.

For example, we know that every element of F_7 can be written in the form $x^n y^m z^k$. So, just for fun, let's determine the appropriate values for n, m, and k for the element $g = x^{-1}yz\ xz$. We may do this simply by looking at the effect that g has on pattern VII. For convenience, we will pick out a particular R in the pattern and trace the action of g one step at a time. To distinguish this R, we enclose it in a shaded box. Also, we draw the axis of the vertical reflection z as a dashed line segment. See Figure 28.11.

Now, comparing the starting position of the shaded R with its final position, we see that $x^{-1}yzxz = x^{-2}y$. Exercise 7 suggests how one may arrive at the same result through purely algebraic manipulation.

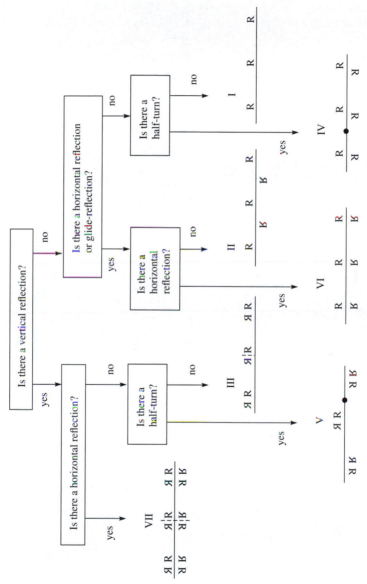

Figure 28.10 Recognition chart for frieze patterns. Adapted from [6, p. 83].

THE CRYSTALLOGRAPHIC GROUPS

The seven frieze groups catalog all symmetry groups that leave a design invariant under all multiples of just one translation. However, there are 17 additional kinds of discrete plane symmetry groups that arise from infinitely repeating designs in a plane. These groups are the symmetry groups of plane patterns whose subgroups of translations are isomorphic to $Z \oplus Z$. Consequently, the patterns are invariant under linear combinations of two linearly independent translations. These 17 groups were first studied by 19th-century crystallographers and are often called the *plane crystallographic groups*. Another term occasionally used for these groups is *wallpaper groups*.

Our approach to the crystallographic groups will be geometric. It is adapted from the excellent article by Schattschneider [5] and the monograph by Crowe [1]. Our goal is to enable the reader to determine which of the 17 plane symmetry groups corresponds to a given periodic pattern. We begin with some examples.

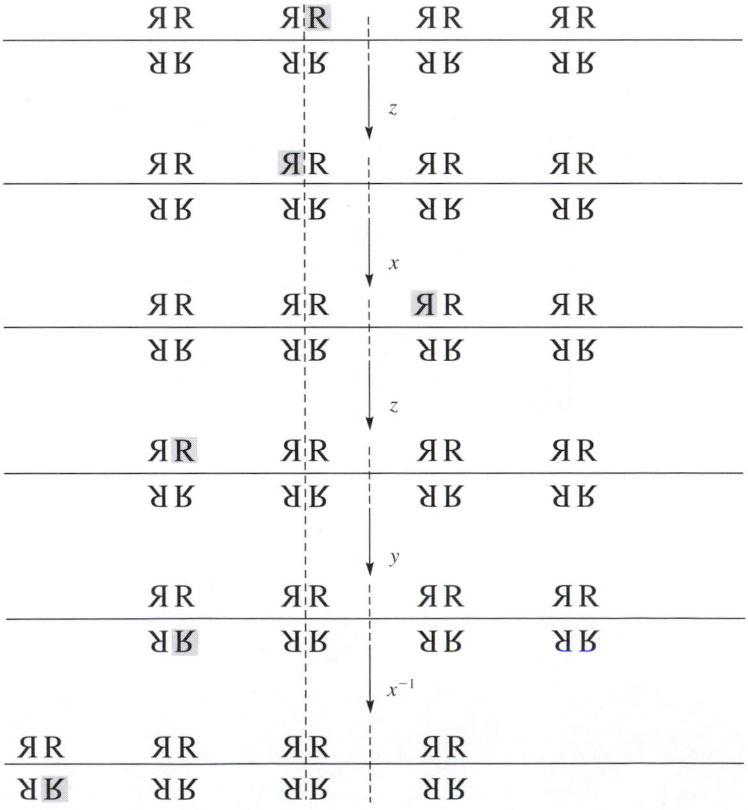

Figure 28.11

The simplest of the 17 crystallographic groups contains translations only. In Figure 28.12, we present an illustration of a representative pattern for this group (imagine the pattern repeated to fill the entire plane). The crystallographic notation for it is $p1$. (This notation is explained in [5].)

The symmetry group of the pattern in Figure 28.13 contains translations and glide-reflections. This group has no (nonzero) rotational or reflective symmetry. The crystallographic notation for it is pg.

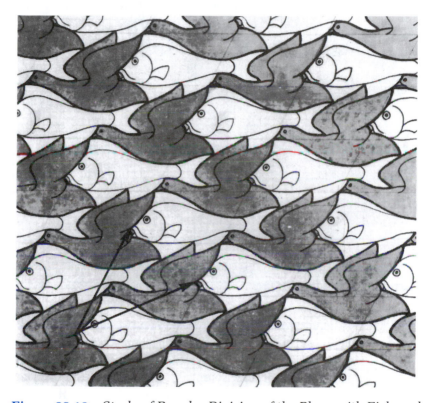

Figure 28.12 *Study of Regular Division of the Plane with Fish and Birds,* 1938. Escher graphic with symmetry group $p1$. The arrows are translation vectors.

Figure 28.14 has translational symmetry and threefold rotational symmetry (that is, the figure can be rotated 120° about certain points and be brought into coincidence with itself). The notation for this group is $p3$.

Representative patterns for all 17 plane crystallographic groups, together with their notations, are given in Figures 28.15 and 28.16. Figure 28.17 uses a triangle motif to illustrate the 17 classes of symmetry patterns.

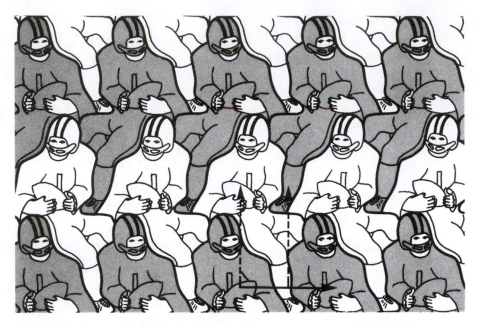

Figure 28.13   Escher-like tessellation by J. L. Teeters, with symmetry group *pg* (disregarding shading). The solid arrow is a translation vector. The dashed arrows are glide-reflection vectors.

IDENTIFICATION OF PLANE PERIODIC PATTERNS

To decide which of the 17 classes any particular plane periodic pattern belongs to, we may use the flowchart presented in Figure 28.18. This is done by determining the rotational symmetry and whether or not the pattern has reflection symmetry or nontrivial glide-reflection symmetry. These three pieces of information will narrow the list of candidates to at most two. The final test, if necessary, is to determine the locations of the centers of rotation.

For example, consider the two patterns in Figure 28.19 generated in a hockey stick motif. Both patterns have a smallest positive rotational symmetry of 120°; both have reflectional and nontrivial glide-reflectional symmetry. Now, according to Figure 28.18, these patterns must be of type *p3m1* or *p31m*. But notice that the pattern on the left has all its threefold centers of rotation on the reflection axis, whereas in the pattern on the right the points where the three blades meet are not on a reflection axis. Thus, the left pattern is *p3m1*, and the right pattern is *p31m*.

Figure 28.14　*Study of Regular Division of the Plane with Human Figures,* 1938. Escher graphic with symmetry *p3* (disregarding shading). The inserted arrows are translation vectors.

Table 28.1 (reproduced from [5, p. 443]) can also be used to determine the type of periodic pattern and contains two other features that are often useful. A *lattice of points* of a pattern is a set of images of any particular point acted on by the translation group of the pattern. A *lattice unit* of a pattern whose translation subgroup is generated by *u* and *v* is a parallelogram formed by a point of the pattern and its image under *u*, *v*, and *u* + *v*. The possible lattices for periodic patterns

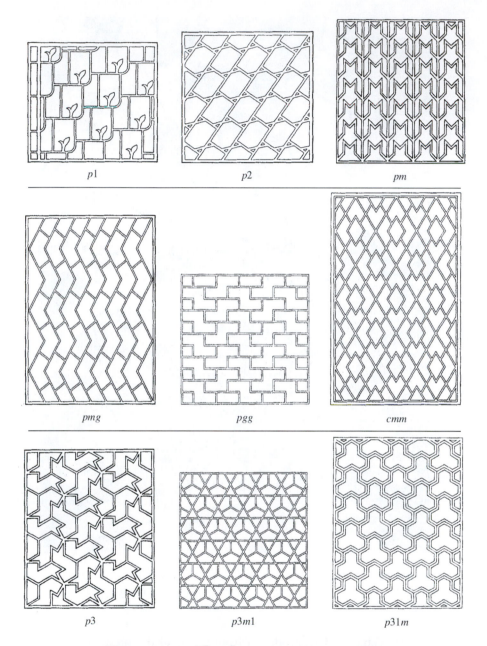

Figure 28.15 The plane symmetry groups.

All designs in Figures 28.15 and 28.16 except *pm*, *p3*, and *pg* are found in [2]. The designs for *p3* and *pg* are based on elements of Chinese lattice designs found in [2]; the design for *pm* is based on a weaving pattern from the Sandwich Islands, found in [3].

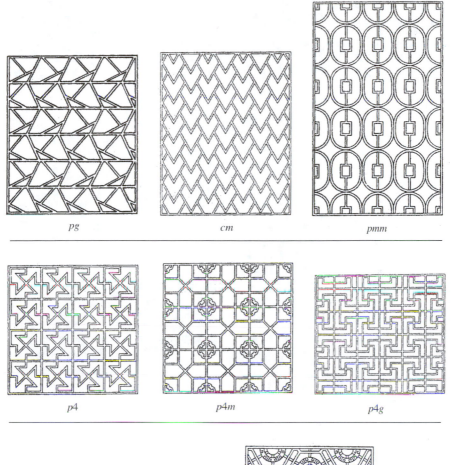

pg　　　　　　　　cm　　　　　　　　pmm

p4　　　　　　　　p4m　　　　　　　　p4g

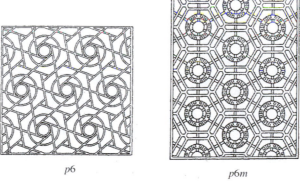

p6　　　　　　　　p6m

Figure 28.16　The plane symmetry groups.

Figure 28.17   The 17 plane periodic patterns formed using a triangle motif.

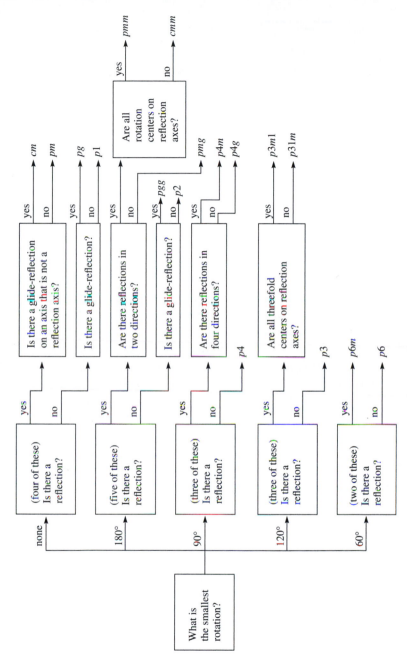

Figure 28.18 Identification flowchart for plane periodic patterns.

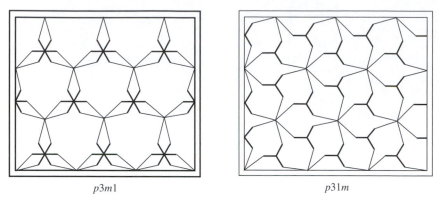

p3m1   *p31m*

Figure 28.19   Patterns generated in a hockey stick motif.

in a plane, together with lattice units, are shown in Figure 28.20. A *generating region* (or *fundamental region*) of a periodic pattern is the smallest portion of the lattice unit whose images under the full symmetry group of the pattern cover the plane. Examples of generating regions for the patterns represented in Figures 28.12, 28.13, and 28.14 are given in Figure 28.21. In Figure 28.21, the portion of the lattice unit with vertical bars is the generating region. The only symmetry pattern in which the lattice unit and the generating region coincide is the $p1$ pattern illustrated in Figure 28.12. Table 28.1 tells what proportion of the lattice unit constitutes the generating region of each plane periodic pattern.

Notice that Table 28.1 reveals that the only possible n-fold rotational symmetries occur when $n = 1, 2, 3, 4,$ and 6. This fact is commonly called the *crystallographic restriction.* The first proof of this was given by the Englishman W. Barlow. The information in Table 28.1 can also be used in reverse to create patterns with a specific symmetry group. The patterns in Figure 28.19 were made in this way.

In sharp contrast to the situation for finite symmetry groups, the transition from two-dimensional crystallographic groups to three-dimensional crystallographic groups introduces a great many more possibilities, since the motif is repeated indefinitely by three independent translations. Indeed, there are 230 three-dimensional crystallographic groups (often called *space groups*). These were independently determined by Fedorov, Schönflies, and Barlow in the 1890s. David Hilbert, one of the leading mathematicians of this century, focused attention on the crystallographic groups in his famous lecture in 1900 at the International Congress of Mathematicians in Paris. One of 23 problems he posed was whether or not the number of crystallographic groups in n dimensions is always finite. This was answered affirmatively by L. Bieberbach in 1910. We mention in passing that in four dimensions, there are 4783 symmetry groups for infinitely repeating patterns.

TABLE 28.1   Identification Chart for Plane Periodic Patterns[a]

Type	Lattice	Highest Order of Rotation	Reflections	Nontrivial Glide-Reflections	Generating Region	Helpful Distinguishing Properties
*p*1	Parallelogram	1	No	No	1 unit	
*p*2	Parallelogram	2	No	No	$\frac{1}{2}$ unit	
pm	Rectangular	1	Yes	No	$\frac{1}{2}$ unit	
pg	Rectangular	1	No	Yes	$\frac{1}{2}$ unit	
cm	Rhombic	1	Yes	Yes	$\frac{1}{2}$ unit	
pmm	Rectangular	2	Yes	No	$\frac{1}{4}$ unit	
pmg	Rectangular	2	Yes	Yes	$\frac{1}{4}$ unit	Parallel reflection axes
pgg	Rectangular	2	No	Yes	$\frac{1}{4}$ unit	
cmm	Rhombic	2	Yes	Yes	$\frac{1}{4}$ unit	Perpendicular reflection axes
*p*4	Square	4	No	No	$\frac{1}{4}$ unit	
*p*4*m*	Square	4	Yes	Yes	$\frac{1}{8}$ unit	Fourfold centers on reflection axes
*p*4*g*	Square	4	Yes	Yes	$\frac{1}{8}$ unit	Fourfold centers not on reflection axes
*p*3	Hexagonal	3	No	No	$\frac{1}{3}$ unit	
*p*3*m*1	Hexagonal	3	Yes	Yes	$\frac{1}{6}$ unit	All threefold centers on reflection axes
*p*31*m*	Hexagonal	3	Yes	Yes	$\frac{1}{6}$ unit	Not all threefold centers on reflection axes
*p*6	Hexagonal	6	No	No	$\frac{1}{6}$ unit	
*p*6*m*	Hexagonal	6	Yes	Yes	$\frac{1}{12}$ unit	

[a]A rotation through an angle of 360°/*n* is said to have order *n*. A glide-reflection is nontrivial if its component translation and reflection are not symmetries of the pattern.

As one might expect, the crystallographic groups are fundamentally important in the study of crystals. In fact, a crystal is defined as a rigid body in which the component particles are arranged in a pattern that repeats in three directions (the repetition is caused by the chemical bonding). A grain of salt and a grain of sugar are two examples of common crystals. In crystalline materials, the motif units are atoms, ions, ionic groups, clusters of ions, or molecules.

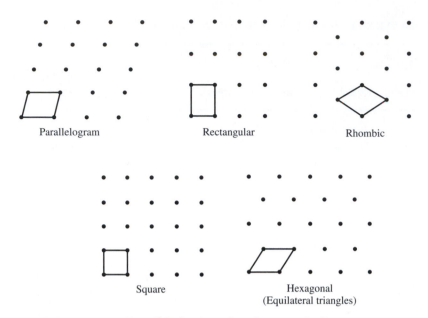

Figure 28.20   Possible lattices for plane periodic patterns.

Perhaps it is fitting to conclude this chapter by recounting two episodes in the history of science in which an understanding of symmetry groups was crucial to a great discovery. In 1912, Max von Laue, a young German physicist, hypothesized that a narrow beam of x-rays directed onto a crystal with a photographic film behind it would be deflected (the technical term is "diffracted") by the unit cell (made up of atoms or ions) and would show up on the film as spots. (See Figure 1.3.) Shortly thereafter, two British scientists, Sir William Henry Bragg and his 22-year-old son William Lawrence Bragg, who was a student, noted that von Laue's diffraction spots together with the known information about crystallographic space groups could be used to calculate the shape of the internal array of atoms. This discovery marked the birth of modern mineralogy. From the first crystal structures deduced by the Braggs to the present, x-ray diffraction has been the means by which the internal structures of crystals are determined. Von Laue was awarded the Nobel Prize in physics in 1914, and the Braggs were jointly awarded the Nobel Prize in physics in 1915.

Our second episode took place in the early 1950s, when a handful of scientists were attempting to learn the structure of the DNA molecule—the basic genetic material. One of these was a graduate student named Francis Crick; another was an x-ray crystallographer, Rosalind Franklin. On one occasion, Crick was shown one of Franklin's research reports and an x-ray diffraction photograph of DNA. At this point, we let Horace Judson [4, pp. 165–166], our source, continue the story.

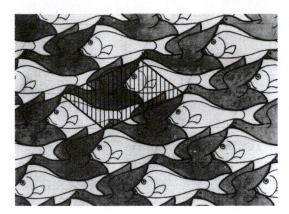

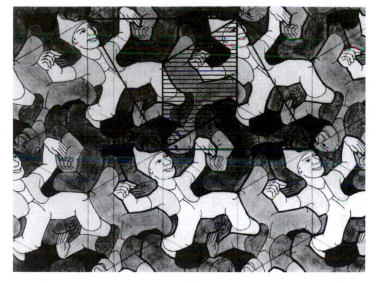

Figure 28.21 A lattice unit and generating region for the patterns in Figures 28.12, 28.13, and 28.14. Generating regions are shaded with vertical bars.

Crick saw in Franklin's words and numbers something just as important, indeed eventually just as visualizable. There was drama, too: Crick's insight began with an extraordinary coincidence. Crystallographers distinguish 230 different space groups, of which the face-centered monoclinic cell with its curious properties of symmetry is only one—though in biological substances a fairly common one. The principal experimental subject of Crick's dissertation, however, was the x-ray diffraction of the crystals of a protein that was of exactly the same space group as DNA. So Crick saw at once the symmetry that neither Franklin nor Wilkins had comprehended, that Perutz, for that matter, hadn't noticed, that had escaped the theoretical crystallographer in Wilkins' lab, Alexander Stokes—namely, that the molecule of DNA, rotated a half turn, came back to congruence with itself. The structure was dyadic, one half matching the other half in reverse.

This was a crucial fact. Shortly thereafter, James Watson and Crick built an accurate model of DNA. In 1962, Watson, Crick, and Maurice Wilkins received the Nobel Prize in medicine and physiology for their discovery. The opinion has been expressed that, had Franklin correctly recognized the symmetry of the DNA molecule, she might have been the one to unravel the mystery and receive the Nobel Prize [4, p. 172].

EXERCISES

You can see a lot just by looking.

YOGI BERRA

1. Show that the frieze group F_6 is isomorphic to $Z \oplus Z_2$.
2. How many nonisomorphic frieze groups are there?
3. In the frieze group F_7, write x^2yzxz in the form $x^ny^mz^k$.
4. In the frieze group F_7, write $x^{-3}zxyz$ in the form $x^ny^mz^k$.
5. In the frieze group F_7, show that $yz = zy$ and $xy = yx$.
6. In the frieze group F_7, show that $zxz = x^{-1}$.
7. Use the results of Exercises 5 and 6 to do Exercises 3 and 4 through symbol manipulation only (that is, without referring to the pattern). (This exercise is referred to in this chapter.)
8. Prove that in F_7 the cyclic subgroup generated by x is a normal subgroup.
9. Quote a previous result that tells why the subgroups $\langle x, y \rangle$ and $\langle x, z \rangle$ must be normal in F_7.
10. Look up the word *frieze* in an ordinary dictionary. Explain why the frieze groups are appropriately named.
11. Determine which of the seven frieze groups is the symmetry group of each of the following patterns.

    a.

b.

c.

d.

e.

f.

12. Determine the frieze group corresponding to each of the following patterns:

    a. $y = \sin x$,

    b. $y = |\sin x|$,

    c. $|y| = \sin x$,

    d. $y = \tan x$,

    e. $y = \csc x$.

13. Determine the symmetry group of the tessellation of the plane exemplified by the brickwork shown.

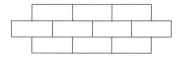

14. Determine the plane symmetry group for each of the patterns in Figure 28.17.

15. Determine which of the 17 crystallographic groups is the symmetry group of each of the following patterns.

a.

b.
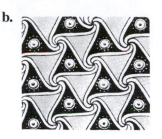

c.

d.
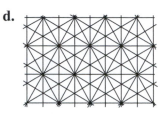

16. In the following figure, there is a point labeled 1. Let α be the translation of the plane that carries the point labeled 1 to the point labeled α, and let β be the translation of the plane that carries the point labeled 1 to the point labeled β. The image of 1 under the composition of α and β is labeled $\alpha\beta$. In the corresponding fashion, label the remaining points in the figure in the form $\alpha^i\beta^j$.

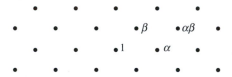

17. The patterns made by automobile tire treads in the snow are frieze patterns. An extensive study of automobile tires revealed that only five of the seven frieze patterns occur. Speculate on which two patterns do not occur and a possible reason why they do not.

18. Locate a nontrivial glide-reflection axis of symmetry in the *cm* pattern in Figure 28.16.

19. Determine which of the frieze groups is the symmetry group of each of the following patterns.
a. $\cdots$ D D D D $\cdots$
b. $\cdots$ V $\wedge$ V $\wedge$ $\cdots$
c. $\cdots$ L L L L $\cdots$

d. $\cdots$ V V V V $\cdots$
e. $\cdots$ N N N N $\cdots$
f. $\cdots$ H H H H $\cdots$
g. $\cdots$ L ⌐ L ⌐ $\cdots$

20. Locate a nontrivial glide-reflection axis of symmetry in the pattern third from the left in the bottom row in Figure 28.17.

References

1. Donald Crowe, *Symmetry, Rigid Motions, and Patterns,* Arlington, Va.: COMAP, 1986.

2. Daniel S. Dye, *A Grammar of Chinese Lattice,* Harvard-Yenching Institute Monograph Series, vol. VI, Cambridge, Mass.: Harvard University Press, 1937. (Reprinted as *Chinese Lattice Designs,* New York: Dover, 1974.)

3. Owen Jones, *The Grammar of Ornament,* New York: Van Nostrand Reinhold, 1972. (Reproduction of the same title, first published in 1856 and reprinted in 1910 and 1928.)

4. Horace Freeland Judson, *The Eighth Day of Creation,* New York: Simon and Schuster, 1979.

5. D. Schattschneider, "The Plane Symmetry Groups: Their Recognition and Notation," *The American Mathematical Monthly* 85 (1978): 439–450.

6. D. K. Washburn and D. W. Crowe, *Symmetries of Culture: Theory and Practice of Plane Pattern Analysis,* Seattle: University of Washington Press, 1988.

Suggested Readings

Vladimir Dubrovsky, "Ornamental Groups," *Quantum,* November/December (1991): 32–35.

An introduction to crystallographic groups illustrated with Escher prints.

S. Garfunkel et al., *For All Practical Purposes,* 5th ed., New York: W. H. Freeman, 2000.

This book has a well-written, richly illustrated chapter on symmetry in art and nature.

W. G. Jackson, "Symmetry in Automobile Tires and the Left-Right Problem," *Journal of Chemical Education* 69 (1992): 624–626.

This article uses automobile tires as a tool for introducing and explaining the symmetry terms and concepts important in chemistry.

C. MacGillivray, *Fantasy and Symmetry—The Periodic Drawings of M. C. Escher,* New York: Harry N. Abrams, 1976.
> This is a collection of Escher's periodic drawings together with a mathematical discussion of each one.

D. Schattschneider, *Visions of Symmetry,* New York: Freeman, 1990.
> A loving, lavish, encyclopedic book on the drawings of M. C. Escher.

H. von Baeyer, "Impossible Crystals," *Discover* 11(2) (1990): 69–78.
> This article tells how the discovery of nonperiodic tilings of the plane led to the discovery of quasicrystals. The x-ray diffraction patterns of quasicrystals exhibit fivefold symmetry—something that had been thought to be impossible.

Suggested Video

Maurits Escher: Painter of Fantasies, 1970, 27 minutes, Cinema Guild, 1697 Broadway, Ste 506, New York, NY 10019, Phone (800) 723–5522.
> An interview with Escher conducted in his home.

Suggested Websites

http://www.mcescher.com/
> This is the official website for the artist M. C. Escher. It features many of his prints and most of his 136 symmetry drawings.

http://ccins.camosun.bc.ca/~jbritton/jbsymteslk.htm
> This spectacular website on symmetry and tessellations has numerous activities and links to many other sites on related topics. It is a wonderful website for K-12 teachers and students.

M. C. Escher

I never got a pass mark in math. The funny thing is I seem to latch on to mathematical theories without realizing what is happening.

M. C. ESCHER

M. C. ESCHER was born on June 17, 1898, in the Netherlands. His artistic work prior to 1937 was dominated by the representation of visible reality, such as landscapes and buildings. Gradually, he became less and less interested in the visible world and became increasingly absorbed in an inventive approach to space. He studied the abstract space-filling patterns used in the Moorish mosaics in the Alhambra in Spain. He also studied the mathematician George Pólya's paper on the 17 plane crystallographic groups. Instead of the geometric motifs used by the Moors and Pólya, Escher preferred to use animals, plants, or people in his space-filling prints.

Escher was fond of incorporating various mathematical ideas into his works. Among these are infinity, Möbius bands, stellations, deformations, reflections, Platonic solids, spirals, and the hyperbolic plane.

Although Escher originals are now quite expensive, it was not until 1951 that he derived a significant portion of his income from his prints. Today, Escher is widely known and appreciated as a graphic artist. His prints have been used to illustrate ideas in hundreds of scientific works. Despite this popularity among scientists, however, Escher has never been held in high esteem in traditional art circles. Escher died on March 27, 1972, in Holland.

To find more information about Escher and his art, visit:

http://www-groups.dcs.st-and.ac.uk/~history/

George Pólya

Thank you Professor Pólya for all your beautiful contributions to mathematics, to science, to education, and to humanity.
    A toast from FRANK HARARY on the occasion of Pólya's 90th birthday

In 1924, the mathematician George Pólya published a paper in a crystallography journal in which he classified the plane symmetry groups and provided a full-page illustration of the corresponding 17 periodic patterns. B. G. Escher, a geologist, sent a copy of the paper to his artist brother, M. C. Escher, who used Pólya's black-and-white geometric patterns as a guide for making his own interlocking colored patterns featuring birds, reptiles, and fish.

George Pólya was born in Budapest, Hungary, on December 13, 1887. He received a teaching certificate from the University of Budapest in languages before turning to philosophy, mathematics, and physics.

In 1912, he was awarded a Ph.D. in mathematics. Horrified by Hitler and World War II, Pólya came to the United States in 1940. After two years at Brown University, he went to Stanford University, where he remained until his death in 1985 at the age of 97.

Pólya contributed to many branches of mathematics, and his collected papers fill four large volumes. Pólya is also famous for his books on problem solving and for his teaching. One of his books has sold more than 1,000,000 copies. The Society for Industrial and Applied Mathematics, the London Mathematical Society, and the Mathematical Association of America each have prizes named after Pólya.

Pólya taught courses and lectured around the country into his 90s. He never learned to drive a car and took his first plane trip at age 75. He was married for 67 years and had no children.

For more information about Pólya, visit:

http://www-groups.dcs.st-and.ac.uk/~history/

John H. Conway

"He's definitely world class, yet he has this kind of childlike enthusiasm."

RONALD GRAHAM speaking of John H. Conway

JOHN H. CONWAY ranks among the most original and versatile contemporary mathematicians. Conway was born in Liverpool, England, in 1938 and grew up in a rough neighborhood. As a youngster, he was often beaten up by older boys and did not do well in high school. Nevertheless, his mathematical ability earned him a scholarship to Cambridge University, where he excelled.

A pattern that uses repeated shapes to cover a flat surface without gaps or overlaps is called a *tiling.* In 1975, Oxford physicist Roger Penrose invented an important new way of tiling the plane with two shapes. Unlike patterns whose symmetry group is one of the 17 plane crystallographic groups, Penrose patterns can be neither translated nor rotated to coincide with themselves. Many of the remarkable properties of the Penrose patterns were discovered by Conway. In 1993, Conway discovered a new prism that can be used to fill three-dimensional space without gaps or overlaps.

Conway has made many significant contributions to number theory, group theory, game theory, knot theory, and combinatorics. Among his most important discoveries are three simple groups, which are now named after him. (Simple groups are the basic building blocks of all groups.) Conway is fascinated by games and puzzles. He invented the game Life and the game Sprouts.

For more information about Conway, visit:

http://www-groups.dcs.st-and.ac.uk/~history/

Symmetry and Counting

Let us pause to slake our thirst one last time at symmetry's bubbling spring.

TIMOTHY FERRIS, *Coming of Age in the Milky Way*

MOTIVATION

Permutation groups naturally arise in many situations involving symmetrical designs or arrangements. Consider, for example, the task of coloring the six vertices of a regular hexagon so that three are black and three are white. Figure 29.1 shows the $\binom{6}{3} = 20$ possibilities. However, if these designs appeared on one side of hexagonal ceramic tiles, it would be nonsensical to count the designs shown in Figure 29.1(a) as different, since all six designs shown there can be obtained from one of them by rotating. (A manufacturer would make only one of the six.) In this case, we say that the designs in Figure 29.1(a) are *equivalent* under the group of rotations of the hexagon. Similarly, the designs in Figure 29.1(b) are equivalent under the group of rotations, as are the designs in Figure 29.1(c) and those in Figure 29.1(d). And, since no design from Figure 29.1(a)–(d) can be obtained from a design in a different part by rotation, we see that the designs within each part of the figure are equivalent to each other but nonequivalent to any design in another figure. However, the designs in Figure 29.1(b) and (c) are equivalent under the dihedral group D_6, since the designs in Figure 29.1(b) can be reflected to yield the designs in Figure 29.1(c). For example, for purposes of arranging three black beads and three white beads to form a necklace, the designs shown in Figure 29.1(b) and (c) would be considered equivalent.

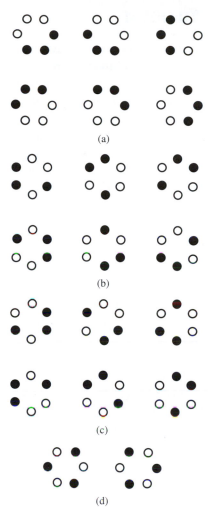

Figure 29.1

   In general, we say that two designs (arrangements) A and B are *equivalent under a group G* of permutations if there is an element ϕ in G such that $\phi(A) = B$. That is, two designs are equivalent under G if they are in the same orbit of G. It follows, then, that the number of non-equivalent designs under G is simply the number of orbits of G. (The set being permuted is the set of all possible designs or arrangements.)

   Notice that the designs in Figure 29.1 divide into four orbits under the group of rotations but only three orbits under the group D_6, since the designs in Figure 29.1(b) and (c) form a single orbit under D_6. Thus, we could obtain all 20 tile designs from just four tiles, but we could obtain all 20 necklaces from just three of them.

BURNSIDE'S THEOREM

Although the problems we have just posed are simple enough to solve by observation, more complicated ones require a more sophisticated approach. Such an approach was provided by Georg Frobenius in 1887. Frobenius's theorem did not become widely known until it appeared in the classic book on group theory by William Burnside in 1911. By an accident of history, Frobenius's theorem has come to be known as "Burnside's Theorem." Before stating this theorem, we recall some notation introduced in Chapter 7 and introduce new notation. If G is a group of permutations on a set S and $i \in S$, then $\text{stab}_G(i) = \{\phi \in G \mid \phi(i) = i\}$ and $\text{orb}_G(i) = \{\phi(i) \mid \phi \in G\}$. For any set X, we use $|X|$ to denote the number of elements in X.

DEFINITION   *Elements Fixed by ϕ*

> For any group G of permutations on a set S and any ϕ in G, we let $\text{fix}(\phi) = \{i \in S \mid \phi(i) = i\}$. This set is called the *elements fixed by ϕ* (or more simply, "fix of ϕ").

Theorem 29.1   **(Burnside)**

> *If G is a finite group of permutations on a set S, then the number of orbits of G on S is*
>
> $$\frac{1}{|G|} \sum_{\phi \in G} |\text{fix}(\phi)|.$$

PROOF.   Let n denote the number of pairs (ϕ, i), with $\phi \in G$, $i \in S$, and $\phi(i) = i$. We begin by counting these pairs in two ways. First, for each particular ϕ in G, the number of such pairs is exactly $|\text{fix}(\phi)|$, as i runs over S. So,

$$n = \sum_{\phi \in G} |\text{fix}(\phi)|. \tag{1}$$

Second, for each particular i in S, observe that $|\text{stab}_G(i)|$ is exactly the number of pairs of the form (ϕ, i), as ϕ runs over G. So,

$$n = \sum_{i \in S} |\text{stab}_G(i)|. \tag{2}$$

It follows from Exercise 29 in Chapter 7 that if s and t are in the same orbit of G, then $\text{orb}_G(s) = \text{orb}_G(t)$ and $|\text{stab}_G(s)| = |\text{stab}_G(t)|$. So, if we choose $s \in S$, sum over $\mathrm{orb}_G(s)$, and appeal to the Orbit-Stabilizer Theorem (Theorem 7.3), we have

$$\sum_{t \,\in\, \mathrm{orb}_G(s)} |\mathrm{stab}_G(t)| \;=\; |\mathrm{orb}_G(s)|\,|\mathrm{stab}_G(s)| \;=\; |G|. \tag{3}$$

Finally, by summing over all the elements of G, one orbit at a time, it follows from Equations (1), (2), and (3) that

$$\sum_{\phi \in G} |\mathrm{fix}(\phi)| \;=\; \sum_{i \in S} |\mathrm{stab}_G(i)| \;=\; |G| \cdot (\text{number of orbits})$$

and the result follows.   ▨ ▩

APPLICATIONS

To illustrate how to apply Burnside's Theorem, let us return to the ceramic tile and necklace problems. In the case of counting hexagonal tiles with three black vertices and three white vertices, the objects being permutated are the 20 possible designs, whereas the group of permutations is the group of six rotational symmetries of a hexagon. Obviously, the identity fixes all 20 designs. We see from Figure 29.1 that rotations of 60°, 180°, or 300° fix none of the 20 designs. Finally, Figure 29.2 shows fix(ϕ) for the rotations of 120° and 240°. These data are collected in Table 29.1.

So, applying Burnside's Theorem, we obtain the number of orbits under the group of rotations as

$$\frac{1}{6}(20 + 0 + 2 + 0 + 2 + 0) = 4.$$

Now let's use Burnside's Theorem to count the number of necklace arrangements consisting of three black beads and three white beads. (For the purposes of analysis, we may arrange the beads in the shape of a regular hexagon.) For this problem, two arrangements are equivalent if they are in the same orbit under D_6. Figure 29.3 shows the arrangements fixed by a reflection across a diagonal. Table 29.2 summarizes the information needed to apply Burnside's Theorem.

Figure 29.2   Tile designs fixed by 120° rotation and 240° rotation.

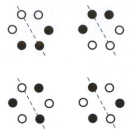

Figure 29.3   Bead arrangements fixed by the reflection across a diagonal.

TABLE 29.1

Element	Number of Designs Fixed by Element
Identity	20
Rotation of 60°	0
Rotation of 120°	2
Rotation of 180°	0
Rotation of 240°	2
Rotation of 300°	0

So, there are

$$\frac{1}{12} \left(1 \cdot 20 + 1 \cdot 0 + 2 \cdot 2 + 2 \cdot 0 + 3 \cdot 4 + 3 \cdot 0\right) = 3$$

nonequivalent ways to string three black beads and three white beads on a necklace.

Now that we have gotten our feet wet on a few easy problems, let's try a more difficult one. Suppose that we have the colors red (R), white (W), and blue (B) that can be used to color the edges of a regular tetrahedron (see Figure 5.1). First, observe that there are $3^6 = 729$ colorings without regard to equivalence. How shall we decide when two colorings of the tetrahedron are nonequivalent? Certainly, if we were to pick up a tetrahedron colored in a certain manner, rotate it, and put it back down, we would think of the tetrahedron as being positioned differently rather than as being colored differently (just as if we picked up a die labeled in the usual way and rolled it, we would not say that the die is now differently labeled). So, our permutation group for this problem is just the group of 12 rotations of the tetrahedron shown in Figure 5.1 and is isomorphic to A_4. (The group consists of the identity;

TABLE 29.2

Type of Element	Number of Elements of This Type	Number of Arrangements Fixed by Type of Element
Identity	1	20
Rotation of order 2 (180°)	1	0
Rotation of order 3 (120° or 240°)	2	2
Rotation of order 6 (60° or 300°)	2	0
Reflection across diagonal	3	4
Reflection across side bisector	3	0

eight elements of order 3, each of which fixes one vertex; and three elements of order 2, each of which fixes no vertex.) Every rotation permutes the 729 colorings, and to apply Burnside's Theorem we must determine the size of fix(ϕ) for each of the 12 rotations of the group.

Clearly, the identity fixes all 729 colorings. Next, consider the element (234) of order 3, shown in the bottom row, second from the left in Figure 5.1. Suppose that a specific coloring is fixed by this element (that is, the tetrahedron appears to be colored the same before and after this rotation). Since (234) carries edge 12 to edge 13, edge 13 to edge 14, and edge 14 to edge 12, these three edges must agree in color (edge ij is the edge joining vertex i and vertex j). The same argument shows that the three edges 23, 34, and 42 also must agree in color. So, |fix(234)| = 3^2, since there are three choices for each of these two sets of three edges. The nine columns in Table 29.3 show the possible colorings of the two sets of three edges. The analogous analysis applies to the other seven elements of order 3.

Now consider the rotation (12)(34) of order 2. (See the second tetrahedron in the top row in Figure 5.1.) Since edges 12 and 34 are fixed, they may be colored in any way and will appear the same after the rotation (12)(34). This yields 3 · 3 choices for those two edges. Since edge 13 is carried to edge 24, these two edges must agree in color. Similarly, edges 23 and 14 must agree. So, we have three choices for the pair of edges 13 and 24 and three choices for the pair of edges 23 and 14. This means that we have 3 · 3 · 3 · 3 ways to color the tetrahedron that will be equivalent under (12)(34). (Table 29.4 gives the complete list of 81 colorings.) So, |fix((12)(34))| = 3^4, and the other two elements of order 2 yield the same results.

TABLE 29.3 Nine Colorings Fixed by (234)

Edge	Colorings								
12	R	R	R	W	W	W	B	B	B
13	R	R	R	W	W	W	B	B	B
14	R	R	R	W	W	W	B	B	B
23	R	W	B	W	R	B	B	R	W
34	R	W	B	W	R	B	B	R	W
24	R	W	B	W	R	B	B	R	W

Now that we have analyzed the three types of group elements, we can apply Burnside's Theorem. In particular, the number of distinct colorings of the edges of a tetrahedron with 3 colors is

$$\frac{1}{12}(1 \cdot 3^6 + 8 \cdot 3^2 + 3 \cdot 3^4) = 87.$$

Surely it would be a difficult task to solve this problem without Burnside's Theorem.

Just as surely, you are wondering who besides mathematicians are interested in counting problems such as the ones we have discussed. Well, chemists are. Indeed, a benzene molecule can be viewed as six carbon atoms arranged in a hexagon with one of the three radicals NH_2, COOH, or OH attached at each carbon atom. See Figure 29.4 for one example.

So Burnside's Theorem enables a chemist to determine the number of benzene molecules (see Exercise 4). Another kind of molecule considered by chemists is visualized as a regular tetrahedron with a carbon atom at the center and any of the four radicals $HOCH_2$

TABLE 29.4 81 Colorings Fixed by (12)(34) (X and Y can be any of R, W, and B)

Edge	Colorings								
12	X	X	X	X	X	X	X	X	X
34	Y	Y	Y	Y	Y	Y	Y	Y	Y
13	R	R	R	W	W	W	B	B	B
24	R	R	R	W	W	W	B	B	B
23	R	W	B	W	R	B	B	R	W
14	R	W	B	W	R	B	B	R	W

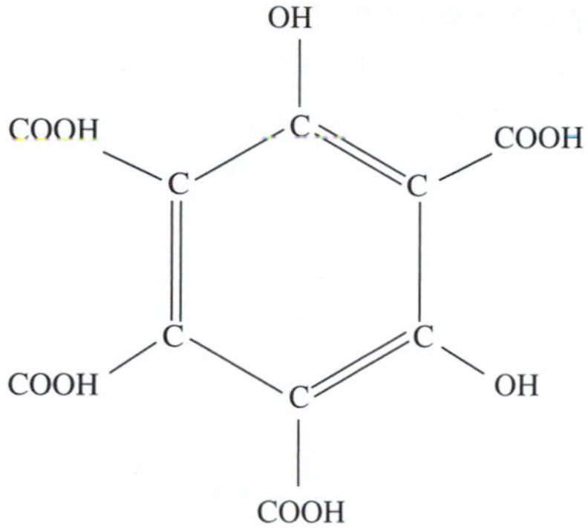

Figure 29.4   A benzene molecule.

(hydroxymethyl), C_2H_5 (ethyl), Cl (chlorine), or H (hydrogen) at the four vertices. Again, the number of such molecules can be easily counted using Burnside's Theorem.

GROUP ACTION

Our informal approach to counting the number of objects that are considered nonequivalent can be made formal as follows. If G is a group and S is a set of objects, we say that G *acts on* S if there is a homomorphism ϕ from G to sym(S), the group of all permutations on S. (The homomorphism is sometimes called the *group action*.) For convenience, we denote the image of g under ϕ as ϕ_g. Then two objects x and y in S are viewed as equivalent under the action of G if and only if $\phi_g(x) = y$ for some g in G. Notice that when ϕ is one-to-one, the elements of G may be regarded as permutations on S. On the other hand, when ϕ is not one-to-one, the elements of G may still be regarded as permutations on S, but there are distinct elements g and h in G such that ϕ_g and ϕ_h induce the same permutation on S [that is, $\phi_g(x) = \phi_h(x)$ for all x in S]. Thus, a group acting on a set is a natural generalization of the permutation group concept.

As an example of group action, let S be the two diagonals of a square and let G be D_4, the group of symmetries of the square. Then ϕ_{R_0}, $\phi_{R_{180}}$, ϕ_D, $\phi_{D'}$ are the identity; $\phi_{R_{90}}$, $\phi_{R_{270}}$, ϕ_H, ϕ_V interchange the two diagonals; and the mapping $g \rightarrow \phi_g$ from D_4 to sym(S) is a group homomorphism. As a second example, note that $GL(n, F)$, the group of invertible $n \times n$ matrices with entries from a field F, acts on the set S of $n \times 1$ column vectors with entries from F by multiplying the vectors on the left by the matrices. In this case, the mapping $g \rightarrow \phi_g$ from $GL(n, F)$ to sym(S) is a one-to-one homomorphism.

We have used group actions several times in this text without calling them that. The proof of Cayley's Theorem (Theorem 6.1) has a group G acting on the elements of G; the proofs of Sylow's Second Theorem and Third Theorem (Theorems 24.4 and 24.5) have a group acting on the set of conjugates of a Sylow p-subgroup; and the proof of the Generalized Cayley Theorem (Theorem 25.3) has G acting on the left cosets of a subgroup H.

EXERCISES ■■

The greater the difficulty, the more glory in surmounting it.

EPICURUS

1. Determine the number of ways in which the four corners of a square can be colored with two colors. (It is permissible to use a single color on all four corners.)

2. Determine the number of different necklaces that can be made using 13 white beads and three black beads.

3. Determine the number of ways in which the vertices of an equilateral triangle can be colored with five colors so that at least two colors are used.

4. A benzene molecule can be viewed as six carbon atoms arranged in a regular hexagon in a plane. At each carbon atom, one of three radicals (NH_2, $COOH$, or OH) can be attached. How many such compounds are possible?

5. Suppose that in Exercise 4 we permit only NH_2 and $COOH$ for the radicals. How many compounds are possible?

6. Determine the number of ways in which the faces of a regular dodecahedron (regular 12-sided solid) can be colored with three colors.

7. Determine the number of ways in which the edges of a square can be colored with six colors so that no color is used on more than one edge.

8. Determine the number of ways in which the edges of a square can be colored with six colors with no restriction placed on the number of times a color can be used.

9. Determine the number of different 11-bead necklaces that can be made using two colors.

10. Determine the number of ways in which the faces of a cube can be colored with three colors.

11. Suppose a cake is cut into 6 identical pieces. How many ways can we color the cake with n colors assuming that each piece receives one color?

12. How many ways can the five points of a five-pointed crown be painted if three colors of paint are available?

13. Let G be a finite group and let sym(G) be the group of all permutations on G. For each g in G, let ϕ_g denote the element of sym(G) defined by $\phi_g(x) = gxg^{-1}$ for all x in G. Show that G acts on itself under the action $g \to \phi_g$. Give an example in which the mapping $g \to \phi_g$ is not one-to-one.

14. Let G be a finite group, let H be a subgroup of G, and let S be the set of left cosets of H in G. For each g in G, let ϕ_g denote the element of sym(S) defined by $\phi_g(xH) = gxH$. Show that G acts on S under the action $g \to \phi_g$.

15. For a fixed square, let L_1 be the perpendicular bisector of the top and bottom of the square and let L_2 be the perpendicular bisector of the two sides. Show that D_4 acts on $\{L_1, L_2\}$ and determine the kernel of the mapping $g \to \phi_g$.

Suggested Readings

Norman Biggs, *Discrete Mathematics,* Oxford: Clarendon Press, 1989.
> Chapter 20 of this book presents a more detailed treatment of the subject of symmetry and counting.

Doris Schattschneider, "Escher's Combinational Patterns," *Electronic Journal of Combinatorics,* 4(2) (1997): R17.
> This article discusses a combinatorial problem concerning generating periodic patterns that the artist M. C. Escher posed and solved in an algorithmic way. The problem can also be solved by using Burnside's Theorem. The article can be downloaded free from the website http://www.combinatorics.org/.

Suggested Video

The Counting Theorem, Films for the Humanities & Sciences, FFH 6368, 24 minutes, Box 2053, Princeton, NJ 08543-2053, 1-800-257-5126.
> This video uses models of a chessboard, a cube, and a necklace together with Burnside's Theorem to solve various counting problems.

William Burnside

In one of the most abstract domains of thought, he [Burnside] has systematized and amplified its range so that, there, his work stands as a landmark in the widening expanse of knowledge. Whatever be the estimate of Burnside made by posterity, contemporaries salute him as a Master among the mathematicians of his own generation.

   A. R. FORSYTH

WILLIAM BURNSIDE was born on July 2, 1852, in London. After graduating from Cambridge University in 1875, Burnside was appointed lecturer at Cambridge, where he stayed until 1885. He then accepted a position at the Royal Naval College at Greenwich and spent the rest of his career in that post.

Burnside wrote more than 150 research papers in many fields. He is best remembered, however, for his pioneering work in group theory and his classic book *Theory of Groups*, which first appeared in 1897. Because of Burnside's emphasis on the abstract approach, many consider him to be the first pure group theorist.

One mark of greatness in a mathematician is the ability to pose important and challenging problems—problems that open up new areas of research for future generations. Here, Burnside excelled. It was he who first conjectured that a group G of odd order has a series of normal subgroups, $G = G_0 \geq G_1 \geq G_2 \geq \cdots \geq G_n = \{e\}$, such that G_i/G_{i+1} is Abelian. This extremely important conjecture was finally proved more than 50 years later by Feit and Thompson in a 255-page paper (see Chapter 25 for more on this). Another of Burnside's conjectures concerns the unique group $B_{m,n}$ with the following three properties: (1) The group has m generators; (2) $x^n = e$ for all x in the group; and (3) every group satisfying properties 1 and 2 is a factor group of $B_{m,n}$. In 1902, Burnside conjectured that $B_{m,n}$ is finite for all m and n. In 1994, Efim Zelmanov received the Fields Medal for his work on a variation of Burnside's conjecture.

Burnside was elected a Fellow of the Royal Society and awarded two Royal medals. He served as president of the Council of the London Mathematical Society and received its De Morgan medal. Burnside died on August 21, 1927.

To find more information about Burnside, visit:

http://www-groups.dcs.st-and.ac.uk/~history/

Cayley Digraphs of Groups

The important thing in science is not so much to obtain new facts as to discover new ways of thinking about them.

SIR WILLIAM LAWRENCE BRAGG, *Beyond Reductionism*

MOTIVATION

In this chapter, we introduce a graphical representation of a group given by a set of generators and relations. The idea was originated by Cayley in 1878. Although this topic is not usually covered in an abstract algebra book, we include it for five reasons: It provides a method of visualizing a group; it connects two important branches of modern mathematics—groups and graphs; it has many applications to computer science; it gives a review of some of our old friends—cyclic groups, dihedral groups, direct products, and generators and relations; and, most importantly, it is fun!

Intuitively, a directed graph (or digraph) is a finite set of points, called *vertices,* and a set of arrows, called *arcs,* connecting some of the vertices. Although there is a rich and important general theory of directed graphs with many applications, we are interested only in those that arise from groups.

THE CAYLEY DIGRAPH OF A GROUP

DEFINITION   *Cayley Digraph of a Group*

Let G be a finite group and let S be a set of generators for G. We define a digraph Cay(S:G), called the *Cayley digraph of G with generating set S,* as follows.

1. Each element of G is a vertex of Cay(S:G).
2. For x and y in G, there is an arc from x to y if and only if $xs = y$ for some $s \in S$.

To tell from the digraph which particular generator connects two vertices, Cayley proposed that each generator be assigned a color, and that the arrow joining x to xs be colored with the color assigned to s. He called the resulting figure the *color graph of the group*. This terminology is still used occasionally. Rather than use colors to distinguish the different generators, we will use solid arrows, dashed arrows, and dotted arrows. In general, if there is an arc from x to y, there need not be an arc from y to x. An arrow emanating from x and pointing to y indicates that there is an arc from x to y.

Following are numerous examples of Cayley digraphs. Note that there are several ways to draw the digraph of a group given by a particular generating set. However, it is not the appearance of the digraph that is relevant but the manner in which the vertices are connected. These connections are uniquely determined by the generating set. Thus, distances between vertices and angles formed by the arcs have no significance. (In the digraphs below, a headless arrow joining two vertices x and y indicates that there is an arc from x to y and an arc from y to x. This occurs when the generating set contains both an element and its inverse. For example, a generator of order 2 is its own inverse.)

EXAMPLE 1 $Z_6 = \langle 1 \rangle$.

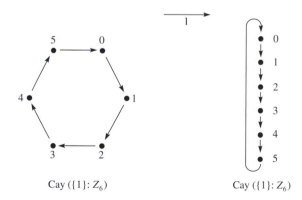

Cay ({1}: Z_6)   Cay ({1}: Z_6)

EXAMPLE 2 $Z_3 \oplus Z_2 = \langle (1, 0), (0, 1) \rangle$.

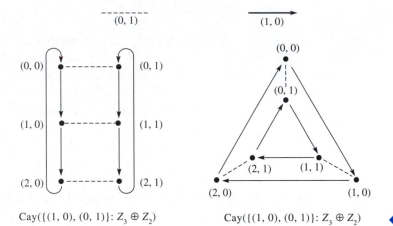

Cay$(\{(1, 0), (0, 1)\}$: $Z_3 \oplus Z_2)$ Cay$(\{(1, 0), (0, 1)\}$: $Z_3 \oplus Z_2)$ ◆

EXAMPLE 3 $D_4 = \langle R_{90}, H \rangle$.

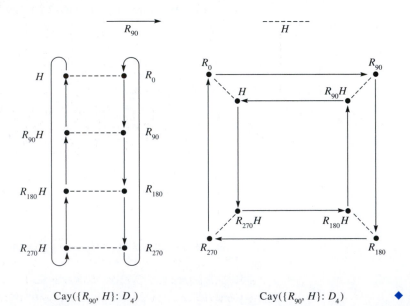

Cay$(\{R_{90}, H\}$: $D_4)$ Cay$(\{R_{90}, H\}$: $D_4)$ ◆

EXAMPLE 4 $S_3 = \langle (12), (123) \rangle$.

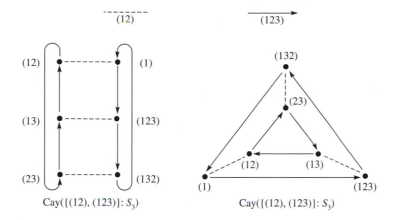

Cay({(12), (123)}: S_3)   Cay({(12), (123)}: S_3)   ◆

EXAMPLE 5 $S_3 = \langle (12), (13) \rangle$.

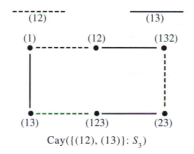

Cay({(12), (13)}: S_3)   ◆

EXAMPLE 6 $A_4 = \langle (12)(34), (123) \rangle$.

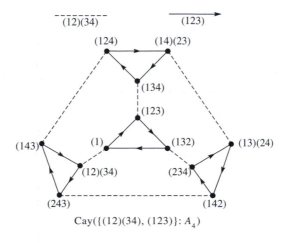

Cay({(12)(34), (123)}: A_4)   ◆

EXAMPLE 7 $Q_4 = \langle a, b \mid a^4 = e, a^2 = b^2, b^{-1}ab = a^3 \rangle.$

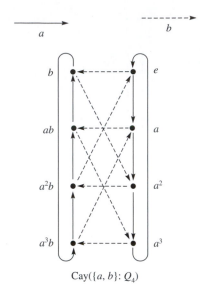

Cay($\{a, b\}$: Q_4)   ◆

EXAMPLE 8 $D_\infty = \langle a, b \mid a^2 = b^2 = e \rangle.$

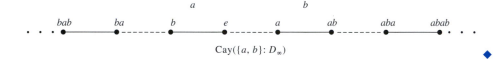

Cay($\{a, b\}$: D_∞)   ◆

The Cayley digraph provides a quick and easy way to determine the value of any product of the generators and their inverses. Consider, for example, the product ab^3ab^{-2} from the group given in Example 7. To reduce this to one of the eight elements used to label the vertices, we need only begin at the vertex e and follow the arcs from each vertex to the next as specified in the given product. Of course, b^{-1} means traverse the b arc in reverse. (Observations such as $b^{-3} = b$ also help.) Tracing the product through, we obtain b. Similarly, one can verify or discover other relations among the generators.

HAMILTONIAN CIRCUITS AND PATHS

Now that we have these directed graphs, what is it that we care to know about them? One question about directed graphs that has been the object of much research was popularized by the Irish mathematician

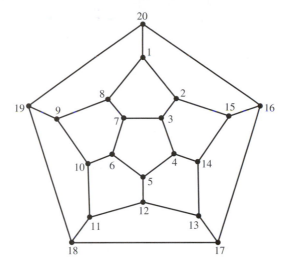

Figure 30.1 Around the world.

Sir William Hamilton in 1859, when he invented a puzzle called "Around the World." His idea was to label the 20 vertices of a regular dodecahedron with the names of famous cities. One solves this puzzle by starting at any particular city (vertex) and traveling "around the world," moving along the arcs in such a way that each other city is visited exactly once before returning to the original starting point. One solution to this puzzle is given in Figure 30.1, where the vertices are visited in the order indicated.

Obviously, this idea can be applied to any digraph; that is, one starts at some vertex and attempts to traverse the digraph by moving along arcs in such a way that each vertex is visited exactly once before returning to the starting vertex. (To go from x to y, there must be an arc from x to y.) Such a sequence of arcs is called a *Hamiltonian circuit* in the digraph. A sequence of arcs that passes through each vertex exactly once without returning to the starting point is called a *Hamiltonian path.* In the rest of this chapter, we concern ourselves with the existence of Hamiltonian circuits and paths in Cayley digraphs.

Figures 30.2 and 30.3 show a Hamiltonian path for the digraph given in Example 2 and a Hamiltonian circuit for the digraph given in Example 7, respectively.

Is there a Hamiltonian circuit in

$$\text{Cay}(\{(1, 0), (0, 1)\}: Z_3 \oplus Z_2)?$$

More generally, let us investigate the existence of Hamiltonian circuits in

$$\text{Cay}(\{(1, 0), (0, 1)\}: Z_m \oplus Z_n),$$

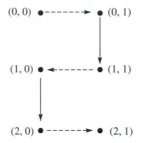

Figure 30.2   Hamiltonian path in $\text{Cay}(\{(1, 0), (0, 1)\}: Z_3 \oplus Z_2)$ from $(0, 0)$ to $(2, 1)$.

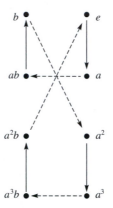

Figure 30.3   Hamiltonian circuit in $\text{Cay}(\{a, b\}: Q_4)$.

where m and n are relatively prime and both are greater than 1. Visualize the Cayley digraph as a rectangular grid coordinatized with $Z_m \oplus Z_n$, as in Figure 30.4. Suppose there is a Hamiltonian circuit in the digraph and (a, b) is some vertex where the circuit exits horizontally. (Clearly, such a vertex exists.) Then the circuit must exit $(a - 1, b + 1)$ horizontally also, for otherwise the circuit passes through $(a, b + 1)$ twice—see Figure 30.5. Repeating this argument again and again, we see that the circuit exits horizontally from each of the vertices $(a, b), (a - 1, b + 1), (a - 2, b + 2), \ldots$, which is just the coset $(a, b) + \langle(-1, 1)\rangle$. But when m and n are relatively prime, $\langle(-1, 1)\rangle$ is the entire group. Obviously, there cannot be a Hamiltonian circuit consisting entirely of horizontal moves. Let us record what we have just proved.

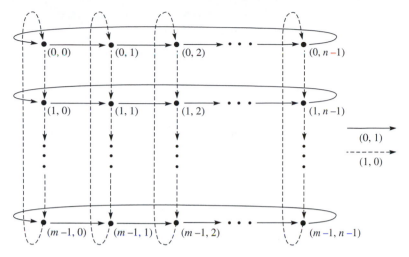

Figure 30.4   Cay($\{(1, 0), (0, 1)\}$: $Z_m \oplus Z_n$).

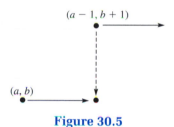

Figure 30.5

Theorem 30.1   **A Necessary Condition**

Cay($\{(1, 0), (0, 1)\}$: $Z_m \oplus Z_n$) does not have a Hamiltonian circuit when m and n are relatively prime and greater than 1.

What about when m and n are not relatively prime? In general, the answer is somewhat complicated, but the following special case is easy to prove.

Theorem 30.2   **A Sufficient Condition**

Cay($\{(1, 0), (0, 1)\}$: $Z_m \oplus Z_n$) has a Hamiltonian circuit when n divides m.

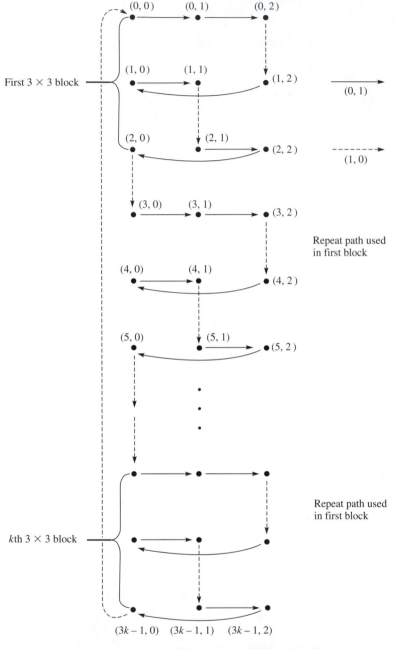

Figure 30.6   Cay($\{(1, 0), (0, 1)\}$: $Z_{3k} \oplus Z_3$).

PROOF.   Say, $m = kn$. Then we may think of $Z_m \oplus Z_n$ as k blocks of size $n \times n$. (See Figure 30.6 for an example.) Start at $(0, 0)$ and cover the vertices of the top block as follows. Use the generator $(0, 1)$ to move horizontally across the first row to the end. Then use the generator $(1, 0)$ to move vertically to the point below, and cover the remaining points in the second row by moving horizontally. Keep this process up until the point $(n - 1, 0)$—the lower left-hand corner of the first block—has been reached. Next, move vertically to the second block and repeat the process used in the first block. Keep this up until the bottom block is covered. Complete the circuit by moving vertically back to $(0, 0)$. ■ ■

Notice that the circuit given in the proof of Theorem 30.2 is easy to visualize but somewhat cumbersome to describe in words. A much more convenient way to describe a Hamiltonian path or circuit is to specify the starting vertex and the sequence of generators in the order in which they are to be applied. In Example 5, for instance, we may start at (1) and alternate the generators (12) and (13) until we return to (1). In Example 3, we may start at R_0 and successively apply R_{90}, R_{90}, R_{90}, H, R_{90}, R_{90}, R_{90}, H. When k is a positive integer and $a, b, \ldots, c$ is a sequence of group elements, we use $k * (a, b, \ldots, c)$ to denote the concatenation of k copies of the sequence $(a, b, \ldots, c)$. Thus, $2 * (R_{90}, R_{90}, R_{90}, H)$ and $2 * (3 * R_{90}, H)$ both mean R_{90}, R_{90}, R_{90}, H, R_{90}, R_{90}, R_{90}, H. With this notation, we may conveniently denote the Hamiltonian circuit given in Theorem 30.2 as

$$m * ((n - 1) * (0, 1), (1, 0)).$$

We leave it as an exercise (Exercise 11) to show that if $x_1, x_2, \ldots, x_n$ is a sequence of generators determining a Hamiltonian circuit starting at some vertex, then the same sequence determines a Hamiltonian circuit for any starting vertex.

From Theorem 30.1, we know that there are some Cayley digraphs of Abelian groups that do not have any Hamiltonian circuits. But Theorem 30.3 shows that each of these Cayley digraphs does have a Hamiltonian path. There are some Cayley digraphs for *non-Abelian* groups that do not even have Hamiltonian paths, but we will not discuss them here.

Theorem 30.3 Abelian Groups Have Hamiltonian Paths

> Let G be a finite Abelian group, and let S be any (nonempty[†]) generating set for G. Then Cay(S:G) has a Hamiltonian path.

[†]If S is the empty set, it is customary to define $\langle S \rangle$ as the identity group. We prefer to ignore this trivial case.

PROOF.   We induct on $|S|$. If $|S| = 1$, say, $S = \{a\}$, then the digraph is just a circle labeled with $e, a, a^2, \ldots, a^{m-1}$, where $|a| = m$. Obviously, there is a Hamiltonian path for this case. Now assume that $|S| > 1$. Choose some $s \in S$. Let $T = S - \{s\}$—that is, T is S with s removed—and set $H = \langle T \rangle$. (Notice that H may be equal to G.)

Because $|T| < |S|$ and H is a finite Abelian group, the induction hypothesis guarantees that there is a Hamiltonian path $(a_1, a_2, \ldots, a_k)$ in Cay$(T{:}H)$. We will show that

$$(a_1, a_2, \ldots, a_k, s, a_1, a_2, \ldots, a_k, s, \ldots, a_1, a_2, \ldots, a_k, s, a_1, a_2, \ldots, a_k),$$

where $a_1, a_2, \ldots, a_k$ occurs $|G|/|H|$ times and s occurs $|G|/|H| - 1$ times, is a Hamiltonian path in Cay$(S{:}G)$.

Because $S = T \cup \{s\}$ and T generates H, the coset Hs generates the factor group G/H. (Since G is Abelian, this group exists.) Hence, the cosets of H are $H, Hs, Hs^2, \ldots, Hs^n$, where $n = |G|/|H| - 1$. Starting from the identity element of G, the path given by $(a_1, a_2, \ldots, a_k)$ visits each element of H exactly once [because $(a_1, a_2, \ldots, a_k)$ is a Hamiltonian path in Cay$(T{:}H)$]. The generator s then moves us to some element of the coset Hs. Starting from there, the path $(a_1, a_2, \ldots, a_k)$ visits each element of Hs exactly once. Then, s moves us to the coset Hs^2, and we visit each element of this coset exactly once. Continuing this process, we successively move to $Hs^3, Hs^4, \ldots, Hs^n$, visiting each vertex in each of these cosets exactly once. Because each vertex of Cay$(S{:}G)$ is in exactly one coset Hs^i, this implies that we visit each vertex of Cay$(S{:}G)$ exactly once. Thus we have a Hamiltonian path.   ■ ■

We next look at Cayley digraphs with three generators.

EXAMPLE 9   Let

$$D_3 = \langle r, f \mid r^3 = f^2 = e, rf = fr^2 \rangle.$$

Then a Hamiltonian circuit in

$$\text{Cay}(\{(r, 0), (f, 0), (e, 1)\}{:} D_3 \oplus Z_6)$$

is given in Figure 30.7.   ◆

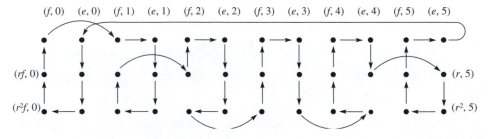

Figure 30.7

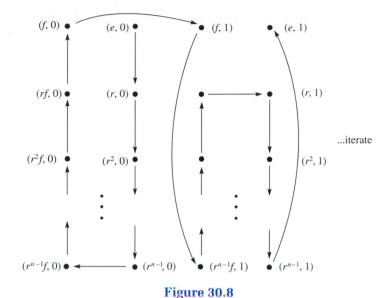

Figure 30.8

Although it is not easy to prove, it is true that

$$\text{Cay}(\{(r, 0), (f, 0), (e, 1)\}: D_n \oplus Z_m)$$

has a Hamiltonian circuit for all n and m. (See [3].) Example 10 shows the circuit for this digraph when m is even.

EXAMPLE 10   Let

$$D_n = \langle r, f \mid r^n = f^2 = e, rf = fr^{-1} \rangle.$$

Then a Hamiltonian circuit in

$$\text{Cay}(\{(r, 0), (f, 0), (e, 1)\}: D_n \oplus Z_m)$$

with m even is traced in Figure 30.8. The sequence of generators that traces the circuit is

$$m * [(n - 1) * (r, 0), (f, 0), (n - 1) * (r, 0), (e, 1)]. \qquad \blacklozenge$$

SOME APPLICATIONS

Cayley graphs are natural models for interconnection networks in computer designs, and Hamiltonicity is an important property in relation to sorting algorithms on such networks. One particular Cayley graph that is used to design and analyze interconnection networks of parallel machines is the symmetric group S_n with the set of all transpositions as the generating set. Hamiltonian paths and circuits in Cayley digraphs arise in a variety of group theory contexts. A Hamiltonian path in a Cayley digraph of a group is simply an ordered listing of the group elements without repetition. The vertices of the digraph are the group elements, and the arcs of the path are generators of the group. In 1948, R. A. Rankin used these ideas (although not the terminology) to prove that certain bell-ringing exercises could not be done by the traditional methods employed by bell ringers. (See [1, Chap. 22] for the group-theoretic aspects of bell ringing.) In 1981, Hamiltonian paths in Cayley digraphs were used in an algorithm for creating computer graphics of Escher-type repeating patterns in the hyperbolic plane [2]. This program can produce repeating hyperbolic patterns in color from among various infinite classes of symmetry groups. The program has now been improved so that the user may choose from many kinds of color symmetry. Two Escher drawings and their computer-drawn counterparts are given in Figures 30.9–30.12.

In this chapter, we have shown how one may construct a directed graph from a group. It is also possible to associate a group—called the *automorphism group*—with every directed graph. In fact, several of the 26 sporadic simple groups were first constructed in this way.

Figure 30.9 M. C. Escher's *Circle Limit I*.

Figure 30.10   A computer duplication of the pattern of M. C. Escher's *Circle Limit I* [2]. The program used a Hamiltonian path in a Cayley digraph of the underlying symmetry group.

Figure 30.11   M. C. Escher's *Circle Limit IV*.

Figure 30.12   A computer drawing inspired by the pattern of M. C. Escher's *Circle Limit IV* [2]. The program used a Hamiltonian path in a Cayley digraph of the underlying symmetry group.

EXERCISES ▪▪

A mathematician is a machine for turning coffee into theorems.

PAUL ERDÖS

1. Find a Hamiltonian circuit in the digraph given in Example 7 different from the one in Figure 30.3.

2. Find a Hamiltonian circuit in

$$\mathrm{Cay}(\{(a, 0), (b, 0), (e, 1)\}: Q_4 \oplus Z_2).$$

3. Find a Hamiltonian circuit in

$$\mathrm{Cay}(\{(a, 0), (b, 0), (e, 1)\}: Q_4 \oplus Z_m)$$

where m is even.

4. Write the sequence of generators for each of the circuits found in Exercises 1, 2, and 3.

5. Use the Cayley digraph in Example 7 to evaluate the product $a^3ba^{-1}ba^3b^{-1}$.

6. Let x and y be two vertices of a Cayley digraph. Explain why two paths from x to y in the digraph yield a group relation.

7. Use the Cayley digraph in Example 7 to verify the relation $aba^{-1}b^{-1}a^{-1}b^{-1} = a^2ba^3$.

8. Identify the following Cayley digraph of a familiar group.

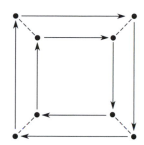

9. Let $D_4 = \langle r, f \mid r^4 = e = f^2, rf = fr^{-1}\rangle$. Verify that
$$6 * [3 * (r, 0), (f, 0), 3 * (r, 0), (e, 1)]$$
is a Hamiltonian circuit in
$$\text{Cay}(\{(r, 0), (f, 0), (e, 1)\}: D_4 \oplus Z_6).$$

10. Draw a picture of $\text{Cay}(\{2, 5\}: Z_8)$.

11. If $s_1, s_2, \ldots, s_n$ is a sequence of generators that determines a Hamiltonian circuit beginning at some vertex, explain why the same sequence determines a Hamiltonian circuit beginning at any point. (This exercise is referred to in this chapter.)

12. Show that the Cayley digraph given in Example 7 has a Hamiltonian path from e to a.

13. Show that there is no Hamiltonian path in
$$\text{Cay}(\{(1, 0), (0, 1)\}: Z_3 \oplus Z_2)$$
from $(0, 0)$ to $(2, 0)$.

14. Draw $\text{Cay}(\{2, 3\}: Z_6)$. Is there a Hamiltonian circuit in this digraph?

15. a. Let G be a group of order n generated by a set S. Show that a sequence $s_1, s_2, \ldots, s_{n-1}$ of elements of S is a Hamiltonian path in $\text{Cay}(S:G)$ if and only if, for all i and j with $1 \le i \le j < n$, we have $s_i s_{i+1} \cdots s_j \ne e$.
   b. Show that the sequence $s_1 s_2 \cdots s_n$ is a Hamiltonian circuit if and only if $s_1 s_2 \cdots s_n = e$, and that whenever $1 \le i \le j < n$, we have $s_i s_{i+1} \cdots s_j \ne e$.

16. Let $D_4 = \langle a, b \mid a^2 = b^2 = (ab)^4 = e\rangle$. Draw $\text{Cay}(\{a, b\}: D_4)$. Why is it reasonable to say that this digraph is undirected?

17. Let D_n be as in Example 10. Show that $2 * [(n - 1) * r, f]$ is a Hamiltonian circuit in $\text{Cay}(\{r, f\}: D_n)$.

18. Let $Q_8 = \langle a, b \mid a^8 = e, a^4 = b^2, b^{-1}ab = a^{-1} \rangle$. Find a Hamiltonian circuit in Cay($\{a, b\}$: Q_8).

19. Let Q_8 be as in Exercise 18. Find a Hamiltonian circuit in
$$\text{Cay}(\{(a, 0), (b, 0), (e, 1)\}: Q_8 \oplus Z_5).$$

20. Prove that the Cayley digraph given in Example 6 does not have a Hamiltonian circuit. Does it have a Hamiltonian path?

21. Find a Hamiltonian circuit in
$$\text{Cay}(\{(R_{90}, 0), (H, 0), (R_0, 1)\}: D_4 \oplus Z_3).$$
Does this circuit generalize to the case $D_{n+1} \oplus Z_n$ for all $n \geq 3$?

22. Let Q_8 be as in Exercise 18. Find a Hamiltonian circuit in
$$\text{Cay}(\{(a, 0), (b, 0), (e, 1)\}: Q_8 \oplus Z_m) \text{ for all even } m.$$

23. Find a Hamiltonian circuit in
$$\text{Cay}(\{(a, 0), (b, 0), (e, 1)\}: Q_4 \oplus Z_3).$$

24. Find a Hamiltonian circuit in
$$\text{Cay}(\{(a, 0), (b, 0), (e, 1)\}: Q_4 \oplus Z_m) \text{ for all odd } m \geq 3.$$

25. Write the sequence of generators that describes the Hamiltonian circuit in Example 9.

26. Let D_n be as in Example 10. Find a Hamiltonian circuit in
$$\text{Cay}(\{(r, 0), (f, 0), (e, 1)\}: D_4 \oplus Z_5).$$
Does your circuit generalize to the case $D_n \oplus Z_{n+1}$ for all $n \geq 4$?

27. Prove that Cay($\{(0, 1), (1, 1)\}$: $Z_m \oplus Z_n$) has a Hamiltonian circuit for all m and n greater than 1.

28. Suppose that a Hamiltonian circuit exists for Cay($\{(1, 0), (0, 1)\}$: $Z_m \oplus Z_n$) and that this circuit exits from vertex (a, b) vertically. Show that the circuit exits from every member of the coset $(a, b) + \langle (1, -1) \rangle$ vertically.

29. Let $D_2 = \langle r, f \mid r^2 = f^2 = e, rf = fr^{-1} \rangle$. Find a Hamiltonian circuit in Cay($\{(r, 0), (f, 0), (e, 1)\}$: $D_2 \oplus Z_3$).

30. Let Q_8 be as in Exercise 18. Find a Hamiltonian circuit in Cay($\{(a, 0), (b, 0), (e, 1)\}$: $Q_8 \oplus Z_3$).

31. In Cay($\{(1, 0), (0, 1)\}$: $Z_4 \oplus Z_5$) find a sequence of generators that visits exactly one vertex twice and all others exactly once and returns to the starting vertex.

32. In Cay($\{(1, 0), (0, 1)\}$: $Z_4 \oplus Z_5$) find a sequence of generators that visits exactly two vertices twice and all others exactly once and returns to the starting vertex.

33. Find a Hamiltonian circuit in Cay($\{(1, 0), (0, 1)\}$: $Z_4 \oplus Z_6$).

34. (Factor Group Lemma) Let S be a generating set for a group G, let N be a cyclic normal subgroup of G, and let

$$\overline{S} = \{sN \mid s \in S\}.$$

If $(a_1N, \ldots, a_rN)$ is a Hamiltonian circuit in $\text{Cay}(\overline{S}{:}G/N)$ and the product $a_1 \cdots a_r$ generates N, prove that

$$|N| * (a_1, \ldots, a_r)$$

is a Hamiltonian circuit in $\text{Cay}(S{:}G)$.

35. A finite group is called *Hamiltonian* if all of its subgroups are normal. (One non-Abelian example is Q_4.) Show that Theorem 30.3 can be generalized to include all Hamiltonian groups.

References

1. F. J. Budden, *The Fascination of Groups,* Cambridge: Cambridge University Press, 1972.

2. Douglas Dunham, John Lindgren, and David Witte, "Creating Repeating Hyperbolic Patterns," *Computer Graphics* 15 (1981): 215–223.

3. David Witte, Gail Letzter, and Joseph A. Gallian, "On Hamiltonian Circuits in Cartesian Products of Cayley Digraphs," *Discrete Mathematics* 43 (1983): 297–307.

Suggested Readings

Frank Budden, "Cayley Graphs for Some Well-Known Groups," *The Mathematical Gazette* 69 (1985): 271–278.

> This article contains the Cayley graphs of A_4, Q_4, and S_4 using a variety of generators and relations.

E. L. Burrows and M. J. Clark, "Pictures of Point Groups," *Journal of Chemical Education* 51 (1974): 87–90.

> Chemistry students may be interested in reading this article. It gives a comprehensive collection of the Cayley digraphs of groups important to chemists.

Douglas Dunham, John Lindgren, and David Witte, "Creating Repeating Hyperbolic Patterns," *Computer Graphics* 15 (1981): 215–223.

> In this beautifully illustrated paper, a process for creating repeating patterns of the hyperbolic plane is described. The paper is a blend of group theory, geometry, and art.

Joseph A. Gallian, "Circuits in Directed Grids," *The Mathematical Intelligencer* 13 (1991): 40–43.
> This article surveys research done on variations of the themes discussed in this chapter.

Joseph A. Gallian and David Witte, "Hamiltonian Checkerboards," *Mathematics Magazine* 57 (1984): 291–294.
> This paper gives some additional examples of Hamiltonian circuits in Cayley digraphs.

Paul Hoffman, "The Man Who Loves Only Numbers," *The Atlantic Monthly* 260 (1987): 60–74.
> A charming portrait of Paul Erdös, a prolific and eccentric mathematician who contributed greatly to graph theory.

Henry Levinson, "Cayley Diagrams," in *Mathematical Vistas: Papers from the Mathematics Section,* New York Academy of Sciences, J. Malkevitch and D. McCarthy, eds., 1990: 62–68.
> This richly illustrated article presents Cayley digraphs of many of the groups that appear in this text.

A. T. White, "Ringing the Cosets," *The American Mathematical Monthly* 94 (1987): 721–746.
> This article analyzes the practice of bell ringing by way of Cayley digraphs.

Suggested Video

N is a Number, Mathematical Association of America (www.maa.org), 58 minutes.
> In this documentary, Erdös discusses politics, death, and mathematics. Many of Erdös's collaborators and friends comment on his work and life.

William Rowan Hamilton

After Isaac Newton, the greatest mathematician of the English-speaking peoples is William Rowan Hamilton.

SIR EDMUND WHITTAKER,
Scientific American

This stamp featuring the quaternions was issued in 1983.

WILLIAM ROWAN HAMILTON was born on August 3, 1805, in Dublin, Ireland. At three, he was skilled at reading and arithmetic. At five, he read and translated Latin, Greek, and Hebrew; at 14, he had mastered 14 languages, including Arabic, Sanskrit, Hindustani, Malay, and Bengali.

In 1833, Hamilton provided the first modern treatment of complex numbers. In 1843, he made what he considered his greatest discovery—the algebra of quaternions. The quaternions represent a natural generalization of the complex numbers with three numbers i, j, and k whose squares are -1. With these, rotations in three and four dimensions can be algebraically treated. Of greater significance, however, is the fact that the quaternions are noncommutative under multiplication. This was the first ring to be discovered in which the commutative property does not hold. After 15 years of fruitless thought, the essential idea for the quaternions suddenly came to him.

Today Hamilton's name is attached to several concepts, such as the Hamiltonian function, which represents the total energy in a physical system; the Hamilton-Jacobi differential equations; and the Cayley-Hamilton Theorem from linear algebra. He also coined the terms *vector*, *scalar*, and *tensor*.

In his later years, Hamilton was plagued by alcoholism. He died on September 2, 1865, at the age of 60.

For more information about Hamilton, visit:

http://www-groups.dcs.st-and.ac.uk/~history/

Paul Erdös

Paul Erdös is a socially helpless Hungarian who has thought about more mathematical problems than anyone else in history.

THE ATLANTIC MONTHLY

PAUL ERDÖS (pronounced AIR-dish) was one of the best known and most highly respected mathematicians of the twentieth century. Unlike most of his contemporaries, who have concentrated on theory building, Erdös focused on problem solving and problem posing. The problems and methods of solution of Erdös—like those of Euler, whose solutions to special problems pointed the way to much of the mathematical theory we have today—have helped pioneer new theories, such as combinatorial and probabilistic number theory, combinatorial geometry, probabilistic and transfinite combinatorics, and graph theory.

Erdös was born on March 26, 1913, in Hungary. Both of his parents were high school mathematics teachers. Erdös, a Jew, left Hungary in 1934 at the age of 21 because of the rapid rise of anti-Semitism in Europe. For the rest of his life he traveled incessantly, rarely staying more than a month in any one place, giving lectures for small honoraria and staying with fellow mathematicians. He had little property and no fixed address. All that he owned he carried with him in a medium-sized suitcase, frequently visiting as many as 15 places in a month. His motto was, "Another roof, another proof." Even in his eighties, he put in 19-hour days doing mathematics.

Erdös wrote more than 1500 research papers. He coauthored papers with more than 500 people. These people are said to have Erdös number 1. People who do not have Erdös number 1, but who have written a paper with someone who does, are said to have Erdös number 2, and so on, inductively. Erdös died of a heart attack on September 20, 1996, in Warsaw, Poland.

For more information about Erdös, visit:

http://www-groups.dcs.st-and.ac.uk/~history/

http://www.oakland.edu/~grossman/erdoshp.html

31

Introduction to Algebraic Coding Theory

Damn it, if the machine can detect an error, why can't it locate the position of the error and correct it?

RICHARD W. HAMMING

MOTIVATION

One of the most interesting and important applications of finite fields has been the development of algebraic coding theory. This theory, which originated in the late 1940s, was created in response to practical communication problems. (Algebraic coding has nothing to do with secret codes.) Algebraic codes are now used in compact disc players, fax machines, modems, and bar code scanners, and are essential to computer maintenance.

To motivate this theory, imagine that we wish to transmit one of two possible signals to a spacecraft approaching Mars. If video pictures reveal the conditions of the proposed landing site and they look unfavorable, we will command the craft to orbit the planet; otherwise, we will command the craft to land. The signal for orbiting will be a 0, and the signal for landing will be a 1. But it is possible that some sort of interference (called *noise*) could cause an incorrect message to be received. To decrease the chance of this happening, redundancy is built into the transmission process. For example, if we wish the craft to orbit Mars, we could send five 0s. The craft's onboard computer is programmed to take any five-digit message received and decode the result by majority rule. So, if 00000 is sent and 10001 is received, the computer decides that 0 was the intended message. Notice that, for the computer to make the wrong decision, at least three errors must occur during transmission. If we assume that errors occur independently, it is less likely that three errors will occur than that two or fewer errors will occur. For this reason, this decision process is frequently called the *maximum-likelihood decoding* procedure. Our particular situation is illustrated in Figure 31.1. The general coding procedure is illustrated in Figure 31.2.

In practice, the means of transmission are telephone, radiowave, modem, microwave, or even a magnetic tape or disk. The noise might be human error, crosstalk, lightning, thermal noise, impulse noise, or deterioration of a tape or disk. Throughout this chapter, we assume that errors in transmission occur independently. Different methods are needed when this is not the case.

Now, let's consider a more complicated situation. This time, assume that we wish to send a sequence of 0s and 1s of length 500. Further, suppose that the probability that an error will be made in the transmission of any particular digit is .01. If we send this message directly, without any redundancy, the probability that it will be received error free is $(.99)^{500}$, or approximately .0066.

On the other hand, if we adopt a threefold repetition scheme by sending each digit three times and decoding each block of three digits received by majority rule, we can do much better. For example, the sequence 1011 is encoded as 111000111111. If the received message is 011000001110, the decoded message is 1001. Now, what is the probability that our 500-digit message will be error free? Well, if a 1, say, is sent, it will be decoded as a 0 if and only if the block received is 001, 010, 100, or 000. The probability that this will occur is

$$(.01)(.01)(.99) + (.01)(.99)(.01) + (.99)(.01)(.01) + (.01)(.01)(.01)$$
$$= (.01)^2[3(.99) + .01]$$
$$= .000298 < .0003.$$

Thus, the probability that any particular digit in the sequence will be decoded correctly is greater than .9997, and it follows that the probability that the entire 500-digit message will be decoded correctly is greater than $(.9997)^{500}$, or approximately .86—a dramatic improvement over .0066.

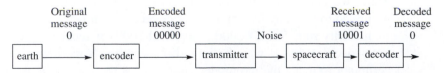

Original message
0

Encoded message
00000

Noise

Received message
10001

Decoded message
0

earth → encoder → transmitter → spacecraft → decoder →

Figure 31.1   Encoding and decoding by fivefold repetition.

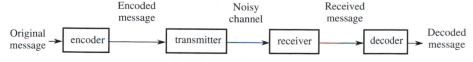

Figure 31.2   General encoding-decoding.

This example illustrates the three basic features of a code. One has a set of messages, a method of encoding these messages, and a method of decoding the received messages. The encoding procedure builds some redundancy into the original messages; the decoding procedure corrects or detects certain prescribed errors. Repetition codes have the advantage of simplicity of encoding and decoding, but they are too inefficient. In a fivefold repetition code, 80% of all transmitted information is redundant. The goal of coding theory is to devise message encoding and decoding methods that are reliable, efficient, and reasonably easy to implement.

Before plunging into the formal theory, it is instructive to look at a sophisticated example.

EXAMPLE 1   The Hamming (7, 4) Code
This time, our message set consists of all possible 4-tuples of 0s and 1s (that is, we wish to send a sequence of 0s and 1s of length 4). Encoding will be done by viewing these messages as 1×4 matrices with entries from Z_2 and multiplying each of the 16 messages on the right by the matrix

$$G = \begin{bmatrix} 1 & 0 & 0 & 0 & 1 & 1 & 0 \\ 0 & 1 & 0 & 0 & 1 & 0 & 1 \\ 0 & 0 & 1 & 0 & 1 & 1 & 1 \\ 0 & 0 & 0 & 1 & 0 & 1 & 1 \end{bmatrix}.$$

(All arithmetic is done modulo 2.) The resulting 7-tuples are called *code words*. (See Table 31.1.)

Notice that the first four digits of each code word constitute just the original message corresponding to the code word. The last three digits of the code word constitute the redundancy features. For this code, we use the *nearest-neighbor* decoding method (which, in the case that the errors occur independently, is the same as the maximum-likelihood decoding procedure). For any received word v, we assume that the word sent is the code word v' that differs from v in the fewest number of positions. If the choice of v' is not unique, we can decide not to decode or arbitrarily choose one of the code words closest to v. (The first option is usually selected when retransmission is practical.)

TABLE 31.1

Message	Encoder G	Code Word	Message	Encoder G	Code Word
0000	$\rightarrow$	0000000	0110	$\rightarrow$	0110010
0001	$\rightarrow$	0001011	0101	$\rightarrow$	0101110
0010	$\rightarrow$	0010111	0011	$\rightarrow$	0011100
0100	$\rightarrow$	0100101	1110	$\rightarrow$	1110100
1000	$\rightarrow$	1000110	1101	$\rightarrow$	1101000
1100	$\rightarrow$	1100011	1011	$\rightarrow$	1011010
1010	$\rightarrow$	1010001	0111	$\rightarrow$	0111001
1001	$\rightarrow$	1001101	1111	$\rightarrow$	1111111

Once we have decoded the received word, we can obtain the message by deleting the last three digits of v'. For instance, suppose that 1000 were the intended message. It would be encoded and transmitted as $u = 1000110$. If the received word were $v = 1100110$ (an error in the second position), it would still be decoded as u, since v and u differ in only one position, whereas v and any other code word would differ in at least two positions. Similarly, the intended message 1111 would be encoded as 1111111. If, instead of this, the word 0111111 were received, our decoding procedure would still give us the intended message 1111. ◆

The code in Example 1 is one of an infinite class of important codes discovered by Richard Hamming in 1948. The Hamming codes are the most widely used codes.

The Hamming (7, 4) encoding scheme can be conveniently illustrated with the use of a Venn diagram, as shown in Figure 31.3. Begin by placing the four message digits in the four overlapping regions I, II, III, and IV, with the digit in position 1 in region I, the digit in position 2 in region II, and so on. For regions V, VI, and VII, assign 0 or 1 so that the total number of 1s in each circle is even.

Consider the Venn diagram of the received word 0001101:

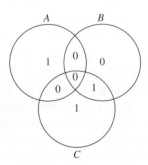

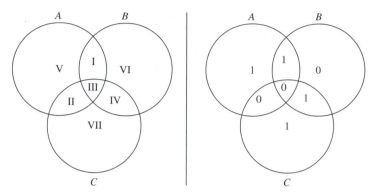

Figure 31.3 Venn diagram of the message 1001 and the encoded message 1001101.

How may we detect and correct an error? Well, observe that each of the circles *A* and *B* has an odd number of 1s. This tells us that something is wrong. At the same time, we note that circle *C* has an even number of 1s. Thus, the portion of the diagram that is in both *A* and *B* but not in *C* is the source of the error. See Figure 31.4.

Quite often, codes are used to detect errors rather than correct them. This is especially appropriate when it is easy to retransmit a message. If a received word is not a code word, we have detected an error. For example, computers are designed to use a parity check for numbers. Inside the computer, each number is represented by a string of 0s and 1s. If there is an even number of 1s in this representation, a 0 is attached to the string; if there is an odd number of 1s in the representation, a 1 is attached to the string. Thus, each number stored in the computer memory has an even number of 1s. Now, when the computer reads a number from memory, it performs a parity check. If the read number has an odd number of 1s, the computer will know

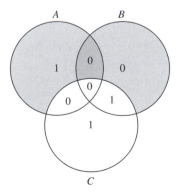

Figure 31.4 Circles *A* and *B* but not *C* have wrong parity.

that an error has been made, and it will reread the number. Note that an even number of errors will not be detected by a parity check.

The methods of error detection introduced in Chapters 0 and 5 are based on the same principle. An extra character is appended to a string of numbers so that a particular condition is satisfied. If we find that such a string does not satisfy that condition, we know that an error has occurred.

LINEAR CODES

We now formalize some of the ideas introduced in the preceding discussion.

DEFINITION   *Linear Code*

An (n, k) *linear code* over a finite field F is a k-dimensional subspace V of the vector space

$$F^n = \underbrace{F \oplus F \oplus \cdots \oplus F}_{n \text{ copies}}$$

over F. The members of V are called the *code words*. When F is Z_2, the code is called *binary*.

One should think of an (n, k) linear code over F as a set of n-tuples from F, where each n-tuple has two parts: the message part, consisting of k digits; and the redundancy part, consisting of the remaining $n - k$ digits. Note that an (n, k) linear code over a finite field F of order q has q^k code words, since every member of the code is uniquely expressible as a linear combination of the k basis vectors with coefficients from F. The set of q^k code words is closed under addition and scalar multiplication by members of F. Also, since errors in transmission may occur in any of the n positions, there are q^n possible vectors that can be received. Where there is no possibility of confusion, it is customary to denote an n-tuple $(a_1, a_2, \ldots, a_n)$ more simply as $a_1 a_2 \cdots a_n$, as we did in Example 1.

EXAMPLE 2   The set

$\{0000000, 0010111, 0101011, 1001101,$
$\qquad\qquad\qquad 1100110, 1011010, 0111100, 1110001\}$

is a $(7, 3)$ binary code.   ◆

EXAMPLE 3   The set $\{0000, 0101, 1010, 1111\}$ is a (4, 2) binary code.   ◆

Although binary codes are by far the most important ones, other codes are occasionally used.

EXAMPLE 4   The set

$$\{0000, 0121, 0212, 1022, 1110, 1201, 2011, 2102, 2220\}$$

is a (4, 2) linear code over Z_3. A linear code over Z_3 is called a *ternary code*.   ◆

To facilitate our discussion of the error-correcting and error-detecting capability of a code, we introduce the following terminology.

DEFINITIONS   *Hamming Distance, Hamming Weight*

> The *Hamming distance* between two vectors of a vector space is the number of components in which they differ. The *Hamming weight* of a vector is the number of nonzero components of the vector. The *Hamming weight* of a linear code is the minimum weight of any nonzero vector in the code.

We will use $d(u, v)$ to denote the Hamming distance between the vectors u and v, and $\text{wt}(u)$ for the Hamming weight of the vector u.

EXAMPLE 5   Let $s = 0010111$, $t = 0101011$, $u = 1001101$, and $v = 1101101$. Then, $d(s, t) = 4$, $d(s, u) = 4$, $d(s, v) = 5$, $d(u, v) = 1$; and $\text{wt}(s) = 4$, $\text{wt}(t) = 4$, $\text{wt}(u) = 4$, $\text{wt}(v) = 5$.   ◆

The Hamming distance and Hamming weight have the following important properties.

Theorem 31.1   **Properties of Hamming Distance and Hamming Weight**

> *For any vectors u, v, and w, $d(u, v) \leq d(u, w) + d(w, v)$ and $d(u, v) = \text{wt}(u - v)$.*

PROOF.   To prove that $d(u, v) = \text{wt}(u - v)$, simply observe that both $d(u, v)$ and $\text{wt}(u - v)$ equal the number of positions in which u and v differ. To prove that $d(u, v) \leq d(u, w) + d(w, v)$, note that if u and v differ in the ith position and u and w agree in the ith position, then w and v differ in the ith position.   ■■

With the preceding definitions and Theorem 31.1, we can now explain why the codes given in Examples 1, 2, and 4 will correct any single error, but the code in Example 3 will not.

Theorem 31.2   **Correcting Capability of a Linear Code**

> *If the Hamming weight of a linear code is at least $2t + 1$, then the code can correct any t or fewer errors. Alternatively, the same code can detect any $2t$ or fewer errors.*

PROOF.   We will use nearest-neighbor decoding; that is, for any received vector v, we will assume that the corresponding code word sent is a code word v' such that the Hamming distance $d(v, v')$ is a minimum. (If there is more than one such v', we do not decode.) Now, suppose that a transmitted code word u is received as the vector v and that at most t errors have been made in transmission. Then, by the definition of distance between u and v, we have $d(u, v) \leq t$. If w is any code word other than u, then $w - u$ is a nonzero code word. Thus, by assumption,

$$2t + 1 \leq \mathrm{wt}(w - u) = d(w, u) \leq d(w, v) + d(v, u) \leq d(w, v) + t,$$

and it follows that $t + 1 \leq d(w, v)$. So, the code word closest to the received vector v is u, and therefore v is correctly decoded as u.

To show that the code can detect $2t$ errors, we suppose that a transmitted code word u is received as the vector v and that at least one error, but no more than $2t$ errors, was made in transmission. Because only code words are transmitted, an error will be detected whenever a received word is not a code word. But v cannot be a code word, since $d(v, u) \leq 2t$, whereas we know that the minimum distance between distinct code words is at least $2t + 1$.   ■ ■

Theorem 31.2 is often misinterpreted to mean that a linear code with Hamming weight $2t + 1$ can correct any t errors *and* detect any $2t$ or fewer errors simultaneously. This is not the case. The user must choose one or the other role for the code. Consider, for example, the Hamming $(7, 4)$ code given in Table 31.1. By inspection, the Hamming weight of the code is $3 = 2 \cdot 1 + 1$, so we may elect either to correct any single error or to detect any one or two errors. To understand why we can't do both, consider the received word 0001010. The intended message could have been 0000000, in which case two errors were made (likewise for the intended messages 1011010 and 0101110), or the intended message could have been 0001011, in which case one error was made. But there is no way for us to know which of these possibilities occurred. If our choice were error correction, we would assume—perhaps mistakenly—that 0001011 was the intended message. If our choice were error detection, we simply would not decode. (Typically, one would request retransmission.)

On the other hand, if we write the Hamming weight of a linear code in the form $2t + s + 1$, we can correct any t errors *and* detect any $t + s$ or fewer errors. Thus, for a code with Hamming weight 5, our options include the following:

1. Detect any four errors ($t = 0, s = 4$).
2. Correct any one error and detect any two or three errors ($t = 1$, $s = 2$).
3. Correct any two errors ($t = 2, s = 0$).

EXAMPLE 6 Since the Hamming weight of the linear code given in Example 2 is 4, it will correct any single error and detect any two errors ($t = 1, s = 1$) or detect any three errors ($t = 0, s = 3$). ◆

It is natural to wonder how the matrix G used to produce the Hamming code in Example 1 was chosen. Better yet, in general, how can one find a matrix G that carries a subspace V of F^k to a subspace of F^n in such a way that for any k-tuple v in V, the vector vG will agree with v in the first k components and build in some redundancy in the last $n - k$ components? Such a matrix is a $k \times n$ matrix of the form

$$
\begin{bmatrix}
1 & 0 & \cdots & 0 & a_{11} & \cdots & a_{1n-k} \\
0 & 1 & \cdots & 0 & \cdot & & \cdot \\
\cdot & \cdot & & \cdot & \cdot & & \cdot \\
\cdot & \cdot & & \cdot & \cdot & & \cdot \\
\cdot & \cdot & & \cdot & \cdot & & \cdot \\
0 & 0 & \cdots & 1 & a_{k1} & \cdots & a_{kn-k}
\end{bmatrix}
$$

where the a_{ij}'s belong to F. A matrix of this form is called the *standard generator matrix* (or *standard encoding matrix*) for the resulting code.

Any $k \times n$ matrix whose rows are linearly independent will transform F^k to a k-dimensional subspace of F^n that could be used to build redundancy, but using the standard generator matrix has the advantage that the original message constitutes the first k components of the transformed vectors. An (n, k) linear code in which the k information digits occur at the beginning of each code word is called a *systematic code*. Schematically, we have

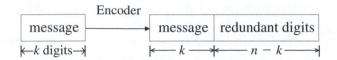

Notice that, by definition, a standard generator matrix produces a systematic code.

EXAMPLE 7   From the set of messages

$$\{000, 001, 010, 100, 110, 101, 011, 111\},$$

we may construct a (6, 3) linear code over Z_2 with the standard generator matrix

$$G = \begin{bmatrix} 1 & 0 & 0 & 1 & 1 & 0 \\ 0 & 1 & 0 & 1 & 0 & 1 \\ 0 & 0 & 1 & 1 & 1 & 1 \end{bmatrix}.$$

The resulting code words are given in Table 31.2. Since the minimum weight of any nonzero code word is 3, this code will correct any single error or detect any double error. ◆

TABLE 31.2

Message	Encoder G	Code Word
000	→	000000
001	→	001111
010	→	010101
100	→	100110
110	→	110011
101	→	101001
011	→	011010
111	→	111100

EXAMPLE 8   Here we take a set of messages as

$$\{00, 01, 02, 10, 11, 12, 20, 21, 22\},$$

and we construct a (4, 2) linear code over Z_3 with the standard generator matrix

$$G = \begin{bmatrix} 1 & 0 & 2 & 1 \\ 0 & 1 & 2 & 2 \end{bmatrix}.$$

TABLE 31.3

Message	Encoder G	Code Word
00	$\rightarrow$	0000
01	$\rightarrow$	0122
02	$\rightarrow$	0211
10	$\rightarrow$	1021
11	$\rightarrow$	1110
12	$\rightarrow$	1202
20	$\rightarrow$	2012
21	$\rightarrow$	2101
22	$\rightarrow$	2220

The resulting code words are given in Table 31.3. This code will correct any single error or detect any double error.   ◆

PARITY-CHECK MATRIX DECODING

Now that we can conveniently encode messages with a standard generator matrix, we need a convenient method for decoding the received messages. Unfortunately, this is not as easy to do; however, in the case where at most one error per code word has occurred, there is a fairly simple method for decoding. (When more than one error occurs in a code word, our decoding method fails.)

To describe this method, suppose that V is a systematic linear code over the field F given by the standard generator matrix $G = [I_k \mid A]$, where I_k represents the $k \times k$ identity matrix and A is the $k \times (n - k)$ matrix obtained from G by deleting the first k columns of G. Then, the $n \times (n - k)$ matrix

$$H = \left[\frac{-A}{I_{n-k}}\right],$$

where $-A$ is the negative of A and I_{n-k} is the $(n - k) \times (n - k)$ identity matrix, is called the *parity-check matrix* for V. (In the literature, the transpose of H is called the parity-check matrix, but H is much more convenient for our purposes.) The decoding procedure is:

1. For any received word w, compute wH.
2. If wH is the zero vector, assume that no error was made.

3. If there is exactly one instance of a nonzero element $s \in F$ and a row i of H such that wH is s times row i, assume that the sent word was $w - (0 \ldots s \ldots 0)$, where s occurs in the ith component. If there is more than one such instance, do not decode.

3′. When the code is binary, category 3 reduces to the following. If wH is the ith row of H for exactly one i, assume that an error was made in the ith component of w. If wH is more than one row of H, do not decode.

4. If wH does not fit into either category 2 or category 3, we know that at least two errors occurred in transmission and we do not decode.

EXAMPLE 9 Consider the Hamming $(7, 4)$ code given in Example 1. The generator matrix is

$$G = \begin{bmatrix} 1 & 0 & 0 & 0 & 1 & 1 & 0 \\ 0 & 1 & 0 & 0 & 1 & 0 & 1 \\ 0 & 0 & 1 & 0 & 1 & 1 & 1 \\ 0 & 0 & 0 & 1 & 0 & 1 & 1 \end{bmatrix},$$

and the corresponding parity-check matrix is

$$H = \begin{bmatrix} 1 & 1 & 0 \\ 1 & 0 & 1 \\ 1 & 1 & 1 \\ 0 & 1 & 1 \\ 1 & 0 & 0 \\ 0 & 1 & 0 \\ 0 & 0 & 1 \end{bmatrix}.$$

Now, if the received vector is $v = 0000110$, we find $vH = 110$. Since this is the first row of H and no other row, we assume that an error has been made in the first position of v. Thus, the transmitted code word is assumed to be 1000110, and the corresponding message is assumed to be 1000. Similarly, if $w = 1011111$ is the received word, then $wH = 101$, and we assume that an error has been made in the second position. So, we assume that 1111111 was sent and that 1111 was the intended message. If the encoded message 1001101 is received as $z = 1001011$ (with errors in the fifth and sixth positions), we find $zH = 110$. Since this matches the first row of H, we decode z as 0001011 and incorrectly assume that the message 0001 was intended. On the other hand, nearest-neighbor decoding would yield the same incorrect result.   ◆

Notice that when only one error was made in transmission, the parity-check decoding procedure gave us the originally intended message. We will soon see under what conditions this is true, but first we need an important fact relating a code given by a generator matrix and its parity-check matrix.

Lemma **Orthogonality Relation**

> *Let C be an (n, k) linear code over F with generator matrix G and parity-check matrix H. Then, for any vector v in F^n, we have vH = 0 (the zero vector) if and only if v belongs to C.*

PROOF. First note that, since H has rank $n - k$, we may think of H as a linear transformation from F^n onto F^{n-k}. Therefore, it follows from the dimension theorem for linear transformations that $n = n - k + \dim (\text{Ker } H)$, so that Ker H has dimension k. (Alternatively, one can use a group theory argument to show that $|\text{Ker } H| = |F|^k$.) Then, since the dimension of C is also k, it suffices to show that $C \subseteq \text{Ker } H$. To do this, let $G = [I_k \mid A]$, so that $H = \begin{bmatrix} -A \\ \hline I_{n-k} \end{bmatrix}$. Then,

$$GH = [I_k \mid A] \begin{bmatrix} -A \\ \hline I_{n-k} \end{bmatrix} = -A + A = [0] \qquad \text{(the zero matrix)}.$$

Now, by definition, any vector v in C has the form mG, where m is a message vector. Thus, $vH = (mG)H = m(GH) = m[0] = 0$ (the zero vector). ∎

Because of the way H was defined, the parity-check matrix method correctly decodes any received word in which no error has been made. But it will do more.

Theorem 31.3 **Parity-Check Matrix Decoding**

> *Parity-check matrix decoding will correct any single error if and only if the rows of the parity-check matrix are nonzero and no one row is a scalar multiple of any other row.*

PROOF. For simplicity's sake, we prove only the binary case. In this special situation, the condition on the rows is that they are nonzero and distinct. So, let H be the parity-check matrix, and let's assume that this condition holds for the rows. Suppose that the transmitted code word w was received with only one error, and that this error occurred in the ith position. Denoting the vector that has a 1 in the ith position and 0s elsewhere by e_i, we may write the received word as $w + e_i$. Now, using the Orthogonality Lemma, we obtain

$$(w + e_i)H = wH + e_iH = 0 + e_iH = e_iH.$$

But this last vector is precisely the ith row of H. Thus, if there was exactly one error in transmission, we can use the rows of the parity-check matrix to identify the location of the error, provided that these rows are distinct. (If two rows, say, the ith and jth, are the same, we know that the error occurred in either the ith position or the jth position, but we do not know in which.)

Conversely, suppose that the parity-check matrix method correctly decodes all received words in which at most one error has been made in transmission. If the ith row of the parity-check matrix H were the zero vector and if the code word $u = 0 \cdots 0$ were received as e_i, we would find $e_iH = 0 \cdots 0$, and we would erroneously assume that the vector e_i was sent. Thus, no row of H is the zero vector. Now, suppose that the ith row of H and the jth row of H are equal and $i \neq j$. Then, if some code word w is transmitted and the received word is $w + e_i$ (that is, there is a single error in the ith position), we find

$$(w + e_i)H = wH + e_iH = i\text{th row of } H = j\text{th row of } H.$$

Thus, our decoding procedure tells us not to decode. This contradicts our assumption that the method correctly decodes all received words in which at most one error has been made. ■ ■

COSET DECODING

There is another convenient decoding method that utilizes the fact that an (n, k) linear code C over a finite field F is a subgroup of the additive group of $V = F^n$. This method was devised by David Slepian in 1956 and is called *coset decoding* (or *standard decoding*). To use this method, we proceed by constructing a table, called a *standard array*. The first row of the table is the set C of code words, beginning in column 1 with the identity $0 \cdots 0$. To form additional rows of the table, choose an element v of V not listed in the table thus far. Among all the elements of the coset $v + C$, choose one of minimum weight, say, v'. Complete the next row of the table by placing under the column headed by the code word c the vector $v' + c$. Continue this process until all the vectors in V have been listed in the table. [Note that an (n, k) linear code over a field with q elements will have $|V{:}C| = q^{n-k}$ rows.] The words in the first column are called the *coset leaders*. The decoding procedure is simply to decode any received word w as the code word at the head of the column containing w.

TABLE 31.4 A Standard Array for a (6, 3) Linear Code

Coset Leaders	Words						
000000	100110	010101	001011	110011	101101	011110	111000
100000	000110	110101	101011	010011	001101	111110	011000
010000	110110	000101	011011	100011	111101	001110	101000
001000	101110	011101	000011	111011	100101	010110	110000
000100	100010	010001	001111	110111	101001	011010	111100
000010	100100	010111	001001	110001	101111	011100	111010
000001	100111	010100	001010	110010	101100	011111	111001
100001	000111	110100	101010	010010	001100	111111	011001

EXAMPLE 10 Consider the (6, 3) binary linear code

$$C = \{000000, 100110, 010101, 001011, 110011, 101101, 011110, 111000\}.$$

The first row of a standard array is just the elements of C. Obviously, 100000 is not in C and has minimum weight among the elements of $100000 + C$, so it can be used to lead the second row. Table 31.4 is the completed table.

If the word 101001 is received, it is decoded as 101101, since 101001 lies in the column headed by 101101. Similarly, the received word 011001 is decoded as 111000. ◆

Recall that the first method of decoding that we introduced was the nearest-neighbor method; that is, any received word w is decoded as the code word c such that $d(w, c)$ is a minimum, provided that there is only one code word c such that $d(w, c)$ is a minimum. The next result shows that in this situation, coset decoding is the same as nearest-neighbor decoding.

Theorem 31.4 Coset Decoding Is Nearest-Neighbor Decoding

> In coset decoding, a received word w is decoded as a code word c such that $d(w, c)$ is a minimum.

PROOF. Let C be a linear code and let w be any received word. Suppose that v is the coset leader for the coset $w + C$. Then, $w + C = v + C$, so $w = v + c$ for some c in C. Thus, using coset decoding, w is decoded as c. Now, if c' is any code word, then $w - c' \in w + C = v + C$, so that $\text{wt}(w - c') \geq \text{wt}(v)$, since the coset leader v was chosen as a vector of minimum weight among the members of $v + C$.

Therefore,

$$d(w, c') = \text{wt}(w - c') \geq \text{wt}(v) = \text{wt}(w - c) = d(w, c).$$

So, using coset decoding, w is decoded as a code word c such that $d(w, c)$ is a minimum.   ■ ■

Recall that in our description of nearest-neighbor decoding, we stated that if the choice for the nearest neighbor of a received word v is not unique, then we can decide not to decode or to decode v arbitrarily from among those words closest to v. In the case of coset decoding, the decoded value of v is always uniquely determined by the coset leader of the row containing the received word. Of course, this decoded value may not be the word that was sent.

When we know a parity-check matrix for a linear code, coset decoding can be considerably simplified.

DEFINITION   *Syndrome*

If an (n, k) linear code over F has parity-check matrix H, then, for any vector u in F^n, the vector uH is called the *syndrome*[†] of u.

The importance of syndromes stems from the following property.

Theorem 31.5   **Same Coset—Same Syndrome**

Let C be an (n, k) linear code over F with a parity-check matrix H. Then, two vectors of F^n are in the same coset of C if and only if they have the same syndrome.

PROOF.   Two vectors u and v are in the same coset of C if and only if $u - v$ is in C. So, by the Orthogonality Lemma, u and v are in the same coset if and only if $0 = (u - v)H = uH - vH$.   ■ ■

We may now use syndromes for decoding any received word w:

1. Calculate wH, the syndrome of w.
2. Find the coset leader v such that $wH = vH$.
3. Assume that the vector sent was $w - v$.

With this method, we can decode any received word with a table that has only two rows—one row of coset leaders and another row with the corresponding syndromes.

[†]This term was coined by D. Hagelbarger in 1959.

EXAMPLE 11 Consider the code given in Example 10. The parity-check matrix for this code is

$$H = \begin{bmatrix} 1 & 1 & 0 \\ 1 & 0 & 1 \\ 0 & 1 & 1 \\ 1 & 0 & 0 \\ 0 & 1 & 0 \\ 0 & 0 & 1 \end{bmatrix}.$$

The list of coset leaders and corresponding syndromes is

Coset leader	000000	100000	010000	001000	000100	000010	000001	100001
Syndromes	000	110	101	011	100	010	001	111

So, to decode the received word $v = 101001$, we compute $vH = 100$. Since the coset leader 000100 has 100 as its syndrome, we assume that $v - 000100 = 101101$ was sent. If the received word is $w = 011001$, we compute $wH = 111$ and assume that $w - 100001 = 111000$ was sent. Notice that these answers are in agreement with those obtained by using the standard-array method of Example 10. ◆

The term *syndrome* is a descriptive term. In medicine, it is used to designate a collection of symptoms that typify a disorder. In coset decoding, the syndrome typifies an error pattern.

In this chapter, we have presented algebraic coding theory in its simplest form. A more sophisticated treatment would make substantially greater use of group theory, ring theory, and especially finite-field theory. For example, Gorenstein (see Chapter 25 for a biography) and Zierler, in 1961, made use of the fact that the multiplicative subgroup of a finite field is cyclic. They associated each digit of certain codes with a field element in such a way that an algebraic equation whose zeros determined the locations of the errors could be derived.

In some instances, two error-correcting codes are employed. The European Space Agency space probe Giotto, which came within 370 miles of the nucleus of Halley's Comet in 1986, had two error-correcting codes built into its electronics. One code checked for independently occurring errors, and another—a so-called Reed-Solomon code—checked for bursts of errors. Giotto achieved an error-detection rate of 0.999999. Reed-Solomon codes are also used on compact discs. They can correct thousands of consecutive errors.

HISTORICAL NOTE: REED-SOLOMON CODES

We conclude this chapter with an adapted version of an article by Barry A. Cipra about the Reed-Solomon codes [1]. It was the first in a series of articles called "Mathematics that Counts" in *SIAM News,* the news journal of the Society for Industrial and Applied Mathematics. The articles highlight developments in mathematics that have led to products and processes of substantial benefit to industry and the public.

The Ubiquitous Reed-Solomon Codes

Irving Reed and Gustave Solomon monitor the encounter of Voyager II with Neptune at the Jet Propulsion Laboratory in 1989.

In this "Age of Information," no one need be reminded of the importance not only of speed but also of accuracy in the storage, retrieval, and transmission of data. Machines *do* make errors, and their non-man-made mistakes can turn otherwise flawless programming into worthless, even dangerous, trash. Just as architects design buildings that will remain standing even through an earthquake, their computer counterparts have come up with sophisticated techniques capable of counteracting digital disasters.

The idea for the current error-correcting techniques for everything from computer hard disk drives to CD players was first introduced in 1960 by Irving Reed and Gustave Solomon, then staff members at MIT's Lincoln Laboratory. . . .

"When you talk about CD players and digital audio tape and now digital television, and various other digital imaging systems that are coming—all of those need Reed-Solomon [codes] as an integral part of the system," says Robert McEliece, a coding theorist in the electrical engineering department at Caltech.

Why? Because digital information, virtually by definition, consists of strings of "bits"—0's and 1's—and a physical device, no matter how capably manufactured, may occasionally confuse the two. Voyager II, for example, was transmitting data at incredibly low power—barely a whisper—over tens of millions of miles. Disk drives pack data so densely that a read/write head can (almost) be excused if it can't tell where one bit stops and the next 1 (or

0) begins. Careful engineering can reduce the error rate to what may sound like a negligible level—the industry standard for hard disk drives is 1 in 10 billion—but given the volume of information processing done these days, that "negligible" level is an invitation to daily disaster. Error-correcting codes are a kind of safety net—mathematical insurance against the vagaries of an imperfect material world.

In 1960, the theory of error-correcting codes was only about a decade old. The basic theory of reliable digital communication had been set forth by Claude Shannon in the late 1940s. At the same time, Richard Hamming introduced an elegant approach to single-error correction and double-error detection. Through the 1950s, a number of researchers began experimenting with a variety of error-correcting codes. But with their SIAM journal paper, McEliece says, Reed and Solomon "hit the jackpot."

The payoff was a coding system based on groups of bits—such as bytes—rather than individual 0's and 1's. That feature makes Reed-Solomon codes particularly good at dealing with "bursts" of errors: six consecutive bit errors, for example, can affect at most two bytes. Thus, even a double-error-correction version of a Reed-Solomon code can provide a comfortable safety factor. . . .

Mathematically, Reed-Solomon codes are based on the arithmetic of finite fields. Indeed, the 1960 paper begins by defining a code as "a mapping from a vector space of dimension m over a finite field K into a vector space of higher dimension over the same field." Starting from a "message" $(a_0, a_1, \ldots, a_{m-1})$, where each a_k is an element of the field K, a Reed-Solomon code produces $(P(0), P(g), P(g^2), \ldots, P(g^{N-1}))$, where N is the number of elements in K, g is a generator of the (cyclic) group of nonzero elements in K, and $P(x)$ is the polynomial $a_0 + a_1x + \cdots + a_{m-1}x^{m-1}$. If N is greater than m, then the values of P overdetermine the polynomial, and the properties of finite fields guarantee that the coefficients of P—i.e., the original message—can be recovered from any m of the values . . . .

In today's byte-sized world, for example, it might make sense to let K be the field of order

2^8, so that each element of K corresponds to a single byte (in computerese, there are four bits to a nibble and two nibbles to a byte). In that case, $N = 2^8 = 256$, and hence messages up to 251 bytes long can be recovered even if two errors occur in transmitting the values $P(0)$, $P(g)$, $\ldots$, $P(g^{255})$. That's a lot better than the 1255 bytes required by the say-everything-five-times approach.

Despite their advantages, Reed-Solomon codes did not go into use immediately—they had to wait for the hardware technology to catch up. "In 1960, there was no such thing as fast digital electronics"—at least not by today's standards, says McEliece. The Reed-Solomon paper "suggested some nice ways to process data, but nobody knew if it was practical or not, and in 1960 it probably wasn't practical."

But technology did catch up, and numerous researchers began to work on implementing the codes. . . . Many other bells and whistles (some of fundamental theoretic significance) have also been added. Compact discs, for example, use a version of a Reed-Solomon code.

Reed, now a professor of electrical engineering at the University of Southern California, is still working on problems in coding theory. Solomon, recently retired from the Hughes Aircraft Company, consults for the Jet Propulsion Laboratory. Reed was among the first to recognize the significance of abstract algebra as the basis for error-correcting codes.

"In hindsight it seems obvious," he told *SIAM News*. However, he added, "coding theory was not a subject when we published that paper." The two authors knew they had a nice result; they didn't know what impact the paper would have.

Three decades later, the impact is clear. The vast array of applications, both current and pending, has settled the question of the practicality and significance of Reed-Solomon codes. "It's clear they're practical, because everybody's using them now," says Berlekamp. Billions of dollars in modern technology depend on ideas that stem from Reed and Solomon's original work. In short, says McEliece, "it's been an extraordinarily influential paper."

EXERCISES ■■

The New Testament offers the basis for modern computer coding theory, in the form of an affirmation of the binary number system.

> *"But let your communication be yea, yea; nay, nay: for whatsoever is more than these cometh of evil."*

ANONYMOUS

1. Find the Hamming weight of each code word in Table 31.1.

2. Find the Hamming distance between the following pairs of vectors: $\{1101, 0111\}$, $\{0220, 1122\}$, $\{11101, 00111\}$.

3. Referring to Example 1, use the nearest-neighbor method to decode the received words 0000110 and 1110100.

4. For any vector space V and any u, v, w in V, prove that the Hamming distance has the following properties:
   a. $d(u, v) = d(v, u)$ (symmetry)
   b. $d(u, v) = 0$ if and only if $u = v$
   c. $d(u, v) = d(u + w, v + w)$ (translation invariance)

5. Determine the (6, 3) binary linear code with generator matrix

$$G = \begin{bmatrix} 1 & 0 & 0 & 0 & 1 & 1 \\ 0 & 1 & 0 & 1 & 0 & 1 \\ 0 & 0 & 1 & 1 & 1 & 0 \end{bmatrix}.$$

6. Show that for binary vectors, $\text{wt}(u + v) \geq \text{wt}(u) - \text{wt}(v)$ and equality occurs if and only if for all i the ith component of u is 1 whenever the ith component of v is 1.

7. If the minimum weight of any nonzero code word is 2, what can we say about the error-correcting capability of the code?

8. Suppose that C is a linear code with Hamming weight 3 and that C' is one with Hamming weight 4. What can C' do that C can't?

9. Let C be a binary linear code. Show that the code words of even weight form a subcode of C. (A *subcode* of a code is a subset of the code that is itself a code.)

10. Let

$$C = \{0000000, 1110100, 0111010, 0011101, 1001110,$$
$$0100111, 1010011, 1101001\}.$$

What is the error-correcting capability of C? What is the error-detecting capability of C?

11. Suppose that the parity-check matrix of a binary linear code is

$$H = \begin{bmatrix} 1 & 0 \\ 0 & 1 \\ 1 & 1 \\ 1 & 0 \\ 0 & 1 \end{bmatrix}.$$

Can the code correct any single error?

12. Use the generator matrix

$$G = \begin{bmatrix} 1 & 0 & 1 & 1 \\ 0 & 1 & 2 & 1 \end{bmatrix}$$

to construct a (4, 2) ternary linear code. What is the parity-check matrix for this code? What is the error-correcting capability of this code? What is the error-detecting capability of this code? Use parity-check decoding to decode the received word 1201.

13. Find all code words of the (7, 4) binary linear code whose generator matrix is

$$G = \begin{bmatrix} 1 & 0 & 0 & 0 & 1 & 1 & 1 \\ 0 & 1 & 0 & 0 & 1 & 0 & 1 \\ 0 & 0 & 1 & 0 & 1 & 1 & 0 \\ 0 & 0 & 0 & 1 & 0 & 1 & 1 \end{bmatrix}.$$

Find the parity-check matrix of this code. Will this code correct any single error?

14. Show that in a binary linear code, either all the code words end with 0, or exactly half end with 0. What about the other components?

15. Suppose that a code word v is received as the vector u. Show that coset decoding will decode u as the code word v if and only if $u - v$ is a coset leader.

16. Consider the binary linear code

$C = \{00000, 10011, 01010, 11001, 00101, 10110, 01111, 11100\}.$

Construct a standard array for C. Use nearest-neighbor decoding to decode 11101 and 01100. If the received word 11101 has exactly one error, can we determine the intended code word? If the received word 01100 has exactly one error, can we determine the intended code word?

17. Construct a (6, 3) binary linear code with generator matrix

$$G = \begin{bmatrix} 1 & 0 & 0 & 1 & 1 & 0 \\ 0 & 1 & 0 & 0 & 1 & 1 \\ 0 & 0 & 1 & 1 & 0 & 1 \end{bmatrix}.$$

Decode each of the received words

$$001001, 011000, 000110, 100001$$

by the following methods:
a. nearest-neighbor method,
b. parity-check matrix method,
c. coset decoding using a standard array,
d. coset decoding using the syndrome method.

18. Suppose that the minimum weight of any nonzero code word in linear code is 6. Discuss the possible options for error correction and error detection.

19. Using the code and the parity-check matrix given in Example 9, show that parity-check matrix decoding cannot detect any multiple errors (that is, two or more errors).

20. Suppose that the last row of a standard array for a binary linear code is

    10000   00011   11010   01001   10101   00110   11111   01100.

    Complete the array.

21. How many code words are there in a (6, 4) ternary linear code? How many possible received words are there for this code?

22. If the parity-check matrix for a binary linear code is

    $$H = \begin{bmatrix} 1 & 1 & 0 \\ 0 & 1 & 1 \\ 1 & 0 & 1 \\ 1 & 0 & 0 \\ 0 & 1 & 0 \\ 0 & 0 & 1 \end{bmatrix},$$

    will the code correct any single error? Why?

23. Suppose that the parity-check matrix for a ternary code is

    $$\begin{bmatrix} 2 & 1 \\ 2 & 2 \\ 1 & 2 \\ 1 & 0 \\ 0 & 1 \end{bmatrix}.$$

    Can the code correct all single errors? Give a reason for your answer.

24. Prove that for nearest-neighbor decoding, the converse of Theorem 31.2 is true.

25. Can a (6, 3) binary linear code be double-error-correcting using the nearest-neighbor method? Do not assume that the code is systematic.

26. Prove that there is no 2×5 standard generator matrix G that will produce a (5, 2) linear code over Z_3 capable of detecting all possible triple errors.

27. Why can't the nearest-neighbor method with a (4, 2) binary linear code correct all single errors?

28. Suppose that one row of a standard array for a binary code is

000100 110000 011110 111101 101010 001001 100111 010011.

Determine the row that contains 100001.

29. Use the field $F = Z_2[x]/\langle x^2 + x + 1 \rangle$ to construct a (5, 2) linear code that will correct any single error.

30. Find the standard generator matrix for a (4, 2) linear code over Z_3 that encodes 20 as 2012 and 11 as 1100. Determine the entire code and the parity-check matrix for the code. Will the code correct all single errors?

31. Assume that C is an (n, k) binary linear code and that, for each position $i = 1, 2, \ldots, n$, the code C has at least one vector with a 1 in the ith position. Show that the average weight of a code word is $n/2$.

32. Let C be an (n, k) linear code over F such that the minimum weight of any nonzero code word is $2t + 1$. Show that not every vector of weight $t + 1$ in F^n can occur as a coset leader.

33. Let C be an (n, k) binary linear code over $F = Z_2$. If $v \in F^n$ but $v \notin C$, show that $C \cup (v + C)$ is a linear code.

34. Let C be a binary linear code. Show that either every member of C has even weight or exactly half the members of C have even weight. (Compare with Exercise 19 in Chapter 5.)

35. Let C be an (n, k) linear code. For each i with $1 \leq i \leq n$, let $C_i = \{v \in C \mid \text{the } i\text{th component of } v \text{ is } 0\}$. Show that C_i is a subcode of C.

Reference

1. Barry A. Cipra, "The Ubiquitous Reed-Solomon Codes," *SIAM News* 26 (January 1993): 1, 11.

Suggested Readings

Norman Levinson, "Coding Theory: A Counterexample to G. H. Hardy's Conception of Applied Mathematics," *The American Mathematical Monthly* 77 (1970): 249–258.

The eminent mathematician G. H. Hardy insisted that "real" mathematics was almost wholly useless. In this article, the author argues that coding theory refutes Hardy's notion. Levinson uses the finite field of order 16 to construct a linear code that can correct any three errors.

T. M. Thompson, *From Error-Correcting Codes Through Sphere Packing to Simple Groups.* Washington, D.C.: The Mathematical Association of America, 1983.

Chapter 1 of this award-winning book gives a fascinating historical account of the origins of error-correcting codes.

Richard W. Hamming

For introduction of error-correcting codes, pioneering work in operating systems and programming languages, and the advancement of numerical computation.

CITATION FOR THE PIORE AWARD, 1979

RICHARD W. HAMMING was born in Chicago, Illinois, on February 11, 1915. He graduated from the University of Chicago with a B.S. degree in mathematics. In 1939, he received an M.A. degree in mathematics from the University of Nebraska, and in 1942, a Ph.D. in mathematics from the University of Illinois.

During the latter part of World War II, Hamming was at Los Alamos, where he was involved in computing atomic-bomb designs. In 1946, he joined Bell Telephone Laboratories, where he worked in mathematics, computing, engineering, and science.

In 1950, Hamming published his famous paper on error-detecting and error-correcting codes. This work started a branch of information theory. The Hamming codes are used in many modern computers. Hamming's work in the field of numerical analysis has also been of fundamental importance.

Hamming received numerous prestigious awards, including the Turing Prize from the Association for Computing Machinery, the Piore Award from the Institute of Electrical and Electronics Engineers (IEEE), and the Oender Award from the University of Pennsylvania. IEEE also named its Hamming Medal, which carries a $10,000 prize, after Hamming and made him the first recipient.

Hamming died of a heart attack on January 7, 1998, at age 82.

To find more information about Hamming, visit:

http://www-groups.dcs.st-and.ac.uk/~history/

Jessie MacWilliams

She was a mathematician who was instrumental in developing the mathematical theory of error-correcting codes from its early development and whose Ph.D. thesis includes one of the most powerful theorems in coding theory.

VERA PLESS, *SIAM News*

An important contributor to coding theory was Jessie MacWilliams. She was born in 1917 in England. After studying at Cambridge University, MacWilliams came to the United States in 1939 to attend Johns Hopkins University. After one year at Johns Hopkins, she went to Harvard for a year.

In 1955, MacWilliams became a programmer at Bell Labs, where she learned about coding theory. Although she made a major discovery about codes while a programmer, she could not obtain a promotion to a math research position without a Ph.D. degree. She completed some of the requirements for the Ph.D. while working full-time at Bell Labs and looking after her family. She then returned to Harvard for a year (1961–1962), where she finished her degree. Interestingly, both MacWilliams and her daughter Ann were studying mathematics at Harvard at the same time.

MacWilliams returned to Bell Labs, where she remained until her retirement in 1983. While at Bell Labs, she made many contributions to the subject of error-correcting codes, including *The Theory of Error-Correcting Code,* written jointly with Neil Sloane—a book that is still a leader in the field. One of her results of great theoretical importance is known as the MacWilliams Identity. She died on May 27, 1990, at the age of 73.

Vera Pless

Vera Pless is a leader in the field of coding theory.

VERA PLESS was born on March 5, 1931, to Russian immigrants on the West Side of Chicago. She accepted a scholarship to attend the University of Chicago at age 15. The program at Chicago emphasized great literature but paid little attention to physics and mathematics. At age 18, with no more than one precalculus course in mathematics, she entered the prestigious graduate program in mathematics at Chicago, where, at that time, there were no women on the mathematics faculty or even women colloquium speakers. After passing her master's exam, she took a job as a research associate at Northwestern University while pursuing a Ph.D. there. In 1957, she obtained her degree.

Over the next several years, Pless stayed at home to raise her children while teaching part-time at Boston University. When she decided to work full-time, she found that women were not welcome at most colleges and universities. One person told her outright, "I would never hire a woman." Fortunately, there was an Air Force Lab in the area that had a group working on error-correcting codes. Although she had never even heard of coding theory, she was hired because of her background in algebra. When the lab discontinued basic research, she took a position as a research associate at MIT in 1972. In 1975, she went to the University of Illinois–Chicago, where she remains.

Having written more than 100 research papers and a widely used book on coding theory, Pless is a leader in the field.

32

An Introduction to Galois Theory

Galois theory is a showpiece of mathematical unification, bringing together several different branches of the subject and creating a powerful machine for the study of problems of considerable historical and mathematical importance.

IAN STEWART, *Galois Theory*

FUNDAMENTAL THEOREM OF GALOIS THEORY

The Fundamental Theorem of Galois Theory is one of the most elegant theorems in mathematics. Look at Figures 32.1 and 32.2. Figure 32.1 depicts the lattice of subgroups of the group of field automorphisms of $Q(\sqrt[4]{2}, i)$. The integer along an upward lattice line from a group H_1 to a group H_2 is the index of H_1 in H_2. Figure 32.2 shows the lattice of subfields of $Q(\sqrt[4]{2}, i)$. The integer along an upward line from a field K_1 to a field K_2 is the degree of K_2 over K_1. Notice that the lattice in Figure 32.2 is the lattice of Figure 32.1 turned upside down. This is only one of many relationships between these two lattices. The Fundamental Theorem of Galois Theory relates the lattice of subfields of an algebraic extension E of a field F to the subgroup structure of the group of automorphisms of E that send each element of F to itself. This relationship was discovered in the process of attempting to solve a polynomial equation $f(x) = 0$ by radicals.

Before we can give a precise statement of the Fundamental Theorem of Galois Theory, we need some terminology and notation.

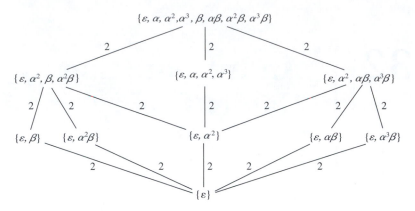

Figure 32.1   Lattice of subgroups of the group of field automorphisms of $Q(\sqrt[4]{2}, i)$, where $\alpha : i \to i$ and $\sqrt[4]{2} \to -i\sqrt[4]{2}$, $\beta : i \to -i$, and $\sqrt[4]{2} \to \sqrt[4]{2}$.

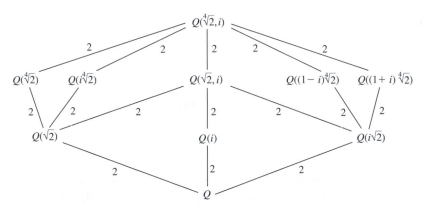

Figure 32.2   Lattice of subfields of $Q(\sqrt[4]{2}, i)$.

DEFINITIONS   *Automorphism, Galois Group, Fixed Field of H*

Let E be an extension field of the field F. An *automorphism of E is* a ring isomorphism from E onto E. The *Galois group* of E over F, Gal(E/F), is the set of all automorphisms of E that take every element of F to itself. If H is a subgroup of Gal(E/F), the set

$$E_H = \{x \in E \mid \phi(x) = x \text{ for all } \phi \in H\}$$

is called the *fixed field of H.*

It is easy to show that the set of automorphisms of E forms a group under composition. We leave as exercises (Exercises 3 and 5) the verifications that the automorphism group of E fixing F is a subgroup of the automorphism group of E and that, for any subgroup H of

Gal(E/F), the fixed field E_H of H is a subfield of E. Be careful not to misinterpret Gal(E/F) as something having to do with factor rings or factor groups. It does not.

The following examples will help you assimilate these definitions. In each example, we simply indicate how the automorphisms are defined. We leave to the reader the verifications that the mappings are indeed automorphisms.

EXAMPLE 1   Consider the extension $Q(\sqrt{2})$ of Q. Since

$$Q(\sqrt{2}) = \{a + b\sqrt{2} \mid a, b \in Q\}$$

and any automorphism of a field containing Q must act as the identity on Q (Exercise 1), an automorphism ϕ of $Q(\sqrt{2})$ is completely determined by $\phi(\sqrt{2})$. Thus,

$$2 = \phi(2) = \phi(\sqrt{2}\sqrt{2}) = (\phi(\sqrt{2}))^2$$

and therefore $\phi(\sqrt{2}) = \pm\sqrt{2}$. This proves that the group Gal($Q(\sqrt{2})/Q$) has two elements, the identity mapping and the mapping that sends $a + b\sqrt{2}$ to $a - b\sqrt{2}$.   ◆

EXAMPLE 2   Consider the extension $Q(\sqrt[3]{2})$ of Q. An automorphism ϕ of $Q(\sqrt[3]{2})$ is completely determined by $\phi(\sqrt[3]{2})$. By an argument analogous to that in Example 1, we see that $\phi(\sqrt[3]{2})$ must be a cube root of 2. Since $Q(\sqrt[3]{2})$ is a subset of the real numbers and $\sqrt[3]{2}$ is the only real cube root of 2, we must have $\phi(\sqrt[3]{2}) = \sqrt[3]{2}$. Thus, ϕ is the identity automorphism and Gal($Q(\sqrt[3]{2})/Q$) has only one element. Obviously, the fixed field of Gal($Q\sqrt[3]{2})/Q$) is $Q(\sqrt[3]{2})$.   ◆

EXAMPLE 3   Consider the extension $Q(\sqrt[4]{2}, i)$ of $Q(i)$. Any automorphism ϕ of $Q(\sqrt[4]{2}, i)$ fixing $Q(i)$ is completely determined by $\phi(\sqrt[4]{2})$. Since

$$2 = \phi(2) = \phi((\sqrt[4]{2})^4) = (\phi(\sqrt[4]{2}))^4,$$

we see that $\phi(\sqrt[4]{2})$ must be a fourth root of 2. Thus, there are at most four possible automorphisms of $Q(\sqrt[4]{2}, i)$ fixing $Q(i)$. If we define an automorphism α such that $\alpha(i) = i$ and $\alpha(\sqrt[4]{2}) = i\sqrt[4]{2}$, then $\alpha \in$ Gal($Q(\sqrt[4]{2}, i)/Q(i)$) and α has order 4. Thus, Gal($Q(\sqrt[4]{2}, i)/Q(i)$) is a cyclic group of order 4. The fixed field of $\{\varepsilon, \alpha^2\}$ (where ε is the identity automorphism) is $Q(\sqrt{2}, i)$. The lattice of subgroups of Gal($Q(\sqrt[4]{2}, i)/Q(i)$) and the lattice of subfields of $Q(\sqrt[4]{2}, i)$ containing $Q(i)$ are shown in Figure 32.3. As in Figures 32.1 and 32.2, the integers along the lines of the group lattice represent the index of a

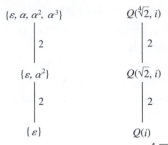

Figure 32.3   Lattice of subgroups of Gal($Q(\sqrt[4]{2}, i)/Q(i)$) and lattice of subfields of $Q(\sqrt[4]{2}, i)$ containing $Q(i)$.

subgroup in the group above it, and the integers along the lines of the field lattice represent the degree of the extension of a field over the field below it.   ◆

EXAMPLE 4   Consider the extension $Q(\sqrt{3}, \sqrt{5})$ of Q. Since

$$Q(\sqrt{3}, \sqrt{5}) = \{a + b\sqrt{3} + c\sqrt{5} + d\sqrt{3}\sqrt{5} \mid a, b, c, d \in Q\},$$

any automorphism ϕ of $Q(\sqrt{3}, \sqrt{5})$ is completely determined by the two values $\phi(\sqrt{3})$ and $\phi(\sqrt{5})$. This time there are four automorphisms:

ε	α	β	$\alpha\beta$
$\sqrt{3} \to \sqrt{3}$	$\sqrt{3} \to -\sqrt{3}$	$\sqrt{3} \to \sqrt{3}$	$\sqrt{3} \to -\sqrt{3}$
$\sqrt{5} \to \sqrt{5}$	$\sqrt{5} \to \sqrt{5}$	$\sqrt{5} \to -\sqrt{5}$	$\sqrt{5} \to -\sqrt{5}$

Obviously, Gal($Q(\sqrt{3}, \sqrt{5})/Q$) is isomorphic to $Z_2 \oplus Z_2$. The fixed field of $\{\varepsilon, \alpha\}$ is $Q(\sqrt{5})$, the fixed field of $\{\varepsilon, \beta\}$ is $Q(\sqrt{3})$, and the fixed field of $\{\varepsilon, \alpha\beta\}$ is $Q(\sqrt{3}\sqrt{5})$. The lattice of subgroups of Gal($Q(\sqrt{3}, \sqrt{5})/Q$) and the lattice of subfields of $Q(\sqrt{3}, \sqrt{5})$ are shown in Figure 32.4.   ◆

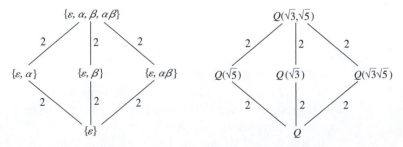

Figure 32.4   Lattice of subgroups of Gal($Q(\sqrt{3}, \sqrt{5})/Q$) and lattice of subfields of $Q(\sqrt{3}, \sqrt{5})$.

Example 5 is a bit more complicated than our previous examples. In particular, the automorphism group is non-Abelian.

EXAMPLE 5   Direct calculations show that $\omega = -1/2 + i\sqrt{3}/2$ satisfies the equations $\omega^3 = 1$ and $\omega^2 + \omega + 1 = 0$. Now, consider the extension $Q(\omega, \sqrt[3]{2})$ of Q. We may describe the automorphisms of $Q(\omega, \sqrt[3]{2})$ by specifying how they act on ω and $\sqrt[3]{2}$. There are six in all:

ε	α	β	β^2	$\alpha\beta$	$\alpha\beta^2$
$\omega \rightarrow \omega$	$\omega \rightarrow \omega^2$	$\omega \rightarrow \omega$	$\omega \rightarrow \omega$	$\omega \rightarrow \omega^2$	$\omega \rightarrow \omega^2$
$\sqrt[3]{2} \rightarrow \sqrt[3]{2}$	$\sqrt[3]{2} \rightarrow \sqrt[3]{2}$	$\sqrt[3]{2} \rightarrow \omega\sqrt[3]{2}$	$\sqrt[3]{2} \rightarrow \omega^2\sqrt[3]{2}$	$\sqrt[3]{2} \rightarrow \omega^2\sqrt[3]{2}$	$\sqrt[3]{2} \rightarrow \omega\sqrt[3]{2}$

Since $\alpha\beta \neq \beta\alpha$, we know that $\text{Gal}(Q(\omega, \sqrt[3]{2})/Q)$ is isomorphic to S_3. (See Theorem 7.2.) The lattices of subgroups and subfields are shown in Figure 32.5.

The lattices in Figure 32.5 have been arranged so that the field occupying the same position as some group is the fixed field of that group. For instance, $Q(\omega\sqrt[3]{2})$ is the fixed field of $\{\varepsilon, \alpha\beta\}$.   ◆

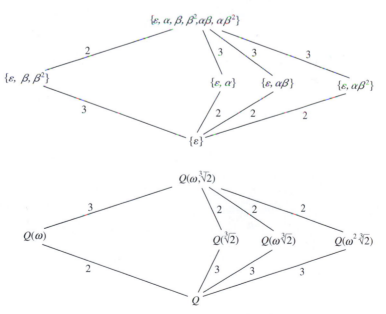

Figure 32.5   Lattice of subgroups of $\text{Gal}(Q(\omega, \sqrt[3]{2})/Q)$ and lattice of subfields of $Q(\omega, \sqrt[3]{2})$, where $\omega = -1/2 + i\sqrt{3}/2$.

The preceding examples show that, in certain cases, there is an intimate connection between the lattice of subfields between E and F and the lattice of subgroups of Gal(E/F). In general, if E is an extension of F, and we let $\mathcal{F}$ be the lattice of subfields of E containing F and let $\mathcal{G}$ be the lattice of subgroups of Gal(E/F), then for each K in $\mathcal{F}$, the group Gal(E/K) is in $\mathcal{G}$, and for each H in $\mathcal{G}$, the field E_H is in $\mathcal{F}$. Thus, we may define a mapping $g: \mathcal{F} \to \mathcal{G}$ by $g(K) = $ Gal(E/K) and a mapping $f: \mathcal{G} \to \mathcal{F}$ by $f(H) = E_H$. It is easy to show that if K and L belong to $\mathcal{F}$ and $K \subseteq L$, then $g(K) \supseteq g(L)$. Similarly, if G and H belong to $\mathcal{G}$ and $G \subseteq H$, then $f(G) \supseteq f(H)$. Thus, f and g are inclusion-reversing mappings between $\mathcal{F}$ and $\mathcal{G}$. We leave it to the reader to show that for any K in $\mathcal{F}$, we have $(fg)(K) \supseteq K$, and for any G in $\mathcal{G}$, we have $(gf)(G) \supseteq G$. When E is an arbitrary extension of F, these inclusions may be strict. However, when E is a suitably chosen extension of F, the Fundamental Theorem of Galois Theory, Theorem 32.1, says that f and g are inverses of each other, so that the inclusions are equalities. In particular, f and g are inclusion-reversing isomorphisms between the lattices $\mathcal{F}$ and $\mathcal{G}$. A stronger result than that given in Theorem 32.1 is true, but our theorem illustrates the fundamental principles involved. The student is referred to [1, p. 285] for additional details and proofs.

Theorem 32.1 **Fundamental Theorem of Galois Theory**

Let F be a field of characteristic 0 or a finite field. If E is the splitting field over F for some polynomial in $F[x]$, then the mapping from the set of subfields of E containing F to the set of subgroups of Gal(E/F) given by $K \to$ Gal(E/K) is a one-to-one correspondence. Furthermore, for any subfield K of E containing F,

1. *$[E:K] = |$Gal(E/K)$|$ and $[K:F] = |$Gal(E/F)$|/|$Gal(E/K)$|$.*
   [The index of Gal(E/K) in Gal(E/F) equals the degree of K over F.]
2. *If K is the splitting field of some polynomial in $F[x]$, then Gal(E/K) is a normal subgroup of Gal(E/F) and Gal(K/F) is isomorphic to Gal(E/F)/Gal(E/K).*
3. *$K = E_{\text{Gal}(E/K)}$. [The fixed field of Gal(E/K) is K.]*
4. *If H is a subgroup of Gal(E/F), then $H = $ Gal(E/E_H).*
   [The automorphism group of E fixing E_H is H.]

Generally speaking, it is much easier to determine a lattice of subgroups than a lattice of subfields. For example, it is usually quite difficult to determine, directly, how many subfields a given field has, and it is often difficult to decide whether or not two field extensions are the same. The corresponding questions about groups are much more tractable. Hence, the Fundamental Theorem of Galois Theory can be a great labor-saving device. Here is an illustration. [Recall from Chapter 20 that if $f(x) \in F[x]$ and the zeros of $f(x)$ in some extension of F are $a_1, a_2, \ldots, a_n$, then $F(a_1, a_2, \ldots, a_n)$ is the splitting field of $f(x)$ over F.]

EXAMPLE 6   Let $\omega = \cos(2\pi/7) + i\sin(2\pi/7)$, so that $\omega^7 = 1$, and consider the field $Q(\omega)$. How many subfields does it have and what are they? First, observe that $Q(\omega)$ is the splitting field of $x^7 - 1$ over Q, so that we may apply the Fundamental Theorem of Galois Theory. A simple calculation shows that the automorphism ϕ that sends ω to ω^3 has order 6. Thus,

$$[Q(\omega){:}Q] = |\text{Gal}(Q(\omega)/Q)| \geq 6.$$

Also, since

$$x^7 - 1 = (x - 1)(x^6 + x^5 + x^4 + x^3 + x^2 + x + 1)$$

and ω is a zero of $x^7 - 1$, we see that

$$|\text{Gal}(Q(\omega)/Q)| = [Q(\omega){:}Q] \leq 6.$$

Thus, $\text{Gal}(Q(\omega)/Q)$ is a cyclic group of order 6. So, the lattice of subgroups of $\text{Gal}(Q(\omega)/Q)$ is trivial to compute. See Figure 32.6.

This means that $Q(\omega)$ contains exactly two proper extensions of Q: one of degree 3 corresponding to the fixed field of $\langle \phi^3 \rangle$ and one of degree 2 corresponding to the fixed field of $\langle \phi^2 \rangle$. To find the fixed field of $\langle \phi^3 \rangle$, we must find a member of $Q(\omega)$ that is not in Q and that is fixed by ϕ^3. Experimenting with various possibilities leads us to discover that $\omega + \omega^{-1}$ is fixed by ϕ^3 (see Exercise 9), and it follows that $Q \subset Q(\omega + \omega^{-1}) \subseteq Q(\omega)_{\langle \phi^3 \rangle}$. Since $[Q(\omega)_{\langle \phi^3 \rangle} : Q] = 3$ and $[Q(\omega + \omega^{-1}) : Q]$ divides $[Q(\omega)_{\langle \phi^3 \rangle} : Q]$, we see that $Q(\omega + \omega^{-1}) = Q(\omega)_{\langle \phi^3 \rangle}$. A similar argument shows that $Q(\omega^3 + \omega^5 + \omega^6)$ is the fixed field of $<\phi^2>$. Thus, we have found all subfields of $Q(\omega)$.   ◆

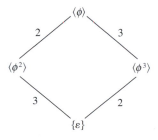

Figure 32.6   Lattice of subgroups of $\text{Gal}(Q(\omega)/Q)$, where $\omega = \cos(2\pi/7) + i\sin(2\pi/7)$.

EXAMPLE 7   Consider the extension $E = GF(p^n)$ of $F = GF(p)$. Let us determine $\text{Gal}(E/F)$. By Corollary 2 of Theorem 22.2, E has the form $F(b)$ for some b where b is the zero of an irreducible polynomial $p(x)$ of the form $x^n + a_{n-1}x^{n-1} + \cdots + a_1 x + a_0$, where $a_{n-1}, a_{n-2}, \ldots, a_0$ belong to F. Since any field automorphism ϕ of E must take 1 to itself, it follows that ϕ acts as the identity on F. Thus, $p(b) = 0$ implies that $p(\phi(b)) = 0$. And because $p(x)$ has at most n zeros, we know that there are at most n possibilities for $\phi(b)$. On the other hand, by Exercise 41 in Chapter 13, we know that the mapping $\sigma(a) = a^p$ for all $a \in E$ is an automorphism of E, and it follows from the fact that E^* is cyclic (Theorem 22.2) that the group $\langle \sigma \rangle$ has order n (see Exercise 7 in Chapter 22). Thus, $\text{Gal}(GF(p^n)/GF(p)) \approx Z_n$.   ◆

SOLVABILITY OF POLYNOMIALS BY RADICALS

For Galois, the elegant correspondence between groups and fields given by Theorem 32.1 was only a means to an end. Galois sought to solve a problem that had stymied mathematicians for centuries. Methods for solving linear and quadratic equations were known thousands of years ago (the Quadratic Formula). In the 16th century, Italian mathematicians developed formulas for solving any third- or fourth-degree equation. Their formulas involved only the operations of addition, subtraction, multiplication, division, and extraction of roots (radicals). For example, the equation

$$x^3 + bx + c = 0$$

has the three solutions

$$A + B,$$
$$-(A + B)/2 + (A - B)\sqrt{-3}/2,$$
$$-(A + B)/2 - (A - B)\sqrt{-3}/2,$$

where

$$A = \sqrt[3]{\frac{-c}{2} + \sqrt{\frac{b^3}{27} + \frac{c^2}{4}}} \quad \text{and} \quad B = \sqrt[3]{\frac{-c}{2} - \sqrt{\frac{b^3}{27} + \frac{c^2}{4}}}.$$

The formulas for the general cubic $x^3 + ax^2 + bx + c = 0$ and the general quartic (fourth-degree polynomial) are even more complicated, but nevertheless can be given in terms of radicals of rational expressions of the coefficients.

Both Abel and Galois proved that there is no general solution of a fifth-degree equation by radicals. In particular, there is no "quintic formula." Before discussing Galois's method, which provided a group-theoretic criterion for the solution of an equation by radicals and led to the modern-day Galois theory, we need a few definitions.

DEFINITION *Solvable by Radicals*

> Let F be a field, and let $f(x) \in F[x]$. We say that $f(x)$ is *solvable by radicals over F* if $f(x)$ splits in some extension $F(a_1, a_2, \ldots, a_n)$ of F and there exist positive integers $k_1, \ldots, k_n$ such that $a_1{}^{k_1} \in F$ and $a_i{}^{k_i} \in F(a_1, \ldots, a_{i-1})$ for $i = 2, \ldots, n$.

So, a polynomial in $F[x]$ is solvable by radicals if we can obtain all of its zeros by adjoining nth roots (for various n) to F. In other words, each zero of the polynomial can be written as an expression (usually a messy one) involving elements of F combined by the operations of addition, subtraction, multiplication, division, and extraction of roots.

EXAMPLE 8   Let $\omega = \cos\left(\frac{2\pi}{8}\right) + i \sin\left(\frac{2\pi}{8}\right) = \frac{\sqrt{2}}{2} + i\frac{\sqrt{2}}{2}$. Then $x^8 - 3$ splits in $Q(\omega, \sqrt[8]{3})$, $\omega^8 \in Q$, and $(\sqrt[8]{3})^8 \in Q \subset Q(\omega)$. Thus, $x^8 - 3$ is solvable by radicals over Q. Although the zeros of $x^8 - 3$ are most conveniently written in the form $\sqrt[8]{3}, \sqrt[8]{3}\,\omega, \sqrt[8]{3}\,\omega^2, \ldots,$ $\sqrt[8]{3}\,\omega^7$, the notion of solvable by radicals is best illustrated by writing them in the form

$$\pm\sqrt[8]{3}, \pm\sqrt{-1}\sqrt[8]{3}, \pm\sqrt[8]{3}\left(\frac{\sqrt{2}}{2} + \frac{\sqrt{-1}\sqrt{2}}{2}\right),$$

$$\pm\sqrt[8]{3}\left(\frac{\sqrt{2}}{2} - \frac{\sqrt{-1}\sqrt{2}}{2}\right). \qquad \blacklozenge$$

Thus, the problem of solving a polynomial equation for its zeros can be transformed into a problem about field extensions. At the same time, we can use the Fundamental Theorem of Galois Theory to transform a problem about field extensions into a problem about groups. This is exactly how Galois showed that there are fifth-degree polynomials that cannot be solved by radicals, and this is exactly how we will do it. Before giving an example of such a polynomial, we need some additional group theory.

DEFINITION   *Solvable Group*

We say that a group G is *solvable* if G has a series of subgroups

$$\{e\} = H_0 \subset H_1 \subset H_2 \subset \cdots \subset H_k = G,$$

where, for each $0 \le i < k$, H_i is normal in H_{i+1} and H_{i+1}/H_i is Abelian.

Obviously, Abelian groups are solvable. So are the dihedral groups and any group whose order has the form p^n, where p is a prime (see Exercises 22 and 23). The monumental Feit-Thompson Theorem (see Chapter 25) says that every group of odd order is solvable. In a certain sense, solvable groups are almost Abelian. On the other hand, it follows directly from the definitions that any non-Abelian simple group is not solvable. In particular, A_5 is not solvable. It follows from Exercise 21 in Chapter 25 that S_5 is not solvable. Our goal is to connect the notion of solvability of polynomials by radicals to that of solvable groups. The next theorem is a step in this direction.

Theorem 32.2   Splitting Field of $x^n - a$

Let F be a field of characteristic 0 and let $a \in F$. If E is the splitting field of $x^n - a$ over F, then the Galois group $\mathrm{Gal}(E/F)$ is solvable.

PROOF.   We first handle the case where F contains a primitive nth root of unity ω. Let b be a zero of $x^n - a$ in E. Then the zeros of $x^n - a$ are $b, \omega b, \omega^2 b, \ldots, \omega^{n-1} b$, and therefore $E = F(b)$. In this case, we claim that $\mathrm{Gal}(E/F)$ is Abelian and hence solvable. To see this, observe that any automorphism in $\mathrm{Gal}(E/F)$ is completely determined by its action on b. Also, since b is a zero of $x^n - a$, we know that any element of $\mathrm{Gal}(E/F)$ sends b to another zero of $x^n - a$. That is, any element of $\mathrm{Gal}(E/F)$ takes b to $\omega^i b$ for some i. Let ϕ and σ be two elements of $\mathrm{Gal}(E/F)$. Then, since $\omega \in F$, ϕ and σ fix ω and $\phi(b) = \omega^j b$ and $\sigma(b) = \omega^k b$ for some j and k. Thus,

$$(\sigma\phi)(b) = \sigma(\phi(b)) = \sigma(\omega^j b) = \sigma(\omega^j)\sigma(b) = \omega^j \omega^k b = \omega^{j+k} b,$$

whereas

$$(\phi\sigma)(b) = \phi(\sigma(b)) = \phi(\omega^k b) = \phi(\omega^k)\phi(b) = \omega^k \omega^j b = \omega^{k+j} b,$$

so that $\sigma\phi$ and $\phi\sigma$ agree on b and fix the elements of F. This shows that $\sigma\phi = \phi\sigma$, and therefore $\mathrm{Gal}(E/F)$ is Abelian.

Now suppose that F does not contain a primitive nth root of unity. Let ω be a primitive nth root of unity and let b be a zero of $x^n - a$ in E.

The case where $a = 0$ is trivial, so we may assume that $b \neq 0$. Since ωb is also a zero of $x^n - a$, we know that both ω and ωb belong to E, and therefore $\omega = \omega b / b$ is in E as well. Thus, $F(\omega)$ is contained in E, and $F(\omega)$ is the splitting field of $x^n - 1$ over F. Analogously to the case above, for any automorphisms ϕ and σ in $\mathrm{Gal}(F(\omega)/F)$ we have $\phi(\omega) = \omega^j$ for some j and $\sigma(\omega) = \omega^k$ for some k. Then,

$$(\sigma\phi)(\omega) = \sigma(\phi(\omega)) = \sigma(\omega^j) = (\sigma(\omega))^j = (\omega^k)^j$$
$$= (\omega^j)^k = (\phi(\omega))^k = \phi(\omega^k) = \phi(\sigma(\omega)) = (\phi\sigma)(\omega).$$

Since elements of $\mathrm{Gal}(F(\omega)/F)$ are completely determined by their action on ω, this shows that $\mathrm{Gal}(F(\omega)/F)$ is Abelian.

Because E is the splitting field of $x^n - a$ over $F(\omega)$ and $F(\omega)$ contains a primitive nth root of unity, we know from the case we have already done that $\mathrm{Gal}(E/F(\omega))$ is Abelian and, by Part 2 of Theorem 32.1, the series

$$\{e\} \subseteq \mathrm{Gal}(E/F(\omega)) \subseteq \mathrm{Gal}(E/F)$$

is a normal series. Finally, since both $\mathrm{Gal}(E/F(\omega))$ and

$$\mathrm{Gal}(E/F)/\mathrm{Gal}(E/F(\omega)) \approx \mathrm{Gal}(F(\omega)/F)$$

are Abelian, $\mathrm{Gal}(E/F)$ is solvable. ∎

To reach our main result about polynomials that are solvable by radicals, we need two important facts about solvable groups.

Theorem 32.3   Factor Group of a Solvable Group Is Solvable

> *A factor group of a solvable group is solvable.*

PROOF.   Suppose that G has a series of subgroups

$$\{e\} = H_0 \subset H_1 \subset H_2 \subset \cdots \subset H_k = G,$$

where, for each $0 \leq i < k$, H_i is normal in H_{i+1} and H_{i+1}/H_i is Abelian. If N is any normal subgroup of G, then

$$\{e\} = H_0 N/N \subset H_1 N/N \subset H_2 N/N \subset \cdots \subset H_k N/N = G/N$$

is the requisite series of subgroups that guarantees that G/N is solvable. (See Exercise 25.) ∎

Theorem 32.4   *N* and *G*/*N* Solvable Implies *G* Is Solvable

> *Let N be a normal subgroup of a group G. If both N and G/N are solvable, then G is solvable.*

PROOF.   Let a series of subgroups of N with Abelian factors be

$$N_0 \subset N_1 \subset \cdots \subset N_t = N$$

and let a series of subgroups of G/N with Abelian factors be

$$N/N = H_0/N \subset H_1/N \subset \cdots \subset H_s/N = G/N.$$

Then the series

$$N_0 \subset N_1 \subset \cdots \subset N_t = H_0 \subset H_1 \subset \cdots \subset H_s = G$$

has Abelian factors (see Exercise 27).   ■ ■

We are now able to make the critical connection between solvability of polynomials by radicals and solvable groups.

***Theorem 32.5 (Galois)*   Solvable by Radicals Implies Solvable Group**

> *Let F be a field of characteristic 0 and let $f(x) \in F[x]$.*
> *Suppose that $f(x)$ splits in $F(a_1, a_2, \ldots, a_t)$, where $a_1^{n_1} \in F$*
> *and $a_i^{n_i} \in F(a_1, \ldots, a_{i-1})$ for $i = 2, \ldots, t$. Let E be the*
> *splitting field for $f(x)$ over F in $F(a_1, a_2, \ldots, a_t)$. Then the*
> *Galois group $\mathrm{Gal}(E/F)$ is solvable.*

PROOF.   We induct on t. For the case $t = 1$, we have $F \subseteq E \subseteq F(a_1)$. Let $a = a_1^{n_1}$ and let L be a splitting field of $x^{n_1} - a$ over F. Then $F \subseteq E \subseteq L$, and both E and L are splitting fields of polynomials over F. By Part 2 of Theorem 32.1, $\mathrm{Gal}(E/F) \approx \mathrm{Gal}(L/F)/\mathrm{Gal}(L/E)$. It follows from Theorem 32.2 that $\mathrm{Gal}(L/F)$ is solvable, and from Theorem 32.3 we know that $\mathrm{Gal}(L/F)/\mathrm{Gal}(L/E)$ is solvable. Thus, $\mathrm{Gal}(E/F)$ is solvable.

Now suppose $t > 1$. Let $a = a_1^{n_1} \in F$, let L be a splitting field of $x^{n_1} - a$ over E, and let $K \subseteq L$ be the splitting field of $x^{n_1} - a$ over F. Then L is a splitting field of $(x^{n_1} - a)f(x)$ over F, and L is a splitting field of $f(x)$ over K. Since $F(a_1) \subseteq K$, we know that $f(x)$ splits in $K(a_2, \ldots, a_t)$, so the induction hypothesis implies that $\mathrm{Gal}(L/K)$ is solvable. Also, Theorem 32.2 asserts that $\mathrm{Gal}(K/F)$ is solvable, which, from Theorem 32.1, tells us that $\mathrm{Gal}(L/F)/\mathrm{Gal}(L/K)$ is solvable. Hence, Theorem 32.4 implies that $\mathrm{Gal}(L/F)$ is solvable. So, by Part 2 of Theorem 32.1 and Theorem 32.3, we know that the factor group $\mathrm{Gal}(L/F)/\mathrm{Gal}(L/E) \approx \mathrm{Gal}(E/F)$ is solvable.   ■ ■

It is worth remarking that the converse of Theorem 32.3 is true also; that is, if E is the splitting field of a polynomial $f(x)$ over a field F of characteristic 0 and $\mathrm{Gal}(E/F)$ is solvable, then $f(x)$ is solvable by radicals over F.

It is known that every finite group is a Galois group over some field. However, one of the major unsolved problems in algebra, first posed by Emmy Noether, is determining which finite groups can occur as Galois groups over Q. Many people suspect that the answer is "all of them." It is known that every solvable group is a Galois group over Q. John Thompson has recently proved that certain kinds of simple groups, including the Monster, are Galois groups over Q. The suggested reading by Ian Stewart provides more information on this topic.

INSOLVABILITY OF A QUINTIC

We will finish our introduction to Galois theory by explicitly exhibiting a polynomial that has integer coefficients and that is not solvable by radicals over Q.

Consider $g(x) = 3x^5 - 15x + 5$. By Eisenstein's Criterion (Theorem 17.4), $g(x)$ is irreducible over Q. Since $g(x)$ is continuous and $g(-2) = -61$ and $g(-1) = 17$, we know that $g(x)$ has a real zero between -2 and -1. A similar analysis shows that $g(x)$ also has real zeros between 0 and 1 and between 1 and 2.

Each of these real zeros has multiplicity 1, as can be verified by long division or by appealing to Theorem 20.6. Furthermore, $g(x)$ has no more than three real zeros, because Rolle's Theorem from calculus guarantees that between each pair of real zeros of $g(x)$ there must be a zero of $g'(x) = 15x^4 - 15$. So, for $g(x)$ to have four real zeros, $g'(x)$ would have to have three real zeros, and it does not. Thus, the other two zeros of $g(x)$ are nonreal complex numbers, say, $a + bi$ and $a - bi$. (See Exercise 59 in Chapter 15.)

Now, let's denote the five zeros of $g(x)$ by a_1, a_2, a_3, a_4, a_5. Since any automorphism of $K = Q(a_1, a_2, a_3, a_4, a_5)$ is completely determined by its action on the a's and must permute the a's, we know that $Gal(K/Q)$ is isomorphic to a subgroup of S_5, the symmetric group on five symbols. Since a_1 is a zero of an irreducible polynomial of degree 5 over Q, we know that $[Q(a_1):Q] = 5$, and therefore 5 divides $[K:Q]$. Thus, the Fundamental Theorem of Galois Theory tells us that 5 also divides $|Gal(K/Q)|$. So, by Cauchy's Theorem (corollary to Theorem 24.3), we may conclude that $Gal(K/Q)$ has an element of order 5. Since the only elements in S_5 of order 5 are the 5-cycles, we know that $Gal(K/Q)$ contains a 5-cycle. The mapping from **C** to **C**, sending $a + bi$ to $a - bi$, is also an element of $Gal(K/Q)$. Since this mapping fixes the three real zeros and interchanges the two complex zeros of $g(x)$, we know that $Gal(K/Q)$ contains a 2-cycle. But, the only subgroup of S_5 that contains both a 5-cycle and a 2-cycle is S_5. (See Exercise 25 in Chapter 25.) So, Gal(K/Q) is isomorphic to S_5. Finally, since S_5 is not solvable (see Exercise 21), we have succeeded in exhibiting a fifth-degree polynomial that is not solvable by radicals.

EXERCISES

Seeing much, suffering much, and studying much are the three pillars of learning.

BENJAMIN DISRAELI

1. Let E be an extension field of Q. Show that any automorphism of E acts as the identity on Q. (This exercise is referred to in this chapter.)

2. Determine the group of field automorphisms of GF(4).

3. Let E be a field extension of the field F. Show that the automorphism group of E fixing F is indeed a group. (This exercise is referred to in this chapter.)

4. Given that the automorphism group of $Q(\sqrt{2}, \sqrt{5}, \sqrt{7})$ is isomorphic to $Z_2 \oplus Z_2 \oplus Z_2$, determine the number of subfields of $Q(\sqrt{2}, \sqrt{5}, \sqrt{7})$ that have degree 4 over Q.

5. Let E be a field extension of a field F and let H be a subgroup of Gal(E/F). Show that the fixed field of H is indeed a field. (This exercise is referred to in this chapter.)

6. Let E be the splitting field of $x^4 + 1$ over Q. Find Gal(E/Q). Find all subfields of E. Find the automorphisms of E that have fixed fields $Q(\sqrt{2})$, $Q(\sqrt{-2})$, and $Q(i)$. Is there an automorphism of E whose fixed field is Q?

7. Let $f(x) \in F[x]$ and let the zeros of $f(x)$ be $a_1, a_2, \ldots, a_n$. If $K = F(a_1, a_2, \ldots, a_n)$, show that Gal($K/F$) is isomorphic to a group of permutations of the a_i's. [When K is the splitting field of $f(x)$ over F, the group Gal(K/F) is called the *Galois group of $f(x)$.*]

8. Show that the Galois group of a polynomial of degree n has order dividing $n!$.

9. Referring to Example 6, show that the automorphism ϕ has order 6. Show that $\omega + \omega^{-1}$ is fixed by ϕ^3 and $\omega^3 + \omega^5 + \omega^6$ is fixed by ϕ^2. (This exercise is referred to in this chapter.)

10. Let $E = Q(\sqrt{2}, \sqrt{5})$. What is the order of the group Gal(E/Q)? What is the order of Gal($Q(\sqrt{10})/Q$)?

11. Suppose that F is a field of characteristic 0 and E is the splitting field for some polynomial over F. If Gal(E/F) is isomorphic to A_4, show that there is no subfield K of E such that $[K:F] = 2$.

12. Determine the Galois group of $x^3 - 1$ over Q and $x^3 - 2$ over Q. (See Exercise 7 for the definition.)

13. Suppose that K is the splitting field of some polynomial over a field F of characteristic 0. If $[K:F] = p^2q$, where p and q are distinct primes, show that K has subfields L_1, L_2, and L_3 such that $[K:L_1] = p$, $[K:L_2] = p^2$, and $[K:L_3] = q$.

14. Suppose that E is the splitting field of some polynomial over a field F of characteristic 0. If $\text{Gal}(E/F)$ is isomorphic to D_5, draw the subfield lattice for the fields between E and F.

15. Suppose that $F \subset K \subset E$ are fields and E is the splitting field of some polynomial in $F[x]$. Show, by means of an example, that K need not be the splitting field of some polynomial in $F[x]$.

16. Suppose that E is the splitting field of some polynomial over a field F of characteristic 0. If $[E:F]$ is finite, show that there is only a finite number of fields between E and F.

17. Suppose that E is the splitting field of some polynomial over a field F of characteristic 0. If $\text{Gal}(E/F)$ is an Abelian group of order 10, draw the subfield lattice for the fields between E and F.

18. Let ω be a nonreal complex number such that $\omega^5 = 1$. If ϕ is the automorphism of $Q(\omega)$ that carries ω to ω^4, find the fixed field of $\langle \phi \rangle$.

19. Determine the isomorphism class of the group $\text{Gal}(GF(64)/GF(2))$.

20. Determine the isomorphism class of the group $\text{Gal}(GF(729)/GF(9))$.

21. Show that S_5 is not solvable. (This exercise is referred to in this chapter.)

22. Show that the dihedral groups are solvable.

23. Show that a group of order p^n, where p is prime, is solvable.

24. Show that S_n is solvable when $n \leq 4$.

25. Complete the proof of Theorem 32.3 by showing that the given series of groups satisfies the definition for solvability.

26. Show that a subgroup of a solvable group is solvable.

27. Let N be a normal subgroup of G and let K/N be a normal subgroup of G/N. Prove that K is a normal subgroup of G. (This exercise is referred to in this chapter.)

28. Show that any automorphism of $GF(p^n)$ acts as the identity on $GF(p)$.

Reference

1. G. Ehrlich, *Fundamental Concepts of Abstract Algebra,* Boston: PWS-Kent, 1991.

Suggested Readings

Tony Rothman, "The Short Life of Évariste Galois," *Scientific American,* April (1982): 136–149.

> This article gives an elementary discussion of Galois's proof that the general fifth-degree equation cannot be solved by radicals. The article also goes into detail about Galois's controversial life and death. In this regard, Rothman refutes several accounts given by other Galois biographers.

Ian Stewart, "The Duelist and the Monster," *Nature* 317 (1985): 12–13.

> This nontechnical article discusses recent work of John Thompson pertaining to the question of "which groups can occur as Galois groups."

Suggested Website

To find more information about the history of quadratic, cubic, and quartic equations, visit:

http://www-groups.dcs.st-and.ac.uk/~history/

Philip Hall

He [Hall] was preeminent as a group theorist and made many fundamental discoveries; the conspicuous growth of interest in group theory in this century owes much to him.

J. E. ROSEBLADE

PHILIP HALL was born on April 11, 1904, in London. Abandoned by his father shortly after birth, Hall was raised by his mother, a dressmaker. He demonstrated academic prowess early by winning a scholarship to Christ's Hospital, where he had several outstanding mathematics teachers. At Christ's Hospital, Hall won a medal for the best English essay, the gold medal in mathematics, and a scholarship to King's College, Cambridge.

Although abstract algebra was a field neglected at King's College, Hall studied Burnside's book *Theory of Groups* and some of Burnside's later papers. After graduating in 1925, he stayed on at King's College for further study and was elected to a fellowship in 1927. That same year, Hall discovered a major "Sylow-like" theorem about solvable groups: If a solvable group has order mn, where m and n are relatively prime, then every subgroup whose order divides m is contained in a group of order m and all subgroups of order m are conjugate. Over the next three decades, Hall developed a general theory of finite solvable groups that had a profound influence on John Thompson's spectacular achievements of the 1960s. In the 1930s, Hall also developed a general theory of groups of prime-power order that has become a foundation of modern finite group theory. In addition to his fundamental contributions to finite groups, Hall wrote many seminal papers on infinite groups.

Among the concepts that have Hall's name attached to them are Hall subgroups, Hall algebras, Hall-Littlewood polynomials, Hall divisors, the marriage theorem from graph theory, and the Hall commutator collecting process. Beyond his own discoveries, Hall had an enormous influence on algebra through his research students. No fewer than one dozen have become eminent mathematicians in their own right. Hall died on December 30, 1982.

To find more information about Hall, visit:

http://www-groups.dcs.st-and.ac.uk/~history/

33

Cyclotomic Extensions

". . . to regard old problems from a new angle requires creative imagination and marks real advances in science."

ALBERT EINSTEIN

MOTIVATION

For the culminating chapter of this book, it is fitting to choose a topic that ties together results about groups, rings, fields, geometric constructions, and the history of mathematics. The so-called *cyclotomic extensions* is such a topic. We begin with the history.

The ancient Greeks knew how to construct regular polygons of 3, 4, 5, 6, 8, 10, 12, 15, and 16 sides with a straightedge and compass. And, given a construction of a regular n-gon, it is easy to construct a regular $2n$-gon. The Greeks attempted to fill in the gaps (7, 9, 11, 13, 14, 17, . . .) but failed. More than 2200 years passed before anyone was able to advance our knowledge of this problem beyond that of the Greeks. Incredibly, Gauss, at age 19, showed that a regular 17-gon is constructible, and shortly thereafter he completely solved the problem of exactly which n-gons are constructible. It was this discovery of the constructibility of the 17-sided regular polygon that induced Gauss to dedicate his life to the study of mathematics. Gauss was so proud of this accomplishment that he requested that a regular 17-sided polygon be engraved on his tombstone.

Gauss was led to his discovery of the constructible polygons through his investigation of the factorization of polynomials of the form $x^n - 1$ over Q. In this chapter, we examine the factors of $x^n - 1$ and show how Galois theory can be used to determine which regular n-gons are constructible with a straightedge and compass. The irreducible factors of $x^n - 1$ are important in number theory and combinatorics.

CYCLOTOMIC POLYNOMIALS

Recall from Example 2 in Chapter 16 that the complex zeros of $x^n - 1$ are 1, $\omega = \cos(\frac{2\pi}{n}) + i \sin(\frac{2\pi}{n})$, ω^2, ω^3, ..., ω^{n-1}. Thus, the splitting field of $x^n - 1$ over Q is $Q(\omega)$. This field is called the *nth cyclotomic extension* of Q, and the irreducible factors of $x^n - 1$ over Q are called the *cyclotomic polynomials*.

Since $\omega = \cos(\frac{2\pi}{n}) + i \sin(\frac{2\pi}{n})$ generates a cyclic group of order n under multiplication, we know from Corollary 2 of Theorem 4.2 that the generators of $\langle \omega \rangle$ are the elements of the form ω^k, where $1 \leq k \leq n$ and $\gcd(n, k) = 1$. These generators are called the *primitive nth roots of unity*. Recalling that we use $\phi(n)$ to denote the number of positive integers less than or equal to n and relatively prime to n, we see that for each positive integer n there are precisely $\phi(n)$ primitive nth roots of unity. The polynomials whose zeros are the $\phi(n)$ primitive nth roots of unity have a special name.

DEFINITION

For any positive integer n, let $\omega_1, \omega_2, \ldots, \omega_{\phi(n)}$ denote the primitive nth roots of unity. The *nth cyclotomic polynomial over Q* is the polynomial $\Phi_n(x) = (x - \omega_1)(x - \omega_2) \cdots (x - \omega_{\phi(n)})$.

In particular, note that $\Phi_n(x)$ is monic and has degree $\phi(n)$. In Theorem 33.2 we will prove that $\Phi_n(x)$ has integer coefficients.

EXAMPLE 1 $\Phi_1(x) = x - 1$, since 1 is the only zero of $x - 1$. $\Phi_2(x) = x + 1$, since the zeros of $x^2 - 1$ are 1 and -1, and -1 is the only primitive root. $\Phi_3(x) = (x - \omega)(x - \omega^2)$, where $\omega = \cos(\frac{2\pi}{3}) + i \sin(\frac{2\pi}{3}) = (-1 + i\sqrt{3})/2$, and direct calculations show that $\Phi_3(x) = x^2 + x + 1$. Since the zeros of $x^4 - 1$ are ± 1 and $\pm i$ and only i and $-i$ are primitive, $\Phi_4(x) = (x - i)(x + i) = x^2 + 1$. ◆

In practice, one does not use the definition of $\Phi_n(x)$ to compute it. Instead, one uses the formulas given in the exercises and makes recursive use of the following result.

Theorem 33.1

For every positive integer n, $x^n - 1 = \prod_{d|n} \Phi_d(x)$, where the product runs over all positive divisors d of n.

Before proving this theorem, let us be sure that the statement is clear. For $n = 6$, for instance, the theorem asserts that $x^6 - 1 = \Phi_1(x)\Phi_2(x)\Phi_3(x)\Phi_6(x)$, since 1, 2, 3, and 6 are the positive divisors of 6.

PROOF. Since both polynomials in the statement are monic, it suffices to show that they have the same zeros and that all zeros have multiplicity 1. Let $\omega = \cos(\frac{2\pi}{n}) + i\sin(\frac{2\pi}{n})$. Then $\langle\omega\rangle$ is a cyclic group of order n, and $\langle\omega\rangle$ contains all the nth roots of unity. From Theorem 4.3 we know that for each j, $|\omega^j|$ divides n so that $(x - \omega^j)$ appears as a factor in $\Phi_{|\omega^j|}(x)$. On the other hand, if $x - \alpha$ is a linear factor of $\Phi_d(x)$ for some divisor d of n, then $\alpha^d = 1$, and therefore $\alpha^n = 1$. Thus, $x - \alpha$ is a factor of $x^n - 1$. Finally, since no zero of $x^n - 1$ can be a zero of $\Phi_d(x)$ for two different d's, the result is proved. ∎ ∎

Before we illustrate how Theorem 33.1 can be used to calculate $\Phi_n(x)$ recursively, we state an important consequence of the theorem.

Theorem 33.2

> *For every positive integer n, $\Phi_n(x)$ has integer coefficients.*

PROOF. The case $n = 1$ is trivial. By induction, we may assume that $g(x) = \prod_{\substack{d\mid n \\ d<n}}\Phi_d(x)$ has integer coefficients. From Theorem 33.1 we know that $x^n - 1 = \Phi_n(x)g(x)$, and, because $g(x)$ is monic, we may carry out the division in $Z[x]$ (see Exercise 43 in Chapter 16). Thus, $\Phi_n(x) \in Z[x]$. ∎ ∎

Now let us do some calculations. If p is a prime, we have from Theorem 33.1 that $x^p - 1 = \Phi_1(x)\Phi_p(x) = (x - 1)\Phi_p(x)$, so that $\Phi_p(x) = (x^p - 1)/(x - 1) = x^{p-1} + x^{p-2} + \cdots + x + 1$. From Theorem 33.1 we have

$$x^6 - 1 = \Phi_1(x)\Phi_2(x)\Phi_3(x)\Phi_6(x),$$

so that $\Phi_6(x) = (x^6 - 1)/((x - 1)(x + 1)(x^2 + x + 1))$. So, by long division, $\Phi_6(x) = x^2 - x + 1$. Similarly, $\Phi_{10}(x) = (x^{10} - 1)/((x - 1)(x + 1)(x^4 + x^3 + x^2 + x + 1)) = x^4 - x^3 + x^2 - x + 1$.

The exercises provide shortcuts that often make long division unnecessary. The values of $\Phi_n(x)$ for all n up to 15 are shown in Table 33.1. The software for Computer Exercise 1 provides the values for $\Phi_n(x)$ for all values of n up to 1000. Judging from Table 33.1, one might be led to conjecture that 1 and -1 are the only nonzero

TABLE 33.1 The Cyclotomic Polynomials $\Phi_n(x)$ up to $n = 15$

n	$\Phi_n(x)$
1	$x - 1$
2	$x + 1$
3	$x^2 + x + 1$
4	$x^2 + 1$
5	$x^4 + x^3 + x^2 + x + 1$
6	$x^2 - x + 1$
7	$x^6 + x^5 + x^4 + x^3 + x^2 + x + 1$
8	$x^4 + 1$
9	$x^6 + x^3 + 1$
10	$x^4 - x^3 + x^2 - x + 1$
11	$x^{10} + x^9 + x^8 + x^7 + x^6 + x^5 + x^4 + x^3 + x^2 + x + 1$
12	$x^4 - x^2 + 1$
13	$x^{12} + x^{11} + x^{10} + x^9 + x^8 + x^7 + x^6 + x^5 + x^4 + x^3 + x^2 + x + 1$
14	$x^6 - x^5 + x^4 - x^3 + x^2 - x + 1$
15	$x^8 - x^7 + x^5 - x^4 + x^3 - x + 1$

coefficients of the cyclotomic polynomials. However, it has been shown that every integer is a coefficient of some cyclotomic polynomial.

The next theorem reveals why the cyclotomic polynomials are important.

Theorem 33.3 (Gauss)

> *The cyclotomic polynomials $\Phi_n(x)$ are irreducible over Z.*

PROOF. Let $f(x) \in Z[x]$ be a monic irreducible factor of $\Phi_n(x)$. Because $\Phi_n(x)$ is monic and has no multiple zeros, it suffices to show that every zero of $\Phi_n(x)$ is a zero of $f(x)$.

Since $\Phi_n(x)$ divides $x^n - 1$ in $Z[x]$, we may write $x^n - 1 = f(x)g(x)$, where $g(x) \in Z[x]$. Let ω be a primitive nth root of unity that is a zero of $f(x)$. Then $f(x)$ is the minimal polynomial for ω over Q. Let p be any prime that does not divide n. Then, by Corollary 2 of Theorem 4.2, ω^p is also a primitive nth root of unity, and therefore $0 = (\omega^p)^n - 1 = f(\omega^p)g(\omega^p)$, and so $f(\omega^p) = 0$ or $g(\omega^p) = 0$. Suppose $f(\omega^p) \neq 0$. Then $g(\omega^p) = 0$, and so ω is a zero of $g(x^p)$. Thus, from Theorem 21.3, $f(x)$ divides $g(x^p)$ in $Q[x]$. Since $f(x)$ is monic, $f(x)$ actually divides $g(x^p)$ in $Z[x]$ (see Exercise 43 in Chapter 16). Say $g(x^p) = f(x)h(x)$, where

$h(x) \in Z[x]$. Now let $\bar{g}(x)$, $\bar{f}(x)$, and $\bar{h}(x)$ denote the polynomials in $Z_p[x]$ obtained from $g(x)$, $f(x)$, and $h(x)$, respectively, by reducing each coefficient modulo p. Since this reduction process is a ring homomorphism from $Z[x]$ to $Z_p[x]$ (see Exercise 9 in Chapter 16), we have $\bar{g}(x^p) = \bar{f}(x)\bar{h}(x)$ in $Z_p[x]$. From Exercise 29 in Chapter 16 and Corollary 5 of Theorem 7.1, we then have $(\bar{g}(x))^p = \bar{g}(x^p) = \bar{f}(x)\bar{h}(x)$, and since $Z_p[x]$ is a unique factorization domain, it follows that $\bar{g}(x)$ and $\bar{f}(x)$ have an irreducible factor in $Z_p[x]$ in common; call it $m(x)$. Thus, we may write $\bar{f}(x) = k_1(x)m(x)$ and $\bar{g}(x) = k_2(x)m(x)$, where $k_1(x)$, $k_2(x) \in Z_p[x]$. Then, viewing $x^n - 1$ as a member of $Z_p[x]$, we have $x^n - 1 = \bar{f}(x)\bar{g}(x) = k_1(x)k_2(x)(m(x))^2$. In particular, $x^n - 1$ has a multiple zero in some extension of $Z_p[x]$. But because p does not divide n, the derivative nx^{n-1} of $x^n - 1$ is not 0, and so nx^{n-1} and $x^n - 1$ do not have a common factor of positive degree in $Z_p[x]$. Since this contradicts Theorem 20.5, we must have $f(\omega^p) = 0$.

We reformulate what we have thus far proved as follows: If β is any primitive nth root of unity that is a zero of $f(x)$ and p is any prime that does not divide n, then β^p is a zero of $f(x)$. Now let k be any integer between 1 and n that is relatively prime to n. Then we can write $k = p_1 p_2 \cdots p_t$, where each p_i is a prime that does not divide n (repetitions are permitted). It follows then that each of ω, ω^{p_1}, $(\omega^{p_1})^{p_2}$, ..., $(\omega^{p_1 p_2 \cdots p_{t-1}})^{p_t} = \omega^k$ is a zero of $f(x)$. Since every zero of $\Phi_n(x)$ has the form ω^k, where k is between 1 and n and is relatively prime to n, we have proved that every zero of $\Phi_n(x)$ is a zero of $f(x)$. This completes the proof. ■ ■

Of course, Theorems 33.3 and 33.1 give us the factorization of $x^n - 1$ as a product of irreducible polynomials over Q. But Theorem 33.1 is also useful for finding the irreducible factorization of $x^n - 1$ over Z_p. The next example provides an illustration. Irreducible factors of $x^n - 1$ over Z_p are used to construct error-correcting codes.

EXAMPLE 2   We determine the irreducible factorization of $x^6 - 1$ over Z_2 and Z_3. From Table 33.1, we have $x^6 - 1 = (x - 1)(x + 1)$ $(x^2 + x + 1)(x^2 - x + 1)$. Taking all the coefficients on both sides mod 2, we obtain the same expression, but we must check that these factors are irreducible over Z_2. Since $x^2 + x + 1$ has no zeros in Z_2, it is irreducible over Z_2 (see Theorem 17.1). Finally, since $-1 = 1$ in Z_2, we have the irreducible factorization $x^6 - 1 = (x + 1)^2(x^2 + x + 1)^2$. Over Z_3, we again start with the factorization $x^6 - 1 = (x - 1)$ $(x + 1)(x^2 + x + 1)(x^2 - x + 1)$ over Z and view the coefficients mod 3. Then 1 is a zero of $x^2 + x + 1$ in Z_3, and by long division we obtain $x^2 + x + 1 = (x - 1)(x + 2) = (x + 2)^2$. Similarly,

$x^2 - x + 1 = (x - 2)(x + 1) = (x + 1)^2$. So, the irreducible factorization of $x^6 - 1$ over Z_3 is $(x + 1)^3(x + 2)^3$.   ◆

We next determine the Galois group of the cyclotomic extensions of Q.

Theorem 33.4

> *Let ω be a primitive nth root of unity. Then $Gal(Q(\omega)/Q) \approx U(n)$.*

PROOF.   Since $1, \omega, \omega^2, \ldots, \omega^{n-1}$ are all the nth roots of unity, $Q(\omega)$ is the splitting field of $x^n - 1$ over Q. For each k in $U(n)$, ω^k is a primitive nth root of unity, and by the lemma preceding Theorem 20.4, there is a field automorphism of $Q(\omega)$, which we denote by ϕ_k, that carries ω to ω^k and acts as the identity on Q. Moreover, these are all the automorphisms of $Q(\omega)$, since any automorphism must map a primitive nth root of unity to a primitive nth root of unity. Next, observe that for every $r, s \in U(n)$,

$$(\phi_r\phi_s)(\omega) = \phi_r(\omega^s) = (\phi_r(\omega))^s = (\omega^r)^s = \omega^{rs} = \phi_{rs}(\omega).$$

This shows that the mapping from $U(n)$ onto $Gal(Q(\omega)/Q)$ given by $k \rightarrow \phi_k$ is a group homomorphism. Clearly, the mapping is an isomorphism, since $\omega^r \neq \omega^s$ when $r, s \in U(n)$ and $r \neq s$.   ■ ■

The next example uses Theorem 33.4 and the results of Chapter 8 to demonstrate how to determine the Galois group of cyclotomic extensions.

EXAMPLE 3   Let $\alpha = \cos(\frac{2\pi}{9}) + i \sin(\frac{2\pi}{9})$ and let $\beta = \cos(\frac{2\pi}{15}) + i \sin(\frac{2\pi}{15})$. Then

$$Gal(Q(\alpha)/Q) \approx U(9) \approx Z_6$$

and

$$Gal(Q(\beta)/Q) \approx U(15) \approx U(5) \oplus U(3) \approx Z_4 \oplus Z_2.$$ ◆

THE CONSTRUCTIBLE REGULAR *n*-GONS

As an application of the theory of cyclotomic extensions and Galois theory, we determine exactly which regular *n*-gons are constructible with a straightedge and compass. But first we prove a technical lemma.

Lemma

> Let n be a positive integer and let $\omega = \cos(\frac{2\pi}{n}) + i \sin(\frac{2\pi}{n})$. Then $Q(\cos(\frac{2\pi}{n})) \subseteq Q(\omega)$.

PROOF. Observe that from $(\cos(\frac{2\pi}{n}) + i \sin(\frac{2\pi}{n}))(\cos(\frac{2\pi}{n}) - i \sin(\frac{2\pi}{n})) = \cos^2(\frac{2\pi}{n}) + \sin^2(\frac{2\pi}{n}) = 1$, we have $\cos(\frac{2\pi}{n}) - i \sin(\frac{2\pi}{n}) = 1/\omega$. Moreover, $(\omega + 1/\omega)/2 = (2\cos(\frac{2\pi}{n}))/2 = \cos(\frac{2\pi}{n})$. Thus, $\cos(\frac{2\pi}{n}) \in Q(\omega)$. ■ ■

Theorem 33.5 (Gauss, 1796)

> It is possible to construct the regular n-gon with a straightedge and compass if and only if n has the form $2^k p_1 p_2 \cdots p_t$, where $k \geq 0$ and the p_i's are distinct primes of the form $2^m + 1$.

PROOF. If it is possible to construct a regular n-gon, then we can construct the angle $2\pi/n$ and therefore the number $\cos(\frac{2\pi}{n})$. By the results of Chapter 23, we know that $\cos(\frac{2\pi}{n})$ is constructible only if $[Q(\cos(\frac{2\pi}{n})) : Q]$ is a power of 2. To determine when this is so, we will use Galois theory.

Let $\omega = \cos(\frac{2\pi}{n}) + i \sin(\frac{2\pi}{n})$. Then $|\text{Gal}(Q(\omega)/Q)| = [Q(\omega) : Q] = \phi(n)$. By the lemma on the preceding page, $Q(\cos(\frac{2\pi}{n})) \subseteq Q(\omega)$, and by Theorem 32.1 we know that

$$[Q(\cos(\tfrac{2\pi}{n})) : Q] = |\text{Gal}(Q(\omega)/Q)|/|\text{Gal}(Q(\omega)/Q(\cos(\tfrac{2\pi}{n})))|$$
$$= \phi(n)/|\text{Gal}(Q(\omega)/Q(\cos(\tfrac{2\pi}{n})))|.$$

Recall that the elements σ of $\text{Gal}(Q(\omega)/Q)$ have the property that $\sigma(\omega) = \omega^k$ for $1 \leq k \leq n$. That is, $\sigma(\cos(\frac{2\pi}{n}) + i \sin(\frac{2\pi}{n})) = \cos(\frac{2\pi k}{n}) + i \sin(\frac{2\pi k}{n})$. If such a σ belongs to $\text{Gal}(Q(\omega)/Q(\cos(\frac{2\pi}{n})))$, then we must have $\cos(\frac{2\pi k}{n}) = \cos(\frac{2\pi}{n})$. Clearly, this holds only when $k = 1$ and $k = n - 1$. So, $|\text{Gal}(Q(\omega)/Q(\cos(\frac{2\pi}{n})))| = 2$, and therefore $[Q(\cos(\frac{2\pi}{n})) : Q] = \phi(n)/2$. Thus, if an n-gon is constructible, then $\phi(n)/2$ must be a power of 2. Of course, this implies that $\phi(n)$ is a power of 2.

Write $n = 2^k p_1{}^{n_1} p_2{}^{n_2} \cdots p_t{}^{n_t}$, where $k \geq 0$, the p_i's are distinct odd primes, and the n_i's > 0. Then $\phi(n) = |U(n)| = |U(2^k)||U(p_1{}^{n_1})||U(p_2{}^{n_2})| \cdots |U(p_t{}^{n_t})| = 2^{k-1} p_1{}^{n_1-1}(p_1 - 1) p_2{}^{n_2-1}(p_2 - 1) \cdots p_t{}^{n_t-1}(p_t - 1)$ must be a power of 2. Clearly, this implies that each $n_i = 1$ and each $p_i - 1$ is a power of 2. This completes the proof that the condition in the statement is necessary.

To prove that the condition given in Theorem 33.5 is also sufficient, suppose that n has the form $2^k p_1 p_2 \cdots p_t$, where the p_i's are distinct odd primes of the form $2^m + 1$, and let $\omega = \cos\left(\frac{2\pi}{n}\right) + i\sin\left(\frac{2\pi}{n}\right)$. By Theorem 33.3, $Q(\omega)$ is a splitting field of an irreducible polynomial over Q, and therefore, by the Fundamental Theorem of Galois Theory, $\phi(n) = [Q(\omega) : Q] = |\text{Gal}(Q(\omega)/Q)|$. Since $\phi(n)$ is a power of 2 and $\text{Gal}(Q(\omega)/Q)$ is an Abelian group, it follows by induction (see Exercise 15) that there is a series of subgroups

$$H_0 \subset H_1 \subset \cdots \subset H_t = \text{Gal}(Q(\omega)/Q)$$

where H_0 is the identity, H_1 is the subgroup of $\text{Gal}(Q(\omega)/Q)$ of order 2 that fixes $\cos\left(\frac{2\pi}{n}\right)$, and $|H_{i+1} : H_i| = 2$ for $i = 0, 1, 2, \ldots, t - 1$. By the Fundamental Theorem of Galois Theory, we then have a series of subfields of the real numbers

$$Q = E_{H_t} \subset E_{H_{t-1}} \subset \cdots \subset E_{H_1} = Q(\cos\left(\frac{2\pi}{n}\right)),$$

where $[E_{H_{i-1}} : E_{H_i}] = 2$. So, for each i, we may choose $\beta_i \in E_{H_{i-1}}$ such that $E_{H_{i-1}} = E_{H_i}(\beta_i)$. Then β_i is a zero of a polynomial of the form $x^2 + b_i x + c_i \in E_{H_i}[x]$, and it follows that $E_{H_{i-1}} = E_{H_i}(\sqrt{b_i^2 - 4c_i})$. Thus, it follows from Exercise 3 in Chapter 23 that every element of $Q(\cos\left(\frac{2\pi}{n}\right))$ is constructible. ∎ ∎

It is interesting to note that Gauss did not use Galois theory in his proof. In fact, Gauss gave his proof 15 years before Galois was born.

EXERCISES

Difficulties should act as a tonic. They should spur us to greater exertion.

B.C. FORBES

1. Determine the minimal polynomial for $\cos\left(\frac{\pi}{3}\right) + i\sin\left(\frac{\pi}{3}\right)$ over Q.
2. Factor $x^{12} - 1$ as a product of irreducible polynomials over Z.
3. Factor $x^8 - 1$ as a product of irreducible polynomials over Z_2, Z_3, and Z_5.
4. For any $n > 1$, prove that the sum of all the nth roots of unity is 0.
5. For any $n > 1$, prove that the product of the nth roots of unity is $(-1)^{n+1}$.
6. Let ω be a primitive 12th root of unity over Q. Find the minimal polynomial for ω^4 over Q.
7. Let F be a finite extension of Q. Prove that there are only a finite number of roots of unity in F.
8. For any $n > 1$, prove that the irreducible factorization over Z of $x^{n-1} + x^{n-2} + \cdots + x + 1$ is $\Pi\Phi_d(x)$, where the product runs over all positive divisors d of n greater than 1.
9. If $2^n + 1$ is prime for some $n \geq 1$, prove that n is a power of 2. (Primes of the form $2^n + 1$ are called *Fermat primes*.)

10. Prove that $\Phi_n(0) = 1$ for all $n > 1$.

11. Prove that if a field contains the nth roots of unity for n odd, then it also contains the $2n$th roots of unity.

12. Let m and n be relatively prime positive integers. Prove that the splitting field of $x^{mn} - 1$ over Q is the same as the splitting field of $(x^m - 1)(x^n - 1)$ over Q.

13. Prove that $\Phi_{2n}(x) = \Phi_n(-x)$ for all odd integers $n > 1$.

14. Prove that if p is a prime and k is a positive integer, then $\Phi_{p^k}(x) = \Phi_p(x^{p^{k-1}})$. Use this to find $\Phi_8(x)$ and $\Phi_{27}(x)$.

15. Prove the assertion made in the proof of Theorem 33.5 that there exists a series of subgroups $H_0 \subset H_1 \subset \cdots \subset H_t$ with $|H_{i+1} : H_i| = 2$ for $i = 0, 1, 2, \ldots, t - 1$. (This exercise is referred to in this chapter.)

16. Prove that $x^9 - 1$ and $x^7 - 1$ have isomorphic Galois groups over Q. (See Exercise 7 in Chapter 32 for the definition.)

17. Let p be a prime that does not divide n. Prove that $\Phi_{pn}(x) = \Phi_n(x^p)/\Phi_n(x)$.

18. Prove that the Galois groups of $x^{10} - 1$ and $x^8 - 1$ over Q are not isomorphic.

19. Let E be the splitting field of $x^5 - 1$ over Q. Show that there is a unique field K with the property that $Q \subset K \subset E$.

20. Let E be the splitting field of $x^6 - 1$ over Q. Show that there is no field K with the property that $Q \subset K \subset E$.

21. Let $\omega = \cos(\frac{2\pi}{15}) - i\sin(\frac{2\pi}{15})$. Find the three elements of $\text{Gal}(Q(\omega)/Q)$ of order 2.

COMPUTER EXERCISES ▪▪

Mathematics is not a deductive science—that's a cliche. When you try to prove a theorem, you don't just list the hypotheses, and then start to reason. What you do is trial and error, experimentation, guesswork.

PAUL HALMOS, I Want to Be a Mathematician.

Software for the first computer exercise in this chapter is available at the website:

http://www.d.umn.edu/~jgallian

1. This program returns the nth cyclotomic polynomial. Enter several choices for n and the program returns $\Phi_n(x)$.

2. Use computer software such as *Mathematica, Maple,* or *GAP* to find the irreducible factorization over Z of all polynomials of the form $x^n - 1$, where n is between 2 and 100. On the basis of this information, make a conjecture about the nature of coefficients of the irreducible factors of $x^n - 1$ for all n. Then test your conjecture for $n = 105$.

Carl Friedrich Gauss

He [Gauss] lives everywhere in mathematics.

E. T. BELL, Men of Mathematics

This stamp was issued by East Germany in 1977. It commemorates Gauss's construction of a regular 17-sided polygon with a straight-edge and compass.

CARL FRIEDRICH GAUSS, considered by many to be the greatest mathematician who has ever lived, was born in Brunswick, Germany, on April 30, 1777. While still a teenager, he made many fundamental discoveries. Among these were the method of "least squares" for handling statistical data, and a proof that a 17-sided regular polygon can be constructed with a straightedge and compass (this result was the first of its kind since discoveries by the Greeks 2000 years earlier). In his Ph.D. dissertation in 1799, he proved the Fundamental Theorem of Algebra.

Throughout his life, Gauss largely ignored the work of his contemporaries and, in fact, made enemies of many of them. Young mathematicians who sought encouragement from him were usually rebuffed. Despite this fact, Gauss had many outstanding students, including Eisenstein, Riemann, Kummer, Dirichlet, and Dedekind.

Gauss died in Göttingen at the age of 77 on February 23, 1855. At Brunswick, there is a statue of him. Appropriately, the base is in the shape of a 17-point star. In 1989, Germany issued a bank note (see page 110) depicting Gauss and the Gaussian distribution.

To find more information about Gauss, visit:

http://www-groups.dcs.st-and.ac.uk/~history/

Manjul Bhargava

We are watching him [Bhargava] very closely.
He is going to be a superstar.
He's amazingly mature mathematically.
He is changing the subject in a fundamental way.

PETER SARNAK

MANJUL BHARGAVA was born in Canada on August 8, 1974 and grew up in Long Island, New York. After graduating from Harvard in 1996, Bhargava went to Princeton to pursue his Ph.D. under the direction of Andrew Wiles (see page 338). Bhargava investigated a "composition law" first formulated by Gauss in 1801 for combining two quadratic equations (equations in a form such as $x^2 + 3xy + 6y^2 = 0$) in a way that was very different from normal addition and revealed a lot of information about number systems. Bhargava tackled an aspect of the problem in which no progress had been made in more than 200 years. He not only broke new ground in that area but also discovered 13 more composition laws and developed a coherent mathematical framework to explain them. He then applied his theory of composition to solve a number of fundamental problems concerning the distribution of extension fields of the rational numbers and of other related algebraic objects. What made Bhargava's work especially remarkable is that he was able to explain all his revolutionary ideas using only elementary mathematics. In commenting on Bhargava's results, Wiles said, "He did it in a way that Gauss himself could have understood and appreciated."

In 2000, Bhargava received the first Clay Mathematics Institute Long-Term Prize Fellowship given to mathematicians under 30 years of age who have already contributed profound ideas and major achievements to the discipline of mathematics. In 2002, he was named as one of Popular Science magazine's "Brilliant 10," in celebration of scientists who are shaking up their fields. In 2003, Bhargava accepted a full professorship with tenure at Princeton at the age of 28.

In addition to doing mathematics, Bhargava is an accomplished tabla player who has studied with the world's most distinguished tabla masters. He performs extensively in the New York and Boston areas.

SUPPLEMENTARY EXERCISES FOR CHAPTERS 24–33 ▬ ▪▪

The road to wisdom?—Well it's plain and simple to express:
Err
and err
and err again
but less
and less
and less.

PIET HEIN, "The Road to Wisdom," Grooks (1966)

True/false questions for Chapters 24–33 are available on the Web at

http://www.d.umn.edu/~jgallian/TF

1. Let $G = \langle x, y \mid x = (xy)^3, y = (xy)^4 \rangle$. To what familiar group is G isomorphic?

2. Let $G = \langle z \mid z^6 = 1 \rangle$ and $H = \langle x, y \mid x^2 = y^3 = 1, xy = yx \rangle$. Show that G and H are isomorphic.

3. Show that a group of order $315 = 3^2 \cdot 5 \cdot 7$ has a subgroup of order 45.

4. Let G be a group of order $p^2 q^2$, where p and q are primes and $p > q$. If $|G| \neq 36$, prove that G has a normal Sylow p-subgroup.

5. Let H denote a Sylow 7-subgroup of a group G and K a Sylow 5-subgroup of G. Assume that $|H| = 49$, $|K| = 5$, and K is a subgroup of $N(H)$. Show that H is a subgroup of $N(K)$.

6. Prove that there is no finite group of order greater than 6 that can have exactly three conjugacy classes.

7. Suppose that K is a normal Sylow p-subgroup of H and that H is a normal subgroup of G. Prove that K is a normal subgroup of G. (Compare this with Exercise 49 in Chapter 9.)

8. Show that the polynomial $x^5 - 6x + 3$ over Q is not solvable by radicals.

9. Let H and K be subgroups of G. Prove that HK is a subgroup of G if $H \leq N(K)$.

10. Suppose that H is a subgroup of a finite group G and that H contains $N(P)$, where P is some Sylow p-subgroup of G. Prove that $N(H) = H$.

11. Prove that a simple group G of order 168 cannot contain an element of order 21.

12. Prove that the only group of order 561 is Z_{561}.

13. Prove that the center of a non-Abelian group of order 105 has order 5.

14. Let n be an odd integer that is at least 3. Prove that every Sylow subgroup of D_n is cyclic.

15. Let G be the digraph obtained from $\text{Cay}(\{(1, 0), (0, 1)\}: Z_3 \oplus Z_5)$ by deleting the vertex $(0, 0)$. [Also, delete each arc to or from $(0, 0)$.] Prove that G has a Hamiltonian circuit.

16. Prove that the digraph obtained from $\text{Cay}(\{(1, 0), (0, 1)\}: Z_4 \oplus Z_7)$ by deleting the vertex $(0, 0)$ has a Hamiltonian circuit.

17. Let G be a finite group generated by a and b. Let $s_1, s_2, \ldots, s_n$ be the arcs of a Hamiltonian circuit in the digraph $\text{Cay}(\{a, b\}: G)$. We say that the vertex $s_1 s_2 \cdots s_i$ *travels by* a if $s_{i+1} = a$. Show that if a vertex x travels by a, then every vertex in the coset $x\langle ab^{-1} \rangle$ travels by a.

18. Recall that the dot product $u \cdot v$ of two vectors $u = (u_1, u_2, \ldots, u_n)$ and $v = (v_1, v_2, \ldots, v_n)$ from F^n is

$$u_1 v_1 + u_2 v_2 + \cdots + u_n v_n$$

(where the addition and multiplication are those of F). Let C be an (n, k) linear code. Show that

$$C^\perp = \{v \in F^n \mid v \cdot u = 0 \text{ for all } u \in C\}$$

is an $(n, n - k)$ linear code. This code is called the *dual* of C.

19. Find the dual of each of the following binary codes:
    a. $\{00, 11\}$,
    b. $\{000, 011, 101, 110\}$,
    c. $\{0000, 1111\}$,
    d. $\{0000, 1100, 0011, 1111\}$.

20. Let C be a binary linear code such that $C \subseteq C^\perp$. Show that $\text{wt}(v)$ is even for all v in C.

21. Let C be an (n, k) binary linear code. If v is a binary n-tuple, but $v \notin C^\perp$, show that $v \cdot u = 0$ for exactly half of the elements u in C.

22. Suppose that C is an (n, k) binary linear code and the vector $11 \cdots 1 \in C^\perp$. Show that $\text{wt}(v)$ is even for every v in C.

23. Suppose that C is an (n, k) binary linear code and $C = C^\perp$. (Such a code is called *self-dual*.) Prove that n is even. Prove that $11 \cdots 1$ is a code word.

24. If G is a finite solvable group, show that there exist subgroups of G

$$\{e\} = H_0 \subset H_1 \subset H_2 \subset \cdots \subset H_n = G$$

such that H_{i+1}/H_i has prime order.

The End.

Title of song by JOHN LENNON AND PAUL MCCARTNEY,
Abbey Road, side 2, October 1969

Selected Answers

Many of the proofs given below are merely sketches. In these cases, the student should supply the complete proof.

CHAPTER 0

To make headway, improve your head. B. C. FORBES

1. $\{1, 2, 3, 4\}$; $\{1, 3, 5, 7\}$; $\{1, 5, 7, 11\}$; $\{1, 3, 7, 9, 11, 13, 17, 19\}$; $\{1, 2, 3, 4, 6, 7, 8, 9, 11, 12, 13, 14, 16, 17, 18, 19, 21, 22, 23, 24\}$

3. 12, 2, 2, 10, 1, 0, 4, 5.

5. 1942, June 18; 1953, December 13.

7. By using 0 as an exponent if necessary, we may write $a = p_1^{m_1} \cdots p_k^{m_k}$ and $b = p_1^{n_1} \cdots p_k^{n_k}$, where the p's are distinct primes and the m's and n's are nonnegative. Then $\text{lcm}(a, b) = p_1^{s_1} \cdots p_k^{s_k}$, where $s_i = \max(m_i, n_i)$, and $\gcd(a, b) = p_1^{t_1} \cdots p_k^{t_k}$, where $t_i = \min(m_i, n_i)$. Then $\text{lcm}(a, b) \cdot \gcd(a, b) = p_1^{m_1 + n_1} \cdots p_k^{m_k + n_k} = ab$.

9. Write $a = nq_1 + r_1$ and $b = nq_2 + r_2$, where $0 \le r_1, r_2 < n$. We may assume that $r_1 \ge r_2$. Then $a - b = n(q_1 - q_2) + (r_1 - r_2)$, where $r_1 - r_2 \ge 0$. If $a \bmod n = b \bmod n$, then $r_1 = r_2$ and n divides $a - b$. If n divides $a - b$, then by the uniqueness of the remainder, we have $r_1 - r_2 = 0$.

11. Use Exercise 9.

13. Use the "GCD Is a Linear Combination" theorem (Theorem 0.2).

15. Since st divides $a - b$, both s and t divide $a - b$. The converse is true when $\gcd(s, t) = 1$.

17. Use Euclid's Lemma and the Fundamental Theorem of Arithmetic.

19. Use proof by contradiction.

21. Let S be a set with $n + 1$ elements and pick some a in S. By induction, S has 2^n subsets that do not contain a. But there is a one-to-one correspondence between the subsets of S that do not contain a and those that do. So there are $2 \cdot 2^n = 2^{n+1}$ subsets in all.

23. Consider $n = 200! + 2$.

25. Say $p_1 p_2 \cdots p_r = q_1 q_2 \cdots q_s$, where the p's and q's are primes. By the Generalized Euclid's Lemma, p_1 divides some q_i, say q_1 (we may relabel the q's if necessary). Then $p_1 = q_1$ and $p_2 \cdots p_r = q_2 \cdots q_s$. Repeating this argument at each step, we obtain $p_2 = q_2, \cdots, p_r = q_r$ and $r = s$.

27. Suppose that S is a set that contains a and whenever $n \geq a$ belongs to S, then $n + 1 \in S$. We must prove that S contains all integers greater than or equal to a. Let T be the set of all integers greater than a that are not in S and suppose that T is not empty. Let b be the smallest integer in T (if T has no negative integers, b exists because of the Well Ordering Principle; if T has negative integers, it can have only a finite number of them so that there is a smallest one). Then $b - 1 \in S$, and therefore $b = (b - 1) + 1 \in S$.

29. The statement is true for any divisor of $8^4 - 4 = 4092$.

31. 6 P.M.

33. Observe that the number with the decimal representation $a_9 a_8 \cdots a_1 a_0$ is $a_9 \cdot 10^9 + a_8 \cdot 10^8 + \cdots + a_1 \cdot 10 + a_0$. Then use Exercise 11 and the fact that $a_i 10^i \bmod 9 = a_i \bmod 9$ to deduce that the check digit is $(a_9 + a_8 + \cdots + a_1 + a_0) \bmod 9$.

35. For the case in which the check digit is not involved, see the answer to Exercise 33. If a transposition involving the check digit $c = (a_1 + a_2 + \cdots + a_{10}) \bmod 9$ goes undetected, then $a_{10} = (a_1 + a_2 + \cdots + a_9 + c) \bmod 9$. Substitution yields $2(a_1 + a_2 + \cdots + a_9) \bmod 9 = 0$. Therefore, modulo 9, we have $10(a_1 + a_2 + \cdots + a_9) = a_1 + a_2 + \cdots + a_9 = 0$. It follows that $c = a_{10}$. In this case the transposition does not yield an error.

37. Say the number is $a_8 a_7 \ldots a_1 a_0 = a_8 \cdot 10^8 + a_7 \cdot 10^7 + \cdots + a_1 \cdot 10 + a_0$. Then the error is undetected if and only if $(a_i 10^i - a_i' 10^i) \bmod 7 = 0$. Multiplying both sides by 5^i and noting that $50 \bmod 7 = 1$, we obtain $(a_i - a_i') \bmod 7 = 0$.

39. 4

43. Cases where $(2a - b - c) \bmod 11 = 0$ are undetected.

45. The check digit would be the same.

47. 4302311568

49. No. $(1, 0) \in R$ and $(0, -1) \in R$, but $(1, -1) \notin R$.

51. a belongs to the same subset as a. If a and b belong to the subset A, then b and a also belong to A. If a and b belong to the subset A and b and c belong to the subset B, then $A = B$, since the distinct subsets of P are disjoint. So, a and c belong to A.

53. 2. Since β is one-to-one, $\beta(\alpha(a_1)) = \beta(\alpha(a_2))$ implies that $\alpha(a_1) = \alpha(a_2)$ and since α is one-to-one, $a_1 = a_2$.

    3. Let $c \in C$. There is a b in B such that $\beta(b) = c$ and an a in A such that $\alpha(a) = b$. Thus, $(\beta\alpha)(a) = \beta(\alpha(a)) = \beta(b) = c$.

    4. Since α is one-to-one and onto we may define $\alpha^{-1}(x) = y$ if and only if $\alpha(y) = x$. Then $\alpha^{-1}(\alpha(a)) = a$ and $\alpha(\alpha^{-1}(b)) = b$.

CHAPTER 1

> *Think of what you're saying, you can get it wrong and still think that it's all right.*   JOHN LENNON AND PAUL MCCARTNEY, *"We Can Work It Out,"* single

1. Three rotations: $0°$, $120°$, $240°$, and three reflections across lines from vertices to midpoints of opposite sides.

3. no
5. D_n has n rotations of the form $k(360°/n)$, where $k = 0, \cdots, n - 1$. In addition, D_n has n reflections. When n is odd, the axes of reflection are the lines from the vertices to the midpoints of the opposite sides. When n is even, half of the axes of reflection are obtained by joining opposite vertices; the other half, by joining midpoints of opposite sides.
7. A rotation followed by a rotation either fixes every point (and so is the identity) or fixes only the center of rotation. However, a reflection fixes a line.
9. Observe that $1 \cdot 1 = 1$; $1(-1) = -1$; $(-1)1 = -1$; $(-1)(-1) = 1$. These relationships also hold when 1 is replaced by "rotation" and -1 is replaced by "reflection."
11. $HD = DV$ but $H \neq V$.
13. R_0, R_{180}, H, V
15. See answer for Exercise 13.
17. In each case, the group is D_6.
19. cyclic
21. D_{28}
23. It would wobble violently.

CHAPTER 2

The nobelist pleasure is the joy of understanding.
LEONARDO DA VINCI

1. Does not contain the identity; closure fails.
3. Under modulo 4, 2 does not have an inverse. Under modulo 5, each element has an inverse.

5. $\begin{bmatrix} 9 & 9 \\ 10 & 8 \end{bmatrix}$

7. a. $2a + 3b$; **b.** $-2a + 2(-b + c)$; **c.** $-3(a + 2b) + 2c = 0$
9. e
11. Use the fact that det $(AB) = (\det A)(\det B)$.
13. 29
15. $(ab)^n$ need not equal $a^n b^n$ in a non-Abelian group.
17. Use Exercise 16.
19. For the case $n > 0$, use induction. For $n < 0$, note that $e = (a^{-1}ba)^n(a^{-1}ba)^{-n} = (a^{-1}ba)^n(a^{-1}b^{-n}a)$ and solve for $(a^{-1}ba)^n$.
21. $\{1, 3, 5, 9, 13, 15, 19, 23, 25, 27, 39, 45\}$
23. Suppose x appears in a row labeled with a twice. Say $x = ab$ and $x = ac$; then cancellation yields $b = c$. But we use distinct elements to label the columns.
25. Use Exercise 23.
27. $a^{-1}cb^{-1}$; aca^{-1}
29. If $x^3 = e$ and $x \neq e$, then $(x^{-1})^3 = e$ and $x \neq x^{-1}$. So, nonidentity solutions come in pairs. If $x^2 \neq e$, then $x^{-1} \neq x$ and $(x^{-1})^2 \neq e$. So solutions to $x^2 \neq e$ come in pairs.
31. Observe that $a(a^{-1}b^{-1})b = e(a^{-1}b^{-1})(ba)$.
33. Since $a^2 = b^2 = (ab)^2 = e$, we have $aabb = abab$. Now cancel on the left and right.
35. If n is not prime, the set is not closed under multiplication modulo n. If n is prime, the set is closed and for every r in the set there are integers s and t such that $1 = rs + nt = rs$ modulo n.

37. Closure follows from the definition of multiplication. The identity is $\begin{bmatrix} 1/2 & 1/2 \\ 1/2 & 1/2 \end{bmatrix}$.

The inverse of $\begin{bmatrix} a & a \\ a & a \end{bmatrix}$ is $\begin{bmatrix} 1/(4a) & 1/(4a) \\ 1/(4a) & 1/(4a) \end{bmatrix}$.

CHAPTER 3

> *The brain is as strong as its weakest think.*   ELEANOR DOAN

1. $|Z_{12}| = 12$; $|U(10)| = 4$; $|U(12)| = 4$; $|U(20)| = 8$; $|D_4| = 8$
   In Z_{12}, $|0| = 1$; $|1| = |5| = |7| = |11| = 12$; $|2| = |10| = 6$; $|3| = |9| = 4$; $|4| = |8| = 3$; $|6| = 2$.
   In $U(10)$, $|1| = 1$; $|3| = |7| = 4$; $|9| = 2$.
   In $U(12)$, $|1| = 1$; $|5| = 2$; $|7| = 2$; $|11| = 2$.
   In $U(20)$, $|1| = 1$; $|3| = |7| = |13| = |17| = 4$; $|9| = |11| = |19| = 2$.
   In D_4, $|R_0| = 1$; $|R_{90}| = |R_{270}| = 4$; $|R_{180}| = |H| = |V| = |D| = |D'| = 2$.
   In each case, notice that the order of the element divides the order of the group.
3. In Q, $|0| = 1$ and all other elements have infinite order. In Q^*, $|1| = 1$, $|-1| = 2$, and all other elements have infinite order.
5. Each is the inverse of the other.
7. If a has infinite order, then $e, a, a^2, \ldots$ are all distinct and belong to G, so G is infinite. If $|a| = n$, then $e, a, a^2, \ldots, a^{n-1}$ are distinct and belong to G.
9. By brute force, show that $k^4 = 1$ for all k.
11. For any integer $n \geq 3$, D_n contains elements a and b of order 2 with $|ab| = n$. In general, there is no relationship among $|a|$, $|b|$, and $|ab|$.
13. $U_4(20) = \{1, 9, 13, 17\}$; $U_5(20) = \{1, 11\}$; $U_5(30) = \{1, 11\}$; $U_{10}(30) = \{1, 11\}$.
   $U_k(n)$ is closed because $(ab) \bmod k = (a \bmod k)(b \bmod k) = 1 \cdot 1 = 1$. H is not closed.
15. If $x \in Z(G)$, then $x \in C(a)$ for all a, so $x \in \underset{a \in G}{\cap} C(a)$. If $x \in \underset{a \in G}{\cap} C(a)$, then $xa = ax$ for all a in G, so $x \in Z(G)$.
17. a. $C(5) = G$; $C(7) = \{1, 3, 5, 7\}$
   b. $Z(G) = \{1, 5\}$
   c. $|2| = 2$; $|3| = 4$. They divide the order of the group.
19. Mimic the proof of Theorem 3.5.
21. No. In D_4, $C(R_{180}) = D_4$.
23. For the first part, see Example 4. For the second part, use D_4.
25. Since the only elements of finite order in $\mathbf{R}^*$ are 1 and -1, the only finite subgroups are $\{1\}$ and $\{1, -1\}$.
27. 2
29. Note that $\begin{bmatrix} 1 & 1 \\ 0 & 1 \end{bmatrix}^n = \begin{bmatrix} 1 & n \\ 0 & 1 \end{bmatrix}$.
31. For any positive integer n, a rotation of $360°/n$ has order n. A rotation of $\sqrt{2}°$ has infinite order.
33. $\langle R_0 \rangle$, $\langle R_{90} \rangle$, $\langle R_{180} \rangle$, $\langle D \rangle$, $\langle D' \rangle$, $\langle H \rangle$, $\langle V \rangle$. (Note that $\langle R_{90} \rangle = \langle R_{270} \rangle$). The subgroups $\{R_0, R_{180}, D, D'\}$ and $\{R_0, R_{180}, H, V\}$ are not cyclic.
35. Nonidentity elements of odd order come in pairs. So, there must be some element a of even order, say $|a| = 2m$. Then $|a^m| = 2$.

37. Let $|g| = m$ and write $m = nq + r$ where $0 \leq r < n$. Then $g^r = g^{m - nq} = g^m(g^n)^{-q} = (g^n)^{-q}$ belongs to H. So, $r = 0$.

39. $1 \in H$. Let $a, b \in H$. Then $(ab^{-1})^2 = a^2(b^2)^{-1}$, which is the product of two rationals. 2 can be replaced by any positive integer.

41. $|\langle 3 \rangle| = 4$

43. Let $\begin{bmatrix} a & b \\ c & d \end{bmatrix}$ and $\begin{bmatrix} a' & b' \\ c' & d' \end{bmatrix}$ belong to H. It suffices to show that $a - a' + b - b' + c - c' + d - d' = 0$. This follows from $a + b + c + d = 0 = a' + b' + c' + d'$. If 0 is replaced by 1, H is not a subgroup.

45. If 2^a and $2^b \in K$, then $2^a(2^b)^{-1} = 2^{a-b} \in K$, since $a - b \in H$.

47. $\begin{bmatrix} 2 & 0 \\ 0 & 2 \end{bmatrix}^{-1} = \begin{bmatrix} \frac{1}{2} & 0 \\ 0 & \frac{1}{2} \end{bmatrix}$ is not in H.

49. If $a + bi$ and $c + di \in H$, then $(a + bi)(c + di)^{-1} = (ac + bd) + (bc - ad)i$ and $(ac + bd)^2 + (bc - ad)^2 = 1$, so that H is a subgroup. H is the unit circle in the complex plane.

51. a. $\left\{ \begin{bmatrix} a + b & a \\ a & b \end{bmatrix} \middle| ab + b^2 \neq a^2; a, b \in \mathbf{R} \right\}$

b. $\left\{ \begin{bmatrix} a & b \\ b & a \end{bmatrix} \middle| a^2 \neq b^2; a, b \in \mathbf{R} \right\}$

c. $\left\{ \begin{bmatrix} a & 0 \\ 0 & a \end{bmatrix} \middle| a \neq 0; a \in \mathbf{R} \right\}$

53. Use Theorem 0.2.

CHAPTER 4

There will be an answer, let it be.

JOHN LENNON AND PAUL MCCARTNEY, *"Let It Be,"* single

1. For Z_6, generators are 1 and 5; for Z_8, generators are 1, 3, 5, and 7; for Z_{20}, generators are 1, 3, 7, 9, 11, 13, 17, and 19.

3. $\langle 20 \rangle = \{20, 10, 0\}$
$\langle 10 \rangle = \{10, 20, 0\}$

5. $\langle 3 \rangle = \{3, 9, 7, 1\}$
$\langle 7 \rangle = \{7, 9, 3, 1\}$

7. $U(8)$ or D_3.

9. Six subgroups; generators are the divisors of 20.
Six subgroups; generators are a^k, where k is a divisor of 20.

11. Certainly, $a^{-1} \in \langle a \rangle$. So $\langle a^{-1} \rangle \subseteq \langle a \rangle$. So, by symmetry, $\langle a \rangle \subseteq \langle a^{-1} \rangle$ as well.

13. Let $k = \text{lcm}(m, n) \bmod 24$. Then $\langle a^m \rangle \cap \langle a^n \rangle = \langle a^k \rangle$.

15. $|g|$ divides 12 is equivalent to $g^{12} = e$. So, if $a^{12} = e$ and $b^{12} = e$, then $(ab^{-1})^{12} = a^{12}(b^{12})^{-1} = ee^{-1} = e$. The general result is given in Exercise 23 of Chapter 3.

17. is odd or infinite

19. $\langle 1 \rangle, \langle 7 \rangle, \langle 11 \rangle, \langle 17 \rangle, \langle 19 \rangle, \langle 29 \rangle$

21. **a.** $|a|$ divides 12. **b.** $|a|$ divides m. **c.** By Theorem 4.3, $|a| = 1, 2, 3, 4, 6, 8, 12,$ or 24. If $|a| = 2$, then $a^8 = (a^2)^4 = e^4 = e$. A similar argument eliminates all other possibilities except 24.

23. Yes, by Theorem 4.3. The subgroups of Z are of the form $\{0, \pm n, \pm 2n, \pm 3n, \cdots \}$, where n is any integer.

25. For the first part, use Theorem 4.4; D_n has n elements of order 2 when n is odd and $n + 1$ elements of order 2 when n is even.

27. Two: a and a^{-1}

29. 1000000, 3000000, 5000000, 7000000

By Theorem 4.3, $\langle 1000000 \rangle$ is the unique subgroup of order 8, and only those on the list are generators.

31. Let $G = \{a_1, a_2, \ldots, a_k\}$. Now let $|a_i| = n_i$. Consider $n = n_1 n_2 \cdots n_k$.

33. Mimic Exercise 32.

35. Mimic Exercise 34.

37. Suppose a and b are relatively prime positive integers and $\langle a/b \rangle = Q^+$. Then there is some positive integer n such that $(a/b)^n = 2$. Clearly, $n \neq 0, 1,$ or -1. If $n > 1$, $a^n = 2b^n$, so that 2 divides a. But then 2 divides b as well. A similar contradiction occurs if $n < -1$.

39. For 6, use Z_{2^5}. For n, use $Z_{2^{n-1}}$.

41. Let $t = \text{lcm}(m, n)$ and $|ab| = s$. Then $(ab)^t = a^t b^t = e$, and therefore s divides t. Also, $e = (ab)^s = a^s b^s$, so that $a^s = b^{-s}$, and therefore a^s and b^{-s} belong to $\langle a \rangle \cap \langle b \rangle = \{e\}$. Thus, m divides s and n divides s, and, therefore, t divides s. This proves that $s = t$. For the second part, try D_3.

43. An infinite cyclic group does not have an element of prime order. A finite cyclic group can have only one subgroup for each divisor of its order. A subgroup of order p has exactly $p - 1$ elements of order p. Another element of order p would give another subgroup of order p.

45. $4, 3 \cdot 4, 7 \cdot 4, 9 \cdot 4$

47. 1 of order 1; 33 of order 2; 2 of order 3; 10 of order 11; 20 of order 33

49. 1, 2, 10, 20

51. Say a and b are distinct elements of order 2. If a and b commute, then ab is a third element of order 2. If a and b do not commute, then aba is a third element of order 2.

53. Use Exercise 14 of Chapter 3 and Theorem 4.3.

55. 1 and 2

57. In a cyclic group there are at most n solutions to the equation $x^n = e$.

59. 12 or 60; 48

61. Say b is a generator of the group. Since p and $p^n - 1$ are relatively prime, we know by Corollary 2 of Theorem 4.2 that b^p also generates the group. Finally, observe that $(b^p)^k = (b^k)^p$.

63. Use the fact that a cyclic group of even order has a unique element of order 2.

65. G is a group because it is closed. It is not cyclic because every nonzero element has order 3.

SUPPLEMENTARY EXERCISES FOR CHAPTERS 1–4

Four short words sum up what has lifted most successful individuals above the crowd: a little bit more. They did all that was expected of them and a little bit more.   A. LOU VICKERY

1. a. Let xh_1x^{-1} and xh_2x^{-1} belong to xHx^{-1}. Then $(xh_1x^{-1})(xh_2x^{-1})^{-1} = xh_1h_2^{-1}x^{-1} \in xHx^{-1}$ also.
   b. Let $\langle h \rangle = H$. Then $\langle xhx^{-1} \rangle = xHx^{-1}$.
   c. $(xh_1x^{-1})(xh_2x^{-1}) = xh_1h_2x^{-1} = xh_2h_1x^{-1} = (xh_2x^{-1})(xh_1x^{-1})$.

3. Suppose $\mathrm{cl}(a) \cap \mathrm{cl}(b) \neq \varnothing$. Say $xax^{-1} = yby^{-1}$. Then $(y^{-1}x)a(y^{-1}x)^{-1} = b$. Thus, for any ubu^{-1} in $\mathrm{cl}(b)$, we have $ubu^{-1} = (uy^{-1}x)a(uy^{-1}x)^{-1} \in \mathrm{cl}(a)$. This shows that $\mathrm{cl}(b) \subseteq \mathrm{cl}(a)$. By symmetry, $\mathrm{cl}(a) \subseteq \mathrm{cl}(b)$. Because $a = eae^{-1} \in \mathrm{cl}(a)$, the union of the conjugacy classes is G.

5. Observe that $(xax^{-1})^k = xa^kx^{-1}$. Thus, $(xax^{-1})^k = e$ if and only if $a^k = e$.

7. Try D_4.

9. By Exercise 5, for every x in G, $|xax^{-1}| = |a|$, so that $xax^{-1} = a$ or $xa = ax$.

11. 1 of order 1, 15 of order 2, 8 of order 15, 4 of order 5, 2 of order 3.

13. Let $|G| = 5$. Let $a \neq e$ belong to G. If $|a| = 5$, we are done. If $|a| = 3$, then $\{e, a, a^2\}$ is a subgroup of G. Let b be either of the remaining two elements of G. Then the set $\{e, a, a^2, b, ab, a^2b\}$ consists of six different elements, a contradiction. Thus, $|a| \neq 3$. Similarly, $|a| \neq 4$. We may now assume that every nonidentity element of G has order 2. Pick $a \neq e$ and $b \neq e$ in G with $a \neq b$. Then $\{e, a, b, ab\}$ is a subgroup of G. Let c be the remaining element of G. Then $\{e, a, b, ab, c, ac, bc, abc\}$ is a set of eight distinct elements of G, a contradiction. It now follows that if $a \in G$ and $a \neq e$, then $|a| = 5$.

15. $a^n(b^n)^{-1} = (ab^{-1})^n$, so G^n is a subgroup. For the non-Abelian group, try D_3.

17. Suppose $G = H \cup K$. Pick $h \in H$ with $h \notin K$. Pick $k \in K$, but $k \notin H$. Then, $hk \in G$, but $hk \notin H$ and $hk \notin K$. $U(8)$ is the union of the three subgroups.

19. If $|a| = p^k$ and $|b| = p^r$ with $k \leq r$, say, then $|ab^{-1}|$ divides p^r.

21. Note that $ba^2 = ab$ and $a^3 = b^2 = e$ imply $ba = a^2b$. Thus, every member of the group can be written in the form a^ib^j. Therefore, the group is $\{e, a, a^2, b, ab, a^2b\}$. D_3 satisfies these conditions.

23. $xy = yx$ if and only if $xyx^{-1}y^{-1} = e$. But, $(xy)x^{-1}y^{-1} = x^{-1}(xy)y^{-1} = ee = e$.

25. Let $x \in N(gHg^{-1})$. Then $x(gHg^{-1})x^{-1} = gHg^{-1}$. Thus $g^{-1}xgHg^{-1}x^{-1}g = g^{-1}xgH(g^{-1}xg)^{-1} = H$. This means that $g^{-1}xg \in N(H)$. So $x \in gN(H)g^{-1}$. Reverse the argument to show $gN(H)g^{-1} \subseteq N(gHg^{-1})$.

27. Look at D_{11}.

29. Solution from *Mathematics Magazine*.[†] "Yes. Let a be an arbitrary element of S. The set $\{a^n \mid n = 1, 2, 3, \ldots\}$ is finite, and therefore $a^m = a^n$ for some m, n with $m > n \geq 1$. By cancellation we have $a^{r(a)} = a$, where $r(a) = m - n + 1 > 1$. If x is any element of S, then $aa^{r(a)-1}x = a^{r(a)}x = ax$, and this implies that $a^{r(a)-1}x = x$. Similarly, we see that $xa^{r(a)-1} = x$, and the element $e = a^{r(a)-1}$ is an identity. The identity element is unique, for if e' is another identity, then $e = ee' = e'$. If $r(a) > 2$ then $a^{r(a)-2}$ is an inverse of a, and if $r(a) = 2$ then $a^2 = a = e$ is its own inverse. Thus S is a group."

[†]*Mathematics Magazine* 63 (April 1990): 136.

31. 1^1 is rational so $H \neq \emptyset$. Say a^m and b^n are rational. Then $(ab^{-1})^{mn} = (a^m)^n/(b^n)^m$ is rational.

33. Use $\det (AB) = (\det A)(\det B)$ to prove H is a subgroup. H is not a subgroup when $\det A$ is an integer, since $\det A^{-1}$ need not be an integer.

35. Choose $x \neq e$ and $y \notin \langle x \rangle$. Then $G = \langle x \rangle \cup \langle y \rangle$. But then $xy \in \langle y \rangle$, so that $\langle x \rangle \subseteq \langle y \rangle$ and therefore $G = \langle y \rangle$. To prove that $|G| = pq$ or p^3, use Theorem 4.3.

37. If T and U are not closed, then there are elements x and y in T and w and z in U such that xy is not in T and wz is not in U. It follows that $xy \in U$ and $wz \in T$. Then $xywz = (xy)wz \in U$ and $xywz = xy(wz) \in T$, a contradiction.

39. Let G be the group of all polynomials with integer coefficients under addition. Let H_k be the subgroup of polynomials of degree at most k together with the zero polynomial (the zero polynomial does not have a degree).

41. Take $g = a$.

43. Let $S = \{s_1, s_2, s_3, \ldots, s_k\}$ and let g be any element in G. Then the set $\{gs_1^{-1}, gs_2^{-1}, gs_3^{-1}, \ldots, gs_k^{-1}\}$ and S have at least one element in common. Say $gs_i^{-1} = s_j$. Then $g = s_j s_i$.

45. Let $K = \{x \in G \mid |x| \text{ divides } d\}$. The subgroup test shows that K is a subgroup. Let $x \in H$. By Theorem 4.3, $|x|$ divides d. So, $H \subseteq K$. Let $y \in K$, $|y| = t$, and $d = tq$. By Theorem 4.3, H has a subgroup of order t and G has only one subgroup of order t. So, $\langle y \rangle \subseteq H$.

CHAPTER 5

Mistakes are often the best teachers.   JAMES A. FROUDE

1. a. 2   **b.** 3   **c.** 5

3. a. 3   **b.** 12   **c.** 6   **d.** 6   **e.** 12   **f.** 2

5. 12

7. For S_6, the possible orders are 1, 2, 3, 4, 5, 6; for A_6, 1, 2, 3, 4, 5; for S_7, 1, 2, 3, 4, 5, 6, 7, 10, 12; for A_7, 1, 2, 3, 4, 5, 6, 7.

9. a. even   **b.** odd   **c.** even   **d.** odd   **e.** even

11. even; odd

13. An even number of 2-cycles followed by an even number of 2-cycles gives an even number of 2-cycles in all. So the finite subgroup test is verified.

15. Suppose that α can be written as a product of m 2-cycles and β can be written as a product of n 2-cycles. Then $\alpha\beta$ can be written as a product of $m + n$ 2-cycles. Now observe that $m + n$ is even if and only if m and n are both even or both odd.

17.

a. $\alpha^{-1} = \begin{bmatrix} 1 & 2 & 3 & 4 & 5 & 6 \\ 2 & 1 & 3 & 5 & 4 & 6 \end{bmatrix}$

b. $\beta\alpha = \begin{bmatrix} 1 & 2 & 3 & 4 & 5 & 6 \\ 1 & 6 & 2 & 3 & 4 & 5 \end{bmatrix}$

c. $\alpha\beta = \begin{bmatrix} 1 & 2 & 3 & 4 & 5 & 6 \\ 6 & 2 & 1 & 5 & 3 & 4 \end{bmatrix}$

19. Suppose H contains at least one odd permutation, say, σ. Imitate the proof of Theorem 5.7 with σ in place of (12).

21. No. The identity is even.

23. a. $C(\alpha_3) = \{\alpha_1, \alpha_2, \alpha_3, \alpha_4\}$
   b. $C(\alpha_{12}) = \{\alpha_1, \alpha_7, \alpha_{12}\}$

25. 90

27. $\beta = (2457136)$

29. (124586739), (142568793), (214856379).

31. Let α, $\beta \in \mathrm{stab}(a)$. Then $\alpha\beta(a) = \alpha(\beta(a)) = \alpha(a) = a$. Also, $\alpha(a) = a$ implies $\alpha^{-1}(\alpha(a)) = \alpha^{-1}(a)$ or $a = \alpha^{-1}(a)$.

33. m is a multiple of 6 but not a multiple of 30.

35. $6!/5 = 144$

37. 3, 7, 9

39. Let $\alpha = (123)$ and $\beta = (145)$.

41. $(123)(12) \neq (12)(123)$ in S_n $(n \geq 3)$.

43. Cycle decomposition shows that any nonidentity element of A_5 is a 5-cycle, a 3-cycle, or a product of a pair of disjoint 2-cycles. Then, observe that there are $(5 \cdot 4 \cdot 3 \cdot 2 \cdot 1)/5 = 24$ group elements of the form $(abcde)$, $(5 \cdot 4 \cdot 3)/3 = 20$ group elements of the form (abc), and $(5 \cdot 4 \cdot 3 \cdot 2 \cdot 1)/(2 \cdot 2 \cdot 2) = 15$ group elements of the form $(ab)(cd)$.

45. *Hint:* $(13)(12) = (123)$ and $(12)(34) = (324)(132)$.

47. Verifying that $a * \sigma(b) \neq b * \sigma(a)$ is done by examining all cases. To prove the general case, observe that $\sigma^i(a) * \sigma^{i+1}(b) \neq \sigma^i(b) * \sigma^{i+1}(a)$ can be written in the form $\sigma^i(a) * \sigma(\sigma^i(b)) \neq \sigma^i(b) * \sigma(\sigma^i(a))$, which is the case already done. If a transposition were not detected, then $\sigma(a_1) * \cdots * \sigma^i(a_i) * \sigma^{i+1}(a_{i+1}) * \cdots * \sigma^n(a_n) = \sigma(a_1) * \cdots * \sigma^i(a_{i+1}) * \sigma^{i+1}(a_i) * \cdots * \sigma^n(a_n)$, which implies $\sigma^i(a_i) * \sigma^{i+1}(a_{i+1}) = \sigma^i(a_{i+1}) * \sigma^{i+1}(a_i)$.

49. Observe that $(a_1 a_2 \cdots a_n) = (1a_1)(1a_n)(1a_{n-1}) \cdots (1a_1)$.

51. If α has odd order k and α is an odd permutation, then $\epsilon = \alpha^k$ would be odd.

53. The product of an element of $Z(A_4)$ of order 2 and an element of A_4 of order 3 would have order 6. The product of an element of $Z(A_4)$ of order 3 and an element of A_4 of order 2 would have order 6.

55. Labeling the four tires 1, 2, 3, and 4 in clockwise order starting with 1 being the tire in the upper left-hand corner, we may represent the four patterns as:
   $\alpha = (1324)$ top left-hand pattern
   $\beta = (1423)$ top right-hand pattern
   $\gamma = (14)(23)$ bottom right-hand pattern
   $\delta = (13)(24)$ bottom left-hand pattern
Notice that $\alpha^{-1} = \beta$ and that $\delta = \alpha^2\gamma$. Thus, we need only find the smallest subgroup of S_4 containing α and γ. To this end, observe that the set $\{\varepsilon, \alpha, \alpha^2, \alpha^3, \gamma, \alpha\gamma, \alpha^2\gamma, \alpha^3\gamma\}$ is closed under multiplication on the left and right by both α and γ. This implies that the set is closed under multiplication and is therefore a group. Since $\alpha\gamma \neq \gamma\alpha$, the subgroup is non-Abelian.

CHAPTER 6

Think and you won't sink.   B. C. FORBES, *Epigrams*

1. Try $n \rightarrow 2n$.
3. $\phi(xy) = \sqrt{xy} = \sqrt{x}\sqrt{y} = \phi(x)\phi(y)$
5. Try $1 \rightarrow 1, 3 \rightarrow 5, 5 \rightarrow 7, 7 \rightarrow 11$.
7. D_{12} has elements of order 12 and S_4 does not.
9. Since $T_e(x) = ex = x$ for all x, T_e is the identity. For the second part, observe that $T_g \circ (T_g)^{-1} = T_e = T_{gg^{-1}} = T_g \circ T_{g^{-1}}$ and cancel.
11. For any x in the group, we have $(\phi_g\phi_h)(x) = \phi_g(\phi_h(x)) = \phi_g(hxh^{-1}) = ghxh^{-1}g^{-1} = (gh)x(gh)^{-1} = \phi_{gh}(x)$.
13. $\phi_{R_{90}}$ and ϕ_{R_0} disagree on H; $\phi_{R_{90}}$ and ϕ_H disagree on R_{90}; $\phi_{R_{90}}$ and ϕ_D disagree on R_{90}. The remaining cases are similar.
15. Let $\alpha \in \text{Aut}(G)$. We show that α^{-1} is operation-preserving: $\alpha^{-1}(xy) = \alpha^{-1}(x)\alpha^{-1}(y)$ if and only if $\alpha(\alpha^{-1}(xy)) = \alpha(\alpha^{-1}(x)\alpha^{-1}(y))$, that is, if and only if $xy = \alpha(\alpha^{-1}(x))\alpha(\alpha^{-1}(y)) = xy$. So α^{-1} is operation-preserving. That $\text{Inn}(G)$ is a group follows from the equation $\phi_g\phi_h = \phi_{gh}$.
17. That α is one-to-one follows from the fact that r^{-1} exists modulo n. The operation-preserving condition is Exercise 11 in Chapter 0.
19. The inverse of a one-to-one function is one-to-one. To see that ϕ^{-1} is operation-preserving, let a and b belong to $\overline{G}$. Then $\phi^{-1}(ab) = \phi^{-1}(a) \phi^{-1}(b)$ if and only if $ab = \phi(\phi^{-1}(a))\phi(\phi^{-1}(b)) = ab$ [we obtained the first equality by applying ϕ to both sides of $\phi^{-1}(ab) = \phi^{-1}(a)\phi^{-1}(b)$]. Finally, let $g \in G$. Then $\phi^{-1}(\phi(g)) = g$, so that ϕ^{-1} is onto.
21. $T_g(x) = T_g(y)$ if and only if $gx = gy$ or $x = y$. This shows that T_g is a one-to-one function. Let $y \in G$. Then $T_g(g^{-1}y) = y$, so that T_g is onto.
23. Apply the appropriate definitions.
25. Show that Q is not cyclic.
27. Try $a + bi \rightarrow \begin{bmatrix} a & -b \\ b & a \end{bmatrix}$.
29. Yes, by Cayley's Theorem.
31. Observe that $\phi_g(y) = gyg^{-1}$ and $\phi_{zg}(y) = zgy(zg)^{-1} = zgyg^{-1}z^{-1} = gyg^{-1}$ since $2 \in Z(G)$. So, $\phi_g = \phi_{zg}$.
33. $\phi_g = \phi_h$ implies $gxg^{-1} = hxh^{-1}$ for all x. This implies $h^{-1}gx(h^{-1}g)^{-1} = x$, and therefore $h^{-1}g \in Z(G)$.
35. Say $|a| = n$. Then $\phi_a{}^n(x) = a^nxa^{-n} = x$, so that $\phi_a{}^n$ is the identity. For the example, take $a = R_{90}$ in D_4.
37. Say ϕ is an isomorphism from Q to $\mathbf{R}^+$ and ϕ takes 1 to a. It follows that the integer r maps to a^r and the rational r/s maps to $a^{r/s}$. But $a^{r/s} \neq a^\pi$ for any r/s.
39. $(R_0 R_{90} R_{180} R_{270})(H D'V D)$.
41. Consider the mapping $\phi(x) = x^2$ and note that 2 is not in the image.
43. Use the fact that if $a > 0$, then $a = \sqrt{a}\sqrt{a}$. For the second part, use the first part together with the fact that the inverse of an automorphism is an automorphism.

CHAPTER 7

Use missteps as stepping stones to deeper understanding and greater achievement. SUSAN TAYLOR

1. $H = \{\alpha_1, \alpha_2, \alpha_3, \alpha_4\}$, $\alpha_5 H = \{\alpha_5, \alpha_8, \alpha_6, \alpha_7\}$, $\alpha_9 H = \{\alpha_9, \alpha_{11}, \alpha_{12}, \alpha_{10}\}$

3. $H, 1 + H, 2 + H$

5. a. yes **b.** yes **c.** no

7. $8/2 = 4$, so there are four cosets. Let $H = \{1, 11\}$. The cosets are $H, 7H, 13H, 19H$.

9. Since $|a^4| = 15$, there are two cosets: $\langle a^4 \rangle$ and $a\langle a^4 \rangle$.

11. Suppose that $h \in H$ and $h < 0$. Then $hR^+ \subseteq hH = H$. But hR^+ is the set of all negative real numbers. Thus, $H = R^*$.

13. $1, 2, 3, 4, 5, 6, 10, 12, 15, 20, 30, 60$

15. Use Lagrange's Theorem (Theorem 7.1) and Corollary 3.

17. By Exercise 16, we have $5^6 \bmod 7 = 1$. So, using mod 7, we have $5^{15} = 5^6 \cdot 5^6 \cdot 5^2 \cdot 5 = 1 \cdot 1 \cdot 4 \cdot 5 = 6$; $7^{13} \bmod 11 = 2$.

19. Use Corollary 4 of Lagrange's Theorem (Theorem 7.1) together with Theorem 0.2.

21. Since G has odd order, no element can have order 2. Thus, for each $x \neq e$, we know that $x \neq x^{-1}$. So, we can write the product of all the elements in the form $ea_1 a_1^{-1} a_2 a_2^{-1} \cdots a_n a_n^{-1} = e$.

23. Let H be the subgroup of order p and K be the subgroup of order q. Then $H \cup K$ has $p + q - 1 < pq$ elements. Let a be any element in G that is not in $H \cup K$. By Lagrange's Theorem, $|a| = p, q,$ or pq. But $|a| \neq p$, for if so, then $\langle a \rangle = H$. Similarly, $|a| \neq q$.

25. $1, 3, 11, 33$. If $|x| = 33$, then $|x^{11}| = 3$. Elements of order 11 occur in multiples of 10.

27. Observe that $|G:H| = |G|/|H|$, $|G:K| = |G|/|K|$, and $|K:H| = |K|/|H|$.

29. Certainly, $a \in \mathrm{orb}_G(a)$. Now suppose that $c \in \mathrm{orb}_G(a) \cap \mathrm{orb}_G(b)$. Then $c = \alpha(a)$ and $c = \beta(b)$ for some α and β, and therefore $(\beta^{-1}\alpha)(a) = b$. So, if $x \in \mathrm{orb}_G(b)$, then $x = \gamma(b) = (\gamma\beta^{-1}\alpha)(a)$ for some γ. This proves that $\mathrm{orb}_G(b) \subseteq \mathrm{orb}_G(a)$. By symmetry, $\mathrm{orb}_G(a) \subseteq \mathrm{orb}_G(b)$.

31. a. $\mathrm{stab}_G(1) = \{(1), (24)(56)\}$; $\mathrm{orb}_G(1) = \{1, 2, 3, 4\}$
   b. $\mathrm{stab}_G(3) = \{(1), (24)(56)\}$; $\mathrm{orb}_G(3) = \{3, 4, 1, 2\}$
   c. $\mathrm{stab}_G(5) = \{(1), (12)(34), (13)(24), (14)(23)\}$; $\mathrm{orb}_G(5) = \{5, 6\}$

33. Suppose that $|Z(G)| = p^{n-1}$ and let a be an element of G not in $Z(G)$. Then $C(a)$ contains both a and $Z(G)$. By Lagrange's Theorem, we must have $C(a) = G$. But then $a \in Z(G)$.

35. 2520

37. Suppose $x^2 = y^2$ and let $|G| = 2k + 1$. Then $x = xe = xx^{2k+1} = x^{2k+2} = (x^2)^{k+1} = (y^2)^{k+1} = y^{2k+2} = yy^{2k+1} = ye = y$.

39. Suppose that $B \in G$ and $\det(B) = 2$. Then $\det(A^{-1}B) = 1$, so that $A^{-1}B \in H$ and therefore $B \in AH$. Conversely, for any $Ah \in AH$ we have $\det(Ah) = \det(A)\det(h) = 2 \cdot 1 = 2$.

41. It is the set of all permutations that carry face 2 to face 1.

43. $aH = bH$ if and only if $\det(a) = \pm\det(b)$.

45. 50

47. For any positive integer n, let $\omega_n = \cos(\frac{2\pi}{n}) + i\sin(\frac{2\pi}{n})$. The finite subgroups of C^* are those of the forms $\langle \omega_n \rangle$. To verify this, let H denote any finite subgroup of C^* of order n. Then every element of H is a solution to $x^n = 1$. But the solution set of $x^n = 1$ in C^* is $\langle \omega_n \rangle$.

CHAPTER 8

There is always a right and a wrong way, and the wrong way always seems the more reasonable.   GEORGE MOORE, *The Bending of the Bough*

1. Closure and associativity in the product follow from the closure and associativity in each component. The identity in the product is the n-tuple with the identity in each component. The inverse of $(g_1, g_2, \ldots, g_n)$ is $(g_1^{-1}, g_2^{-1}, \ldots, g_n^{-1})$.

3. Use $g \rightarrow (g, e_H)$ and $h \rightarrow (e_G, h)$.

5. To show that $Z \oplus Z$ is not cyclic, note that $(a, b + 1) \notin \langle (a, b) \rangle$.

7. Try $(g_1, g_2) \rightarrow (g_2, g_1)$.

9. Yes, by Theorem 8.2.

11. Observe by Theorem 4.4 that as long as d divides n, the number of elements of order d in a cyclic group depends only on d. So, in both $Z_{8000000}$ and Z_4 there are $\phi(4) = 2$ elements of order 4 and $\phi(2) = 1$ element of order 2. Similarly for $Z_m \oplus Z_n$.

13. Try $a + bi \rightarrow (a, b)$.

15. Use Exercise 3 and Theorem 4.3.

17. 3

19. Map $\begin{bmatrix} a & b \\ c & d \end{bmatrix}$ to (a, b, c, d). Let $\mathbf{R}^k$ denote $\mathbf{R} \oplus \mathbf{R} \oplus \cdots \oplus \mathbf{R}$ (k factors). Then the group of $m \times n$ matrices under addition is isomorphic to $\mathbf{R}^{mn}$.

21. $(g, g)(h, h)^{-1} = (gh^{-1}, gh^{-1})$
    When $G = \mathbf{R}$, $G \oplus G$ is the plane and H is the line $y = x$.

23. $\langle (3, 0) \rangle, \langle (3, 1) \rangle, \langle (3, 2) \rangle, \langle (0, 1) \rangle$

25. 60

27. $\{0, 400\} \oplus \{0, 50, 100, 150\}$

29. Compare the number of elements of order 2 in each group.

31. Try $3^m 6^n \rightarrow (m, n)$.

33. D_{24} has elements of order 24, whereas $D_3 \oplus D_4$ does not.

35. 12

37. $\text{Aut}(U(25)) \approx \text{Aut}(Z_{20}) \approx U(20) \approx U(4) \oplus U(5) \approx Z_2 \oplus Z_4$.

39. $2^k - 1$; $2^t - 1$, where t is the number of the $n_1, n_2, \ldots, n_k$ that are even.

41. No. $Z_{10} \oplus Z_{12} \oplus Z_6$ has 7 elements of order 2 whereas $Z_{15} \oplus Z_4 \oplus Z_{12}$ has only 3.

43. Using the fact that an isomorphism from Z_{12} is determined by the image of 1 and the fact that a generator must map to a generator, we determine that there are 4 isomorphisms.

45. Since $a \in Z_m$ and $b \in Z_n$, we know that $|a|$ divides m and $|b|$ divides n. So, $|(a, b)| = \text{lcm}(|a|, |b|)$ divides $\text{lcm}(m, n)$.

47. Z, Z_3, Z_4, Z_6

49. Observe that every nonidentity element of $Z_p \oplus Z_p$ has order p and each subgroup of order p contains $p - 1$ of them. So, there are exactly $(p^2 - 1)/(p - 1) = p + 1$ subgroups of order p.

51. $U(165) \approx U(11) \oplus U(15) \approx U(5) \oplus U(33) \approx U(3) \oplus U(55) \approx U(3) \oplus U(5) \oplus U(11)$

53. Mimic the analysis for elements of order 12 in $U(720)$ in this chapter.

55. 60

57. They are both isomorphic to $Z_{10} \oplus Z_4$.

59. That $U(n)^2$ is a subgroup follows from Exercise 15 of Supplementary Exercises for Chapters 1–4. $1^2 = (n - 1)^2$ shows that it is a proper subgroup.

61. 275

63. $U(117) \approx U(9) \oplus U(13) \approx Z_6 \oplus Z_{12}$, which contains $\langle (2, 0) \rangle \oplus \langle (0, 4) \rangle$.

65. Consider $U(49)$.

67. Consider $U(65)$.

69. NO

SUPPLEMENTARY EXERCISES FOR CHAPTERS 5–8

All things are difficult before they are easy.   THOMAS FULLER

1. Consider the finite and infinite cases separately. In the finite case, note that $|H| = |\phi(H)|$. Now use Theorem 4.3. For the infinite case, use Exercise 2 in Chapter 6.

3. Observe that $\phi(x^{-1}y^{-1}xy) = (\phi(x))^{-1}(\phi(y))^{-1}\phi(x)\phi(y)$, so ϕ carries the generators of G' to the generators of G'.

5. All nonidentity elements of G and H have order 3. $G \neq H$.

7. Certainly the set HK has $|H||K|$ symbols. However, not all symbols need represent distinct group elements. That is, we may have $hk = h'k'$ although $h \neq h'$ and $k \neq k'$. We must determine the extent to which this happens. For every t in $H \cap K$, $hk = (ht)(t^{-1}k)$, so each group element in HK is represented by at least $|H \cap K|$ products in HK. But $hk = h'k'$ implies $t = h^{-1}h' = k(k')^{-1} \in H \cap K$, so that $h' = ht$ and $k' = t^{-1}k$. Thus each element in HK is represented by exactly $|H \cap K|$ products. So, $|HK| = |H| |K|/|H \cap K|$.

9. $U(n)$, where $n = 4, 8, 3, 6, 12, 24$.

11. *Hint:* $3 + 2i = \sqrt{13}\left(\dfrac{3}{\sqrt{13}} + \dfrac{2}{\sqrt{13}} i \right)$

13. Suppose $\phi\colon Q \to R$ is an isomorphism. Let $\phi(1) = x_0$. Show that $\phi(a/b) = (a/b)x_0$ for all integers a, b with $b \neq 0$.

15. In Q, the equation $2x = a$ has a solution for all a. The corresponding equation $x^2 = b$ in Q^+ does not have a solution for all b.

17. Suppose $x^{p-2} = 1$. Since $|U(p)| = p - 1$, we have that $x^{p-1} = 1$ for all $x \in U(p)$. So, by cancellation, $x = 1$.

19. $\langle 3 \rangle \oplus \langle 4 \rangle$

21. $Z_{18}, Z_2 \oplus Z_3 \oplus Z_3, D_9, D_3 \oplus Z_3$.

23. Say $\alpha = a_1 a_2 \cdots a_n$ and $\beta = b_1 \cdots b_m$, where the a's and b's are cycles. Then $\alpha \beta^{-1} = a_1 a_2 \cdots a_n b_m^{-1} \cdots b_1^{-1}$ is a finite number of cycles.

25. Count elements of order 2.

27. Count elements of order 2.

29. $x = \phi_a(x) = axa^{-1}$, so that $xa = ax$. Conversely, if G is Abelian, ϕ_a is the identity.

31. $U_{50}(450)$

33. $(4, 10)$

35. Count elements of order 2.

37. 20; $(8, 7, (3251))$

39. $Z_8, Z_4 \oplus Z_2, Z_2 \oplus Z_2 \oplus Z_2, D_4$, and the quaternions.

41. $(12)(34)(56789)$.

43. $\{(1), (345), (354), (12), (12)(345), (12)(354)\}$.

45. $\beta = (17395)(286)$.

47. Say the points in H lie on the line $y = mx$. Then $(a, b) + H = \{(a + x, b + mx) \mid x \in R\}$. This set is the line $y - b = m(x - a)$.

49. $aH = bH$ implies $a^{-1}b \in H$. So $(a^{-1}b)^{-1} = b^{-1}a \in H$. Thus, $Hb^{-1}a = H$ or $Hb^{-1} = Ha^{-1}$. These steps are reversible.

51. Let β have order 2. In disjoint cycle form, β is a product of transpositions, so there must be some i missing from this product. Thus, $\beta(i) = i$. Pick j such that $\beta(j) \neq j$. Since σ is an n-cycle, some power of σ, say σ^t, takes i to j. If β commutes with σ, it commutes with σ^t as well. Then $(\sigma^t\beta)(i) = \sigma^t(\beta(i)) = \sigma^t(i) = j$, whereas $(\beta\sigma^t)(i) = \beta(\sigma^t(i)) = \beta(j) \neq j$. This proves that $\sigma^t\beta \neq \beta\sigma^t$.

CHAPTER 9

> *There's a mighty big difference between good, sound reasons and reasons that sound good.*   BURTON HILLIS

1. no

3. Say $i < j$ and let $h \in H_i \cap H_j$. Then $h \in H_1 H_2 \cdots H_i \cdots H_{j-1} \cap H_j = \{e\}$.

5. Recall that if A and B are matrices, then $\det(ABA^{-1}) = (\det A)(\det B)(\det A)^{-1}$.

7. Let $x \in G$. If $x \in H$, then $xH = H = Hx$. If $x \notin H$, then xH is the set of elements in G, not in H. But Hx is also the set of elements in G, not in H.

9. $G/H \approx Z_4$
$G/K \approx Z_2 \oplus Z_2$

11. No, look at D_3.

13. This follows directly from $(ab)h = a(bh)$ for all $h \in H$.

15. 2

17. $H = \{0 + \langle 20 \rangle, 4 + \langle 20 \rangle, 8 + \langle 20 \rangle, 12 + \langle 20 \rangle, 16 + \langle 20 \rangle\}$.
$G/H = \{0 + \langle 20 \rangle + H, 1 + \langle 20 \rangle + H, 2 + \langle 20 \rangle + H, 3 + \langle 20 \rangle + H\}$.

19. $40/10 = 4$

21. By Theorem 9.5, the group has an element a of order 3 and an element b of order 11. Then $|ab| = 33$.

23. ∞; no, $(6, 3) + \langle(4, 2)\rangle$ has order 2.

25. Z_8

27. yes; no

29. Since $\det(AB^{-1}) = (\det A)(\det B)^{-1}$, H is a subgroup.
Since $\det(CAC^{-1}) = (\det C)(\det A)(\det C)^{-1} = \det A$, H is normal.

31. Certainly, every nonzero real number is of the form $\pm r$, where r is a positive real number. Real numbers commute, and $\mathbf{R}^+ \cap \{1, -1\} = \{1\}$.

33. No. If $G = H \times K$, then $|g| = \text{lcm}(|h|, |k|)$ provided that $|h|$ and $|k|$ are finite. If $|h|$ or $|k|$ is infinite, so is $|g|$.

35. For the first question, note that $\langle 3 \rangle \cap \langle 6 \rangle = \{1\}$ and $\langle 3 \rangle\langle 6 \rangle \cap \langle 10 \rangle = \{1\}$. For the second question, observe that $12 = 3^{-1}6^2$.

37. Say $|g| = n$. Then $(gH)^n = g^nH = eH = H$. Now use Corollary 2 to Theorem 4.1.

39. Let $x \in C(H)$, $g \in G$, and $h \in H$. We must show that $gxg^{-1}h = hgxg^{-1}$. Note that in the expression $(gxg^{-1})h(gxg^{-1})^{-1} = gxg^{-1}hgx^{-1}g^{-1}$ the terms x and x^{-1} cancel since $g^{-1}hg \in H$ and x commutes with every element of H. Then we have $(gxg^{-1})h(gxg^{-1})^{-1} = gxg^{-1}hgx^{-1}g^{-1} = gg^{-1}hgg^{-1} = h$. So, $gxg^{-1} \in C(H)$.

41. Take $G = Z_6, H = \{0, 3\}, a = 1$, and $b = 4$.

43. Use the "*G/Z* Theorem."

45. If H is normal in G, then $xNhN(xN)^{-1} = xhx^{-1}N \in H/N$, so H/N is normal in G/N. Now assume H/N is normal in G/N. Then $xhx^{-1}N = xNhN(xN)^{-1} \in H/N$. Thus, $xhx^{-1}N = h'N$ for $h' \in H$. So, $xhx^{-1} = h'n$ for some $n \in N$.

47. Say H has index n. Then $(\mathbf{R}^*)^n = \{x^n \mid x \in \mathbf{R}^*\} \subseteq H$. If n is odd, then $(\mathbf{R}^*)^n = \mathbf{R}^*$; if n is even, then $(\mathbf{R}^*)^n = \mathbf{R}^+$. So, $H = \mathbf{R}^*$ or $H = \mathbf{R}^+$.

49. Use Exercise 7 and observe that $VK \neq KV$.

51. Suppose n_1h_1 and $n_2h_2 \in NH$. Then $n_1h_1n_2h_2 = n_1n'h_1h_2 \in NH$. Also $(n_1h_1)^{-1} = h_1^{-1}n_1^{-1} = nh_1^{-1} \in NH$.

53. Let $N = \langle a \rangle, H = \langle a^k \rangle$, and $x \in G$. Then, $x(a^k)^m x^{-1} = (xa^m x^{-1})^k = (a^r)^k = (a^k)^r \in H$.

55. $\gcd(|x|, |G/H|) = 1$ implies $\gcd(|xH|, |G/H|) = 1$. But $|xH|$ divides $|G/H|$. Thus $|xH| = 1$ and therefore $xH = H$.

57. Note that G/N is a group and use Corollary 4 of Theorem 7.1.

59. Use Theorems 9.4 and 9.3.

61. Say $|gH| = n$. Then $|g| = nt$ (by Exercise 37) and $|g^t| = n$. For the second part, consider $Z/\langle k \rangle$.

63. It is not a group table. No, because $\mathcal{H}$ is not normal in D_4.

65. Use Theorem 9.3 and Theorem 7.2.

67. By Exercise 66, A_5 would have an element of the form $(ab)(cd)$ that commutes with every element of A_5. Try (abc).

69. Observe that $xg^2x^{-1} = (xgx^{-1})^2$.

71. Mimic the argument given in Example 13 in this chapter.

CHAPTER 10

Sixty minutes of thinking of any kind is bound to lead to confusion and unhappiness. JAMES THURBER

1. Note that $\det(AB) = (\det A)(\det B)$.

3. Note that $(f + g)' = f' + g'$.

5. Observe that $(xy)^r = x^r y^r$. Odd values of r yield an isomorphism.

7. $(\sigma\phi)(g_1g_2) = \sigma(\phi(g_1g_2)) = \sigma(\phi(g_1)\phi(g_2)) = \sigma(\phi(g_1))\sigma(\phi(g_2)) = (\sigma\phi)(g_1)(\sigma\phi)(g_2)$.

9. $\phi((g, h)(g', h')) = \phi((gg', hh')) = gg' = \phi((g, h))\phi((g', h'))$. The kernel is $\{(e, h)|h \in H\}$.

11. Consider $\phi: Z \oplus Z \to Z_a \oplus Z_b$ given by $\phi((x, y)) = (x \bmod a, y \bmod b)$ and use Theorem 10.3.

13. $(a, b) \to b$ is a homomorphism from $A \oplus B$ onto B with kernel $A \oplus \{e\}$.

15. 3, 13, 23

17. Suppose ϕ is such a homomorphism. By Theorem 10.3, Ker $\phi = \langle (8, 1) \rangle, \langle (0, 1) \rangle$ or $\langle (8, 0) \rangle$. In these cases, $(1, 0)$ + Ker ϕ has order either 16 or 8. So, $(Z_{16} \oplus Z_2)/$ Ker ϕ is not isomorphic to $Z_4 \oplus Z_4$.

19. Since |Ker ϕ| is not 1 and divides 17, ϕ is the trivial map.

21. $\langle 5 \rangle$

23. a. The possible images are isomorphic to Z_1, Z_2, Z_3, Z_4, Z_6, and Z_{12}.
 b. $\langle 1 \rangle \approx Z_{36}, \langle 2 \rangle \approx Z_{18}, \langle 3 \rangle \approx Z_{12}, \langle 4 \rangle \approx Z_9, \langle 6 \rangle \approx Z_6$, and $\langle 12 \rangle \approx Z_3$.

25. 4 onto; 10 to

27. For each k with $0 \le k \le n - 1$, the mapping $1 \to k$ determines a homomorphism.

29. Use Theorem 10.3 and properties 5, 7, and 8 of Theorem 10.2.

31. $\phi^{-1}(7) = 7$ Ker $\phi = \{7, 17\}$

33. 11 Ker ϕ

35. $\phi((a, b) + (c, d)) = \phi((a + c, b + d)) = (a + c) - (b + d) = a - b + c - d = \phi((a, b)) + \phi((c, d))$. Ker $\phi = \{(a, a) \mid a \in Z\}$. $\phi^{-1}(3) = \{(a + 3, a) \mid a \in Z\}$.

37. $\phi(xy) = (xy)^6 = x^6 y^6 = \phi(x)\phi(y)$. Ker $\phi = \langle \cos 60° + i \sin 60° \rangle$.

39. Show that the mapping from K to KN/N given by $k \to kN$ is an onto homomorphism with kernel $K \cap N$.

41. For each divisor d of k there is a unique subgroup of Z_k of order d, and this subgroup is generated by $\phi(d)$ elements. A homomorphism from Z_n to a subgroup of Z_k must carry 1 to a generator of the subgroup. Furthermore, the order of the image of 1 must divide n, so we need consider only those divisors d of k that also divide n.

43. $D_4, \{e\}, Z_2, Z_2 \oplus Z_2$

45. It is divisible by 10.

47. It is infinite. Z

49. Let γ be the natural homomorphism from G onto G/N. Let $\overline{H}$ be a subgroup of G/N and let $\gamma^{-1}(\overline{H}) = H$. Then H is a subgroup of G and $H/N = \gamma(H) = \gamma(\gamma^{-1}(\overline{H})) = \overline{H}$.

51. The mapping $g \to \phi_g$ is a homomorphism with kernel $Z(G)$.

53. $(f + g)(3) = f(3) + g(3)$. The kernel is the set of elements in $Z[x]$ whose graphs pass through the point $(3, 0)$.

55. Use Exercise 50 in Chapter 9 and Exercise 39 above to prove the first assertion. To verify that $G/(H \cap K)$ is not cyclic, observe that it has two subgroups of order 2.

57. Mimic Example 16.

59. Suppose that H is a proper subgroup of G that is not properly contained in a proper subgroup of G. Then G/H has no nontrivial, proper subgroup. It follows from Exercise 22 in Chapter 7 that G/H is isomorphic to Z_p for some prime p. But then for every coset $g + H$ we have $p(g + H) = H$ so that $pg \in H$ for all $g \in G$. But then $G = pG \subseteq H$. Both Q and R satisfy the hypothesis.

CHAPTER 11

Think before you think!   STANISLAW J. LEC, *Unkempt Thoughts*

1. $n = 4$

$Z_4, Z_2 \oplus Z_2$

3. $n = 36$

$Z_9 \oplus Z_4, Z_3 \oplus Z_3 \oplus Z_4, Z_9 \oplus Z_2 \oplus Z_2, Z_3 \oplus Z_3 \oplus Z_2 \oplus Z_2$

5. The only Abelian groups of order 45 are Z_{45} and $Z_3 \oplus Z_3 \oplus Z_5$. In the first group, $|3| = 15$; in the second one, $|(1, 1, 1)| = 15$. $Z_3 \oplus Z_3 \oplus Z_5$ does not have an element of order 9.

7. $Z_9 \oplus Z_3 \oplus Z_4$; $Z_9 \oplus Z_3 \oplus Z_2 \oplus Z_2$

9. $Z_4 \oplus Z_2 \oplus Z_3 \oplus Z_5$

11. By the Fundamental Theorem, any finite Abelian group G is isomorphic to some direct product of cyclic groups of prime-power order. Now go across the direct product and, for each distinct prime you have, pick off the largest factor of the prime-power. Next, combine all of these into one factor (you can do this, since the subscripts are relatively prime). Let us call the order of this new factor n_1. Now repeat this process with the remaining original factors and call the order of the resulting factor n_2. Then n_2 divides n_1, since each prime-power divisor of n_2 is also a prime-power divisor of n_1. Continue in this fashion. Example: If

$$G \approx Z_{27} \oplus Z_3 \oplus Z_{125} \oplus Z_{25} \oplus Z_4 \oplus Z_2 \oplus Z_2,$$

then

$$G \approx Z_{27 \cdot 125 \cdot 4} \oplus Z_{3 \cdot 25 \cdot 2} \oplus Z_2.$$

Now note that 2 divides $3 \cdot 25 \cdot 2$ and $3 \cdot 25 \cdot 2$ divides $27 \cdot 125 \cdot 4$.

13. $Z_2 \oplus Z_2$

15. a. 1   **b.** 1   **c.** 1   **d.** 1   **e.** 1   **f.** There is a unique Abelian group of order n if and only if n is not divisible by the square of any prime.

17. $Z_2 \oplus Z_2$

19. $Z_3 \oplus Z_3$

21. n is square-free (no prime factor of n occurs more than once).

23. Among the first 11 elements in the table, there are nine elements of order 4. None of the other isomorphism classes has this many.

25. $Z_4 \oplus Z_2 \oplus Z_2$; one internal direct product is $\langle 7 \rangle \times \langle 101 \rangle \times \langle 199 \rangle$.

27. 3; 6; 12

29. $Z_4 \oplus Z_4$

31. Use Theorems 11.1, 8.1, and 4.3.

33. $|\langle a \rangle K| = |a||K|/|\langle a \rangle \cap K| = |a||K| = |\bar{a}||\bar{K}|p = |\bar{G}|p = |G|.$

35. By the Fundamental Theorem of Finite Abelian Groups, it suffices to show that every group of the form $Z_{p_1^{n_1}} \oplus Z_{p_2^{n_2}} \oplus \cdots \oplus Z_{p_k^{n_k}}$ is a subgroup of a U-group. Consider first a group of the form $Z_{p_1^{n_1}} \oplus Z_{p_2^{n_2}}$ (p_1 and p_2 need not be distinct). By Dirichlet's Theorem, for some s and t there are distinct primes q and r such that $q = tp_1^{n_1} + 1$ and $r = sp_2^{n_2} + 1$. Then $U(qr) = U(q) \oplus U(r) \approx Z_{tp_1^{n_1}} \oplus Z_{sp_2^{n_2}}$, and this latter group contains a subgroup isomorphic to $Z_{p_1^{n_1}} \oplus Z_{p_2^{n_2}}$. The general case follows in the same way.

SUPPLEMENTARY EXERCISES FOR CHAPTERS 9–11

You cannot have success without the failures.

H.G. HASLER, *The Observer*

1. Say $aH = Hb$. Then $a = hb$ for some h in H. Then $Ha = Hhb = Hb = aH$.

3. Suppose diag(G) is normal. Then $(e, a)(b, b)(e, a)^{-1} = (b, aba^{-1}) \in$ diag(G). Thus $b = aba^{-1}$. If G is Abelian, $(g, h)(b, b)(g, h)^{-1} = (gbg^{-1}, hbh^{-1}) = (b, b)$. The index of diag($G$) is $|G|$.

5. Let $\alpha \in$ Aut(G) and $\phi_a \in$ Inn(G). Then $(\alpha \phi_a \alpha^{-1})(x) = (\alpha \phi_a)(\alpha^{-1}(x)) = \alpha(a\alpha^{-1}(x)a^{-1}) = \alpha(a)x(\alpha(a))^{-1} = \phi_{\alpha(a)}(x).$

7. R^* (See Example 2 in Chapter 10.)

9. a.

$$Z(H) = \left\{ \begin{bmatrix} 1 & 0 & b \\ 0 & 1 & 0 \\ 0 & 0 & 1 \end{bmatrix} \middle| b \in Q \right\}$$

The mapping

$$\begin{bmatrix} 1 & 0 & b \\ 0 & 1 & 0 \\ 0 & 0 & 1 \end{bmatrix} \to b$$

is an isomorphism.

b. The mapping

$$\begin{bmatrix} 1 & a & b \\ 0 & 1 & c \\ 0 & 0 & 1 \end{bmatrix} \to (a, c)$$

is a homomorphism with $Z(H)$ as the kernel.

c. The proofs are valid with $\mathbf{R}$ and Z_p.

11. $b(a/b + Z) = a + Z = Z$

13. Use Exercise 5 of the Supplementary Exercises for Chapters 1–4. Such a set is possible only when n is prime. For the first example, consider D_p, where p is a prime. For the second example, try D_4.

15. Observe that $hkh^{-1}k^{-1} = (hkh^{-1})k^{-1} \in K$ and $hkh^{-1}k^{-1} = h(kh^{-1}k^{-1}) \in H$.

17. Use Theorem 7.3 and Exercise 7 of Chapter 9.

19. First observe that $\phi((4, 0, 0)) = \phi(4(1, 0, 0)) = 4\phi(1, 0, 0) = (0, 0)$, so that Ker $\phi = \{(0, 0, 0), (4, 0, 0)\}$. But then $(Z_8 \oplus Z_2 \oplus Z_2)/\text{Ker } \phi$ has more than three elements of order 2, whereas $Z_4 \oplus Z_4$ has only three.

21. Use Theorem 7.2 together with the fact that S_4 has no element of order 6.

23. The number is m in all cases.

25. The mapping $g \to g^n$ is a homomorphism from G onto G^n with kernel G_n.

27. Let $|H| = p$. Exercise 7 of Supplementary Exercises for Chapters 5–8 shows that H is the only subgroup of order p. But xHx^{-1} is also a subgroup of order p. So, $xHx^{-1} = H$.

29. Say a and b are integers and $a/b + Z$ has order n in Q/Z. Then $na/b = m$ for some integer m. Thus, $a/b + Z = m/n + Z = m(1/n + Z) \in \langle 1/n + Z \rangle$.

31. If $(1, 0) \to a$ and $(0, 1) \to b$, then $(x, 0) \to ax$ and $(0, y) \to by$.

33. First note that by Exercise 11 every element in Q/Z has finite order. For each positive integer n, let B_n denote the set of elements of order n and suppose that ϕ is an isomorphism from Q/Z to itself. Then, by property 5 of Theorem 6.2, $\phi(B_n) \subseteq B_n$. By Exercise 29 we know that B_n is finite, and since ϕ preserves orders and is one-to-one, we must have $\phi(B_n) = B_n$. Since it follows from Exercise 11 and Exercise 29 that $Q/Z = \cup B_n$, where the union is taken over all positive integers n, we have $\phi(Q/Z) = Q/Z$.

35. If the group is not Abelian, for any element a not in the center, the inner automorphism induced by a is not the identity; if the group is Abelian and contains an element a with $|a| > 2$, then $x \to x^{-1}$ works; if every nonidentity element has order 2, then G is isomorphic to a group of the form $Z_2 \oplus Z_2 \oplus \cdots \oplus Z_2$. In this case, the mapping that takes $(a_1, a_2, a_3, \ldots a_k)$ to $(a_2, a_1, a_3, \ldots, a_k)$ is not the identity.

CHAPTER 12

> *Think for yourself.* Title of a song by GEORGE HARRISON,
> *Rubber Soul*

1. For any $n > 1$, the ring $M_2(Z_n)$ of 2×2 matrices with entries from Z_n is a finite non-commutative ring. The set $M_2(2Z)$ of 2×2 matrices with even integer entries is an infinite noncommutative ring that does not have a unity.

3. The only property that is not immediate is closure under multiplication:

$$(a + b\sqrt{2})(c + d\sqrt{2}) = (ac + 2bd) + (ad + bc)\sqrt{2}.$$

5. The proofs given for a group apply to a ring as well.

7. In Z_p, nonzero elements have multiplicative inverses. Use them.

9. If a and b belong to the intersection, then they belong to each member of the intersection. Thus $a - b$ and ab belong to each member of the intersection. So, $a - b$ and ab belong to the intersection.

11. Part 3: $0 = 0(-b) = (a + (-a))(-b) = a(-b) + (-a)(-b) = -(ab) + (-a)(-b)$. So, $ab = (-a)(-b)$.
Part 4: $a(b - c) = a(b + (-c)) = ab + a(-c) = ab + (-(ac)) = ab - ac$.
Part 5: Use part 2.
Part 6: Use part 3.

13. *Hint:* Z is a cyclic group under addition, and every subgroup of a cyclic group is cyclic.

15. For positive m and n, observe that $(m \cdot a)(n \cdot b) = (a + a + \cdots + a)(b + b + \cdots + b) = (ab + ab + \cdots + ab)$, where the last term has mn summands. Similar arguments apply in the remaining cases.

17. From Exercise 15, we have $(n \cdot a)(m \cdot a) = (nm) \cdot a^2 = (mn) \cdot a^2 = (m \cdot a)(n \cdot a)$.

19. Let a, b belong to the center. Then $(a - b)x = ax - bx = xa - xb = x(a - b)$. Also, $(ab)x = a(bx) = a(xb) = (ax)b = (xa)b = x(ab)$.

21. $(x_1, \ldots, x_n)(a_1, \ldots, a_n) = (x_1, \ldots, x_n)$ for all x_i in R_i if and only if $x_i a_i = x_i$ for all x_i in R_i and $i = 1, \ldots, n$.

23. $\{1, -1, i, -i\}$

25. $f(x) = 1$ and $g(x) = -1$.

27. If a is a unit, then $b = a(a^{-1}b)$.

29. Consider $a^{-1} - a^{-2}b$.

31. Try the ring $M_2(Z)$.

33. Note that $2x = (2x)^3 = 8x^3 = 8x$.

35. For Z_6 use $n = 3$. For Z_{10} use $n = 5$. Say $m = p^2 t$ where p is a prime. Then $(pt)^n = 0$ in Z_m since m divides $(pt)^n$.

37. Every subgroup of Z_n is closed under multiplication.

39. $ara - asa = a(r - s)a$. $(ara)(asa) = ara^2 sa = arsa$. $a1a = a^2 = 1$, so $1 \in S$.

41. The subring test is satisfied.

43. Look at $(1, 0, 1)$ and $(0, 1, 1)$.

45. Observe that $n \cdot 1 - m \cdot 1 = (n - m) \cdot 1$. Also, $(n \cdot 1)(m \cdot 1) = (nm) \cdot ((1)(1)) = (nm) \cdot 1$.

47. $\{m/2^n \mid m \in Z, n \in Z^+\}$

49. $(a + b)(a - b) = a^2 + ba - ab - b^2 = a^2 - b^2$ if and only if $ba - ab = 0$.

51. $Z_2 \oplus Z_2$; $Z_2 \oplus Z_2 \oplus \cdots$ (infinitely many copies).

CHAPTER 13

Work now or wince later.   B. C. FORBES, *Epigrams*

1. The verifications for Examples 1–6 follow from elementary properties of real and complex numbers. For Example 7, note that

$$\begin{bmatrix} 1 & 0 \\ 0 & 0 \end{bmatrix}\begin{bmatrix} 0 & 0 \\ 0 & 1 \end{bmatrix} = \begin{bmatrix} 0 & 0 \\ 0 & 0 \end{bmatrix}.$$

For Example 8, note that $(1, 0)(0, 1) = (0, 0)$.

3. Let $ab = 0$ and $a \neq 0$. Then $ab = a \cdot 0$, so $b = 0$.

5. Let $k \in Z_n$. If $\gcd(k, n) = 1$, then k is a unit. If $\gcd(k, n) = d > 1$, write $k = sd$. Then $k(n/d) = sd(n/d) = sn = 0$.

7. Let $s \in R$, $s \neq 0$. Consider the set $S = \{sr \mid r \in R\}$. If $S = R$, then $sr = 1$ (the unity) for some r. If $S \neq R$, then there are distinct r_1 and r_2 such that $sr_1 = sr_2$. In this case, $s(r_1 - r_2) = 0$. To see what happens when the "finite" condition is dropped, consider Z.

9. $(a_1 + b_1\sqrt{d}) - (a_2 + b_2\sqrt{d}) = (a_1 - a_2) + (b_1 - b_2)\sqrt{d}$; $(a_1 + b_1\sqrt{d})(a_2 + b_2\sqrt{d}) = (a_1a_2 + b_1b_2d) + (a_1b_2 + a_2b_1)\sqrt{d}$. Thus the set is a ring. Since $Z[\sqrt{d}]$ is a subring of the ring of complex numbers, it has no zero-divisors.

11. The even integers.

13. $(1 - a)(1 + a + a^2 + \cdots + a^{n-1}) = 1 + a + a^2 + \cdots + a^{n-1} - a - a^2 - \cdots - a^n = 1 - a^n = 1 - 0 = 1$.

15. Suppose $a \neq 0$ and $a^n = 0$ (where we take n to be as small as possible). Then $a \cdot 0 = 0 = a^n = a \cdot a^{n-1}$, so by cancellation, $a^{n-1} = 0$.

17. If $a^2 = a$ and $b^2 = b$, then $(ab)^2 = a^2b^2 = ab$.

19. $(3 + 4i)^2 = 3 + 4i$.

21. $a^2 = a$ implies $a(a - 1) = 0$. So if a is a unit, $a - 1 = 0$ and $a = 1$.

23. See Theorems 3.1 and 12.3.

25. Note that $ab = 1$ implies $aba = a$. Thus $0 = aba - a = a(ba - 1)$. So, $ba - 1 = 0$.

27. A subdomain of an integral domain D is a subset of D that is an integral domain under the operations of D. To show that P is a subdomain, show that it is a subring and contains 1. Every subdomain contains 1 and is closed under addition and subtraction, so every subdomain contains P. $|P| = $ char D when char D is prime and $|P|$ is infinite when char D is 0.

29. Use Theorems 13.3, 13.4, and 7.1 (Lagrange's Theorem).

31. a. Since $a^3 = b^3$, $a^6 = b^6$. Then $a = b$ because we can cancel a^5 from both sides (since $a^5 = b^5$).

b. Use the fact that there exist integers s and t such that $1 = sn + tm$, but remember that you cannot use negative exponents in a ring.

33. $(1 - a)^2 = 1 - 2a + a^2 = 1 - 2a + a = 1 - a$.

35. Z_8

37. Let $S = \{a_1, a_2, \ldots, a_n\}$ be the nonzero elements of the ring. First show that $S = \{a_1a_1, a_1a_2, \ldots, a_1a_n\}$. Thus, $a_1 = a_1a_i$ for some i. Then a_i is the unity, for if a_k is any element of S, we have $a_1a_k = a_1a_ia_k$, so that $a_1(a_k - a_ia_k) = 0$.

39. Say $|x| = n$ and $|y| = m$ with $n < m$. Consider $(nx)y = x(ny)$.

41. a. Use the Binomial Theorem.

b. Use part **a** and induction.

c. Consider $\{0, 3, 6, 9\}$ under addition and multiplication modulo 12.

43. Use Theorems 13.4 and 9.5 and Exercise 39.

45. $n \begin{bmatrix} a & b \\ c & d \end{bmatrix} = \begin{bmatrix} 0 & 0 \\ 0 & 0 \end{bmatrix}$ for all members of $M_2(R)$ if and only if $na = 0$ for all a in R.

47. Use Exercise 46.

49. a. 2 **b.** 2, 3 **c.** 2, 3, 6, 11 **d.** 2, 3, 9, 10

51. 2

53. See Example 10.

55. Use Exercise 23 and part **a** of Exercise 41.

57. Choose $a \neq 0$ and $a \neq 1$ and consider $1 + a$.

59. $\phi(x) = \phi(x \cdot 1) = \phi(x) \cdot \phi(1)$ so $\phi(1) = 1$. Also, $1 = \phi(1) = \phi(xx^{-1}) = \phi(x) \phi(x^{-1})$.

CHAPTER 14

Not one student in a thousand breaks down from overwork.

WILLIAM ALLAN NEILSON

1. Let $r_1 a$ and $r_2 a$ belong to $\langle a \rangle$. Then $r_1 a - r_2 a = (r_1 - r_2)a \in \langle a \rangle$. If $r \in R$ and $r_1 a \in \langle a \rangle$, then $r(r_1 a) = (rr_1)a \in \langle a \rangle$.

3. Clearly, I is not empty. Now observe that $(r_1 a_1 + \cdots + r_n a_n) - (s_1 a_1 + \cdots + s_n a_n) = (r_1 - s_1)a_1 + \cdots + (r_n - s_n)a_n \in I$. Also, if $r \in R$, then $r(r_1 a_1 + \cdots + r_n a_n) = (rr_1)a_1 + \cdots + (rr_n)a_n \in I$. That $I \subseteq J$ follows from closure under addition and multiplication by elements from R.

5. Let $a + bi, c + di \in S$. Then $(a + bi) - (c + di) = a - c + (b - d)i$ and $b - d$ is even. Also, $(a + bi)(c + di) = ac - bd + (ad + cb)i$ and $ad + cb$ is even. Finally, $(1 + 2i)(1 + i) = -1 + 3i \notin S$.

7. $ar_1 - ar_2 = a(r_1 - r_2)$; $(ar_1)r = a(r_1 r)$
$4R = \langle 8 \rangle$

9. If n is prime, use Euclid's Lemma (Chapter 0). If n is not prime, say $n = st$ where $s < n$ and $t < n$, then st belongs to nZ but s and t do not.

11. a. $a = 1$ **b.** $a = 3$ **c.** $a = \gcd(m, n)$

13. a. $a = 12$

b. $a = 48$. To see this, note that every element of $\langle 6 \rangle \langle 8 \rangle$ has the form $6t_1 8k_1 + 6t_2 8k_2 + \cdots + 6t_n 8k_n = 48s \in \langle 48 \rangle$. So, $\langle 6 \rangle \langle 8 \rangle \subseteq \langle 48 \rangle$. Also, since $48 \in \langle 6 \rangle \langle 8 \rangle$, we have $\langle 48 \rangle \subseteq \langle 6 \rangle \langle 8 \rangle$.

c. $a = mn$

15. Let $r \in R$. Then $r = 1r \in A$.

17. Let $u \in I$ be a unit and let $r \in R$. Then $r = r(u^{-1}u) = (ru^{-1})u \in I$.

19. Clearly, I is closed under subtraction. Also, if $b_1, b_2, b_3,$ and b_4 are even, then every entry of $\begin{bmatrix} a_1 & a_2 \\ a_3 & a_4 \end{bmatrix} \begin{bmatrix} b_1 & b_2 \\ b_3 & b_4 \end{bmatrix}$ is even.

21. Use the observation that every member of R can be written in the form

$\begin{bmatrix} 2q_1 + r_1 & 2q_2 + r_2 \\ 2q_3 + r_3 & 2q_4 + r_4 \end{bmatrix}$. Then note that $\begin{bmatrix} 2q_1 + r_1 & 2q_2 + r_2 \\ 2q_3 + r_3 & 2q_4 + r_4 \end{bmatrix} + I = \begin{bmatrix} r_1 & r_2 \\ r_3 & r_4 \end{bmatrix} + I$.

23. $(br_1 + a_1) - (br_2 + a_2) = b(r_1 - r_2) + (a_1 - a_2) \in B$; $r'(br + a) = b(r'r) + r'a \in B$.

25. Use Exercise 17.

27. Since every element of $\langle x \rangle$ has the form $xg(x)$, we have $\langle x \rangle \subseteq I$. If $f(x) \in I$, then $f(x) = a_n x^n + \cdots + a_1 x = x(a_n x^{n-1} + \cdots + a_1) \in \langle x \rangle$.

29. Suppose $f(x) + A \neq A$. Then $f(x) + A = f(0) + A$ and $f(0) \neq 0$. Thus,

$$(f(x) + A)^{-1} = \frac{1}{f(0)} + A.$$

This shows that R/A is a field. Now use Theorem 14.4.

31. Since $(3 + i)(3 - i) = 10$, $10 + \langle 3 + i \rangle = 0 + \langle 3 + i \rangle$. Also, $i + \langle 3 + i \rangle = -3 + \langle 3 + i \rangle = 7 + \langle 3 + i \rangle$. So, $Z[i]/\langle 3 + i \rangle = \{k + \langle 3 + i \rangle \mid k = 0, 1, \ldots, 9\}$, since $1 + \langle 3 + i \rangle$ has additive order 10.

33. Use Theorems 14.3 and 14.4.

35. Since every $f(x)$ in $\langle x, 2 \rangle$ has the form $f(x) = xg(x) + 2h(x)$, we have $f(0) = 2h(0)$, so that $f(x) \in I$. If $f(x) \in I$, then $f(x) = a_n x^n + \cdots + a_1 x + 2k = x(a_n x^{n-1} + \cdots + a_1) + 2k \in \langle x, 2 \rangle$. I is prime and maximal. $Z[x]/I$ has two elements.

37. $3x + 1 + I$

39. Every ideal is a subgroup. Every subgroup of a cyclic group is cyclic.

41. Say $b, c \in \text{Ann}(A)$. Then $(b - c)a = ba - ca = 0 - 0 = 0$. Also, $(rb)a = r(ba) = r \cdot 0 = 0$.

43. a. $\langle 3 \rangle$ **b.** $\langle 3 \rangle$ **c.** $\langle 3 \rangle$

45. Suppose $(x + N(\langle 0 \rangle))^n = 0 + N(\langle 0 \rangle)$. We must show that $x \in N(\langle 0 \rangle)$. We know that $x^n + N(\langle 0 \rangle) = 0 + N(\langle 0 \rangle)$, so that $x^n \in N(\langle 0 \rangle)$. Then, for some m, $(x^n)^m = 0$, and therefore $x \in N(\langle 0 \rangle)$.

47. The set $Z_2[x]/\langle x^2 + x + 1 \rangle$ has only four elements and each of the nonzero ones has a multiplicative inverse. For example,

$$(x + \langle x^2 + x + 1 \rangle)(x + 1 + \langle x^2 + x + 1 \rangle) = 1 + \langle x^2 + x + 1 \rangle.$$

49. $x + 2 + \langle x^2 + x + 1 \rangle$ is not zero, but its square is.

51. If f and $g \in A$, then $(f - g)(0) = f(0) - g(0)$ is even and $(f \cdot g)(0) = f(0) \cdot g(0)$ is even. $f(x) = \sqrt{2} \in R$ and $g(x) = 2 \in A$, but $f(x)g(x) \notin A$.

53. *Hint:* Any ideal of R/I has the form A/I, where A is an ideal of R.

55. Use the fact that R/I is an integral domain to show that $R/I = \{I, 1 + I\}$.

57. $\langle x \rangle \subset \langle x, 2^n \rangle \subset \langle x, 2^{n-1} \rangle \subset \cdots \subset \langle x, 2 \rangle$

59. Taking $r = 1$ and $s = 0$ shows that $a \in I$. Taking $r = 0$ and $s = 1$ shows that $b \in I$. If J is any ideal that contains a and b, then it contains I because of the closure conditions.

SUPPLEMENTARY EXERCISES FOR CHAPTERS 12–14

If at first you don't succeed, try, try, again. Then quit. There's no use being a damn fool about it. W. C. FIELDS

1. In Z_{10} they are 0, 1, 5, and 6. In Z_{20}, they are 0, 1, 5, and 16. In Z_{30}, they are 0, 1, 6, 10, 15, 16, 21, and 25.

3. We must show that $a^n = 0$ implies $a = 0$. First show this for the case when n is a power of 2. If n is not a power of 2, say 13, for example, note that $a^{13} = 0$ implies $a^{16} = 0$.

5. Suppose $A \not\subseteq C$ and $B \not\subseteq C$. Pick $a \in A$ and $b \in B$ such that $a, b \notin C$. But $ab \in C$ and C is prime.

7. $\{0\} \oplus \{0\}$, $\mathbf{R} \oplus \mathbf{R}$, $\mathbf{R} \oplus \{0\}$, and $\{0\} \oplus \mathbf{R}$. The ideals of $F \oplus F$ are $\{0\} \oplus \{0\}$, $F \oplus F$, $F \oplus \{0\}$, and $\{0\} \oplus F$.

9. Observe that $(Z \oplus Z)/A = \{(b, 0) \mid b = 0, 1, \ldots, p - 1\} \approx Z_p$, a field.

11. Suppose $a_1, a_2 \in A$ but $a_1 \notin B$ and $a_2 \notin C$. Use $a_1 + a_2$ to derive a contradiction.

13. Clearly $\langle a \rangle$ contains the right-hand side. Now show that the right-hand side contains a and is an ideal.

15. Since A is an ideal, $ab \in A$. Since B is an ideal, $ab \in B$. So $ab \in A \cap B = \{0\}$.

17. 6

19. Use Exercise 4.

21. Consider $x^2 + 1 + \langle x^4 + x^2 \rangle$.

23. Consider Z_8.

25. Say char $R = p$ (remember p must be prime). Then char $R/A =$ the additive order of $1 + A$. But $|1 + A|$ divides $|1| = p$.

27. Use Theorems 13.2, 14.3, and 14.4.

29. Observe that $A = \left\{ \begin{bmatrix} a & b \\ 0 & 0 \end{bmatrix} \middle| a, b \in Z_2 \right\}$ but $\begin{bmatrix} 1 & 1 \\ 1 & 1 \end{bmatrix} \begin{bmatrix} 1 & 0 \\ 0 & 0 \end{bmatrix} = \begin{bmatrix} 1 & 0 \\ 1 & 0 \end{bmatrix}$ is not in A.

31. $Z[i]/A$ has two elements. (From this it follows that A is maximal. See Theorem 14.4.)

33. A finite subset of a field is a subfield if it contains a nonzero element and is closed under addition and multiplication.

35. Observe that $(a + bi)(a - bi) = a^2 + b^2$.

37. 5

39. The inverse is $2x + 3$.

41. Observe that $Z_5[x, y]/\langle x, y \rangle \approx Z_5$ and use Theorem 14.4.

43. Say $(a,b)^n = (0, 0)$. Then $a^n = 0$ and $b^n = 0$. If $a^m = 0$ and $b^n = 0$, then $(a, b)^{mn} = ((a^m)^n, (b^n)^m) = (0, 0)$.

45. If $a^2 = a$, then $p^k | a(a - 1)$. Since a and $a - 1$ are relatively prime, $p^k | a$ or $p^k | (a - 1)$. So, $a = 0$ or $a = 1$.

47. In $Z_3[\sqrt{2}]$, $(a + b\sqrt{2})^{-1} = (a - b\sqrt{2})/(a^2 - 2b^2) = (a - b\sqrt{2})/(a^2 + b^2)$. In $Z_7[\sqrt{2}]$, $(1 + 2\sqrt{2})(1 + 5\sqrt{2}) = 0$.

CHAPTER 15

For every problem there is a solution which is simple, clean and wrong. H. L. MENCKEN

1. Part 3: $\phi(A)$ is a subgroup because ϕ is a group homomorphism. Let $s \in S$ and $\phi(r) = s$. Then $s\phi(a) = \phi(r)\phi(a) = \phi(ra)$ and $\phi(a)s = \phi(a)\phi(r) = \phi(ar)$.
Part 4: Let a and b belong to $\phi^{-1}(B)$ and r belong to R. Then $\phi(a)$ and $\phi(b)$ are in B. So, $\phi(a) - \phi(b) = \phi(a) + \phi(-b) = \phi(a - b) \in B$. Thus, $a - b \in B$. Also, $\phi(ra) = \phi(r)\phi(a) \in B$ and $\phi(ar) = \phi(a)\phi(r) \in B$. So, ra and $ar \in \phi^{-1}(B)$.

3. We already know $R/\text{Ker } \phi \approx \phi(R)$ as groups. Let $\Phi(x + \text{Ker } \phi) = \phi(x)$. Note that $\Phi((r + \text{Ker } \phi)(s + \text{Ker } \phi)) = \Phi(rs + \text{Ker } \phi) = \phi(rs) = \phi(r)\phi(s) = \Phi(r + \text{Ker } \phi) \Phi(s + \text{Ker } \phi)$.

5. $\phi(2 + 4) = \phi(1) = 5$, whereas $\phi(2) + \phi(4) = 0 + 0 = 0$.

7. a. No. Suppose $2 \to a$. Now consider $2 + 2$ and $2 \cdot 2$.
b. No.

9. If a and b ($b \neq 0$) belong to every member of the collection, then so do $a - b$ and ab^{-1}. Thus, by Exercise 23 in Chapter 13, the intersection is a subfield.

11. Apply the definition.

13. Multiplication is not preserved.

15. yes

17. The set of all polynomials passing through the point $(1, 0)$.

19. The group A/B is cyclic of order 4. The ring A/B has no unity.

21. The zero map and the identity map.

23. Use Exercise 22.

25. Say 1 is the unity of R. Let $s = \phi(r)$ be any element of S. Then $\phi(1)s = \phi(1)\phi(r) = \phi(1r) = \phi(r) = s$. Similarly, $s\phi(1) = s$.

27. Observe that an idempotent must map to an idempotent. It follows that $(a, b) \to a$, $(a, b) \to b$, and $(a, b) \to 0$ are the only ring homomorphisms.

29. Say $m = a_k a_{k-1} \cdots a_1 a_0$ and $n = b_k b_{k-1} \cdots b_1 b_0$. Then $m - n = (a_k - b_k)10^k + (a_{k-1} - b_{k-1})10^{k-1} + \cdots + (a_1 - b_1)10 + (a_0 - b_0)$. Now use the test for divisibility by 9.

31. Use the appropriate divisibility tests.

33. Mimic Example 8.

35. Observe that $(2 \cdot 10^{75} + 2) \bmod 3 = 1$ and $(10^{100} + 1 = 2) \bmod 3 = -1 \bmod 3$.

37. This follows directly from Theorem 13.3 and Theorem 10.1, part 3.

39. No. The kernel must be an ideal.

41. a. Suppose $ab \in \phi^{-1}(A)$. Then $\phi(a)\phi(b) \in A$, so that $a \in \phi^{-1}(A)$ or $b \in \phi^{-1}(A)$.
   b. Consider the natural homomorphism from R to S/A. Then use Theorems 15.3 and 14.4.

43. a. $\phi((a, b) + (a', b')) = \phi((a + a', b + b')) = a + a' = \phi((a, b)) + \phi((a', b'))$ so ϕ preserves addition. Also, $\phi((a, b)(a', b')) = \phi((aa', bb')) = aa' = \phi((a, b))\phi((a', b'))$.
   b. $\phi(a) = \phi(b)$ implies that $(a, 0) = (b, 0)$, which implies that $a = b$. $\phi(a + b) = (a + b, 0) = (a, 0) + (b, 0) = \phi(a) + \phi(b)$. Also, $\phi(ab) = (ab, 0) = (a, 0)(b, 0) = \phi(a)\phi(b)$.
   c. Use $(r, s) \to (s, r)$.

45. Observe that $x^4 = 1$ has two solutions in $\mathbf{R}$ but four in $\mathbf{C}$.

47. Use Exercises 40 and 46.

49. If $a/b = a'/b'$ and $c/d = c'/d'$, then $ab' = ba'$ and $cd' = dc'$. So, $acb'd' = (ab')(cd') = (ba')(dc') = bda'c'$. Thus, $ac/bd = a'c'/b'd'$ and therefore $(a/b)(c/d) = (a'/b')(c'/d')$.

51. First note that any field containing Z and i must contain $Q[i]$. Then prove that $(a + bi)/(c + di) \in Q[i]$.

53. The subfield of E is $\{ab^{-1} \mid a, b \in D, b \neq 0\}$.

55. Reflexive and symmetric properties follow from the commutativity of D. For transitivity, assume $a/b \equiv c/d$ and $c/d \equiv e/f$. Then $adf = (bc)f = b(cf) = bde$, and cancellation yields $af = be$.

57. Try $ab^{-1} \to a/b$.

59. The mapping $a + bi \to a - bi$ is a ring isomorphism of $\mathbf{C}$.

61. Certainly the unity 1 is contained in every subfield. So, if a field has characteristic p, the subfield $\{0, 1, \ldots, p - 1\}$ is contained in every subfield. If a field has characteristic 0, then $\{(m \cdot 1)(n \cdot 1)^{-1} \mid m, n \in Z, n \neq 0\}$ is a subfield contained in every subfield. This subfield is isomorphic to Q [map $(m \cdot 1)(n \cdot 1)^{-1}$ to m/n].

63. Observe that $(x + y)/1 = x/1 + y/1$ and $(xy)/1 = x/1\, y/1$.

CHAPTER 16

You know my methods. Apply them! SHERLOCK HOLMES, *The Hound of the Baskervilles*

1. $f + g = 3x^4 + 2x^3 + 2x + 2$
   $f \cdot g = 2x^7 + 3x^6 + x^5 + 2x^4 + 3x^2 + 2x + 2$
3. 1, 2, 4, 5
5. Write $f(x) = (x - a)q(x) + r(x)$. Since deg $(x - a) = 1$, deg $r(x) = 0$ or $r(x) = 0$. So $r(x)$ is a constant. Also, $f(a) = r(a)$.
7. Use Corollary 1 of Theorem 16.2.
9. Let $f(x), g(x) \in R[x]$. By inserting terms with the coefficient 0, we may write

$$f(x) = a_n x^n + \cdots + a_0 \quad \text{and} \quad g(x) = b_n x^n + \cdots + b_0.$$

Then

$$\overline{\phi}(f(x) + g(x)) = \phi(a_n + b_n)x^n + \cdots + \phi(a_0 + b_0)$$
$$= (\phi(a_n) + \phi(b_n))x^n + \cdots + \phi(a_0) + \phi(b_0)$$
$$= (\phi(a_n)x^n + \cdots + \phi(a_0)) + (\phi(b_n)x^n + \cdots + \phi(b_0))$$
$$= \overline{\phi}(f(x)) + \overline{\phi}(g(x)).$$

Multiplication is done similarly.
11. Quotient, $2x^2 + 2x + 1$; remainder, 2
13. It is its own inverse.
15. No. See Exercise 17.
17. If $f(x) = a_n x^n + \cdots + a_0$ and $g(x) = b_m x^m + \cdots + b_0$, then $f(x) \cdot g(x) = a_n b_m x^{m+n} + \cdots + a_0 b_0$.
19. Use Corollary 3 of Theorem 16.2.
21. If $f(x) \neq g(x)$, then $\deg[f(x) - g(x)] < \deg p(x)$. But the minimum degree of any member of $\langle p(x) \rangle$ is deg $p(x)$.
23. Start with $(x - 1/2)(x + 1/3)$ and clear fractions.
25. "Long divide" $x - a$ into $f(x)$ and induct on deg $f(x)$.
27. By Theorem 16.4, $I = \langle x - 1 \rangle$.
29. Use Corollary 2 of Theorem 15.5 and Exercise 9 in this chapter.
31. For any a in $U(p)$, $a^{p-1} = 1$, so every member of $U(p)$ is a zero of $x^{p-1} - 1$. Now use the Factor Theorem and a degree argument.
33. Use Exercise 32.
35. Observe that, modulo 101, $(50!)^2 = (50!)(-1)(-2) \cdots (-50) = (50!)(100)(99) \cdots (51) = 100!$ and use Exercise 32.
37. Take $R = Z$ and $I = \langle 2 \rangle$.
39. *Hint:* $F[x]$ is a PID. So $\langle f(x), g(x) \rangle = \langle a(x) \rangle$ for some $a(x) \in F[x]$. Thus $a(x)$ divides both $f(x)$ and $g(x)$. This means that $a(x)$ is a constant.
41. Write $f(x) = (x - a)g(x)$. Use the product rule to compute $f'(x)$.
43. Say deg $g(x) = m$, deg $h(x) = n$, and $g(x)$ has leading coefficient a. Let $k(x) = g(x) - ax^{m-n}h(x)$. Then deg $k(x) < $ deg $g(x)$ and $h(x)$ divides $k(x)$ in $Z[x]$ by induction. So, $h(x)$ divides $k(x) + ax^{m-n}h(x) = g(x)$ in $Z[x]$.
45. Consider the remainder when x^{43} is divided by $x^2 + x + 1$.
47. Observe that every term of $f(a)$ has the form $c_i a^i$ and $c_i a^i \bmod m = c_i b^i \bmod m$. To prove the second statement, assume that there is some integer k such that $f(k) = 0$.

If k is even, then because $k \bmod 2 = 0 \bmod 2$, we have by the first statement $0 = f(k)$ mod $2 = f(0)$ mod 2 so that $f(0)$ is even. This shows that k is not even. If k is odd, then $k \bmod 2 = 1 \bmod 2$, so by the first statement $f(k) = 0$ is odd. This contradiction completes the proof.

49. A solution to $x^{25} - 1 = 0$ in Z_{37} is a solution to $x^{25} = 1$ in $U(37)$. So, by Corollary 2 of Theorem 4.1, $|x|$ divides 25. Moreover, we must also have that $|x|$ divides $|U(37)| = 36$.

CHAPTER 17

> *Experience enables you to recognize a mistake when you make it again.*   FRANKLIN P. JONES

1. Use Theorem 17.1.

3. If $f(x)$ is not primitive, then $f(x) = ag(x)$, where a is an integer greater than 1. Then a is not a unit in $Z[x]$ and $f(x)$ is reducible.

5. a. If $f(x) = g(x)h(x)$, then $af(x) = ag(x)h(x)$.
    b. If $f(x) = g(x)h(x)$, then $f(ax) = g(ax)h(ax)$.
    c. If $f(x) = g(x)h(x)$, then $f(x + a) = g(x + a)h(x + a)$.
    d. Try $a = 1$.

7. Find an irreducible polynomial $p(x)$ of degree 2 over Z_5. Then $Z_5[x]/\langle p(x) \rangle$ is a field of order 25.

9. Note that -1 is a zero. No, since 4 is not a prime.

11. Let $f(x) = x^4 + 1$ and $g(x) = f(x + 1) = x^4 + 4x^3 + 6x^2 + 4x + 2$. Then $f(x)$ is irreducible over Q if $g(x)$ is. Eisenstein's Criterion shows that $g(x)$ is irreducible over Q. To see that $x^4 + 1$ is reducible over $\mathbf{R}$, observe that

$$x^8 - 1 = (x^4 + 1)(x^4 - 1)$$

so any complex zero of $x^4 + 1$ is a complex zero of $x^8 - 1$. Also note that the complex zeros of $x^4 + 1$ must have order 8 (when considered as an element of $\mathbf{C}$). Let $\omega = \sqrt{2}/2 + i\sqrt{2}/2$. Then Example 2 in Chapter 16 tells us that the complex zeros of $x^4 + 1$ are ω, ω^3, ω^5, and ω^7, so

$$x^4 + 1 = (x - \omega)(x - \omega^3)(x - \omega^5)(x - \omega^7).$$

But we may pair these factors up as

$$((x - \omega)(x - \omega^7))((x - \omega^3)(x - \omega^5)) = (x^2 - \sqrt{2}\,x + 1)(x^2 + \sqrt{2}\,x + 1)$$

to factor using reals (see DeMoivre's Theorem, Example 8 in Chapter 0).

13. $(x + 3)(x + 5)(x + 6)$

15. a. Consider the number of distinct expressions of the form $(x - c)(x - d)$.
    b. Reduce the problem to the case considered in part **a**.

17. Use Exercise 16, and imitate Example 10.

19. Map $Z_3[x]$ onto $Z_3[i]$ by $f(x) \to f(i)$. This is a ring homomorphism with kernel $\langle x^2 + 1 \rangle$.

21. $x^2 + 1$, $x^2 + x + 2$, $x^2 + 2x + 2$

23. 1 has multiplicity 1, 3 has multiplicity 2.

25. We know that $a_n(r/s)^n + a_{n-1}(r/s)^{n-1} + \cdots + a_0 = 0$. So $a_n r^n + s a_{n-1} r^{n-1} + \cdots + s^n a_0 = 0$. This shows that $s \mid a_n r^n$ and $r \mid s^n a_0$. Now use Euclid's Lemma and the fact that r and s are relatively prime.

27. Use induction and Corollary 2 of Theorem 17.5.

29. If there is an a in Z_p such that $a^2 = -1$, then $x^4 + 1 = (x^2 + a)(x^2 - a)$.
If there is an a in Z_p such that $a^2 = 2$, then $x^4 + 1 = (x^2 + ax + 1)(x^2 - ax + 1)$.
If there is an a in Z_p such that $a^2 = -2$, then $x^4 + 1 = (x^2 + ax - 1)(x^2 - ax - 1)$.
To show that one of these three cases must occur, consider the homomorphism from Z_p^* to itself given by $x \to x^2$. Since the kernel is $\{1, -1\}$, the image H has index 2 (we may assume that $p \neq 2$). Suppose that neither -1 nor 2 belongs to H. Then, since there is only one coset other than H, we have $-1H = 2H$. Also, since Z_p^*/H has order 2, any element in Z_p^*/H is the square of the identity coset H. Thus, $H = (-1H)(-1H) = (-1H)(2H) = -2H$, so that -2 is in H.

31. Since $(f + g)(a) = f(a) + g(a)$ and $(f \cdot g)(a) = f(a)g(a)$, the mapping is a homomorphism. Clearly, $p(x)$ belongs to the kernel. By Theorem 17.5, $\langle p(x) \rangle$ is a maximal ideal, so the kernel is $\langle p(x) \rangle$.

33. The mapping $a \to a + \langle p(x) \rangle$ is an isomorphism.

35. The analysis is identical except that $0 \leq q, r, t, u \leq n$. Now, just as when $n = 2$, we have $q = r = t = 1$, but this time $0 \leq u \leq n$. However, when $u > 2$, $P(x) = x(x + 1) \cdot (x^2 + x + 1)(x^2 - x + 1)^u$ has $(-u + 2)x^{2u+3}$ as one of its terms. Since the coefficient of x^{2u+3} represents the number of dice with the label $2u + 3$, the coefficient cannot be negative. Thus, $u \leq 2$, as before.

37. Although the probability of rolling any particular sum is the same with either pair of dice, the probability of rolling doubles is different (1/6 with ordinary dice, 1/9 with Sicherman dice). Thus the probability of going to jail is different. Other probabilities are also affected. For example, if in jail one cannot land on Virginia by rolling a pair of 2s with Sicherman dice, but one is twice as likely to land on St. James with a pair of 3s with the Sicherman dice as with ordinary dice.

CHAPTER 18

He thinks things through very carefully, before going off half-cocked. GENERAL CARL SPAATZ, in *Presidents Who Have Known Me,* GEORGE E. ALLEN

1. 1. $|a^2 - db^2| = 0$ implies $a^2 = db^2$. Thus $a = 0 = b$, since otherwise $d = 1$ or d is divisible by the square of a prime.
2. $N((a + b\sqrt{d})(a' + b'\sqrt{d})) = N(aa' + dbb' + (ab' + a'b)d) = |(aa' + dbb')^2 - d(ab' + a'b)^2| = |a^2a'^2 + d^2b^2b'^2 - da^2b'^2 - da'^2b^2| = |a^2 - db^2||a'^2 - db'^2| = N(a + b\sqrt{d}) N(a' + b'\sqrt{d})$.
3. If $xy = 1$, then $1 = N(1) = N(xy) = N(x)N(y)$ and $N(x) = 1 = N(y)$. If $N(a + b\sqrt{d}) = 1$, then $\pm 1 = a^2 - db^2 = (a + b\sqrt{d})(a - b\sqrt{d})$ and $a + b\sqrt{d}$ is a unit.
4. This property follows directly from properties 2 and 3.

3. Let $I = \cup I_i$. Let $a, b \in I$ and $r \in R$. Then $a \in I_i$ for some i and $b \in I_j$ for some j. Thus $a, b \in I_k$, where $k = \max\{i, j\}$. So, $a - b \in I_k \subseteq I$ and ra and $ar \in I_k \subseteq I$.

5. Clearly, $\langle ab \rangle \subseteq \langle b \rangle$. If $\langle ab \rangle = \langle b \rangle$, then $b = rab$, so that $1 = ra$ and a is a unit.

7. Say $x = a + bi$ and $y = c + di$. Then
$$xy = (ac - bd) + (bc + ad)i.$$

So
$$d(xy) = (ac - bd)^2 + (bc + ad)^2 = (ac)^2 + (bd)^2 + (bc)^2 + (ad)^2.$$

On the other hand,
$$d(x)d(y) = (a^2 + b^2)(c^2 + d^2) = a^2c^2 + b^2d^2 + b^2c^2 + a^2d^2.$$

9. Suppose $a = bu$, where u is a unit. Then $d(b) \le d(bu) = d(a)$. Also, $d(a) \le d(au^{-1}) = d(b)$.

11. $m = 0$ and $n = -1$ give $q = -i, r = -2 - 2i$.

13. $3 \cdot 7$ and $(1 + 2\sqrt{-5})(1 - 2\sqrt{-5})$. Mimic Example 8 to show that these are irreducible.

15. Observe that $10 = 2 \cdot 5$ and $10 = (2 - \sqrt{-6})(2 + \sqrt{-6})$ and mimic Example 8. A PID is a UFD.

17. Suppose $3 = \alpha\beta$, where $\alpha, \beta \in Z[i]$ and neither is a unit. Then $9 = d(3) = d(\alpha)d(\beta)$, so that $d(\alpha) = 3$. But there are no integers such that $a^2 + b^2 = 3$. Observe that $2 = -i(1 + i)^2$ and $5 = (1 + 2i)(1 - 2i)$.

19. Use Exercise 1 with $d = -1.5$ and $1 + 2i$; 13 and $3 + 2i$; 17 and $4 + i$.

21. Mimic Example 1.

23. $(-1+\sqrt{5})(1+\sqrt{5}) = 4 = 2 \cdot 2$. Now use Exercise 22.

25. Use the fact that x is a unit if and only if $N(x) = 1$.

27. See Example 3.

29. $(a_1a_2 \cdots a_{n-1})a_n \in \langle p \rangle$ implies $a_1a_2 \cdots a_{n-1} \in \langle p \rangle$ or $a_n \in \langle p \rangle$. Thus, by induction, p divides some a_i.

31. Use Exercise 10 and Theorem 14.4.

33. Suppose R satisfies the ascending chain condition and there is an ideal I of R that is not finitely generated. Then pick $a_1 \in I$. Since I is not finitely generated, $\langle a_1 \rangle$ is a proper subset of I, so we may choose $a_2 \in I$ but $a_2 \notin \langle a_1 \rangle$. As before, $\langle a_1, a_2 \rangle$ is proper, so we may choose $a_3 \in I$ but $a_3 \notin \langle a_1, a_2 \rangle$. Continuing in this fashion, we obtain a chain of infinite length $\langle a_1 \rangle \subset \langle a_1, a_2 \rangle \subset \langle a_1, a_2, a_3 \rangle \subset \cdots$.

Now suppose every ideal of R is finitely generated and there is a chain $I_1 \subset I_2 \subset I_3 \subset \cdots$. Let $I = \cup I_i$. Then $I = \langle a_1, a_2, \cdots, a_n \rangle$. Since $I = \cup I_i$, each a_i belongs to some member of the union, say $I_{i'}$. Letting $k = \max \{i' \mid i = 1, \ldots, n\}$, we see that all $a_i \in I_k$. Thus, $I \subseteq I_k$ and the chain has length at most k.

35. Say $I = \langle a + bi \rangle$. Then $a^2 + b^2 + I = (a + bi)(a - bi) + I = I$ and $a^2 + b^2 \in I$. For any $c, d \in Z$, let $c = q_1(a^2 + b^2) + r_1$ and $d = q_2(a^2 + b^2) + r_2$, where $0 \le r_1, r_2 < a^2 + b^2$. Then $c + di + I = r_1 + r_2i + I$.

37. $N(6 + 2\sqrt{-7}) = 64 = N(1 + 3\sqrt{-7})$. For the other part, use Exercise 25.

39. Theorem 18.1 shows that primes are irreducible. So, assume that a is an irreducible in a UFD R and that $a|bc$ in R. We must show that $a|b$ or $a|c$. Since $a|bc$ there is an element d in R such that $bc = ad$. Now replace b, c, and d by their factorizations as a product of irreducibles and use uniqueness.

SUPPLEMENTARY EXERCISES FOR CHAPTERS 15–18

> *Errors, like straws, upon the surface flow;*
> *He who would search for pearls must dive below.*
> JOHN DRYDEN

1. Use Theorem 15.3, Supplementary Exercise 8 for Chapters 12–14, Theorem 14.4, and Example 13 in Chapter 14.

3. To show the isomorphism, use the First Isomorphism Theorem.

5. Use the First Isomorphism Theorem.

7. Consider the obvious homomorphism from $Z[x]$ onto $Z_2[x]$. Then use the First Isomorphism Theorem and Theorem 14.3.

9. As in Example 7 in Chapter 6, the mapping is onto, is one-to-one, and preserves multiplication. Also, $a(x + y)a^{-1} = axa^{-1} + aya^{-1}$, so that it preserves addition as well.

11. $Z[i]/\langle 2 + i \rangle = \{0 + \langle 2 + i \rangle, 1 + \langle 2 + i \rangle, 2 + \langle 2 + i \rangle, 3 + \langle 2 + i \rangle, 4 + \langle 2 + i \rangle\}$. Note that

$$5 + \langle 2 + i \rangle = (2 + i)(2 - i) + \langle 2 + i \rangle = 0 + \langle 2 + i \rangle.$$

13. Observe that $(3 + 2\sqrt{2})(3 - 2\sqrt{2}) = 1$.

15. In Z_n we are given $(k + 1)^2 = k + 1$. So, $k^2 + 2k + 1 = k + 1$ or $k^2 = -k = n - k$. Also, $(n - k)^2 = n^2 - 2nk + k^2 = k^2$, so $(n - k)^2 = n - k$.

17. Observe that for any integer a, $a^2 \bmod 4 = 0$ or 1.

19. Use the Mod 2 Irreducibility Test.

21. Use Theorem 14.4. The factor ring has two elements.

23. Use Theorem 14.4.

25. Say $a/b, c/d \in R$. Then $(ad - bc)/(bd)$ and $ac/(bd) \in R$ by Euclid's Lemma. The field of quotients is Q.

27. $Z[i]/\langle 3 \rangle$ is a field and $Z_3 \oplus Z_3$ is not.

29. Consider the mapping from $R[x]$ to $(R/I)[x]$ given by $a_n x^n + \cdots + a_0 \rightarrow (a_n + I)x^n + \cdots + (a_0 + I)$.

31. Let $I = \langle 2 \rangle$. Then $Z_8[x]/I$ is isomorphic to $Z_2[x]$.

33. $\langle x, 3 \rangle$.

CHAPTER 19

> *When I was young I observed that nine out of every ten things*
> *I did were failures, so I did ten times more work.*
> GEORGE BERNARD SHAW

1. $\mathbf{R}^n$ has basis $\{(1, 0, \ldots, 0), (0, 1, 0, \ldots, 0), \ldots, (0, 0, \ldots, 1)\}$;

$$M_2(Q) \text{ has basis } \left\{ \begin{bmatrix} 1 & 0 \\ 0 & 0 \end{bmatrix}, \begin{bmatrix} 0 & 1 \\ 0 & 0 \end{bmatrix}, \begin{bmatrix} 0 & 0 \\ 1 & 0 \end{bmatrix}, \begin{bmatrix} 0 & 0 \\ 0 & 1 \end{bmatrix} \right\};$$

$Z_p[x]$ has basis $\{1, x, x^2, \ldots\}$; $\mathbf{C}$ has basis $\{1, i\}$.

3. $(a_2 x^2 + a_1 x + a_0) + (a_2' x^2 + a_1' x + a_0') = (a_2 + a_2')x^2 + (a_1 + a_1')x + (a_0 + a_0')$ and $a(a_2 x^2 + a_1 x + a_0) = aa_2 x^2 + aa_1 x + aa_0$. A basis is $\{1, x, x^2\}$. Yes.

5. Linearly dependent, since $-3(2, -1, 0) - (1, 2, 5) + (7, -1, 5) = (0, 0, 0)$.

7. Suppose $au + b(u + v) + c(u + v + w) = 0$. Then $(a + b + c)u + (b + c)v + cw = 0$. Since $\{u, v, w\}$ are linearly independent, we obtain $c = 0$, $b + c = 0$, and $a + b + c = 0$. So, $a = b = c = 0$.

9. If the set is linearly independent, it is a basis. If not, then delete one of the vectors that is a linear combination of the others (see Exercise 8). This new set still spans V. Repeat this process until you obtain a linearly independent subset. This subset will still span V, since you deleted only vectors that are linear combinations of the remaining ones.

11. Let u_1, u_2, u_3 be a basis for U and w_1, w_2, w_3 be a basis for W. Use the fact that $u_1, u_2, u_3, w_1, w_2, w_3$ are linearly dependent over F. In general, if $\dim U + \dim W > \dim V$, then $U \cap W \neq \{0\}$.

13. no

15. yes; 2

17.
$$\begin{bmatrix} a & a + b \\ a + b & b \end{bmatrix} + \begin{bmatrix} a' & a' + b' \\ a' + b' & b' \end{bmatrix} = \begin{bmatrix} a + a' & a + b + a' + b' \\ a + b + a' + b' & b + b' \end{bmatrix}$$

and $c \begin{bmatrix} a & a + b \\ a + b & b \end{bmatrix} = \begin{bmatrix} ac & ac + bc \\ ac + bc & bc \end{bmatrix}$.

19. Suppose B is a basis. Then every member of V is some linear combination of elements of B. If $a_1v_1 + \cdots + a_nv_n = a_1'v_1 + \cdots + a_n'v_n$, where $v_i \in B$, then $(a_1 - a_1')v_1 + \cdots + (a_n - a_n')v_n = 0$ and $a_i - a_i' = 0$ for all i. Conversely, if every member of V is a unique linear combination of elements of B, certainly B spans V. Also, if $a_1v_1 + \cdots + a_nv_n = 0$, then $a_1v_1 + \cdots + a_nv_n = 0v_1 + \cdots + 0v_n$ and $a_i = 0$ for all i.

21. Since $w_1 = a_1u_1 + a_2u_2 + \cdots + a_nu_n$ and $a_1 \neq 0$, we have $u_1 = a_1^{-1}(w_1 - a_2u_2 - \cdots - a_nu_n)$, and therefore $u_1 \in \langle w_1, u_2, \ldots, u_n \rangle$. Clearly, $u_2, \ldots, u_n \in \langle w_1, u_2, \ldots, u_n \rangle$. Hence every linear combination of $u_1, \ldots, u_n$ is in $\langle w_1, u_2, \ldots, u_n \rangle$.

23. $\{(1, 0, 1, 1), (0, 1, 0, 1)\}$.

25. Study the proof of Theorem 19.1.

27. If V and W are vector spaces over F, then the mapping must preserve addition and scalar multiplication. That is, $T: V \to W$ must satisfy $T(u + v) = T(u) + T(v)$ for all vectors u and v in V, and $T(au) = aT(u)$ for all vectors u in V and scalars a in F. A vector space isomorphism from V to W is a one-to-one linear transformation from V onto W.

29. Suppose v and u belong to the kernel and a is a scalar. Then $T(v + u) = T(v) + T(u) = 0 + 0 = 0$ and $T(av) = aT(u) = a \cdot 0 = 0$.

31. Let $\{v_1, v_2, \ldots, v_n\}$ be a basis for V. Map $a_1v_1 + a_2v_2 + \cdots + a_nv_n$ to $(a_1, a_2, \ldots, a_n)$.

CHAPTER 20

> *Well here's another clue for you all.*   JOHN LENNON AND PAUL
> MCCARTNEY, "Glass Onion," *The White Album*

1. Compare with Exercise 24 in the Supplementary Exercises for Chapters 12–14.

3. $Q(\sqrt{-3})$

5. $Q(\sqrt{-3})$

7. Note that $x = \sqrt{1 + \sqrt{5}}$ implies $x^4 - 2x^2 - 4 = 0$.

9. $a^5 = a^2 + a + 1$; $a^{-2} = a^2 + a + 1$; $a^{100} = a^2$

11. The set of all expressions of the form
$$(a_n \pi^n + a_{n-1} \pi^{n-1} + \cdots + a_0)/(b_m \pi^m + b_{m-1} \pi^{m-1} + \cdots + b_0),$$
where $b_m \neq 0$.

13. $x^7 - x = x(x^6 - 1) = x(x^3 + 1)(x^3 - 1) = x(x - 1)^3(x + 1)^3$; $x^{10} - x = x(x^9 - 1) = x(x - 1)^9$ (see Exercise 41 in Chapter 13).

15. *Hint:* Use Exercise 41 in Chapter 13.

17. $a = 4/3, b = 2/3, c = 5/6$

19. Use the fact that $1 + i = -(4 - i) + 5$ and $4 - i = 5 - (1 + i)$.

21. If the zeros of $f(x)$ are $a_1, a_2, \ldots, a_n$, then the zeros of $f(x + a)$ are $a_1 - a, a_2 - a, \ldots, a_n - a$. Now use Exercise 20.

23. Q and $Q(\sqrt{2})$

25. Let $F = Z_3[x]/\langle x^3 + 2x + 1 \rangle$ and denote the cosets $x + \langle x^3 + 2x + 1 \rangle$ by β and $2 + \langle x^3 + 2x + 1 \rangle$ by 2. Then $x^3 + 2x + 1 = (x - \beta)(x - \beta - 1)(x + 2\beta + 1)$.

27. Suppose that $\phi: Q(\sqrt{-3}) \to Q(\sqrt{3})$ is an isomorphism. Since $\phi(1) = 1$, we have $\phi(-3) = -3$. Then $-3 = \phi(-3) = \phi(\sqrt{-3}\sqrt{-3}) = (\phi(\sqrt{-3}))^2$. This is impossible, since $\phi(\sqrt{-3})$ is a real number.

29. Use long division.

31. Use Theorem 20.5.

33. Use Theorem 20.5.

CHAPTER 21

Work is the greatest thing in the world, so we should always save some of it for tomorrow. DON HERALD

1. It follows from Theorem 21.1 that if $p(x)$ and $q(x)$ are both monic irreducible polynomials in $F[x]$ with $p(a) = q(a) = 0$, then deg $p(x) = $ deg $q(x)$. If $p(x) \neq q(x)$, then $(p - q)(a) = p(a) - q(a) = 0$ and deg $(p(x) - q(x)) < $ deg $p(x)$, contradicting Theorem 21.1. To prove Theorem 21.3, use the Division Algorithm (Theorem 16.2).

3. Note that $[Q(\sqrt[n]{2}): Q] = n$ and use Theorem 21.5.

5. Use Exercise 4.

7. Suppose $Q(\sqrt{a}) = Q(\sqrt{b})$. If $\sqrt{b} \in Q$, then $\sqrt{a} \in Q$ and we may take $c = \sqrt{a}/\sqrt{b}$. If $\sqrt{b} \notin Q$, then $\sqrt{a} \notin Q$. Write $\sqrt{a} = r + s\sqrt{b}$. It follows that $r = 0$ and $a = bs^2$. The other direction follows from Exercise 20 in Chapter 20.

9. Observe that $[F(a):F]$ must divide $[E:F]$.

11. Note that $[F(a, b):F]$ is divisible by both $m = [F(a):F]$ and $n = [F(b):F]$ and $[F(a,b):F] \leq mn$.

13. Note that a is a zero of $x^3 - a^3$ over $F(a^3)[x]$. For the second part, take $F = Q, a = 1$; $F = Q, a = (-1 + i\sqrt{3})/2$; $F = Q, a = \sqrt[3]{2}$.

15. Suppose $E_1 \cap E_2 \neq F$. Then $[E_1:E_1 \cap E_2][E_1 \cap E_2:F] = [E_1:F]$ implies $[E_1:E_1 \cap E_2] = 1$, so that $E_1 = E_1 \cap E_2$. Similarly, $E_2 = E_1 \cap E_2$.

17. E must be an algebraic extension of $\mathbf{R}$, so that $E \subseteq \mathbf{C}$. But then $[\mathbf{C}:E][E:\mathbf{R}] = [\mathbf{C}:\mathbf{R}] = 2$.

19. Let a be a zero of $p(x)$ in some extension of F. First note $[E(a):E] \leq [F(a):F] = \deg p(x)$. Then observe that $[E(a):F(a)][F(a):F] = [E(a):E][E:F]$. This implies that $\deg p(x)$ divides $[E(a):E]$, so that $\deg p(x) = [E(a):E]$.

21. *Hint:* If $\alpha + \beta$ and $\alpha\beta$ are algebraic, then so is $\sqrt{(\alpha + \beta)^2 - 4\alpha\beta}$.

23. $\sqrt{b^2 - 4ac}$

25. Use the Factor Theorem.

27. Say a is a generator of F^*. Then $F = Z_p(a)$, and it suffices to show that a is algebraic over Z_p. If $a \in Z_p$, we are done. Otherwise, $1 + a = a^k$ for some $k \neq 0$. If $k > 0$, we are done. If $k < 0$, then $a^{-k} + a^{1-k} = 1$ and we are done.

29. If $[K:F] = n$, then there are elements $v_1, v_2, \ldots, v_n$ in K that constitute a basis for K over F. The mapping $a_1v_1 + \ldots + a_nv_n \to (a_1, \ldots, a_n)$ is a vector space isomorphism from K to F^n. If K is isomorphic to F^n, then the n elements in K corresponding to $(1, 0, \ldots, 0), (0, 1, \ldots, 0), \ldots, (0, 0, \ldots, 1)$ in F^n constitute a basis for K over F.

31. Observe that $[F(a,b) : F(a)] = [F(a)(b) : F(a)] \leq [F(b) : F] \leq [F(a)(b) : F(b)] [F(b) : F] = [F(a)(b) : F] = [F(a,b) : F]$.

33. Mimic Example 5.

35. Mimic Example 6.

CHAPTER 22

Tell me tell me tell me come on tell me the answer. JOHN LENNON AND PAUL MCCARTNEY, "Helter Skelter," *The White Album*

1. $[GF(729):GF(9)] = 3$
$[GF(64):GF(8)] = 2$

3. Use Theorem 22.2.

5. The only possibilities for $f(x)$ are $x^3 + x + 1$ and $x^3 + x^2 + 1$. See Exercise 8 in Chapter 20 for the first case. For the second case, let a be a zero of $x^3 + x^2 + 1$ in some extension of Z_2. Then use the fact that $a^3 = a^2 + 1$ to test the elements of $Z_2(a)^* = \{1, a, a^2, \ldots, a^6\}$ for zeros of $f(x)$.

7. Use Exercise 38 in Chapter 15 and Corollary 4 of Lagrange's Theorem (Theorem 7.1).

9. Use the fact that if $g(x)$ is an irreducible factor of $x^8 - x$ over Z_2 and $\deg g(x) = m$, then the field $Z_2[x]/\langle g(x)\rangle$ has order 2^m and is a subfield of $GF(8)$. Now use Theorem 22.3.

11. Direct calculations show that given $x^3 + 2x + 1 = 0$, we have $x^2 \neq 1$ and $x^{13} \neq 1$.

13. Direct calculations show that $x^{13} = 1$, whereas $(2x)^2 \neq 1$ and $(2x)^{13} \neq 1$. Thus, $2x$ is a generator.

15. First observe that for any field F the set F^* is a group under multiplication. Now use Theorem 22.2 and Theorem 4.3.

17. Find a quadratic irreducible polynomial $p(x)$ over Z_3; then $Z_3[x]/\langle p(x)\rangle$ is a field of order 9.

19. Let $a, b \in K$. Then, by Exercise 41b in Chapter 13, $(a - b)^{p^m} = a^{p^m} - b^{p^m} = a - b$. Also, $(ab)^{p^m} = a^{p^m} b^{p^m} = ab$. So, K is a subfield.

21. Consider $x^{p^n-1} - 1$ and use Corollary 3 of Lagrange's Theorem (Theorem 7.1).

23. identical

25. Consider $g(x) = x^2 - a$. Note that $|GF(p)[x]/\langle g(x)\rangle| = p^2$, so that $g(x)$ has a zero in $GF(p^2)$. Now use Theorem 22.3.

27. Use Exercise 7.

29. Since F^* is a cyclic group of order 124, it has a unique element of order 2.

31. Use Exercise 41 in Chapter 13.

33. Consider the field of quotients of $Z_p[x]$. The polynomial $f(x) = x$ is not the image of any element.

CHAPTER 23

> *Why, sometimes I've believed as many as six impossible things before breakfast.*   LEWIS CARROLL

1.

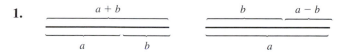

3. Let y denote the length of the hypotenuse of the right triangle with base 1 and let x denote the length of the hypotenuse of the right triangle with base $|c|$. Then $y^2 = 1 + d^2$, $y^2 + x^2 = (1 + |c|)^2$, and $|c|^2 + d^2 = x^2$. So, $1 + 2|c| + |c|^2 = 1 + d^2 + |c|^2 + d^2$, which simplifies to $|c| = d^2$.

5. Use $\sin^2 \theta + \cos^2 \theta = 1$.

7. Use $\cos 2\theta = 2 \cos^2 \theta - 1$.

9. Use $\sin(\alpha - \beta) = \sin \alpha \cos \beta - \cos \alpha \sin \beta$.

11. Solving two linear equations with coefficients from F involves only the operations of F.

13. Use Theorem 17.1 and Exercise 25 in Chapter 17.

15. If so, then an angle of $40°$ is constructible. Now use Exercise 10.

17. This amounts to showing that $\sqrt{\pi}$ is not constructible. But if $\sqrt{\pi}$ is constructible, so is π. However, $[Q(\pi) : Q]$ is infinite.

19. No, since $[Q(\sqrt[3]{3}):Q] = 3$.

21. No, since $[Q(\sqrt[3]{\pi}):Q]$ is infinite.

SUPPLEMENTARY EXERCISES FOR CHAPTERS 19–23

> *The things taught in colleges and schools are not an education, but the means of education.*   RALPH WALDO EMERSON, *Journals*

1. Use Theorem 20.5.

3. Suppose b is one solution of $x^n = a$. Since F^* is a cyclic group of order $q - 1$, it has a cyclic subgroup of order n, say $\langle c \rangle$. Then each member of $\langle c \rangle$ is a solution to the equation $x^n = 1$. It follows that $b\langle c \rangle$ is the solution set of $x^n = a$.

5. $(5a^2 + 2)/a = 5a + 2a^{-1}$. Now observe that since $a^2 + a + 1 = 0$, we know that $a(-a - 1) = 1$, and so $a^{-1} = -a - 1$. Thus, $(5a^2 + 2)/a = -2 + 3a$.

7. 5

9. Since $F(a) = F(a^{-1})$, we have degree of $a = [F(a):F] = [F(a^{-1}):F] = $ degree of a^{-1}.

11. If ab is a zero of $c_n x^n + \cdots + c_1 x + c_0 \in F[x]$, then a is a zero of $c_n b^n x^n + \cdots + c_1 bx + c_0 \in F(b)[x]$.

13. Every element of $F(a)$ can be written in the form $f(a)/g(a)$, where $f(x), g(x) \in F[x]$. If $f(a)/g(a)$ is algebraic and not in F, then there is some $h(x) \in F[x]$ such that $h(f(a)/g(a)) = 0$. By clearing fractions and collecting like powers of a, we obtain a polynomial in a with coefficients from F equal to 0. But then a would be algebraic over F.

15. Use Corollary 2 to Theorem 22.2.

17. If the basis elements commute, then so would any combination of basis elements. However, the entire space is not commutative.

19. $\{x, x^2, x^3\}$

21. Use Exercise 41 in Chapter 13.

23. By Theorem 20.5, the zeros of $x^n - a$ are distinct, say $\alpha_1, \alpha_2, \ldots, \alpha_n$. Then $\beta_i = \alpha_1/\alpha_i$ for $i = 1, 2, \ldots, n$ are all the nth roots of unity.

CHAPTER 24

Difficulty, my brethren, is the nurse of greatness.
WILLIAM CULLEN BRYANT

1. $a = eae^{-1}$; $cac^{-1} = b$ implies $a = c^{-1}bc = c^{-1}b(c^{-1})^{-1}$; $a = xbx^{-1}$ and $b = ycy^{-1}$ imply $a = xycy^{-1}x^{-1} = xyc(xy)^{-1}$.

3. Observe that $T(xC(a)) = xax^{-1} = yay^{-1} = T(yC(a))$ if and only if $y^{-1}xa = ay^{-1}x$ if and only if $y^{-1}x \in C(a)$ if and only if $yC(a) = xC(a)$. This proves that T is well defined and one-to-one. T is onto by definition.

5. By way of contradiction, assume that H is the only Sylow 2-subgroup of G and that K is the only Sylow 3-subgroup of G. Then H and K are normal and Abelian (Corollary to Theorem 24.5 and corollary to Theorem 24.2). So, $G = H \times K \approx H \oplus K$ and, from Exercise 4 in Chapter 8, G is Abelian.

7. Use Exercise 7 in Supplementary Exercises for Chapters 5–8.

9. Use Exercise 51 in Chapter 9 and Exercise 7 of the Supplementary Exercises for Chapters 5–8.

11. 8

13. 15

15. By Exercise 14, G has seven subgroups of order 3.

17. 10; $\langle(123)\rangle, \langle(234)\rangle, \langle(134)\rangle, \langle(345)\rangle, \langle(245)\rangle$

21. 21

23. Sylow's Third Theorem implies that the Sylow 3- and Sylow 5-subgroups are unique. Pick any x not in the union of these. Then $|x| = 15$.

25. By Sylow, $n_{17} = 1$ or 35. Assume $n_{17} = 35$. Then the union of the Sylow 17-subgroups has 561 elements. By Sylow, $n_5 = 1$. Thus, we may form a cyclic subgroup of order 85 (Exercise 51 in Chapter 9 and Theorem 24.6). But then there are 64 elements of order 85. This gives too many elements.

27. Use the "G/Z Theorem" (Theorem 9.3).

29. Let H be the Sylow 3-subgroup and suppose that the Sylow 5-subgroups are not normal. By Sylow, there must be six Sylow 5-subgroups, call them $K_1, \ldots, K_6$. These subgroups have 24 elements of order 5. Also, the cyclic subgroups $HK_1, \ldots, HK_6$ each have eight generators. Thus, there are 48 elements of order 15.

31. Mimic the proof of Sylow's First Theorem.

33. Pick $x \in Z(G)$ such that $|x| = p$. If $x \in H$, by induction, $N(H/\langle x \rangle) > H/\langle x \rangle$, say $y\langle x \rangle \in N(H/\langle x \rangle)$ but not $H/\langle x \rangle$. Now show $y \in N(H)$ but not H. If $x \notin H$, then $x \in N(H)$, so that $N(H) > H$.

35. Automorphisms preserve order.

37. Since 3 divides $|N(K)|$ we know that $N(K)$ has subgroup H_1 of order 3. Now use the fact that H_1K is cyclic group of order 15 and Exercise 23 in the Supplementary Exercises for Chapters 1–4.

39. Normality of H implies $\text{cl}(h) \subseteq H$ for h in H. Now observe that $h \in \text{cl}(h)$. This is true only when H is normal.

41. The mapping from H to xHx^{-1} given by $h \to xhx^{-1}$ is an isomorphism.

43. Say $\text{cl}(x) = \{x, g_1xg_1^{-1}, g_2xg_2^{-1}, \ldots, g_kxg_k^{-1}\}$. If $x^{-1} = g_ixg_i^{-1}$, then for each $g_jxg_j^{-1}$ in $\text{cl}(x)$ we have $(g_jxg_j^{-1})^{-1} = g_jx^{-1}g_j^{-1} = g_j(g_ixg_i^{-1})g_j^{-1} \in \text{cl}(x)$. Because $|G|$ has odd order, $g_jxg_j^{-1} \neq (g_jxg_j^{-1})^{-1}$. It follows that $|\text{cl}(x)|$ is even. But $|\text{cl}(x)|$ divides $|G|$.

45. Mimic Example 4.

47. Say $\text{cl}(e)$ and $\text{cl}(a)$ are the only two conjugacy classes of a group G of order n. Then $\text{cl}(a)$ has $n - 1$ elements all of the same order, say m. If $m = 2$, then it follows from Exercise 33 in Chapter 2 that G is Abelian. But then $\text{cl}(a) = \{a\}$ and so $n = 2$. If $m > 2$, then $\text{cl}(a)$ has at most $n - 2$ elements since conjugation of a by e, a, and a^2 each yields a.

49. $|C(a)|/|G|$

51. $\text{Pr}(D_4) = 5/8$, $\text{Pr}(S_3) = 1/2$, $\text{Pr}(A_4) = 1/3$

53. Exactly as in the case for a group, we have for a ring $R = \{x_1, x_2, \ldots, x_n\}$, $\text{Pr}(R) = |K|/n^2$, where $K = \{(x, y) | xy = yx, x, y \in R\}$. Also, $|K| = |C(x_1)| + |C(x_2)| + \cdots + |C(x_n)|$. From Exercise 28 in the Supplemental Exercises for Chapters 12–14, we know that $R/C(R)$ is not cyclic. Thus, $|R/C(R)| \geq 4$ and so $|C(R)| \leq |R|/4$. So, for at least 3/4 of the elements x of R, we have $|C(x)| \leq |R|/2$. Then starting with the elements in the center and proceeding to the elements not in the center we have $|K| \leq |R|/4 + (1/2)(3/4)|R| = (5/8)|R|$.

CHAPTER 25

Learn to reason forward and backward on both sides of a question. THOMAS BLANDI

1. Use the $2 \cdot$ odd test.

3. Use the Index Theorem.

5. Suppose G is a simple group of order 525. Let L_7 be a Sylow 7-subgroup of G. It follows from Sylow's theorems that $|N(L_7)| = 35$. Let L be a subgroup of $N(L_7)$ of order 5. Since $N(L_7)$ is cyclic (Theorem 24.6), $N(L) \geq N(L_7)$, so that 35 divides $|N(L)|$. But L is contained in a Sylow 5-subgroup (Theorem 24.4), which is Abelian (see the Corollary to Theorem 24.2). Thus, 25 divides $|N(L)|$ as well. It follows that 175 divides $|N(L)|$. The Index Theorem now yields a contradiction.

7. $n_{11} = 12$. Use the N/C Theorem (Example 15 in Chapter 10) to show that there is an element of order 22, then use the Embedding Theorem and observe that A_{12} has no element of order 22.

9. Use the fact that A_{12} has no element of order 33.

11. If we can find a pair of distinct Sylow 2-subgroups A and B such that $|A \cap B| = 8$, then $N(A \cap B) \geq AB$, so that $N(A \cap B) = G$. Now let H and K be any distinct pair of Sylow 2-subgroups. Then $16 \cdot 16/|H \cap K| = |HK| \leq 112$ (Supplementary Exercise 7

for Chapters 5–8), so that $|H \cap K|$ is at least 4. If $|H \cap K| = 8$, we are done. So, assume $|H \cap K| = 4$. Then $N(H \cap K)$ picks up at least 8 elements from H and at least 8 from K (see Exercise 33 in Chapter 24). Thus, $|N(H \cap K)| \geq 16$ and is divisible by 8. So, $|N(H \cap K)| = 16, 56$, or 112. Since the latter two cases yield normal subgroups, we may assume $|N(H \cap K)| = 16$. If $N(H \cap K) = H$, then $|H \cap K| = 8$, since $N(H \cap K)$ contains at least 8 elements from K. So, we may assume that $N(H \cap K) \neq H$. Then, we may take $A = N(H \cap K)$ and $B = H$.

15. Use the Index Theorem.

17. $n_5 = 6$ and $n_3 = 10$ or 40. If there are two Sylow 2-subgroups L_2 and $L_2{'}$ whose intersection has order 4, show that $N(L_2 \cap L_2{'})$ has index at most 5. Now use the Embedding Theorem. If $n_3 = 40$, the union of all the Sylow subgroups has more than 120 elements. If $n_3 = 10$, use the *N/C* Theorem to show that there is an element of order 6 and then use the Embedding Theorem and observe that A_6 has no element of order 6.

19. Let α be as in the proof of the Generalized Cayley Theorem. Then Ker $\alpha \leq H$ and $|G/\text{Ker } \alpha|$ divides $|G:H|!$ Now show $|\text{Ker } \alpha| = |H|$. A subgroup of index 2 is normal.

21. If H is a proper normal subgroup of S_5, then $H \cap A_5 = A_5$ or $\{\varepsilon\}$. But $H \cap A_5 = A_5$ implies $H = A_5$, whereas $H \cap A_5 = \{\varepsilon\}$ implies $H = \{\varepsilon\}$ or $|H| = 2$. (See Exercise 19 in Chapter 5.) Now use Exercise 66 in Chapter 9 and Exercise 46 in Chapter 5.

23. By direct computation, show that $PSL(2, Z_7)$ has more than four Sylow 3-subgroups, more than one Sylow 7-subgroup, and more than one Sylow 2-subgroup. *Hint:* Observe that $\begin{bmatrix} 1 & 4 \\ 1 & 5 \end{bmatrix}$ has order 3. Now use conjugation to find four other subgroups of order 3; observe that $\left|\begin{bmatrix} 5 & 5 \\ 1 & 4 \end{bmatrix}\right| = 7$ and use conjugation to find another subgroup of order 7; observe that $\left|\begin{bmatrix} 5 & 1 \\ 3 & 5 \end{bmatrix}\right| = 4$ and use conjugation to find six more elements of order 4 (which guarantees that more than one Sylow 2-subgroup exists). Now argue as we did to show that A_5 is simple. In the cases that the supposed normal subgroup N has order 2 or 4, show that in G/N, the Sylow 7-subgroup is normal. But then, G has a normal subgroup of order 14 or 28, which were already ruled out.

25. Mimic Exercise 24.

27. Suppose there is a simple group of order 60 that is not isomorphic to A_5. The Index Theorem implies $n_2 \neq 1$ or 3, and the Embedding Theorem implies $n_2 \neq 5$. Thus, $n_2 = 15$. Counting shows that there must be two Sylow 2-subgroups whose intersection has order 2. Now mimic the argument used in showing that there is no simple group of order 144 to show that the normalizer of this intersection has index 5, 3, or 1, but the Embedding Theorem and the Index Theorem rule these out.

CHAPTER 26

If you make a mistake, make amends.   LOU HOLTZ

1. $u \sim u$ because u is obtained from itself by no insertions; if v can be obtained from u by inserting or deleting words of the form xx^{-1} or $x^{-1}x$, then u can be obtained from v by reversing the procedure; if u can be obtained from v and v can be obtained from w, then u can be obtained from w by first obtaining v from w then u from v.

3. $b(a^2N) = b(aN)a = a^3bNa = a^3b(aN) = a^3a^3bN$
$\qquad = a^6bN = a^6Nb = a^2Nb = a^2bN$
$\quad b(a^3N) = b(a^2N)a = a^2bNa = a^2b(aN) = a^2a^3bN$
$\qquad = a^5bN = a^5Nb = aNb = abN$
$\quad b(bN) = b^2N = N$
$\quad b(abN) = baNb = a^3bNb = a^3b^2N = a^3N$
$\quad b(a^2bN) = ba^2Nb = a^2bNb = a^2b^2N = a^2N$
$\quad b(a^3bN) = ba^3Nb = abNb = ab^2N = aN$

5. Let F be the free group on $\{a_1, a_2, \ldots, a_n\}$. Let N be the smallest normal group containing $\{w_1, w_2, \ldots, w_t\}$ and let M be the smallest normal subgroup containing $\{w_1, w_2, \ldots, w_t, w_{t+1}, \ldots, w_{t+k}\}$. Then $F/N \approx G$ and $F/M \approx \overline{G}$. The homomorphism from F/N to F/M given by $aN \rightarrow aM$ induces a homomorphism from G onto $\overline{G}$.

   To prove the corollary, observe that the theorem shows that K is a homomorphic image of G, so that $|K| \le |G|$.

7. Clearly, a and ab belong to $\langle a, b \rangle$, so $\langle a, ab \rangle \subseteq \langle a, b \rangle$. Now show that a and b belong to $\langle a, ab \rangle$.

9. Show that $|G| \le 2n$ and that D_n satisfies the relations that define G.

11. Since $x^2 = y^2 = e$, we have $(xy)^{-1} = y^{-1}x^{-1} = yx$. Also, $xy = z^{-1}yz$, so that $(xy)^{-1} = (z^{-1}yz)^{-1} = z^{-1}y^{-1}z = z^{-1}yz = xy$.

13. **a.** b^6 **b.** b^7a

15. Center is $\langle x^2 \rangle$. $|xy| = 8$.

17. Use the fact that the mapping from G onto G/N given by $x \rightarrow xN$ is a homomorphism.

19. For H to be a normal subgroup we must have $yxy^{-1} \in H = \{e, y^3, y^6, y^9, x, xy^3, xy^6, xy^9\}$. But $yxy^{-1} = yxy^{11} = (yxy)y^{10} = xy^{10}$.

21. 6; the given relations imply that $a^2 = e$. G is isomorphic to Z_6.

23. 1, 2, and ∞

25. $ab = c \Rightarrow abc^{-1} = e$
$cd = a \Rightarrow (abc^{-1})cd = ae \Rightarrow bd = e \Rightarrow d = b^{-1}$
$da = b \Rightarrow bda = b^2 \Rightarrow ea = b^2 \Rightarrow a = b^2$
$ab = c \Rightarrow b^3 = c$
So $G = \langle b \rangle$.
$bc = d \Rightarrow bb^3 = b^{-1} \Rightarrow b^5 = e$. So $|G| = 1$ or 5.
But Z_5 satisfies the defining relations with $a = 1, b = 3, c = 4$, and $d = 2$.

27. D_4

CHAPTER 27

> *If at first you don't succeed—that makes you about average.*
> BRADENTON, *Florida Herald*

1. If T is a distance-preserving function and the distance between points a and b is positive, then the distance between $T(a)$ and $T(b)$ is positive.

3. See Figure 1.5.

5. 12

7. $4n$

9. **a.** Z_2
   b. $Z_2 \oplus Z_2$
   c. $G \oplus Z_2$, where G is the plane symmetry group of a circle.

11. 6
13. An inversion in $\mathbf{R}^3$ leaves only a single point fixed, while a rotation leaves a line fixed.
15. In $\mathbf{R}^4$, a plane is fixed. In $\mathbf{R}^n$, a hyperplane of dimension $n - 2$ is fixed.
17. Let T be an isometry; p, q, and r the three noncollinear points; and s any other point in the plane. Then the quadrilateral determined by $T(p)$, $T(q)$, $T(r)$, and $T(s)$ is congruent to the one formed by p, q, r, and s. Thus, $T(s)$ is uniquely determined by $T(p)$, $T(q)$, and $T(r)$.
19. a rotation
21. The point of intersection of the perpendicular bisector of the line joining p and p' and the perpendicular bisector of the line joining q and q'.

CHAPTER 28

The thing that counts is not what we know but the ability to use what we know.   LEO L. SPEARS

1. Try $x^n y^m \rightarrow (n, m)$.
3. xy
5. Use Figure 28.9.
7. $x^2 yzxz = x^2 yx^{-1} = x^2 x^{-1} y = xy$
$x^{-3} zxyz = x^{-3} x^{-1} y = x^{-4} y$
9. A subgroup of index 2 is normal.
11. a. V   **b.** I   **c.** II   **d.** VI   **e.** VII   **f.** III
13. *cmm*
15. a. *p4m*   **b.** *p3*   **c.** *p31m*   **d.** *p6m*
17. The principal purpose of tire tread design is to carry water away from the tire. Patterns I and III do not have horizontal reflective symmetry. Thus these designs would not carry water away equally on both halves of the tire.
19. a. VI   **b.** V   **c.** I   **d.** III   **e.** IV   **f.** VII   **g.** IV

CHAPTER 29

With every mistake we must surely be learning.   GEORGE HARRISON, "While My Guitar Gently Weeps," *The White Album*

1. 6
3. 30
5. 13
7. 45
9. 126
11. 29
13. For the first part, see Exercise 11 in Chapter 6. For the second part, try D_4.
15. The kernel is $\{R_0, R_{180}, H, V\}$.

CHAPTER 30

> *I am not bound to please thee with my answers.*
> SHAKESPEARE, *The Merchant of Venice*

1. $4 * (b, a)$

3. $(m/2) * \{3 * [(a, 0), (b, 0)], (a, 0), (e, 1), 3 * (a, 0), (b, 0), 3 * (a, 0), (e, 1)\}$

5. $a^3 b$

7. Both yield paths from e to $a^3 b$.

11. Say we start at x. Then we know the vertices $x, xs_1, xs_1 s_2, \ldots, xs_1 s_2 \cdots s_{n-1}$ are distinct and $x = xs_1 s_2 \cdots s_n$. So if we apply the same sequence beginning at y, then cancellation shows that $y, ys_1, ys_1 s_2, \ldots, ys_1 s_2 \cdots s_{n-1}$ are distinct and $y = ys_1 s_2 \cdots s_n$.

13. If there were a Hamiltonian path from $(0, 0)$ to $(2, 0)$, there would be a Hamiltonian circuit in the digraph, since $(2, 0) + (1, 0) = (0, 0)$.

15. a. If $s_1, s_2, \ldots, s_{n-1}$ traces a Hamiltonian path and $s_i s_{i+1} \cdots s_j = e$, then the vertex $s_1 s_2 \cdots s_{i-1}$ appears twice. Conversely, if $s_i s_{i+1} \cdots s_j \neq e$, then the sequence $e, s_1, s_1 s_2, \ldots, s_1 s_2 \cdots s_{n-1}$ yields the n vertices (otherwise, cancellation gives a contradiction).

b. This follows directly from part **a**.

17. The sequence traces the digraph in a clockwise fashion.

19. Abbreviate $(a, 0)$, $(b, 0)$, and $(e, 1)$ by a, b, and 1, respectively. A circuit is $4 * (4 * 1, a), 3 * a, b, 7 * a, 1, b, 3 * a, b, 6 * a, 1, a, b, 3 * a, b, 5 * a, 1, a, a, b, 3 * a, b, 4 * a, 1, 3 * a, b, 3 * a, b, 3 * a, b$.

21. Abbreviate $(R_{90}, 0)$, $(H, 0)$, and $(R_0, 1)$ by R, H, and 1, respectively. A circuit is $3 * (R, 1, 1), H, 2 * (1, R, R), R, 1, R, R, 1, H, 1, 1$.

23. Abbreviate $(a, 0)$, $(b, 0)$, and $(e, 1)$ by a, b, and 1, respectively. A circuit is $2 * (1, 1, a), a, b, 3 * a, 1, b, b, a, b, b, 1, 3 * a, b, a, a$.

25. Abbreviate $(r, 0)$, $(f, 0)$, and $(e, 1)$ by r, f, and 1, respectively. Then the sequence is $r, r, f, r, r, 1, f, r, r, f, r, 1, r, f, r, r, f, 1, r, r, f, r, r, 1, f, r, r, f, r, 1, r, f, r, r, f, 1$.

27. $m * ((n - 1) * (0, 1), (1, 1))$

29. Abbreviate $(r, 0)$, $(f, 0)$, and $(e, 1)$ by r, f, and 1, respectively. A circuit is $1, r, 1, 1, f, r, 1, r, 1, r, f, 1$.

31. $2 * [3 * (1, 0), (0, 1)], 4 * (1, 0), (0, 1), 2 * [3 * (1, 0), (0, 1)]$

33. $12 * ((1, 0), (0, 1))$.

35. In the proof of Theorem 30.3, we used the hypothesis that G is Abelian in two places: We needed H to satisfy the induction hypothesis, and we needed to form the factor group G/H. Now, if we assume only that G is Hamiltonian, then H also is Hamiltonian and G/H exists.

CHAPTER 31

> *We must view with profound respect the infinite capacity of the human mind to resist the introduction of useful knowledge.* THOMAS R. LOUNSBURY

1. wt(0001011) = 3; wt(0010111) = 4; wt(0100101) = 3; etc.

3. 1000110; 1110100

5. 000000, 100011, 010101, 001110, 110110, 101101, 011011, 111000

7. Not all single errors can be detected.

9. Observe that a vector has even weight if and only if it can be written as a sum of an even number of vectors of weight 1.

11. No, by Theorem 31.3.

13. 0000000, 1000111, 0100101, 0010110, 0001011, 1100010, 1010001, 1001100, 0110011, 0101110, 0011101, 1110100, 1101001, 1011010, 0111000, 1111111;

$$H = \begin{bmatrix} 1 & 1 & 1 \\ 1 & 0 & 1 \\ 1 & 1 & 0 \\ 0 & 1 & 1 \\ 1 & 0 & 0 \\ 0 & 1 & 0 \\ 0 & 0 & 1 \end{bmatrix};$$

yes.

15. Suppose that u is decoded as v and that x is the coset leader of the row containing u. Coset decoding means v is at the head of the column containing u. So, $x + v = u$ and $x = u - v$. Now suppose $u - v$ is a coset leader and u is decoded as y. Then y is at the head of the column containing u. Since v is a code word, $u = u - v + v$ is in the row containing $u - v$. Thus $u - v + y = u$ and $y = v$.

17. 000000, 100110, 010011, 001101, 110101, 101011, 011110, 111000

$$H = \begin{bmatrix} 1 & 1 & 0 \\ 0 & 1 & 1 \\ 1 & 0 & 1 \\ 1 & 0 & 0 \\ 0 & 1 & 0 \\ 0 & 0 & 1 \end{bmatrix}$$

001001 is decoded as 001101 by all four methods.

011000 is decoded as 111000 by all four methods.

000110 is decoded as 100110 by all four methods.

Since there are no code words whose distance from 100001 is 1 and three whose distance is 2, the nearest-neighbor method will not decode or will arbitrarily choose a code word; parity-check matrix decoding does not decode 100001; the standard-array and syndrome methods decode 100001 as 000000, 110101, or 101011, depending on which of 100001, 010100, or 001010 is a coset leader.

19. For any received word w, there are only eight possibilities for wH. But each of these eight possibilities satisfies condition 2 or the first portion of condition 3' of the decoding procedure, so decoding assumes that no error was made or one error was made.

21. There are 3^4 code words and 3^6 possible received words.

23. No; row 3 is twice row 1.

25. No. For if so, nonzero code words would be all words with weight at least 5. But this set is not closed under addition.

27. Use Exercise 24 together with the fact that the set of code words is closed under addition.

29. Abbreviate the coset $a + \langle x^2 + x + 1 \rangle$ with a. The following generating matrix will produce the desired code:

$$\begin{bmatrix} 1 & 0 & 1 & 1 & & x \\ 0 & 1 & x & x+1 & x+1 \end{bmatrix}.$$

31. Use Exercise 14.

33. Let $c, c' \in C$. Then, $c + (v + c') = v + c + c' \in v + C$ and $(v + c) + (v + c') = c + c' \in C$, so the set $C \cup (v + C)$ is closed under addition.

35. If the ith component of both u and v is 0, then so is the ith component of $u - v$ and au, where a is a scalar.

CHAPTER 32

> *Wisdom rises upon the ruins of folly.*   THOMAS FULLER,
> *Gnomologia*

1. Note that $\phi(1) = 1$. Thus $\phi(n) = n$. Also, $1 = \phi(1) = \phi(nn^{-1}) = \phi(n)\phi(n^{-1}) = n\phi(n^{-1})$, so that $1/n = \phi(n^{-1})$.

3. If α and β are automorphisms of E fixing F, so are α^{-1} and $\alpha\beta$.

5. If a and b are fixed by elements of H, so are $a + b$, $a - b$, $a \cdot b$, and a/b.

7. It suffices to show that each member of $\mathrm{Gal}(K/F)$ defines a permutation on the a_i's. Let $\alpha \in \mathrm{Gal}(K/F)$ and write

$$f(x) = c_n x^n + c_{n-1} x^{n-1} + \cdots + c_0 = c_n(x - a_1)(x - a_2) \cdots (x - a_n).$$

Then $f(x) = \alpha(f(x)) = c_n(x - \alpha(a_1))(x - \alpha(a_2)) \cdots (x - \alpha(a_n))$. Thus, $f(a_i) = 0$ implies $a_i = \alpha(a_j)$ for some j, so that α permutes the a_i's.

9. $\phi^6(\omega) = \omega^{729} = \omega$.
$\phi^3(\omega + \omega^{-1}) = \omega^{27} + \omega^{-27} = \omega^{-1} + \omega$.
$\phi^2(\omega^3 + \omega^5 + \omega^6) = \omega^{27} + \omega^{45} + \omega^{54} = \omega^6 + \omega^3 + \omega^5$.

11. Recall that A_4 has no subgroup of order 6. (See Example 13 in Chapter 9.)

13. Use Sylow's Theorem.

15. Let ω be a primitive cube root of 1. Then $Q \subset Q(\sqrt[3]{2}) \subset Q(\omega, \sqrt[3]{2})$ and $Q(\sqrt[3]{2})$ is not the splitting field of a polynomial in $Q[x]$.

17. Use the lattice of Z_{10}.

19. Z_6 (Be sure you know why the group is cyclic.)

21. See Exercise 21 in Chapter 25.

23. Use Exercise 31 in Chapter 24.

25. Use Exercise 40 in Chapter 10.

27. Since $K/N \lhd G/N$, for any $x \in G$ and $k \in K$, there is a $k' \in K$ such that $k'N = (xN)(kN)(xN)^{-1} = xNkNx^{-1}N = xkx^{-1}N$. So, $xkx^{-1} = k'n$ for some $n \in N$. And since $N \subseteq K$, we have $k'n \in K$.

CHAPTER 33

All wish to posses knowledge, but few, comparatively speaking, are willing to pay the price.
JUVENAL

1. $x^2 - x + 1$.

3. Over Z, $x^8 - 1 = (x - 1)(x + 1)(x^2 + 1)(x^4 + 1)$. Over Z_2, $x^2 + 1 = (x + 1)^2$ and $x^4 + 1 = (x + 1)^4$. So, over Z_2, $x^8 - 1 = (x + 1)^8$. Over Z_3, $x^2 + 1$ is irreducible, but $x^4 + 1$ factors into irreducibles as $(x^2 + x + 2)(x^2 - x - 1)$. So, $x^8 - 1 = (x - 1)(x + 1)(x^2 + 1)(x^2 + x + 2)(x^2 - x - 1)$. Over Z_5, $x^2 + 1 = (x - 2)(x + 2)$, $x^4 + 1 = (x^2 + 2)(x^2 - 2)$, and these last two factors are irreducible. So, $x^8 - 1 = (x - 1)(x + 1)(x - 2)(x + 2)(x^2 + 2)(x^2 - 2)$.

5. Let ω be a primitive nth root of unity. We must prove $\omega\omega^2 \cdots \omega^n = (-1)^{n+1}$. Observe that $\omega\omega^2 \cdots \omega^n = \omega^{n(n+1)/2}$. When n is odd, $\omega^{n(n+1)/2} = (\omega^n)^{(n+1)/2} = 1^{(n+1)/2} = 1$. When n is even, $(\omega^{n/2})^{n+1} = (-1)^{n+1} = -1$.

7. If $[F : Q] = n$ and F has infinitely many roots of unity, then there is no finite bound on their multiplicative orders. Let ω be a primitive mth root of unity in F such that $\phi(m) > n$. Then $[Q(\omega) : Q] = \phi(m)$. But $F \supseteq Q(\omega) \supseteq Q$ implies $[Q(\omega) : Q] \leq n$.

9. Let $2^n + 1 = q$. Then $2 \in U(q)$ and $2^n = q - 1 = -1$ in $U(q)$ implies that $|2| = 2n$. So, by Lagrange's Theorem, $2n$ divides $|U(q)| = q - 1 = 2^n$.

11. Let ω be a primitive nth root of unity. Then $2n$th roots of unity are $\pm 1, \pm\omega, \ldots, \pm\omega^{n-1}$. These are distinct, since $-1 = (-\omega^i)^n$, whereas $1 = (\omega^i)^n$.

13. First observe that deg $\Phi_{2n}(x) = \phi(2n) = \phi(n)$ and deg $\Phi_n(-x) = $ deg $\Phi_n(x) = \phi(n)$. Thus, it suffices to show that every zero of $\Phi_n(-x)$ is a zero of $\Phi_{2n}(x)$. But the fact that ω is a zero of $\Phi_n(-x)$ means that $|-\omega| = n$, which in turn implies that $|\omega| = 2n$.

15. Let $G = \text{Gal}(Q(\omega)/Q)$ and H_1 be the subgroup of G of order 2 that fixes $\cos(\frac{2\pi}{n})$. Then, by induction, G/H_1 has a series of subgroups $H_1/H_1 \subset H_2/H_1 \subset \cdots \subset H_t/H_1 = G/H_1$, so that $|H_{i+1}/H_i : H_i/H_1| = 2$. Now observe that $|H_{i+1}/H_i : H_i/H_1| = |H_{i+1}/H_i|$.

17. Instead, prove that $\Phi_n(x)\Phi_{pn}(x) = \Phi_n(x^p)$. Since both sides are monic and have degree $p\phi(n)$, it suffices to show that every zero of $\Phi_n(x)\Phi_{pn}(x)$ is a zero of $\Phi_n(x^p)$. If ω is a zero of $\Phi_n(x)$, then $|\omega| = n$. By Theorem 4.2, $|\omega^p| = n$ also. Thus, ω is a zero of $\Phi_n(x^p)$. If ω is a zero of $\Phi_{pn}(x)$, then $|\omega| = pn$ and therefore $|\omega^p| = n$.

19. Use Theorem 33.4 and Theorem 32.1.

21. $\omega \to \omega^4, \omega \to \omega^{-1}, \omega \to \omega^{-4}$

SUPPLEMENTARY EXERCISES FOR CHAPTERS 24–33

For those who keep trying, failure is temporary.
FRANK TYGER

1. Z_6

3. Let $|G| = 315$ and let H be a Sylow 3-subgroup and K a Sylow 5-subgroup. If $H \lhd G$, then $HK = 45$. If H is not normal, then by Sylow's Third Theorem, $|G/N(H)| = 7$, so that $|N(H)| = 45$.

5. Observe that $K \subseteq N(H)$ implies that HK is a group of order 245. Now, use Sylow's Third Theorem.

7. Note that $gKg^{-1} \subseteq gHg^{-1} = H$. Now use the corollary to Sylow's Third Theorem.

9. Use the same proof as for Exercise 51 in Chapter 9.

11. Since $n_7 = 8$, we know by the Embedding Theorem (Chapter 25) that $G \leq A_8$. But A_8 does not have an element of order 21.

13. Let G be a non-Abelian group of order 105. By Theorem 9.3, $G/Z(G)$ is not cyclic. So $|Z(G)| \neq 3, 7, 15, 21$, or 35. This leaves only 1 or 5 for $|Z(G)|$. Let H, K, and L be Sylow 3-, Sylow 5-, and Sylow 7-subgroups of G, respectively. Now, counting shows that $K \triangleleft G$ or $L \triangleleft G$. Thus, $|KL| = 35$ and KL is a cyclic subgroup of G. But, KL has 24 elements of order 35 (since $|U(Z_{35})| = 24$). Thus, a counting argument shows that $K \triangleleft G$ and $L \triangleleft G$. Now, $|HK| = 15$ and HK is a cyclic subgroup of G. Thus, $HK \subseteq C(K)$ and $KL \subseteq C(K)$. This means that 105 divides $|C(K)|$. So $K \subseteq Z(G)$.

15.

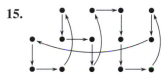

17. It suffices to show that x travels by a implies xab^{-1} travels by a (for we may successively replace x by xab^{-1}). If xab^{-1} traveled by b, then the vertex xa would appear twice in the circuit.

19. a. $\{00, 11\}$
   b. $\{000, 111\}$
   c. $\{0000, 1100, 1010, 1001, 0101, 0110, 0011, 1111\}$
   d. $\{0000, 1100, 0011, 1111\}$

21. The mapping $T_v: F^n \to \{0, 1\}$ given by $T_v(u) = u \cdot v$ is an onto homomorphism. So $|F^n/\text{Ker } T_v| = 2$.

23. It follows from Exercise 18 that if C is an (n, k) linear code, then $C^\perp$ is an $(n, n - k)$ linear code. Thus, in this problem, $k = n - k$. To prove the second claim, use Exercise 18, Exercise 21, the definition of $C^\perp$, and the hypothesis that $C^\perp = C$.

Text Credits for Gallian

Photo Credits

Index of Mathematicians

(Biographies appear on pages in boldface.)

Index of Terms

A49

Essential Theorems in Abstract Algebra

Theorem 3.1 One-Step Subgroup Test
A nonempty subset H of a group G is a subgroup of G if ab^{-1} is in H whenever a and b are in H.

Theorem 4.3 Fundamental Theorem of Cyclic Groups
Every subgroup of a cyclic group is cyclic. If $|\langle a \rangle| = n$, then for each positive divisor k of n, $\langle a \rangle$ has exactly one subgroup of order k and no others.

Theorem 7.1 Lagrange's Theorem
In a finite group the order of a subgroup divides the order of the group.

Theorem 9.1 Normal Subgroup Test
A subgroup H of G is normal in G if and only if $xHx^{-1} \subseteq H$ for all x in G.

Theorem 10.3 First Isomorphism Theorem
If ϕ is a group homomorphism from G to a group, then $G/\text{Ker } \phi \approx \phi(G)$.

Theorem 11.1 Fundamental Theorem of Finite Abelian Groups
Every finite Abelian group is a direct product of cyclic groups of prime-power order.

Theorem 12.3 Subring Test
A nonempty subset S of a ring R is a subring if $a - b$ and ab are in S whenever a and b are in S.

Theorem 13.4 Characteristic of an Integral Domain
The characteristic of an integral domain is 0 or prime.

Theorem 14.1 Ideal Test
A nonempty subset A of a ring R is an ideal if $a - b \in A$ whenever a and b are in A; and ra and ar are in A whenever $a \in A$ and $r \in R$.

Theorem 14.4 R/A is a Field if and only if A is Maximal
Let R be a communitive ring with unity and let A be an ideal of R. Then R/A is a field if and only if A is maximal.

Theorem 15.3 First Isomorphism Theorem for Rings
If ϕ is a ring homomorphism from R to a ring, then $R/\text{Ker } \phi \approx \phi(R)$.

Corollary 1 of Theorem 17.5 $F[x]/\langle p(x) \rangle$ Is a Field
Let F be a field and $p(x)$ an irreducible polynomial over F. Then $F[x]/\langle p(x) \rangle$ is a field.

Theorem 21.5 $[K : F] = [K : E][E : F]$
If K is a finite extension field of the field E and E is a finite extension field of the field F, then $[K : F] = [K : E][E : F]$.

Theorem 22.2 Structure of Finite Fields
The set of nonzero elements of a finite field is a cyclic group under multiplication.

Theorem 24.3 Sylow's First Theorem
Let G be a finite group and p a prime. If p^k divides $|G|$, then G has a subgroup of order p^k.

Theorem 24.5 Sylow's Third Theorem
The number of Sylow p-subgroups of G is equal to 1 modulo p and divides $|G|$. Furthermore, any two Sylow p-subgroups of G are conjugate.